Physical Chemistry

Understand our Chemical World

KU-365-418

Physical Chemistry

Understanding our Chemical World

Paul Monk

Manchester Metropolitan University, UK

John Wiley & Sons, Ltd

Copyright © 2004   John Wiley & Sons Ltd, The Atrium, Southern Gate, Chichester,
West Sussex PO19 8SQ, England

Telephone (+44) 1243 779777

Email (for orders and customer service enquiries): cs-books@wiley.co.uk
Visit our Home Page on www.wileyeurope.com or www.wiley.com

Reprinted March 2007

All Rights Reserved. No part of this publication may be reproduced, stored in a retrieval system or
transmitted in any form or by any means, electronic, mechanical, photocopying, recording, scanning or
otherwise, except under the terms of the Copyright, Designs and Patents Act 1988 or under the terms of
a licence issued by the Copyright Licensing Agency Ltd, 90 Tottenham Court Road, London W1T 4LP,
UK, without the permission in writing of the Publisher. Requests to the Publisher should be addressed
to the Permissions Department, John Wiley & Sons Ltd, The Atrium, Southern Gate, Chichester, West
Sussex PO19 8SQ, England, or emailed to permreq@wiley.co.uk, or faxed to (+44) 1243 770620.

This publication is designed to provide accurate and authoritative information in regard to the subject
matter covered. It is sold on the understanding that the Publisher is not engaged in rendering
professional services. If professional advice or other expert assistance is required, the services of a
competent professional should be sought.

Other Wiley Editorial Offices

John Wiley & Sons Inc., 111 River Street, Hoboken, NJ 07030, USA

Jossey-Bass, 989 Market Street, San Francisco, CA 94103-1741, USA

Wiley-VCH Verlag GmbH, Boschstr. 12, D-69469 Weinheim, Germany

John Wiley & Sons Australia Ltd, 33 Park Road, Milton, Queensland 4064, Australia

John Wiley & Sons (Asia) Pte Ltd, 2 Clementi Loop #02-01, Jin Xing Distripark, Singapore 129809

John Wiley & Sons Canada Ltd, 22 Worcester Road, Etobicoke, Ontario, Canada M9W 1L1

Wiley also publishes its books in a variety of electronic formats. Some content that appears
in print may not be available in electronic books.

Library of Congress Cataloging-in-Publication Data

Monk, Paul M. S.
   Physical chemistry : understanding our chemical world / Paul Monk.
      p. cm.
   Includes bibliographical references and index.
   ISBN 0-471-49180-2 (acid-free paper) – ISBN 0-471-49181-0 (pbk. :
acid-free paper)
   1. Chemistry, Physical and theoretical. I. Title.
   QD453.3.M66 2004
   541 – dc22   2004004224

British Library Cataloguing in Publication Data

A catalogue record for this book is available from the British Library

ISBN 978-0-471-49181-1 (P/B)
ISBN 978-0-471-49180-4 (H/B)

Typeset in 10.5/12.5pt Times by Laserwords Private Limited, Chennai, India

Contents

Preface

This book

Some people make physical chemistry sound more confusing than it really is. One of their best tricks is to define it inaccurately, saying it is 'the physics of chemicals'. This definition is sometimes quite good, since it suggests we look at a chemical system and ascertain how it follows the laws of nature. This is true, but it suggests that chemistry is merely a sub-branch of physics; and the notoriously mathematical nature of physics impels us to avoid this otherwise useful way of looking at physical chemistry.

An alternative and more user-friendly definition tells us that physical chemistry supplies 'the laws of chemistry', and is an addition to the *making* of chemicals. This is a superior lens through which to view our topic because we avoid the bitter aftertaste of pure physics, and start to look more closely at physical chemistry as an *applied* science: we do not look at the topic merely for the sake of looking, but because there are real-life situations requiring a scientific explanation. Nevertheless, most practitioners adopting this approach are still overly mathematical in their treatments, and can make it sound as though the science is fascinating in its own right, but will sometimes condescend to suggest an application of the theory they so clearly relish.

But the definition we will employy here is altogether simpler, and also broader: we merely ask 'why does it happen?' as we focus on the behaviour of each chemical system. Every example we encounter in our everyday walk can be whittled down into small segments of thought, each so simple that a small child can understand. As a famous mystic of the 14th century once said, 'I

> Now published as *Revelations of Divine Love*, by Mother Julian of Norwich.

saw a small hazelnut and I marvelled that everything that exists could be contained within it'. And in a sense she was right: a hazelnut looks brown because of the way light interacts with its outer shell – the topic of spectroscopy (Chapter 9); the hazelnut is hard and solid – the topic of bonding theory (Chapter 2) and phase equilibria (Chapter 5); and the nut is good to eat – we say it is readily metabolized, so we think of kinetics (Chapter 8); and the energetics of chemical reactions (Chapters 2–4). The sensations of taste and sight are ultimately detected within the brain as electrical impulses, which we explain from within the rapidly growing field of electrochemistry (Chapter 7). Even the way a nut sticks to our teeth is readily explained by adsorption science (Chapter 10). Truly, the whole of physical chemistry can be encompassed within a few everyday examples.

So the approach taken here is the opposite to that in most other books of physical chemistry: each small section starts with an example from everyday life, i.e. both the world around us and also those elementary observations that a chemist can be certain to have pondered upon while attending a laboratory class. We then work backwards from the experiences of our hands and eyes toward the cause of why our world is the way it is.

Nevertheless, we need to be aware that physical chemistry is not a closed book in the same way of perhaps classical Latin or Greek. Physical chemistry is a growing discipline, and new experimental techniques and ideas are continually improving the data and theories with which our understanding must ultimately derive.

Inevitably, some of the explanations here have been over-simplified because physical chemistry is growing at an alarming rate, and additional sophistications in theory and experiment have yet to be devised. But a more profound reason for caution is in ourselves: it is all too easy, as scientists, to say 'Now I understand!' when in fact we mean that all the facts before us can be answered by the theory. Nevertheless, if the facts were to alter slightly – perhaps we look at another kind of nut – the theory, as far as we presently understand it, would need to change ever so slightly. Our understanding can never be complete.

So, we need a word about humility. It is said, probably too often, that science is not an emotional discipline, nor is there a place for any kind of reflection on the *human* side of its application. This view is deeply mistaken, because scientists limit themselves if they blind themselves to any contradictory evidence when sure they are right. The laws of physical chemistry can only grow when we have the humility to acknowledge how incomplete is our knowledge, and that our explanation might need to change. For this reason, a simple argument is not necessary the right one; but neither is a complicated one. The examples in this book were chosen to show how the world around us manifests *Physical Chemistry*. The explanation of a seemingly simple observation may be fiendishly complicated, but it may be beautifully simple. It must be admitted that the chemical examples are occasionally artificial. The concept of activity, for example, is widely misunderstood, probably because it presupposes knowledge from so many overlapping branches of physical chemistry. The examples chosen to explain it may be quite absurd to many experienced teachers, but, as an aid to simplification, they can be made to work. Occasionally the science has been simplified to the point where some experienced teachers will maintain that it is technically wrong. But we must start from the beginning if we are to be wise, and only then can we progress via the middle ... and physical chemistry is still a rapidly growing subject, so we don't yet know where it will end.

While this book could be read as an almanac of explanations, it provides students in further and higher education with a *unified* approach to physical chemistry. As a teacher of physical chemistry, I have found the approaches and examples here to be effective with students of HND and the early years of BSc and MChem courses. It has been written for students having the basic chemical and mathematical skills generally expected of university entrants, such as rearrangement of elementary algebra and a little calculus. It will augment the skills of other, more advanced, students.

To reiterate, this book supplies no more than an introduction to physical chemistry, and is not an attempt to cover the whole topic. Those students who have learned some physical chemistry are invited to expand their vision by reading more specialized works. The inconsistencies and simplifications wrought by lack of space and style in this text will be readily overcome by copious background reading. A comprehensive bibliography is therefore included at the end of the book. Copies of the figures and bibliography, as well as live links can be found on the book's website at http://www.wileyeurope.com/go/monkphysical.

Acknowledgements

One of the more pleasing aspects of writing a text such as this is the opportunity to thank so many people for their help. It is a genuine pleasure to thank Professor Séamus Higson of Cranfield University, Dr Roger Mortimer of Loughborough University, and Dr Michele Edge, Dr David Johnson, Dr Chris Rego and Dr Brian Wardle from my own department, each of whom read all or part of the manuscript, and whose comments have been so helpful.

A particular 'thank you' to Mrs Eleanor Riches, formerly a high-school teacher, who read the entire manuscript and made many perceptive and helpful comments.

I would like to thank the many students from my department who not only saw much of this material, originally in the form of handouts, but whose comments helped shape the material into its present form.

Please allow me to thank Michael Kaufman of *The Campaign for a Hydrogen Economy* (formerly the *Hydrogen Association of UK and Ireland*) for helpful discussions to clarify the arguments in Chapter 7, and the *Tin Research Council* for their help in constructing some of the arguments early in Chapter 5.

Concerning permission to reproduce figures, I am indebted to *The Royal Society of Chemistry* for Figures 1.8 and 8.26, the *Open University Press* for Figure 7.10, *Elsevier Science* for Figures 4.7 and 10.3, and *John Wiley & Sons* for Figures 7.19, 10.11 and 10.14. Professor Robin Clarke FRS of University College London has graciously allowed the reproduction of Figure 9.28.

Finally, please allow me to thank Dr Andy Slade, Commissioning Editor of Wiley, and the copy and production editors Rachael Ballard and Robert Hambrook. A special thank you, too, to Pete Lewis.

Paul Monk
Department of Chemistry & Materials
Manchester Metropolitan University
Manchester

Etymological introduction

The hero in *The Name of the Rose* is a medieval English monk. He acts as sleuth, and is heard to note at one point in the story how, 'The study of words is the whole of knowledge'. While we might wish he had gone a little further to mention chemicals, we would have to agree that many of our technical words can be traced back to Latin or Greek roots. The remainder of them originate from the principal scientists who pioneered a particular field of study, known as etymology.

> "Etymology" means the derivation of a word's meaning.

Etymology is our name for the science of *words*, and describes the sometimes-tortuous route by which we inherit them from our ancestors. In fact, most words change and shift their meaning with the years. A classic example describes how King George III, when first he saw the rebuilt St Paul's Cathedral in London, described it as 'amusing, artificial and awful', by which he meant, respectively, it 'pleased him', was 'an artifice' (i.e. grand) and was 'awesome' (i.e. breathtaking).

Any reader will soon discover the way this text has an unusual etymological emphasis: the etymologies are included in the belief that taking a word apart actually helps us to understand it and related concepts. For example, as soon as we know the Greek for 'green' is *chloros*, we understand better the meanings of the proper nouns *chlor*ophyll and *chlor*ine, both of which are green. Incidentally, *phyll* comes from the Greek for 'leaf', and *ine* indicates a substance.

Again, the etymology of the word *oxygen* incorporates much historical information: *oxys* is the Greek for 'sharp', in the sense of an extreme sensory experience, such as the taste of acidic vinegar, and the ending *gen* comes from *gignesthaw* (pronounced 'gin-es-thaw'), meaning 'to be produced'. The classical roots of 'oxygen' reveal how the French scientists who first made the gas thought they had isolated the distinguishing chemical component of acids.

The following tables are meant to indicate the power of this approach. There are several dozen further examples in the text. The bibliography on p. 533 will enable the interested reader to find more examples.

Words derived from a scientist's name

Scientist	Field of study	Present meaning(s)	Derived words[a]
Ampère, André	current, electricity	current	*amp* (unit of current); *amperometry*; *amperometric*; *voltammetry*
Coulomb, Charles	charge, electricity	charge passed	*coulomb* (unit of charge); *coulombic*
Faraday, Michael	capacitance, electrochemistry	electricity	*faraday* (molar electronic charge); *faradaic*; *farad* (unit of capacitance)
Voltaire, Allesandro	electricity, potential	potential	*volt* (unit of potential); *voltaic*; *voltammetry*

[a]Note that derived words do not start with a capital letter.

Words and roots from Latin

Word or root	Original Latin meaning	Modern meaning	Modern examples	Scientific examples
Centi(n)s	hundred times	hundred	century, cent (= $/100)	centi (symbol c) = factor of 10^2
Decie(n)s	ten times	ten	decimate (i.e. to kill 1 in 10); decimal	deci (symbol d) factor of 10; decimetre (= metre $\div$ 10)
Giga(n)s	giant	very large	gigantic	giga (symbol G) = factor of 10^9
Milli	thousand, thousands	thousand, thousandth	millipede, millennium	milli (symbol m) = factor of 10^{-3}
Stratum	bed, couch, coverlet	something beneath	stratify (many layered)	strata of rock (in geology); substrate (chemistry and physics)
Sub	under, beneath, directly under	below, less than	subterfuge, subterranean; sub-standard, subset	substrate (i.e. underlying strata); subscript (in typesetting)
Super	above, on, over, on top of	above, bigger than	superstar, superlative	superscript (in typesetting)

Words and roots from Greek

Word or root	Original Greek meaning	Modern meaning	Modern examples	Scientific examples
Anode (ανοδος)	ascent	positive electrode	anode	anode, anodic, anodize
Baro (βαρος)	weigh down, heaviness	to do with atmosphere	barometer, barometric	barometer, isobar, bar (unit of pressure)
Cathode (καθοδος)	descent	negative electrode	cathode	cathode, cathodic, cathodize
Cyclo (κυκλος)	circle, circular	cycle, circle	bicycle, cylinder, cyclone	cyclotron, cyclization
Di (δις)	two, twice	to do with two	dihedral	to do with two (coordination chemistry)
				dimer, di-stereoisomer (i.e. one of two images)
Iso (ισο)	level, equality	same	Isomer	e.g. isobutane, isomer
Kilo (κιλος)	lots of, many	factor of a thousand	Kilometre	kilo (symbol k) = factor of 10^3
Mega (μεγα-)	great, large	very large	megabyte	mega (symbol M) = factor of 10^6
Mesos (μεσος)	middle, mid	mid, intermediate	mezzanine (mid floor)	mesophase (i.e. phase between two extremes)
Meta (μετα-)	afterwards	after, beyond	metaphor (i.e. beyond the real meaning)	position beyond ortho on a ring
				metathesis (i.e. product of mixing)
Meter (μετρητης)	meter	a meter, to meter	gas meter, metrical	barometer (i.e. measures pressure)
Micro (μιχρος)	small	tiny, small	microscope, micrometer	micro (symbolμ) = factor of 10^{-6}
Mono (μονος)	one	one, alone	monorail, monologue	monomer, e.g. mono-substituted
			monotonous (i.e. on one note)	
Ortho (ορθος)	straight	straight, right	orthodox (i.e. to the standard)	adjacent position on a ring
			othopædic (straightening bones)	
Para (παρα-)	near, beyond, contrary	opposite	paranormal (beyond normal)	position opposite the primary carbon
			paradox (contrary to standard)	
Tetra (τετταρεος)	four, four times	to do with four	tetrahedron	to do with four (coodination chemistry)
Thermo (θερμο)	energy, temperature	Heat	thermos, thermometer	thermos, thermometer

List of Symbols

I_o intensity of incident light beam

I ionic strength

I ionization energy

J rotational quantum number; rotational quantum number of an excited state

J' rotational quantum number of ground state

k force constant of a bond

k proportionality constant

k rate constant

k' pseudo rate constant

k_n rate constant of an nth-order reaction

k_{-n} rate constant for the back reaction of an nth-order reaction

$k_{(n)}$ rate constant of the nth process in a multi-step reaction

k_a rate constant of adsorption

k_d rate constant of desorption

k_H Henry's law constant

K equilibrium constant

K correction constant of an ion-selective electrode

K_a acidity constant ('acid dissociation' constant)

$K_{a(n)}$ acidity constant for the nth dissociation reaction

K_b basicity constant

K_c equilibrium constant formulated in terms of concentration

K_p equilibrium constant formulated in terms of pressure

K_s equilibrium constant of solubility (sometimes called 'solubility product' or 'solubility constant')

K_w autoprotolysis constant of water

$K^{\ddagger}$ equilibrium constant of forming a transition state 'complex'

l length

m gradient of a graph

m mass

M relative molar mass

n number of moles

n number of electrons in a redox reaction

n_m amount of material in an adsorbed monolayer

N number

p pressure

$p_{(i)}$ partial pressure of component i

$p_{(i)}^{\ominus}$ vapour pressure of pure i

$p^{\ominus}$ standard pressure of 10^5 Pa

q heat energy

Q charge

Q reaction quotient

r separation between ions

r radius of a circle or sphere

r bond length

r' bond length in an optically excited species

r_o equilibrium bond length

R electrical resistance

s solubility

s stoichiometric ratio

S entropy

ΔS change in entropy

$S^{\ominus}$ standard entropy

$S^{\ddagger}$ entropy of activation

t time

$t_{\frac{1}{2}}$ half life

T temperature

T optical transmittance

T_o optical transmittance without a sample

T_K Krafft temperature

U internal energy

ΔU change in internal energy, e.g. during reaction

v quantum-number of vibration

v' quantum-number of vibration in an excited-state species

v''	quantum-number of vibration in a ground-state species	v	velocity
V	volume	ν	frequency (the reciprocal of the period of an event)
V	voltage, e.g. of a power pack	ν_0	frequency following transmission (in Raman spectroscopy)
V	Coulomb potential energy		
V_m	molar volume	ξ	extent of reaction
w	work	ρ	density
x	controlled variable on the horizontal axis of a graph	σ	electrical conductivity
		σ	standard deviation
x	deviation of a bond from its equilibrium length	ϕ	electric field strength (electrostatic interaction)
x_i	mole fraction of i	ϕ	work function of a metal
y	observed variable on the vertical of a graph	ϕ	primary quantum yield
		Φ	quantum yield of a reaction
z	charge on ion (so z^+ for a cation and z^- for an anion)	χ	electronegativity
Z	compressibility	ω	wavenumber of a vibration (determined as $\omega = \lambda \div c$)

Symbols for constants

γ	activity coefficient
$\gamma_\pm$	mean ionic activity coefficient
γ	fugacity coefficient
γ	surface tension
δ	small increment
∂	partial differential
Δ	change in a variable (so $\Delta X = X_{(\text{final form})} - X_{(\text{initial form})}$)
ε	extinction coefficient ('molar decadic absorptivity')
ε_r	relative permittivity
ε_0	permittivity of free space
θ	adsorption isotherm
θ	angle
κ	ionic conductivity
λ	wavelength
$\lambda_{(\text{max})}$	the wavelength of a peak in a spectrum
μ	reduced mass
μ_i	chemical potential of i
$\mu_i^{\ominus}$	standard chemical potential of i
ν	stoichiometric constant

A	Debye–Hückel 'A' factor
c	the speed of light *in vacuo*
$c^{\ominus}$	standard concentration
e	charge on an electron, of value 1.6×10^{-19} C
f	mathematical operator ('function of')
F	Faraday constant, of value $96\,485$ C mol^{-1}
k_B	Boltzmann constant, of value 1.38×10^{-23}
L	Avogadro constant, of value 6.022×10^{23} mol^{-1}
N_A	Avogadro number, of value 6.022×10^{23} mol^{-1}
g	acceleration due to gravity, of value 9.81 m s^{-2}
h	Planck constant, of value 6.626×10^{-34} J s
R	gas constant, of value 8.314 J K^{-1} mol^{-1}

Symbols for units

A	ampère
Å	ångström, length of value 10^{-10} m (non-IUPAC)
bar	standard pressure of 10^5 Pa (non-SI unit)
C	coulomb
°C	centigrade (non-SI)
g	gram
Hz	hertz
J	joule
K	kelvin
kg	kilogram
m	metre
mmHg	millimetre of mercury (non-SI unit of pressure)
mol	mole
N	newton
Pa	pascal
s	second (SI unit)
S	siemen
V	volt
W	watt
yr	year
Ω	ohm

Acronyms and abbreviations

CT	charge transfer
d	differential operator (which *never* appears on its own)
HOMO	highest occupied molecular orbital
IQ	intelligence quotient
IR	infrared
IUPAC	International Union of Pure and Applied Chemistry
IVF	*in vitro* fertilization
LCD	liquid crystal display
LHS	left-hand side
LUMO	lowest unoccupied molecular orbital
MLCT	metal-to-ligand charge transfer
MRI	magnetic resonance imaging
NIR	near-infra red
NMR	nuclear magnetic resonance
O	general oxidized form of a redox couple
p	mathematical operator, $-\log_{10}$[variable], so pH $= -\log_{10}$[H$^+$]
PEM	proton exchange membrane
R	general reduced form of a redox couple
RHS	right-hand side
s.t.p.	standard temperature and pressure
SAQ	self-assessment question
SCE	saturated calomel electrode
SCUBA	self-contained underwater breathing apparatus
SHE	standard hydrogen electrode
SHM	simple harmonic motion
SI	*Système Internationale*
S_N1	unimolecular nucleophilic substitution process
S_N2	bimolecular nucleophilic substitution process
SSCE	silver–silver chloride electrode
TS	transition state
TV	television
UPS	UV-photoelectron spectroscopy
UV	ultraviolet
UV–vis	ultraviolet and visible
XPS	X-ray photoelectron spectroscopy

•	radical
+•	radical cation
⊖	standard state

Standard subscripts (other than those where a word or phrase is spelt in full)

ads	adsorption; adsorbed
aq	aqueous
c	combustion
eq	at equilibrium
f	formation
g	gas
l	liquid
LHS	left-hand side of a cell
m	molar
p	at constant pressure
Pt	platinum (usually, as an electrode)
r	reaction
RHS	right-hand side of cell
s	solid
sat'd	saturated
t	at time t (i.e. after a reaction or process has commenced)
V	at constant volume
0	initially (i.e. at time $t = 0$)
∞	measurement taken after an infinite length of time

Chemicals and materials

A	general anion
Bu	butyl
CFC	chlorofluorocarbon
DMF	N,N-dimethylformamide
DMSO	dimethylsulphoxide
DNA	deoxyribonucleic acid
e^-	electron
EDTA	ethylenediamine tetra-acetic acid
HA	general Lowry–Brønsted acid
LPG	liquid petroleum gas
M	general cation
MB	methylene blue
MV	methyl viologen ($1,1'$-dimethyl-$4,4'$-bipyridilium)
O	general oxidized form of a redox couple
PC	propylene carbonate
Ph	phenyl substituent
R	general alkyl substituent
R	general reduced form of a redox couple
SDS	sodium dodecyl sulphate
TFA	tetrafluoroacetic acid
α	particle emitted during radioactive disintegration of nucleus
β	particle emitted during radioactive disintegration of nucleus
γ	high-energy photon (gamma ray)

Standard superscripts (other than those where a word or phrase is spelt in full)

‡	activated quantity
—	anion
+	cation
*	excited state

Standard prefixes powers of ten

Powers of ten: energy in joules

Power	Prefix	Value	Description
10^{-18}	a atto	0.000, 000, 000, 000, 000, 001	1 aJ
10^{-15}	f femto	0.000, 000, 000, 000, 001	1 fJ = energy of a single high-energy photon (γ-ray of $\lambda = 10^{-10}$ m)
10^{-12}	p pico	0.000, 000, 000, 001	1 pJ = energy consumption of a single nerve impulse
10^{-9}	n nano	0.000, 000, 001	1 nJ = energy per beat of a fly's wing
10^{-6}	μ micro	0.000, 001	1 μJ = energy released per second by a single phosphor on a TV screen
10^{-3}	m milli	0.001	1 mJ = energy consumption per second of an LCD watch display
10^{-2}	c centi	0.01	1 cJ = energy per mole of low-energy photons (radio wave of $\lambda = 10$ m)
10^{-1}	d deci	0.1	1 dJ = energy released by passing 1 mA across 1 V for 100 s (energy = Vit)
$10^0 = 1$	1	1	1 J = half the kinetic energy of 0.1 kg (1 N) travelling at a velocity of 1 m s^{-1}
10^3	k kilo	1, 000	1 kJ = half the energy of room temperature (E at RT of 298 K = 2.3 kJ)
10^6	M mega	1, 000, 000	1 MJ = energy of burning $^1/_3$th mole of glucose (about a quarter a Mars bar)
10^9	G giga	1, 000, 000, 000	1 GJ = energy of 1 mole of γ-ray photons ($\lambda = 10^{-10}$ m)
10^{12}	T tera	1, 000, 000, 000, 000	1 TJ = energy (via $E = mc^2$) held in a mass of 100 g
10^{15}	P peta	1, 000, 000, 000, 000, 000	1 PJ = energy released by detonating a very small nuclear bomb

Powers of ten: time in seconds

Power	Prefix	Value	Description
10^{-18}	a atto	0.000,000,000,000,000,001	1 as = the very fastest laser flash
10^{-15}	f femto	0.000,000,000,000,000,001	1 fs = 10 × the time required for photon absorption
10^{-12}	p pico	0.000,000,000,001	1 ps = time required for bond rearrangement
10^{-9}	n nano	0.000,000,001	1 ns = time required for solvent rearrangement e.g. after redox change
10^{-6}	μ micro	0.000,001	1 μs = time for 166 calculations in a Pentium® microprocessor chip
10^{-3}	m milli	0.001	1 ms = time for a nerve impulse
10^{-2}	c centi	0.01	1 cs = fastest human reflexes
10^{-1}	d deci	0.1	1 ds = time required to blink fast
$10^{0}=1$		1	1 s = time required to sneeze
10^{3}	k kilo	1,000	1 ks = average time required to walk 1 mile (ca. 16 minutes)
10^{6}	M mega	1,000,000	1 Ms = life expectancy of a housefly (11.5 days)
10^{9}	G giga	1,000,000,000	1 Gs = average life span of an adult in the third world (31.7 years)
10^{12}	T tera	1,000,000,000,000	1 Ts = age of fossil remains of Neanderthal man (31,700 years)
10^{15}	P peta	1,000,000,000,000,000	1 Ps = time since dinosaurs were last in their prime (31.7 million years ago)

Powers of ten: length in metres

10^{-18}	a atto	0.000,000,000,000,000,001	1 am = diameter of a γ-ray photon (if it is taken to be a particle)
10^{-15}	f femto	0.000,000,000,000,001	1 fm = diameter of an electron (if it is taken to be a particle)
10^{-12}	p pico	0.000,000,000,001	1 pm = ten times the diameter of an atomic nucleus
10^{-9}	n nano	0.000,000,001	1 nm = ten bond lengths
10^{-6}	μ micro	0.000,001	1 μm = wavelength of near infra-red light (i.e. heat)
10^{-3}	m milli	0.001	1 mm = diameter of a grain of sand
10^{-2}	c centi	0.01	1 cm = width of a human finger
10^{-1}	d deci	0.1	1 dm = length of the human tongue
$10^{0}=1$	1	1	1 m = half the height of an adult human
10^{3}	k kilo	1,000	1 km = distance walked in about 12 minutes
10^{6}	M mega	1,000,000	1 Mm = distance between London and Edinburgh
10^{9}	G giga	1,000,000,000	1 Gm = distance travelled by light in 3 minutes
10^{12}	T tera	1,000,000,000,000	1 Tm = ninth of the distance between the sun and the earth
10^{15}	P peta	1,000,000,000,000,000	1 Pm = tenth of a light year (distance travelled by a photon in 1 year)

Powers of ten: mass in grams

Power	Prefix	Value	Example
10^{-18}	a atto	0.000,000,000,000,000,001	1 ag = mass of 1 molecule of polystyrene (of *molar* mass 10^6 g mol^{-1})
10^{-15}	f femto	0.000,000,000,000,001	1 fg = mass of 1 molecule of DNA (of *molar* mass 10^9 g mol^{-1})
10^{-12}	p pico	0.000,000,000,001	1 pg = mass of phosphorous used to dope 1 g of silicon (for a microchip)
10^{-9}	n nano	0.000,000,001	1 ng = mass of a single ferrite particle for use in a computer floppy disc
10^{-6}	µ micro	0.000,001	1µg = mass of ink in a full stop
10^{-3}	m milli	0.001	1 mg = mass of ink on a banknote
10^{-2}	c centi	0.01	1 cg = mass of nicotine absorbed from a single cigarette
10^{-1}	d deci	0.1	1 dg = mass of the active component within 1 aspirin tablet
$10^0=1$	1	1	1 g = mass of a single lentil
10^3	k kilo	1,000	1 kg = mass of a grapefruit
10^6	M mega	1,000,000	1 Mg = mass of a baby elephant
10^9	G giga	1,000,000,000	1 Gg = mass of a large crane (1000 tonnes)
10^{12}	T tera	1,000,000,000,000	1 Tg = mass of a mountain
10^{15}	P peta	1,000,000,000,000,000	1 Pg = mass of a large island such as the Isle of White (10^9 tonnes)

1

Introduction to physical chemistry

Introduction

In this, our introductory chapter, we start by looking at the terminology of physical chemistry. Having decided what physical chemistry actually is, we discuss the nature of variables and relationships. This discussion introduces the way relationships underlying physical chemistry are formulated.

We also introduce the fundamental (base) units of the *Système Internationale* (SI), and discuss the way these units are employed in practice.

We look at the simple gas laws to explore the behaviour of systems with no interactions, to understand the way macroscopic variables relate to microscopic, molecular properties. Finally, we introduce the statistical nature underlying much of the physical chemistry in this book when we look at the Maxwell–Boltzmann relationship.

..1 What is physical chemistry: variables, relationships and laws

Why do we warm ourselves by a radiator?

Cause and effect

We turn on the radiator if we feel cold and warm ourselves in front of it. We become warm because heat travels from the radiator to us, and we absorb its heat energy, causing our own energy content to rise. At root, this explains why we feel more comfortable.

While this example is elementary in the extreme, its importance lies in the way it illustrates the concept of cause and effect. We would not feel warmer if the radiator was at the same temperature as we were. We feel warmer firstly because the radiator is warmer than us, and secondly because some of the heat energy leaves the radiator and we absorb it. A transfer of energy occurs and, therefore, *a change*. Without the *cause*, the *effect* of feeling warmer could not have followed.

A *variable* is an experimental parameter we can change or 'tweak'.

We are always at the mercy of events as they occur around us. The physical chemist could do nothing if nothing happened; *chemists look at changes*. We say a physical chemist alters *variables*, such as pressure or temperature. Typically, a chemist causes one variable to change and looks at the resultant response, if any. Even a lack of a response is a form of result, for it shows us what is and what is not a variable.

Why does water get hot in a kettle?

The fearsome-looking word 'physicochemical' means 'relating to physical chemistry'.

Physicochemical relationships

Putting water into an electric kettle does not cause the water to get hot. The water stays cold until we turn on the power to the kettle element, which converts electrical energy from the mains into heat energy. The heat energy from the kettle element is then absorbed by the water, which gets hot as a direct consequence.

The temperature of the water does not increase much if a small amount of electrical energy is consumed; conversely, the water gets hotter if a greater amount of energy is consumed and thereafter passed to the water. A physical chemist says a 'relationship' exists (in this case) between heat input and temperature, i.e. the temperature of the water depends on the amount of energy consumed.

In words, the symbols $T = f(\text{energy})$ means 'T is a function of energy'. Note how variables are usually printed in *italic* type.

Mathematically, we demonstrate the existence of a relationship by writing $T = f(\text{energy})$, where T is temperature and the small f means 'is a function of'.

So the concept of variables is more powerful than just changing parameters; nor do physical chemists merely vary one parameter and see what happens to others. They search for 'physicochemical' relationships between the variables.

Are these two colours complementary?

Qualitative and quantitative measurements

We often hear this question, either at the clothes shop or at a paint merchant. Either someone wants to know if the pink colour of a sweatshirt matches the mauve of a skirt, or perhaps a decorator needs to know if two shades of green will match when painted on opposing bedroom walls.

But while asking questions concerning whether a series of colours are complementary, we are in fact asking two questions at once: we ask about the colour in relation to how dark or light it is ('What is the *brightness* of the colour?'); but we also ask a more subjective question, saying 'Is the pink more red or more white: what *kind* of pink is it?' We are looking for two types of relationship.

> *Complementary* means 'to make complete'.

In any investigation, we first look for a *qualitative* relationship. In effect, we ask questions like, 'If I change the variable x, is there is a response in a different variable y?' We look at what kind of response we can cause – a scientist wants to know about the *qualities* of the response, hence QUAL-itative. An obvious question relating to qualitative relationships is, 'If I mix solutions of A and B, *does a reaction occur*?'

Only after we know whether or not there *is* a response (and of what general kind) does a physical chemist ask the next question, seeking a *quantitative* assessment. He asks, 'How *much* of the response is caused?' In effect, physical chemists want to know if the magnitude (or quantity) of a response is big, small or intermediate. We say we look for a QUANT-itative aspect of the relationship. An obvious question relating to quantitative relationships is, 'I now know that a reaction occurs when I mix solutions of A and B, but to *what extent* does the reaction occur; *what is the chemical yield*?'

Does my radio get louder if I vary the volume control?

Observed and controlled variables

We want to turn up the radio because it's noisy outside, and we want to hear what is broadcast. We therefore turn the volume knob toward 'LOUD'. At its most basic, the volume control is a variable resistor, across which we pass a current from the battery, acting much like a kettle element. If we turn up the volume control then a larger current is allowed to flow, causing more energy to be produced by the resistor. As a listener, we hear a response because the sound from the speakers becomes louder. The speakers work harder.

But we must be careful about the way we state these relationships. We do not 'turn up the volume' (although in practice we might say these exact words and think in these terms). Rather, we vary the volume control and, *as a response*, our ears experience an increase in the decibels coming through the radio's speakers. The listener controls the magnitude of the noise by deciding how far the volume-control knob needs to be turned. Only then will the volume change. The process does not occur in reverse: we do not change the magnitude of the noise and see how it changes the position of the volume-control knob.

While the magnitude of the noise and the position of the volume knob are both variables, they represent different types, with one depending on the other. The volume control is a *controlled variable* because the listener dictates its position. The amount of noise is the *observed variable* because it only changes in response to variations in the controlled variable, and not before.

> We consciously, carefully, vary the magnitude of the *controlled* variable and look at the response of the *observed* variable.

Relationships and graphs

The x-axis (horizontal) is sometimes called the *abscissa* and the y-axis (vertical) is the *ordinate*. A simple way to remember which axis is which is to say, 'an e**X**panse of road goes horizontally along the x-axis', and 'a **Y**o-Yo goes up and down the y-axis'.

Physical chemists often depict relationships between variables by drawing graphs. The controlled variable is always drawn along the x-axis, and the observed variable is drawn up the y-axis.

Figure 1.1 shows several graphs, each demonstrating a different kind of relationship. Graph (a) is straight line passing through the origin. This graph says: when we vary the controlled variable x, the observed variable y changes in direct proportion. An obvious example in such a case is the colour intensity in a glass of black-currant cordial: the intensity increases in linear proportion to the concentration of the cordial, according to the Beer–Lambert law (see Chapter 9). Graph (a) in Figure 1.1 goes through the origin because there is no purple colour when there is no cordial (its concentration is zero).

Graph (b) in Figure 1.1 also demonstrates the existence of a relationship between the variables x and y, although in this case not a linear relationship. In effect, the graph tells us that the observed variable y increases at a faster rate than does the controlled variable x. A simple example is the distance travelled by a ball as a function of time t as it accelerates while rolling down a hill. Although the graph is not straight, we still say there *is* a relationship, and still draw the controlled variable along the x-axis.

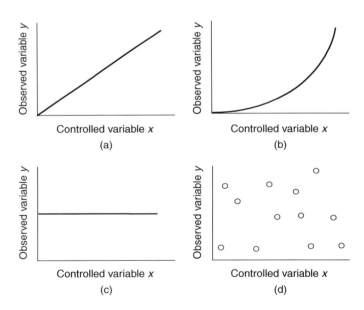

Figure 1.1 Graphs of observed variable (along the y-axis) against controlled variable (along the x-axis). (a) A simple linear proportionality, so $y =$ constant $\times\ x$; (b) a graph showing how y is not a simple function of x, although there is a clear relationship; (c) a graph of the case where variable y is independent of variable x; (d) a graph of the situation in which there is no relationship between y and x, although y does vary

Graph (c) in Figure 1.1 is a straight-line graph, but is horizontal. In other words, whatever we do to the controlled variable x, the observed variable y will not change. In this case, the variable y is not a function of x because changing x will not change y. A simple example would be the position of a book on a shelf as a function of time. In the absence of other forces and variables, the book will not move just because it becomes evening.

Graph (d) in Figure 1.1 shows another situation, this time the data do not demonstrate a straightforward relationship; it might demonstrate there is no relationship at all. The magnitude of the controlled variable x does not have any bearing on the observed variable y. We say the observed variable y is *independent* of the controlled variable x. Nevertheless, there is a range of results for y as x varies. Perhaps x is a *compound variable*, and we are being simplistic in our analysis: an everyday example might be a student's IQ as x and his exam performance as y, suggesting that, while IQ is important, there must be another variable controlling the magnitude of the exam result, such as effort and commitment. Conversely, the value of y might be completely random (so repeating the graphs with the same values of x would generate a different value of y – we say it is *irreproducible*). An example of this latter situation would be the number of people walking along a main road as a function of time.

> *Data* is plural; the singular is *datum*.

> When two variables are multiplied together, we call them a *compound variable*.

Why does the mercury in a barometer go up when the air pressure increases?

Relationships between variables

The pressure p of the air above any point on the Earth's surface relates ultimately to the amount of air above it. If we are standing high up, for example on the top of a tall mountain, there is less air between us and space for gravity to act upon. Conversely, if we stand at the bottom of the Grand Canyon (one of the lowest places on Earth) then more air separates us from space, causing the air pressure p to be much greater.

A barometer is an instrument designed to measure air pressure p. It consists of a pool of liquid mercury in a trough. A long, thin glass tube (sealed at one end) is placed in the centre of the trough with its open-side beneath the surface of the liquid; see Figure 1.2. The pressure of the air acts as a force on the surface of the mercury, forcing it up and into the capillary within the tube. If the air pressure is great, then the force of the air on the mercury is also great, causing much mercury up the tube. A lower pressure is seen as a shorter length h of mercury in the tube.

By performing experiments at different pressures, it is easy to prove the existence of a relationship between the air pressure p and the height h of the mercury column

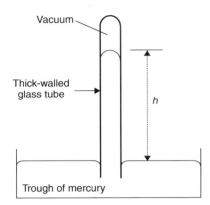

Vacuum

Thick-walled
glass tube

h

Trough of mercury

Figure 1.2 A barometer is a device for measuring pressures. A vacuum-filled glass tube (sealed at one end) is placed in a trough of mercury with its open end beneath the surface of the liquid metal. When the tube is erected, the pressure of the external air presses on the surface and forces mercury up the tube. The height of the mercury column h is directly proportional to the external pressure p

In fact, the value of the constant c in Equation (1.1) comprises several natural constants, including the acceleration due gravity g and the density ρ of the mercury.

in the tube. This relationship follows Equation (1.1):

$$h = c \times p \qquad (1.1)$$

where c is merely a proportionality constant.

In practice, a barometer is merely an instrument on which we look at the length of the column of mercury h and, via Equation (1.1), calculate the air pressure p. The magnitude of h is in direct relation to the pressure p. We ascertain the magnitude of h if we need to know the air pressure p.

While physical chemistry can appear to be horribly mathematical, in fact the mathematics we employ are simply one way (of many) to describe the relationships between variables. Often, we do not know the exact nature of the function until a later stage of our investigation, so the complete form of the relationship has to be discerned in several stages. For example, perhaps we first determine the existence of a linear equation, like Equation (1.1), and only then do we seek to measure an accurate value of the constant c.

We might see this situation written mathematically as, $h \neq f(p)$, where the '$\neq$' means 'is not equal to'. In other words, h is not a function of p in a poor barometer.

But we do know a relationship holds, because there is a *response*. We would say there was no relationship if there was no response. For example, imagine we had constructed a poor-quality barometer (meaning it does not follow Equation (1.1)) and gave it a test run. If we could *independently* verify that the pressure p had been varied over a wide range of values yet the length of the mercury h in the barometer did not change, then we would say no relationship existed between p and h.

Why does a radiator feel hot to the touch when 'on', and cold when 'off'?

Laws and the minus-oneth law of thermodynamics

Feeling the temperature of a radiator is one of the simplest of experiments. No one has ever sat in front of a hot radiator and felt colder. As a qualitative statement, we begin with the excellent generalization, 'heat always travels from the hotter to the colder environment'. We call this observation a *law* because it is universal. Note how such a law is not concerned with magnitudes of change

> A 'law' in physical chemistry relates to a wide range of situations.

but simply relays information about a universal phenomenon: energy in the form of heat will travel from a hotter location or system to a place which is colder. Heat energy never travels in the opposite direction.

We can also notice how, by saying 'hotter' and 'colder' rather than just 'hot' and 'cold', we can make the law wider in scope. The temperature of a radiator in a living room or lecture theatre is typically about 60 °C, whereas a human body has an ideal temperature of about 37 °C. The radiator is hotter than we are, so heat travels *to* us *from* the radiator. It is this heat emitted by the radiator which we absorb in order to feel warmer.

Conversely, now consider placing your hands on a colder radiator having a temperature of 20 °C (perhaps it is broken or has not been switched on). In this second example, although our hands still have the same temperature of 37 °C, this time the heat energy travels *to* the radiator *from* our hands as soon as we touch it. The direction of heat flow has been reversed in response to the reversal of the *relative* difference between the two temperatures. The direction in which the heat energy is transferred is one aspect of why the radiator *feels* cold. We see how the movement of energy not only has a magnitude but also a *direction*.

Such statements concerning the direction of heat transfer are sometimes called the *minus-oneth law of thermodynamics*, which sounds rather daunting. In fact, the word 'thermodynamics' here may be taken apart piecemeal to translate it into everyday English. First the simple bit: 'dynamic' comes from the Greek word *dunamikos*, which means movement. We obtain the conventional English word 'dynamic' from the same root; and a cyclist's

> The 'minus-oneth law of thermodynamics' says, 'heat always travels from hot to cold'.

'dynamo' generates electrical energy from the spinning of a bicycle wheel, i.e. from a moving object. Secondly, *thermo* is another commonly encountered Greek root, and means energy or temperature. We encounter the root *thermo* incorporated into such everyday words as 'thermometer', 'thermal' and 'thermos flask'. A 'thermodynamic' property, therefore, relates to events or processes in which there are 'changes in heat or energy'.

Aside

We need to explain the bizarre name of this law, which is really an accident of history. Soon after the first law of thermodynamics was postulated in the mid nineteenth century, it was realized how the law presupposed a more elementary law, which we now call the zeroth law (see below). We call it the 'zeroth' because zero comes before one. But scientists soon realized how even the zeroth law was too advanced, since it presupposed a yet more elementary law, which explains why the minus-oneth law had to be formulated.

How does a thermometer work?

Thermal equilibrium and the zeroth law of thermodynamics

> The word 'thermometer' has two roots: *meter* denotes a device to measure something, and *thermo* means 'energy' or 'temperature'. Thus, a 'thermometer' is a device for measuring energy as a function of temperature.

A fever is often the first visible sign of someone developing an illness. The body's temperature rises – sometimes dramatically – above its preferred value of 37 °C. As a good generalization, the temperature is hotter when the fever is worse, so it is wise to monitor the temperature of the sick person and thereby check the progress of the illness. A thermometer is the ideal instrument for this purpose.

When measuring a temperature with a thermometer, we place the mercury-containing end into the patient's mouth or armpit and allow the reading to settle. The mercury is encased within a thin-walled glass tube, which itself is placed in contact with the patient. A 'reading' is possible because the mercury expands with increasing temperature: we take the length l of the mercury in the tube to be an accurate function of its temperature T. We read the patient's temperature from the thermometer scale only when the length of the mercury has stopped changing.

But how does the thermometer work in a *thermodynamic* sense, since at no time can the toxic mercury be allowed to touch the patient?

> Bodies together at the same temperature are said to be in 'thermal equilibrium'.

Consider the flow of heat: heat energy first flows from the patient to the glass, and thence flows through the glass into the mercury. Only when all three – mercury, glass and patient – are at the same temperature can the thermometer reading become steady. We say we have *thermal equilibrium* when these three have the same temperature; see Figure 1.3.

Although in some respects a trivial example, a thermometer helps us see a profound truth: only when both (i) the mercury and the glass, *and* (ii) the glass and the patient are at thermal equilibrium can the patient and the mercury truly be said to be at the same temperature. By this means, we have measured the temperature of the patient by

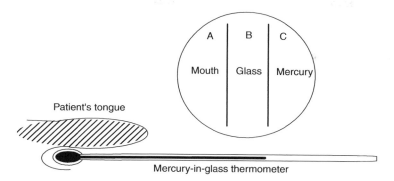

Figure 1.3 The zeroth law states, 'Imagine three bodies, A, B and C. If A and B are in thermal equilibrium, and B and C are in thermal equilibrium, then A and C are also in thermal equilibrium' (see inset). A medic would rephrase the law, 'If mercury is in thermal equilibrium with the glass of a thermometer, and the glass of a thermometer is in thermal equilibrium with a patient, then the mercury and the patient are also in thermal equilibrium'

utilizing a temperature-dependent property of the liquid metal inside the thermometer, yet at no time do we need to expose the patient to the toxic mercury.

We begin to understand the power of thermodynamics when we realize how often this situation arises: in effect, we have made an *indirect* measurement – a frequent occurrence – so we need to formulate another law of thermodynamics, which we call the *zeroth law*. Imagine three bodies, A, B and C. If A and B are in thermal equilibrium, and B and C are in thermal equilibrium, then A and C are also in thermal equilibrium.

> The *zeroth law of thermodynamics* says: imagine three bodies, A, B and C. If A and B are in thermal equilibrium, and B and C are also in thermal equilibrium, then A and C will be in thermal equilibrium.

While sounding overly technical, we have in fact employed the zeroth law with the example of a thermometer. Let us rephrase the definition of the zeroth law and say, 'If mercury is in thermal equilibrium with the glass of a thermometer, and the glass of a thermometer is in thermal equilibrium with a patient, then the mercury and the patient are also in thermal equilibrium'. A medic could not easily determine the temperature of a patient without this, the zeroth law.

From now on we will assume the zeroth law is obeyed each time we use the phrase 'thermal equilibrium'.

.2 The practice of thermodynamic measurement

What is temperature?

Scientific measurement

Although the answer to the simple question 'what *is* temperature?' seems obvious at first sight, it is surprisingly difficult to answer to everyone's satisfaction. In fact, it is

'Corollary' means a deduction following on from another, related, fact or series of facts.

generally easier to state the corollary, 'a body has a higher temperature if it has more energy, and a lower temperature if it has less energy'.

We have been rather glib so far when using words such as 'heat' and 'temperature', and will be more careful in future. Heat is merely one way by which we experience *energy*. Everything contains energy in various amounts, although the exact quantity of the energy is not only unknown but unknowable.

The word 'thermo-chemistry' has two roots: *thermo*, meaning 'temperature or energy', and *chemistry*, the science of the combination of chemicals. We see how 'thermochemistry' studies the energy and temperature changes accompanying *chemical changes*.

Much of the time, we, as physical chemists, will be thinking about energy and the way energetic changes accompany chemical changes (i.e. atoms, ions or whole groups of atoms combine, or add, or are being lost, from molecules). While the total energy cannot be known, we can readily determine the *changes* that occur in tandem with chemical changes. We sometimes give the name *thermochemistry* to this aspect of physical chemistry.

In practice, the concept of temperature is most useful when determining whether two bodies are in thermal equilibrium. Firstly, we need to appreciate how these equilibrium processes are always *dynamic*, which, stated another way, indicates that a body simultaneously emits and absorbs energy, with these respective amounts of energy being equal and opposite. Furthermore, if two bodies participate in a thermal equilibrium then we say that the energy emitted by the first body is absorbed by the second; and the first body also absorbs a similar amount of energy to that emitted by the second body.

A body in 'dynamic equilibrium' with another exchanges energy with it, yet without any *net* change.

Temperature is most conveniently visualized in terms of the senses: we say something is hotter or is colder. The first thermometer for studying changes in temperature was devised in 1631 by the Frenchman Jean Rey, and comprised a length of water in a glass tube, much like our current-day mercury-in-glass thermometers but on a much bigger scale. The controlled variable in this thermometer was temperature T, and the observed variable was the length l of the water in the glass tube.

Rey's thermometer was not particularly effective because the density of water is so low, meaning that the volume of the tube had to be large. And the tube size caused an additional problem. While the water expanded with temperature (as required for the thermometer to be effective), so did the glass encapsulating it. In consequence of both water and glass expanding, although the water expanded in a straightforward way with increasing temperature, the *visible* magnitude of the expansion was not in direct proportion to the temperature rise.

Scientists use the word 'ideal' to mean obeying the laws of science.

Although we could suggest that a relationship existed between the length l and the temperature T (saying one is a function of the other), we could not straightforwardly ascertain the *exact* nature of the function. In an *ideal* thermometer, we write the mathematical relationship, $l = f(T)$. Because Rey's thermometer contained water, Rey was not able to observe a linear dependence of l on T for his thermometer, so he could not write $l = aT + b$ (where a and b are constants).

For a more dense liquid, such as mercury, the relationship between l and T is linear – at least over a relatively *narrow* range of temperatures – so a viable mercury-in-glass thermometer may be constructed. But, because the temperature response is only linear over a narrow range of temperatures, we need to exercise caution.

> 'Narrow' in this case means 50–70 °C at most.

If we assume the existence of a linear response for such a thermometer, then the thermometer is 'calibrated' by correlating the readings of length l using the known properties of the standard, as follows. First, the thermometer is placed in a trough of pure ice at its melting temperature, and the end of the mercury bead marked as 0 °C. The same thermometer is then placed in water at its boiling point and the end of the mercury bead marked as 100 °C. The physical distance between these two extremes is subdivided into 100 equal portions, each representing a temperature increment of 1 °C. This *centigrade* scale is satisfactory for most purposes. (The same scale is sometimes called *Celsius* after a Swedish physicist who championed its use.)

> To 'calibrate' an instrument such as a thermometer, we correlate a physico-chemical property (such as the length l of the mercury) using the temperature-dependent properties of a known standard.

This formulation of the centigrade scale presupposed a linear relationship between length l and temperature T (i.e. the straight line (a) on the graph in Figure 1.4), but we must be aware the relationship might only be approximately linear (e.g. the curved line (b) on the graph in Figure 1.4). The straight and the curved lines only agree at the two temperatures 0 °C and 100 °C merely because they were *defined* that way.

> The 'centigrade' scale was first proposed in 1694 by Renaldi. *Centi* is a Latin prefix meaning 'hundred'.

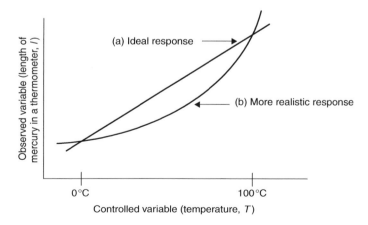

Figure 1.4 In using a thermometer, we assume the existence of a linear response between the length l of the mercury and the controlled variable temperature T. Trace (a) shows such a relationship, and trace (b) shows a more likely situation, in which there is a close approximation to a linear relationship between length l and temperature T

This last paragraph inevitably leads to the questions, 'So how do we know what the exact temperature is?' and 'How do I know if my thermometer follows profile (a) or profile (b) in Figure 1.4?' Usually, we do not know the answer. If we had a single thermometer whose temperature was always accurate then we could use it as a primary standard, and would simply prepare a *calibrated* thermometer against which all others are calibrated.

But there are no ideal (perfect) thermometers in the real world. In practice, we generally experiment a bit until we find a thermometer for which a property X is as close to being a linear function of temperature as possible, and call it a *standard thermometer* (or 'ideal thermometer'). We then calibrate other thermometers in relation to this, the standard. There are several good approximations to a standard thermometers available today: the temperature-dependent (observed) variable in a *gas thermometer* is the volume of a gas V. Provided the pressure of the gas is quite low (say, one-hundredth of atmospheric pressure, i.e. 100 Pa) then the volume V and temperature T do indeed follow a fairly good linear relationship.

A second, popular, standard is the *platinum-resistance thermometer*. Here, the electrical resistance R of a long wire of platinum increases with increased temperature, again with an essentially linear relationship.

Worked Example 1.1   A platinum resistance thermometer has a resistance R of $3.0 \times 10^{-4}\,\Omega$ at $0\,°C$ and $9.0 \times 10^{-4}\,\Omega$ at $100\,°C$. What is the temperature if the resistance R is measured and found to be $4.3 \times 10^{-4}\,\Omega$?

We first work out the exact relationship between resistance R and temperature T. We must assume a *linear* relationship between the two to do so.

> These discussions are expressed in terms of centigrade, although absolute temperatures are often employed – see next section.

The change per degree centigrade is obtained as 'net change in resistance $\div$ net change in temperature'. The resistance R increases by $6.0 \times 10^{-4}\,\Omega$ while the temperature is increased over the $100\,°C$ range; therefore, the increase in resistance per degree centigrade is given by the expression

$$R \text{ per }°C = \frac{6.0 \times 10^{-4}\,\Omega}{100\,°C} = 6 \times 10^{-6}\,\Omega\,°C^{-1}$$

Next, we determine by how much the resistance has increased in going to the new (as yet unknown) temperature. We see how the resistance increases by an amount $(4.3 - 3.0) \times 10^{-4}\,\Omega = 1.3 \times 10^{-4}\,\Omega$.

The increase in temperature is then the rise in resistance divided by the change in resistance increase per degree centigrade.

We obtain

$$\frac{1.3 \times 10^{-4}\,\Omega}{6 \times 10^{-6}\,\Omega\,°C^{-1}}$$

so the new temperature is $21.7\,°C$.

We should note, before proceeding, firstly how the units of Ω on both top and bottom of this fraction cancel; and secondly, how $°C^{-1}$ is in the denominator of the fraction. As a consequence of it being on the bottom of the fraction, it is inverted and so becomes $°C$. In summary, we see how a simple analysis of the units in this sum automatically allows the eventual answer to be expressed in terms of $°C$. We are therefore delighted, because the answer we want is a temperature, and the units tell us it is indeed a temperature.

Aside

This manipulation of units is sometimes called *dimensional analysis*. Strictly speaking, though, dimensional analysis is independent of the units used. For example, the units of speed may be in metres per second, miles per hour, etc., but the dimensions of speed are always a length [L] divided by a time [T]:

$$[\text{speed}] = [L] \div [T]$$

Dimensional analysis is useful in two respects. (1) It can be used to determine the units of a variable in an equation. (2) Using the normal rules of algebra, it can be used to determine whether an equation is *dimensionally correct*, i.e. the units should balance on either side of the equation. *All* equations in any science discipline are dimensionally correct or they are wrong!

A related concept to dimensional analysis is *quantity calculus*, a method we find particularly useful when it comes to setting out table header rows and graph axes. Quantity calculus is the handling of physical quantities and their units using the normal rules of algebra. A *physical quantity* is defined by a *numerical value* and a *unit*:

$$\text{physical quantity} = \text{number} \times \text{unit}$$

e.g.

$$\Delta H = 40.7 \text{ kJ mol}^{-1}$$

which rearranges to

$$\Delta H / \text{kJ mol}^{-1} = 40.7$$

SAQ 1.1 A temperature is measured with the same platinum-resistance thermometer used in Worked Example 1.1, and a resistance $R = 11.4 \times 10^{-4}\,\Omega$ determined. What is the temperature?

The word 'philosophical' comes from the Greek words *philos* meaning 'love' and *sophia* meaning 'wisdom'. *Philosophy* is therefore the love of wisdom. This same usage of 'wisdom' is seen with the initials PhD, which means a 'philosophy doctorate'.

Some people might argue that none of the discussion above actually answers the *philosophical* question, 'What *is* temperature?' We will never come to a completely satisfactory answer; but we can suppose a body has a higher temperature if it contains more energy, and that it has a lower temperature if it has less energy. More importantly, a body will show a rise in temperature if its energy content rises, and it will show a lower temperature if its energy content drops. This is why we sit in front of a fire: we want to absorb energy, which we experience as a higher temperature.

How long is a piece of string?

The SI unit of length

This definition of 'more energy means hotter' needs to be handled with care: consider two identical weights at the same temperature. The higher weight has a greater potential energy.

A common problem in Anglo Saxon England, as well as much of contemporary Europe, was the way cloth merchants could so easily cheat the common people. At a market, it was all too easy to ask for a yard of cloth, to see it measured against the merchant's yardstick, and pay for the cloth only to get home to learn just how short the merchant's stick was. Paying for 10 yards and coming home with only 9 yards was common, it seems; and the problem was not restricted to just cloth, but also to leather and timber.

According to legend, the far-sighted English King Edgar (AD 959–975) solved the problem of how to stop such cheating by standardizing the length. He took 100 foot soldiers and measured the length of the right foot of each, one after the other, as they stood in line along the floor of his threshing hall. This overall length was then subdivided into 100 equal parts to yield the standard length, the foot. The foot is still commonly employed as a unit of length in Britain to this day. Three of these feet made up 1 yard. The king was said to keep in his treasury a rod of gold measuring exactly 1 yard in length. This is one theory of how the phrase 'yardstick' originated. Any merchant accused of cheating was required to bring his yardstick and to compare its length against that of the king. Therefore, a merchant whose stick was shorter was a cheat and paid the consequences. A merchant whose stick was longer was an idiot.

SI units are self-consistent, with all units being defined in terms a basis of seven fundamental units. The SI unit of length *l* is the metre (m).

While feet and yards are still used in Britain and other countries, the usual length is now the metre. At the time of the French Revolution in the 18th century and soon after, the French Academy of Sciences sought to systemize the measurement of all scientific quantities. This work led eventually to the concept of the *Système Internationale*, or SI for short. Within this system, *all* units and definitions are self-consistent. The SI unit of length is the metre.

The original metre rule was kept in the *International Bureau of Weights and Measures* in Sèvres, near Paris, and was a rod of platinum–iridium alloy on which two deep marks were scratched 1 m apart. It was used in exactly the same way as King Edgar's yardstick 10 centuries earlier.

Unfortunately, platinum–iridium alloy was a poor choice, for it has the unusual property of shrinking (albeit microscopically) with time. This SI metre rule is now about 0.3 per cent too short. King Edgar's yardstick, being made of gold, would still be the same length today as when it was made, but gold is too ductile, and could have been stretched, bent or re-scored.

In 1960, the SI unit of length was redefined. While keeping the metre as the unit of length, it is now defined as 1 650 763.73 wavelengths of the light emitted *in vacuo* by krypton-86. This is a sensible standard, because it can be reproduced in any laboratory in the world.

> 'Ductile' means the ability of a metal to be drawn to form a wire, or to be worked. Ductile is the opposite of 'brittle'.

> *In vacuo* is Latin for 'in a vacuum'. Many properties are measured in a vacuum to avoid the complication of interference effects.

How fast is 'greased lightning'?

Other SI standards

In comic books of the 1950s, one of the favourite phrases of super-heroes such as Superman was 'greased lightning!' The idea is one of extreme speed. The lightning we see, greased or otherwise, is a form of light and travels very, very fast. For example, it travels through a vacuum at 3×10^8 m s^{-1}, which we denote with the symbol c. But while the speed c is constant, the actual speed of light may not be: in fact, it alters very slightly depending on the medium through which it travels. We see how a definition of time involving the speed of light is inherently risky, explaining why we now choose to define time in terms of the duration (or fractions and multiples thereof) between static events. And by 'static' we mean unchanging.

SI 'base units'

Time is one of the so-called 'base units' within the SI system, and so is length. Whereas volume can be expressed in terms of a length (for example, a cube has a volume l^3 and side of area l^2), we cannot define length in terms of something simpler. Similarly, whereas a velocity is a length per unit time, we cannot express time in terms of something simpler. In fact, just as compounds are made up of elements, so all scientific units are made up from seven base units: length, time, mass, temperature, current, amount of material and luminous intensity.

> There are seven base SI units: length, time, mass, temperature, current, luminous intensity and amount of material.

Table 1.1 summarizes the seven base (or 'fundamental') SI physical quantities and their units. The last unit, *luminous intensity*, will not require our attention any further.

The SI unit of 'time' t is the second. The second was originally defined as 1/86 400th part of a mean solar day. This definition is

> The SI unit of 'time' t is the second (s).

Table 1.1   The seven fundamental SI physical quantities and their units

Physical quantity	Symbol[a]	SI unit	Abbreviation
Length	l	metre	m
Mass	m	kilogram	kg
Time	t	second	s
Electrical current	I	ampère	A
Thermodynamic temperature	T	kelvin	K
Amount of substance	n	mole	mol
Luminous intensity	I_v	candela	cd

[a]Notice how the abbreviation for each quantity, being a variable, is always italicized, whereas the abbreviation for the unit, which is not a variable, is printed with an upright typeface. None of these unit names starts with a capital.

again quite sensible because it can be reproduced in any laboratory in the world. While slight changes in the length of a solar year do occur, the word 'mean' in our definition obviates any need to consider them. Nevertheless, it was felt necessary to redefine the second; so, in the 1960s, the second was redefined as 9 192 631 770 periods of the radiation corresponding to the transition between two of the hyperfine levels in the ground state of the caesium-133 atom. Without discussion, we note how the heart of a so-called 'atomic clock' contains some caesium-133.

The SI unit of 'temperature' T is the kelvin (K).

In a similar way, the *Système Internationale* has 'defined' other common physicochemical variables. The SI unit of 'temperature' T is the kelvin. We define the kelvin as 1/273.16th part of the thermodynamic temperature difference between absolute zero (see Section 1.4) and the triple point of water, i.e. the temperature at which liquid water is at equilibrium with solid water (ice) and gaseous water (steam) provided that the pressure is 610 Pa.

The SI unit of 'current' I is the ampère (A).

The SI unit of 'current' I is the ampère (A). An ampère was first defined as the current flowing when a charge of 1 C (coulomb) passed per second through a perfect (i.e. resistance-free) conductor. The SI definition is more rigorous: 'the ampère is that constant current which, if maintained in two parallel conductors (each of negligible resistance) and placed *in vacuo* 1 m apart, produces a force between of exactly 2×10^{-7} N per metre of length'. We will not employ this latter definition.

The SI unit of 'amount of substance' n is the mole (mol).

The SI unit of the 'amount of substance' n is the mole. Curiously, the SI General Conference on Weights and Measures only decided in 1971 to incorporate the mole into its basic set of fundamental parameters, thereby filling an embarrassing loophole. The mole is the amount of substance in a system that contains as many elementary entities as does 0.012 kg (12 g) of carbon-12. The amount of substance must be stated in terms of the elementary entities chosen, be they photons, electrons, protons, atoms, ions or molecules.

The number of elementary entities in 1 mol is an experimentally determined quantity, and is called the 'Avogadro constant' L, which has the value 6.022×10^{23} mol^{-1}. The Avogadro constant is also (incorrectly) called the 'Avogadro number'. It is

Table 1.2   Several of the more common units that are not members of the *Système Internationale*

Quantity	Non-SI unit	Abbreviation	Conversion from non-SI to SI
Energy	calorie	cal	$1 \text{ cal} = 4.184 \text{ J}$
Length	ångström	Å	$1 \text{ Å} = 10^{-10} \text{ m}$
Pressure	atmosphere	atm	$1 \text{ atm} = 101\,325 \text{ Pa}$
Pressure	bar	bar	$1 \text{ bar} = 10^5 \text{ Pa}$
Volume	litre	dm^3	$1 \text{ dm}^3 = 10^{-3} \text{ m}^3$

increasingly common to see the Avogadro constant given a different symbol than L. The most popular alternative symbol at present seems to be N_A.

Non-SI units

It is important to be consistent with units when we start a calculation. An enormously expensive spacecraft crashed on the surface of the planet Mars in 1999 because a distance was calculated by a NASA scientist in terms of inches rather than centimetres.

Several non-SI units persist in modern usage, the most common being listed in Table 1.2. A calculation performed wholly in terms of SI units will be self-consistent. Provided we know a suitable way to interchange between the SI and non-SI units, we can still employ our old non-SI favourites.

Aside

In addition to the thermodynamic temperature T there is also the Celsius temperature t, defined as

$$t = T - T_0$$

where $T_0 = 273.15 \text{ K}$.

Sometimes, to avoid confusion with the use of t as the symbol for time, the Greek symbol θ (theta) is substituted for the Celsius temperature t instead.

Throughout this book we adopt T to mean temperature. The context will make clear whether T is required to be in degrees Celsius or kelvin. Beware, though, that most formulae require the use of temperature in kelvin.

Why is the SI unit of mass the kilogram?

Multiples and the SI unit of mass m

The definition of mass in the *Système Internationale* scheme departs from the stated aim of formulating a rigorous, self-consistent set of standards. The SI unit of 'mass'

Table 1.3 Selection of a few physicochemical parameters that comprise combinations of the seven SI fundamental quantities

Quantity	Symbol	SI units
Acceleration	a	$m\,s^{-2}$
Area	A	m^2
Density	ρ	$kg\,m^{-3}$
Force	F	$kg\,m\,s^{-2}$
Pressure	p	$kg\,m^{-1}\,s^{-2}$
Velocity	v	$m\,s^{-1}$
Volume	V	m^3

> The SI unit of 'mass' m is the kilogram (kg).

> In the SI system, 1 g is defined as the mass of 5.02×10^{22} atoms of carbon-12. This number comes from $L/12$, where L is the Avogadro number.

m is the 'kilogram'. Similar to the metre, the original SI standard of mass was a block of platinum metal in Sèvres, near Paris, which weighted exactly 1 kg. The current SI definition is more complicated: because 12.000 g in the SI system represents exactly 1 mol of carbon-12, then 1 g is one-twelfth of a mole of carbon-12.

The problem with the SI base unit being a kilogram is the 'kilo' part. The philosophical idea behind the SI system says any parameter (physical, chemical, mechanical, etc.) can be derived from a suitable combination of the others. For example, the SI unit of velocity is metres per second ($m\,s^{-1}$), which is made up of the two SI fundamental units of length (the metre) and time (the second). A few of these combinations are cited in Table 1.3.

Why is 'the material of action so variable'?

Writing variables and phrases

The classical author Epictetus (*ca* 50–*ca* 138 AD) once said, 'The materials of action are variable, but the use we make of them should be constant'. How wise.

> We give the name 'compound unit' to several units written together. We leave a space between each constituent unit when we write such a compound unit.

When we build a house, we only require a certain number of building materials: say, bricks, tubes and window panes. The quantity surveyor in charge of the building project decides which materials are needed, and writes a quantity beside each on his order form: 10 000 bricks, 20 window panes, etc. Similarly, when we have a velocity, we have the units of 'm' and 's^{-1}', and then quantify it, saying something like, 'The man ran fast, covering a distance of 10 metres per second'. By this means, any parameter is defined both qualitatively (in terms of its units) and quantitatively (in terms of a number). With symbols, we would write $v = 10\ m\,s^{-1}$.

A variable (mass, length, velocity, etc.) is written in a standard format, according to Equation (1.2):

$$\text{Variable or physicochemical quantity} = \text{number} \times \text{units} \qquad (1.2)$$

We sometimes call it a 'phrase'. Because some numbers are huge and others tiny, the SI system allows us a simple and convenient shorthand. We do not need to write out all the zeros, saying the velocity of light c is $300\,000\,000$ m s^{-1}: we can write it as $c = 3 \times 10^8$ m s^{-1} or as 0.3 Gm s^{-1}, where the capital 'G' is a shorthand for 'giga', or $1\,000\,000\,000$. The symbol G (for giga) in this context is called a 'factor'. In effect, we are saying $300\,000\,000$ m s^{-1} = 0.3 Gm s^{-1}. The standard factors are listed on pp. xxviii–xxxi.

> 'Giga' comes from the Latin *gigas*, meaning 'giant' or 'huge'. We also get the everyday words 'giant' and 'gigantic' from this root.

Most people find that writing $300\,000\,000$ m s^{-1} is a bit long winded. Some people do not like writing simple factors such as G for giga, and prefer so-called *scientific notation*. In this style, we write a number followed by a factor expressed as ten raised to an appropriate power. The number above would be 3.0×10^8 m s^{-1}.

> In physical chemistry, a 'factor' is a number by which we multiply the numerical value of a variable. Factors are usually employed with a shorthand notation.

Worked Example 1.2   Identify the variable, number, factor and unit in the phrase, 'energy = 12 kJ mol^{-1}'.

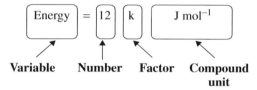

Variable   Number   Factor   Compound unit

Reasoning

Variable – in simple mathematical 'phrases' such as this, we almost always write the variable on the left. A variable is a quantity whose value can be altered.

Number – the easy part! It will be made up of numbers $1, 2, 3, \ldots, 0$.

Factor – if we need a factor, it will always be written between the number and the units (compound or single). A comprehensive list of the simple factors is given on pp. xxviii–xxxi.

Units – the units are always written on the right of a phrase such as this. There are two units here, joules (J) and moles (as 'mol^{-1}', in this case). We should leave a space between them.

A factor is simply shorthand, and is dispensable. We could have dispensed with the *factor* and written the *number* differently, saying energy = $12\,000$ J mol^{-1}. This same energy in scientific notation would be 12×10^3 J mol^{-1}. But units are *not* dispensable.

SAQ 1.2 Identify the variable, number, factor and unit in the phrase, 'length = 3.2 km'.

1.3   Properties of gases and the gas laws

Why do we see eddy patterns above a radiator?

The effects of temperature on density

The air around a hot radiator soon acquires heat. We explain this observation from the 'minus oneth law of thermodynamics' (see Section 1.1), since heat travels from hot to cold.

The density of a gas depends quite strongly on its temperature, so hot air has a smaller density than does cold air; colder air is more dense than hot air. From everyday experience, we know that something is dense if it tries to drop, which is why a stone drops to the bottom of a pond and a coin sinks to the bottom of a pan of water. This relative motion occurs because both the stone and the coin have higher densities than does water, so they drop. Similarly, we are more dense than air and will drop if we fall off a roof.

Just like the coin in water, cold air sinks because it is denser than warmer air. We sometimes see this situation stated as warm air 'displaces' the cold air, which subsequently takes its place. Alternatively, we say 'warm air rises', which explains why we place our clothes *above* a radiator to dry them, rather than below it.

Light entering the room above the radiator passes through these pockets of warm air as they rise through colder air, and therefore passes through regions of different density. The rays of light bend in transit as they pass from region to region, much in the same way as light twists when it passes through a glass of water. We say the light is *refracted*. The eye responds to light, and interprets these refractions and twists as different intensities.

So we see swirling eddy (or 'convective') patterns above a radiator because the density of air is a function of temperature. If all the air had the same temperature, then no such difference in density would exist, and hence we would see no refraction and no eddy currents – which is the case in the summer when the radiator is switched off. Then again, we can sometimes see a 'heat haze' above a hot road, which is caused by exactly the same phenomenon.

Why does a hot-air balloon float?

The effect of temperature on gas volume

A hot-air balloon is one of the more graceful sights of summer. A vast floating ball, powered only by a small propane burner, seems to defy gravity as it floats effortlessly above the ground. But what is it causing the balloon to fly, despite its considerable weight?

The small burner at the heart of the balloon heats the air within the canvas hood of the balloon. The densities of all materials – solid, liquid or gas – alter with temperature. Almost universally, we find the density ρ increases with cooling. Density ρ is defined as the ratio of mass m to volume V, according to

> 'Density' ρ is defined as mass per unit volume.

$$\text{density } \rho = \frac{\text{mass, } m}{\text{volume, } V} \qquad (1.3)$$

It is not reasonable to suppose the mass m of a gas changes by heating or cooling it (in the absence of chemical reactions, that is), so the changes in ρ caused by heating must have been caused by changes in volume V. On the other hand, if the volume were to *decrease* on heating, then the density would *increase*.

So the reason why the balloon floats is because the air inside its voluminous hood has a lower density than the air outside. The exterior air, therefore, sinks lower than the less-dense air inside. And the sinking of the cold air and the rising of the warm air is effectively the same thing: it is movement of the one relative to the other, so the balloon floats above the ground. Conversely, the balloon descends back to earth when the air it contains cools to the same temperature as the air outside the hood.

How was the absolute zero of temperature determined?

Charles's law

J. A. C. Charles (1746–1823) was an aristocratic amateur scientist of the 18th century. He already knew that the volume V of a gas increased with increasing temperature T, and was determined to find a relationship between these variables. The law that now bears his name can be stated as, 'The ratio of volume and temperature for a fixed mass of gas remains constant', provided the external pressure is not altered.

> According to 'Charles's law', a linear relationship exists between V and T (at constant pressure p).

Stated mathematically, Charles demonstrated

$$\frac{V}{T} = \text{constant} \qquad (1.4)$$

where the value of the constant depends on both the amount and the identity of the gas. It also depends on the pressure, so the data are obtained at constant pressure p.

This is one form of 'Charles's law'. (Charles's law is also called 'Gay–Lussac's law'.) Alternatively, we could have multiplied both sides of Equation (1.4) by T, and rewritten it as

> A 'straight line' will always have an equation of the type $y = mx + c$, where m is the gradient and c is the intercept on the y-axis (i.e. when the value of $x = 0$).

$$V = \text{constant} \times T \qquad (1.5)$$

Lord Kelvin (1824–1907) was a great thermodynamicist whom we shall meet quite often in these pages. He noticed how the relationship in Equation (1.5) resembles the equation of a straight line, i.e. takes the form

$$y = mx + c$$

observed variable   gradient   controlled variable   constant

(1.6)

except without an intercept, i.e. $c = 0$. Kelvin obtained good-quality data for the volume of a variety of gases as a function of temperature, and plotted graphs of volume V (as y) against temperature T (as x) for each; curiously, however, he was unable to draw a graph with a zero intercept for any of them.

Kelvin then replotted his data, this time extrapolating each graph till the volume of the gas was zero, which he found to occur at a temperature of $-273.15\,°C$; see Figure 1.5. He then devised a new temperature scale in which this, the coldest of temperatures, was the zero. He called it *absolute zero*, and each subsequent degree was equal to $1\,°C$. This new scale of temperature is now called the *thermodynamic* (or *absolute*) scale of temperature, and is also sometimes called the *Kelvin scale*.

> *Note*: degrees in the Kelvin scale do not have the degree symbol. The units have a capital K, but the noun 'kelvin' has a small letter.

The relationship between temperatures T on the centigrade and the absolute temperature scales is given by

$$T \text{ in } °C = T \text{ in } K - 273.15$$

(1.7)

Equation (1.7) demonstrates how $1\,°C = 1\,K$.

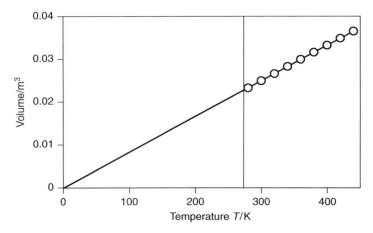

Figure 1.5 A graph of the volume V of a gas (as y) against temperature T (as x) is linear. Extrapolating the gas's volume suggests its volume will be zero if the temperature is $-273.15\,°C$ (which we call $0\,K$, or absolute zero)

SAQ 1.3 What is the temperature T expressed in kelvin if the temperature is 30 °C?

SAQ 1.4 What is the centigrade temperature corresponding to 287.2 K?

SAQ 1.5 The data in the table below relate to gaseous helium. Demonstrate the linear relationship between the volume V and the temperature T.

> We divide each temperature, both kelvin and centigrade, by its respective unit to obtain a *number*, rather than the temperature.

Temperature T/K	280	300	320	340	360	380	400	420	440
Volume V/m^3	0.023	0.025	0.027	0.028	0.030	0.032	0.033	0.035	0.037

Charles's law is often expressed in a slightly different form than Equation (1.4), as

$$\frac{V_1}{T_1} = \frac{V_2}{T_2} \tag{1.8}$$

which is generally regarded as superior to Equation (1.4) because we do not need to know the value of the constant.

Equation (1.8) is also preferred in situations where the volume of a fixed amount of gas changes in response to temperature changes (but at constant pressure). The subscripts refer to the two situations; so, for example, the volume at temperature T_1 is V_1 and the volume at temperature T_2 is V_2.

> Note how we write the controlled variable along the *top* row of a table, with the observed following. (If the table is vertical, we write the controlled variable on the far left.)

SAQ 1.6 The gas inside a balloon has a volume V_1 of 1 dm^3 at 298 K. It is warmed to 350 K. What is the volume following warming? Assume the pressure remained constant.

> The subscripts written to the right of a variable are called 'descriptors'. They are always written as a *sub*script, because a *super*scripted number means a power, i.e. V^2 means $V \times V$.

Why pressurize the contents of a gas canister?

The effect of pressure on gas volume: Boyle's law

It is easy to buy canisters of gas of many sizes, e.g. as fuel when we wish to camp in the country, or for a portable welding kit. The gas will be *n*-butane if the gas is for heating purposes, but might be oxygen or acetylene if the gas is to achieve the higher temperatures needed for welding.

Typically, the components within the can are gaseous at most temperatures. The typical volume of an aerosol can is about 0.3 dm^3 (3×10^{-4} m^3), so it could contain very little gas if stored at normal pressure. But if we purchase a canister of gas and release its entire contents at once, the gas would occupy a volume similar that of

Care: a small *p* indicates pressure, yet a big P is the symbol for the element phosphorus. Similarly, a big *V* indicates volume and a small *v* is the symbol for velocity.

an entire living room. To ensure the (small) can contains this (large) amount of gas, we *pressurize* it to increase its capacity. We see how volume and pressure are interrelated in a reciprocal way: the volume *decreases* as the pressure *increases*.

Robert Boyle was the first to formulate a relationship between *p* and *V*. Boyle was a contemporary of the greatest scientist the world has ever seen, the 17th-century physicist Sir Isaac Newton. Boyle's law was discovered in 1660, and states

$$pV = \text{constant} \tag{1.9}$$

where the numerical value of the constant on the right-hand side of the equation depends on both the identity and amount of the gas, as well as its temperature *T*.

An 'isotherm' is a line on a graph representing values of a variable obtained at constant temperature.

Figure 1.6 shows a graph of pressure *p* (as *y*) against volume *V* (as *x*) for 1 mol of neon gas. There are several curves, each representing data obtained at a different temperature. The temperature per curve was constant, so we call each curve an *isotherm*. The word *isotherm* has two Greek roots: *iso* means 'same' and *thermo* means temperature or energy. An isotherm therefore means at the same energy.

The actual shape of the curves in Figure 1.6 are those of reciprocals. We can prove this mathematical form if we divide both sides of Equation (1.9) by *V*, which yields

'Reciprocal' means to turn a fraction upside down. *X* can be thought of as '*X* ÷ 1', so its reciprocal is 1/*X* (i.e. 1 ÷ *X*).

$$p = \frac{1}{V} \times \text{constant} \tag{1.10}$$

Figure 1.7 shows a graph of volume *p* (as *y*) against 1/volume *V* (as *x*), and has been constructed with the same data as used for

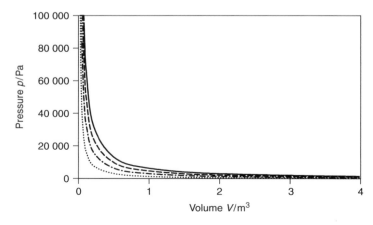

Figure 1.6 Graph of pressure *p* (as *y*) against volume *V* (as *x*) for 1 mol of an ideal gas as a function of temperature: (·····) 200 K; (–·–·–) 400 K; (–––) 600 K; (———) 800 K

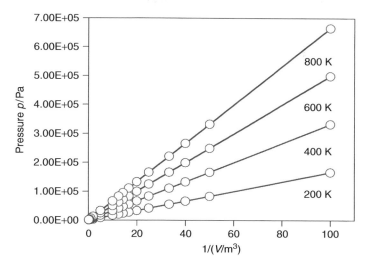

Figure 1.7 Graph of pressure p (as y) against reciprocal volume $1 \div V$ (as x) for 1 mol of an ideal gas as a function of temperature. The data are the same as those from Figure 1.6. The temperatures are indicated. We need to appreciate how plotting the same data on a different set of axes yields a *linear* graph, thereby allowing us to formulate a *relationship* between p and $1 \div V$

Figure 1.6. Each of the lines on the graph is now linear. Again, we find these data are temperature dependent, so each has the gradient of the respective value of 'constant'.

At constant temperature T, an *increase* in pressure (so $p_2 > p_1$) causes a *decrease* in volume (so $V_2 < V_1$). This observation explains why the graph for the gas at the higher temperatures has a smaller value for the constant.

An alternative way of writing Equation (1.9) is

> At constant temperature T, an *increase* in pressure ($p_2 > p_1$) causes a *decrease* in volume ($V_2 < V_1$).

$$p_1 V_1 = p_2 V_2 \tag{1.11}$$

SAQ 1.7 The usual choice of propellant within an aerosol of air freshener is propane gas. What is the volume of propane following compression, if 1 dm^3 of gaseous propane is compressed from a pressure of 1 atm to a pressure of 2.5 atm? Assume the temperature is kept constant during the compression.

Why does thunder accompany lightning?

Effect of changing both temperature and pressure on gas volume

Lightning is one of the most impressive and yet frightening manifestations of nature. It reminds us just how powerful nature can be.

'Experiential' means the way we notice something exists following an experience or sensation.

Lightning is quite a simple phenomenon. Just before a storm breaks, perhaps following a period of hot, fine weather, we often note how the air feels 'tense'. In fact, we are expressing an experiential truth: the air contains a great number of *ions* – charged particles. The existence of a large charge on the Earth is mirrored by a large charge in the upper atmosphere. The only difference between these two charges is that the Earth bears a positive charge and the atmosphere bears a negative charge.

Accumulation of a charge difference between the Earth and the upper atmosphere cannot proceed indefinitely. The charges must eventually equalize somehow: in practice, negative charge in the upper atmosphere passes through the air to neutralize the positive charge on the Earth. The way we see this charge conducted between the Earth and the sky is lightning: in effect, air is ionized to make it a conductor, allowing electrons in the clouds and upper atmosphere to conduct through the air to the Earth's surface. This movement of electrical charge is a *current*, which we see as lightning. Incidentally, ionized air emits light, which explains why we *see* lightning (see Chapter 9). Lightning comprises a massive amount of energy, so the local air through which it conducts tends to heat up to as much as a few thousand degrees centigrade.

And we have already seen how air expands when warmed, e.g. as described mathematically by Charles's law (Equation (1.6)). In fact, the air through which the lightning passes increases in volume to an almost unbelievable extent because of its rise in temperature. And the expansion is very rapid.

SAQ 1.8 Show, using the version of Charles's law in Equation (1.8), how a rise in temperature from 330 K to 3300 K is accompanied by a tenfold increase in volume.

We hear the sensation of sound when the ear drum is moved by compression waves travelling through the air; we hear people because their speech is propagated by subtle pressure changes in the surrounding air. In a similar way, the huge increase in air volume is caused by huge changes in air pressure, itself manifested as sound: we hear the thunder caused by the air expanding, itself in response to lightning.

And the reason why we see the lightning first and hear the thunder later is because light travels faster than sound. The reason why thunder accompanies lightning, then, is because pressure p, volume V and temperature T are interrelated.

How does a bubble-jet printer work?

The ideal-gas equation

A bubble-jet printer is one of the more useful and versatile inventions of the last decade. The active component of the printer is the 'head' through which liquid ink passes before striking the page. The head moves from side to side over the page. When the 'head' is positioned above a part of the page to which an image is required, the computer tells the head to eject a tiny bubble of ink. This jet of ink strikes the page to leave an indelible image. We have printing.

The head is commonly about an inch wide, and consists of a row of hundreds of tiny pores (or 'capillaries'), each connecting the ink reservoir (the cartridge) and the page. The signals from the computer are different for each pore, allowing different parts of the page to receive ink at different times. By this method, images or letters are formed by the printer.

The pores are the really clever bit of the head. Half-way along each pore is a minute heater surrounded by a small pocket of air. In front of the heater is a small bubble of ink, and behind it is the circuitry of the printer, ultimately connecting the heater to the computer. One such capillary is shown schematically in Figure 1.8.

Just before the computer instructs the printer to eject a bubble of ink, the heater is activated, causing the air pocket to increase in temperature T at quite a rapid rate. The temperature increase causes the air to expand to a greater volume V. This greater volume increases the pressure p within the air pocket. The enhanced air pressure p is sufficient to eject the ink bubble from the pore and onto the page. This pressure-activated ejection is similar to spitting.

This ejection of ink from a bubble-jet printer ingeniously utilizes the interconnectedness of pressure p, volume V and temperature T. Experiments with simple gases show how p, T and V are related by the relation

$$\frac{pV}{T} = \text{constant} \qquad (1.12)$$

which should remind us of both Boyle's law and Charles's law.

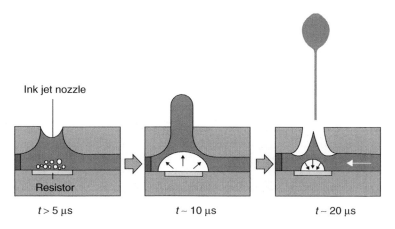

Figure 1.8 Schematic diagram of a capillary (one of hundreds) within the printing 'head' of a bubble-jet printer. The resistor heats a small portion of solution, which boils thereby increasing the pressure. Bubbles form within 5 μs of resistance heating; after 10 μs the micro-bubbles coalesce to force liquid from the aperture; and a bubble is ejected a further 10 μs later. The ejected bubble impinges on the paper moments afterwards to form a written image. Reproduced by permission of Avecia

If there is exactly 1 mol of gas, the pressure is expressed in pascals (Pa), the temperature is in kelvin and the volume is in cubic metres (both SI units), then the value of the constant is $8.314 \, \mathrm{J\,K^{-1}\,mol^{-1}}$. We call it the *gas constant* and give it the symbol R. (Some old books may call R the 'universal gas constant', 'molar gas constant' or just 'the gas constant'. You will find a discussion about R on p. 54)

More generally, Equation (1.12) is rewritten as

Equation (1.13) tells us the constant in Boyle's law is 'nRT' and the (different) constant in Charles's law is '$nR \div p$'.

$$pV = nRT \tag{1.13}$$

where n is the number of moles of gas. Equation (1.13) is called the *ideal-gas equation* (or, sometimes, in older books the 'universal gas equation'). The word 'ideal' here usually suggests that the gas in question obeys Equation (1.13).

Worked Example 1.3   What is the volume of 1 mol of gas at a room temperature of $25\,^{\circ}\mathrm{C}$ at an atmospheric pressure of 10^5 Pa?

First, we convert the data into the correct SI units. In this example, only the temperature needs to be converted. From Equation (1.7), the temperature is 298 K.

Secondly, we rearrange Equation (1.13) to make V the subject, by dividing both sides by p:

$$V = \frac{nRT}{p}$$

and then insert values:

$$V = \frac{1 \text{ mol} \times 8.314 \text{ J K}^{-1}\text{mol}^{-1} \times 298 \text{ K}}{10^5 \text{ Pa}}$$

So the volume $V = 0.0248 \text{ m}^3$.

If we remember how there are 1000 $\mathrm{dm^3}$ in 1 $\mathrm{m^3}$, we see how 1 mol of gas at room temperature and standard pressure has a volume of 24.8 $\mathrm{dm^3}$.

SAQ 1.9   2 mol of gas occupy a volume $V = 0.4 \text{ m}^3$ at a temperature $T = 330$ K. What is the pressure p of the gas?

An alternative form of Equation (1.13) is given as

$$\frac{p_1 V_1}{T_1} = \frac{p_2 V_2}{T_2} \tag{1.14}$$

and is used when we have to start with a constant number of moles of gas n housed in a volume V_1. Its initial pressure is p_1 when the temperature is T_1. Changing one variable causes at least one of the two to change. We say the new temperature is T_2, the new pressure is p_2 and the new volume is V_2. Equation (1.14) then holds provided the number of moles does not vary.

Worked Example 1.4   Nitrogen gas is housed in a sealed, hollow cylinder at a pressure of 10^5 Pa. Its temperature is 300 K and its volume is 30 dm^3. The volume within the cylinder is increased to 45 dm^3, and the temperature is increased at the same time to 310 K. What is the new pressure, p_2?

We first rearrange Equation (1.14) to make the unknown volume p_2 the subject, writing

$$p_2 = \frac{p_1 V_1 T_2}{T_1 V_2}$$

We insert values into the rearranged equation:

$$p_2 = \frac{10^5 \text{ Pa} \times 30 \text{ dm}^3 \times 310 \text{ K}}{300 \text{ K} \times 45 \text{ dm}^3}$$

> Note how the units of volume cancel, meaning we can employ any unit of volume provided the units of V_1 and V_2 are the same.

so $p_2 = 0.69 \times 10^5$ Pa. The answer demonstrates how the pressure drops by about a third on expansion.

SAQ 1.10   The pressure of some oxygen gas is doubled from 1.2×10^5 Pa to 2.4×10^5 Pa. At the same time, the volume of the gas is decreased from 34 dm^3 to 29 dm^3. What is the new temperature T_2 if the initial temperature T_1 was 298 K?

Justification Box 1.1

We start with n moles of gas at a temperature T_1, housed in a volume V_1 at a pressure of p_1. Without changing the amount of material, we change the volume and temperature to V_2 and T_2 respectively, therefore causing the pressure to change to p_2.

The number of moles remains unaltered, so we rearrange Equation (1.13) to make n the subject:

$$n = p_1 V_1 \div R T_1$$

Similarly, the same number of moles n under the second set of conditions is

$$n = p_2 V_2 \div R T_2$$

Again, although we changed the physical conditions, the number of moles n remains constant, so these two equations must be the same. We say

$$\frac{p_1 V_1}{R T_1} = n = \frac{p_2 V_2}{R T_2}$$

As the value of R (the gas constant) does not vary, we can simplify the equation by multiplying both sides by R, to obtain

$$\frac{p_1 V_1}{T_1} = \frac{p_2 V_2}{T_2}$$

which is Equation (1.14).

What causes pressure?

Motion of particles in the gas phase

The question, 'What *is* pressure?' is another odd question, but is not too difficult to answer.

The constituent particles of a substance each have energy. In practice, the energy is manifested as *kinetic* energy – the energy of movement – and explains why all molecules and atoms move continually as an expression of that kinetic energy. This energy decreases as the temperature decreases. The particles only stop moving when cooled to a temperature of absolute zero: 0 K or $-273.15\,°C$.

The particles are not free to move throughout a solid substance, but can vibrate about their mean position. The frequency and amplitude of such vibration increases as the temperature rises. In a liquid, lateral motion of the particles *is* possible, with the motion becoming faster as the temperature increases. We call this energy *translational energy*. Furthermore, as the particles acquire energy with increased temperature, so the interactions (see Chapter 2) between the particles become comparatively smaller, thereby decreasing the viscosity of the liquid and further facilitating rapid motion of the particles. When the interactions become negligible (comparatively), the particles can break free and become gaseous.

And each particle in the gaseous state can move at amazingly high speeds; indeed, they are often supersonic. For example, an average atom of helium travels at a mean speed of 1204 m s^{-1} at 273.15 K. Table 1.4 lists the mean speeds of a few other gas molecules at 273.15 K. Notice how heavier molecules travel more slowly, so carbon dioxide has a mean speed of 363 m s^{-1} at the same temperature. This high speed of atomic and molecular gases as they move is a manifestation of their enormous kinetic energy. It would not be possible to travel so fast in a liquid or solid because they are so much denser – we call them *condensed phases*.

The gas particles are widely separated.

Particles of gas travel fast and in straight lines, unless they collide.

The separation between each particle in gas is immense, and usually thousands of times greater than the diameter of a single gas particle. In fact, more than 99 per cent of a gas's volume is empty space. The simple calculation in Worked Example 1.5 demonstrates this truth.

Table 1.4 The average speeds of gas molecules at 273.15 K, given in order of increasing molecular mass. The speeds $\bar{c}$ are in fact root-mean-square speeds, obtained by squaring each velocity, taking their mean and then taking the square root of the sum

Gas	Speed $\bar{c}/\mathrm{m\,s^{-1}}$
Monatomic gases	
Helium	1204.0
Argon	380.8
Mercury	170.0
Diatomic gases	
Hydrogen	1692.0
Deuterium	1196.0
Nitrogen	454.2
Oxygen	425.1
Carbon monoxide	454.5
Chlorine	285.6
Polyatomic gases	
Methane	600.6
Ammonia	582.7
Water	566.5
Carbon dioxide	362.5
Benzene	272.8

Worked Example 1.5   What is the molar volume of neon, assuming it to be a straightforward solid?

We must first note how the neon must be extremely cold if it is to be a solid – probably no colder than about 20 K.

> The 'molar volume' is the name we give to the volume 'per mole'.

We know that the radius of a neon atom from tables of X-ray crystallographic data is about 10^{-10} m, so the volume of one atom (from the equation of a sphere, $V = \frac{4}{3}\pi r^3$) is 4.2×10^{-30} m^3. If we assume the neon to be a simple solid, then 1 mol of neon would occupy a volume of 4.2×10^{-30} m^3 per atom $\times$ 6.022×10^{23} atoms per mole = 2.5×10^{-6} m^3 mol^{-1}. This volume represents 2.5 cm^3 mol^{-1}.

A volume of 2.5 cm^3 mol^{-1} is clearly much smaller than the value we calculated earlier in Worked Example 1.3 with the ideal-gas equation, Equation (1.13). It is also smaller than the volume of solid neon made in a cryostat, suggesting the atoms in a solid are also separated by much empty space, albeit not so widely separated as in a gas.

> By corollary, if the gas particles move fast and the gas is ideal, the gas particles must travel in straight lines between collisions.

In summary, we realize how each particle of gas has enormous kinetic energy and are separated widely. Yet, like popcorn in a popcorn maker, these particles cannot be classed as wholly independent, one from another, because they collide. They collide

Newton's first law states that every *action* has an equal but opposite *reaction*. His second law relates the force acting on an object to the product of its mass multiplied by its acceleration.

The pressure of a gas is a 'macroscopic' manifestation of the 'microscopic' gas particles colliding with the internal walls of the container.

The surface area inside a cylinder of radius r and height h is $2\pi rh$. Don't forget to include the areas of the two ends, each of which is πr^2.

firstly with each other, and secondly with the internal walls of the container they occupy.

Just like the walls in a squash court, against which squash balls continually bounce, the walls of the gas container experience a force each time a gas particle collides with them. From Newton's laws of motion, the force acting on the wall due to this incessant collision of gas particles is equal and opposite to the force applied to it. If it were not so, then the gas particles would not bounce following a collision, but instead would go *through* the wall.

We see how each collision between a gas particle and the internal walls of the container causes the same result as if we had applied a force to it. If we call the area of the container wall A and give the symbol F to the sum of the forces of all the particles in the gas, then the pressure p exerted by the gas-particle collisions is given by

$$\text{pressure, } p = \frac{\text{force, } F}{\text{area, } A} \qquad (1.15)$$

In summary, the pressure caused by a container housing a gas is simply a manifestation of the particles moving fast and colliding with the container walls.

SAQ 1.11 A cylindrical can contains gas. Its height is 30 cm and its internal diameter is 3 cm. It contains gas at a pressure of 5×10^5 Pa. First calculate the area of the cylinder walls (you will need to know that 1 m = 100 cm, so 1 m^2 = 10^4 cm^2), and then calculate the force necessary to generate this pressure.

Aside

A popular misconception says a molecule in the gas phase travels faster than when in a liquid. In fact, the molecular velocities will be the same in the gas and liquid phases if the temperatures are the same. Molecules only *appear* to travel slower in a liquid because of the large number of collisions between its particles, causing the overall distance travelled per unit time to be quite short.

Why is it unwise to incinerate an empty can of air freshener?

The molecular basis of the gas laws

The writing printed on the side of a can of air freshener contains much information. Firstly, it cites the usual sort of advertising prose, probably saying it's a better product than anyone else's, and smells nicer. Few people seem to bother reading these bits. But in most countries, the law says the label on the can should also gives details of the can's contents, both in terms of the net mass of air freshener it contains and also perhaps a few details concerning its chemical composition. Finally, a few words of instruction say how to dispose safely of the can. In this context, the usual phrase printed on the can is, 'Do not incinerate, even when empty'. But why?

It is common for the can to contain a propellant in addition to the actual components of the air freshener mixture. Commonly, butane or propane are chosen for this purpose, although CFCs were the favoured choice in the recent past.

Such a can is thrown away when it contains no more air freshener, although it certainly still contains much propellant. Incineration of the can leads to an increase in the kinetic energy of the remaining propellant molecules, causing them to move faster and faster. And as their kinetic energy increases, so the frequency with which they strike the internal walls of the can increases. The force of each collision also increases. In fact, we rediscover the ideal gas equation, Equation (1.13), and say that the pressure of the gas (in a constant-volume system) increases in proportion to any increase in its temperature. In consequence, we should not incinerate an old can of air freshener because the internal pressure of any residual propellant increases hugely and the can explodes. Also note the additional scope for injury afforded by propane's flammability.

> CFC stands for chlorofluorocarbon. Most CFCs have now been banned because of their ability to damage the ozone layer in the upper atmosphere.

> Pressure increases with increasing temperature because the collisions between the gas particles and the container wall are more energetic and occur more frequently.

.4   Further thoughts on energy

Why is the room warm?

The energy of room temperature

Imagine coming into a nice, warm room after walking outside in the snow. We instantly feel warmer, because the room is warmer. But what exactly is the energy content of the room? Stated another way, how much energy do we get from the air in the room by virtue of it being at its own particular temperature?

For simplicity, we will consider only the molecules of gas. Each molecule of gas will have kinetic energy (the energy of movement) unless the temperature is absolute zero. This energy may be transferred through inelastic molecules collisions. But how much kinetic energy does the gas have?

At a temperature T, 1 mol of gas has a kinetic energy of $\frac{3}{2}RT$, where T is the thermodynamic temperature and R is the gas constant. This energy is directly proportional to the thermodynamic temperature, explaining why we occasionally call the kinetic energy 'thermal motion energy'. This simple relationship says that temperature is merely a measure of the average kinetic energy of gas molecules moving chaotically.

It is important to appreciate that this energy relates to the *average* energy of 1 mol of gas molecules. The concept of temperature has no meaning when considering a single molecule or atom. For example, the velocity (and hence the kinetic energy) of a single particle changes with time, so in principle its temperature also changes. Temperature only acquires any thermodynamic meaning when we consider *average* velocities for a large number of particles.

Provided we know the temperature of the gas, we know its energy – the energy it has simply by existing at the temperature T.

Worked Example 1.6   What is the energy of 1 mol of gas in a warm room at 310 K?

> The 'room energy' $\frac{3}{2}RT$ derives from the kinetic (movement) energy of a gas or material.

The energy per mole is $\frac{3}{2} \times R \times T$; so, inserting values, energy = $\frac{3}{2} \times 8.314 \text{ J K}^{-1} \text{ mol}^{-1} \times 310$ K.

$$\text{Energy} = 3866 \text{ J mol}^{-1} \approx 3.9 \text{ kJ mol}^{-1}$$

The molar energy of these molecules is about 4 kJ mol^{-1}, which is extremely slight compared with the energy of the bonds connecting the respective atoms within a molecule (see Chapters 2 and 3). There is little chance of this room energy causing bonds to break or form.

SAQ 1.12   What is the room energy per mole on a cold winter's day, at $-8\,^{\circ}$C (265 K)?

What do we mean by 'room temperature'?

Standard temperature and pressure

Suppose two scientists work on the same research project, but one resides in the far north of the Arctic Circle and the other lives near the equator. Even if everything else is the same – such as the air pressure, the source of the chemicals and the manufacturers of the equipment – the difference between the temperatures in the two laboratories will cause their results to differ widely. For example, the 'room energy' RT will differ. One scientist will not be able to repeat the experiments of the other, which is always bad science.

> An experiment should always be performed at a known, fixed temperature.

An experiment should always be performed at known temperature. Furthermore, the temperature should be constant throughout the course of the experiment, and should be noted in the laboratory notebook.

But to enable complete consistency, we devise what is called a set of *standard conditions*. 'Standard pressure' is given the symbol $p^{\ominus}$, and has a value of 10^5 Pa. We sometimes call it '1 bar'. Atmospheric pressure has a value of 101 325 Pa, so it is larger than $p^{\ominus}$. We often give atmospheric pressure the symbol 'atm'.

'Standard temperature' has the value of 298 K exactly, which equates to just below 25 °C. If both the pressure and the temperature are maintained at these standard conditions, then we say the measurement was performed at 'standard temperature and pressure', which is universally abbreviated to 's.t.p.' If the scientists at the equator and the Arctic Circle perform their work in thermostatically controlled rooms, both at s.t.p., then the results of their experiments will be identical.

A 'thermostat' is a device for maintaining a temperature. *Thermo* is Greek for 'energy' or 'temperature', and 'stat' derives from the Greek root *statikos*, meaning 'to stand', i.e. not move or alter.

Why do we get warmed-through in front of a fire, rather than just our skins?

The Maxwell–Boltzmann distribution of energies

If no heat was distributed, then our faces and those parts closest to the fire would quickly become unbearably hot, while the remainder of our flesh would continue to feel cold. Heat *conducts* through the body principally by the fire warming the blood on the surface of the skin, which is then pumped to other parts of the body through the circulatory system. The energy in the warmed blood is distributed within cooler, internal tissues.

It is important to note how the heat energy is *distributed* around the body, i.e. shared and equalized. Nature does not like diversity in terms of energetic content, and provides many mechanisms by which the energy can be 'shared'. We shall discuss this aspect of thermochemistry in depth within Chapter 4.

We can be certain that molecules do not each have the same energy, but a *distribution* of energies. The graph in Figure 1.9 concerns the energies in a body. The x-axis gives the range of energies possible, and the y-axis represents the number of particles in the body (molecules, atoms, etc.) having that energy. The graph clearly shows how few particles possess a large energy and how a few particles have a tiny energy, but the majority have lesser energies. We call this spread of energies the 'Maxwell–Boltzmann distribution'.

All speeds are found at all temperatures, but more molecules travel at faster speeds at the higher temperatures.

The distribution law depicted in Figure 1.9 may be modelled mathematically, to describe the proportions of molecules of molar mass M with energies E in the range E to $E + dE$ that exist in

We often see this relationship called merely the 'Boltzmann distribution', after the Austrian Physicist Ludwig Boltzmann (1844–1906), who played a pivotal role in marrying thermodynamics with statistical and molecular physics.

The thermodynamic temperature is the sole variable required to define the Maxwell–Boltzmann distribution: raising the temperature increases the spread of energies.

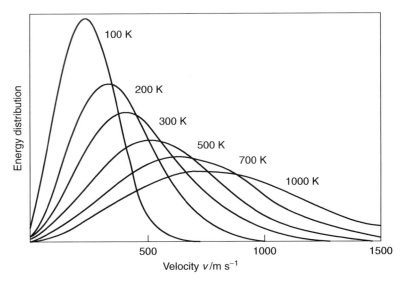

Figure 1.9   Molecular energies follow the Maxwell–Boltzmann distribution: energy distribution of nitrogen molecules (as y) as a function of the kinetic energy, expressed as a molecular velocity (as x). Note the effect of raising the temperature, with the curve becoming flatter and the maximum shifting to a higher energy

thermal equilibrium with each other at a temperature T:

$$f(E) = 4\pi \left(\frac{M}{2\pi RT} \right)^{3/2} E^2 \exp \left(-\frac{Ms^2}{2RT} \right) \tag{1.16}$$

where f on the far left indicates the 'function' that is to be applied to the variable E: the mathematical nature of this function is given by the right-hand side of the equation.

So, in summary, we feel warmer in front of a fire because energy is distributed between those parts facing the flames and the more hidden tissues within.

2

Introducing interactions and bonds

Introduction

We look first at deviations from the ideal-gas equation, caused by inter-particle interactions. Having described induced dipoles (and hydrogen bonds) the interaction strengths are quantified in terms of the van der Waals and virial equations of state.

Next, formal bonds are described, both *covalent* (with electrons shared between participating atoms) and *ionic* (in which electrons are swapped to form charged ions; these ions subsequently associate in response to electrostatic forces). Several underlying factors are expounded, such as ionization energy I and electron affinity $E_{(ea)}$. The energy changes occurring while forming these interactions are alluded to, but are treated properly in Chapter 3.

2.1 Physical and molecular interactions

What is 'dry ice'?

Deviations from the ideal-gas equation

We call solid carbon dioxide (CO_2) 'dry ice'. To the eye, it looks just like normal ice, although it sometimes appears to 'smoke'; see below. Carbon dioxide is a gas at room temperature and only solidifies (at atmospheric pressure) if the temperature drops to about $-78\,^{\circ}C$ or less, so we make dry ice by cooling gaseous CO_2 below its freezing temperature. We call it *dry* ice because, unlike normal ice made with water, warming it above its melting temperature leaves no puddle of liquid, because the CO_2 converts directly to a gas. We say it *sublimes*.

> Substances sublime if they pass directly from a solid to form a gas without being a liquid as an intermediate phase; see Chapter 5.

Gases become denser as we lower their temperature. If CO_2 was still a gas at $-90\,^{\circ}C$, then its molar volume would be $15\,200$ cm^3. In fact, the molar volume of *solid* CO_2 at this temperature is about 30 cm^3. We deduce that CO_2 does not obey the ideal-gas equation (Equation (1.13)) below its freezing temperature, for the very obvious reason that it is no longer a gas.

SAQ 2.1   Show that the volume of 1 mol of CO_2 would be 15 200 cm^3 at $p^\ominus$ and $-90\,°C$ (183 K). [Hint: use the ideal-gas equation. To express this answer in cubic metres, you will need to remember that 1 m^3 = 10^3 dm^3 and 10^6 cm^3.]

Although solidifying CO_2 is an extreme example, it does show how deviations from the ideal-gas equation occur.

How is ammonia liquefied?

Intermolecular forces

Compressing ammonia gas under high pressure forces the molecules into close proximity. In a normal gas, the separation between each molecule is generally large – approximately 1000 molecular diameters is a good generalization. By contrast, the separation between the molecules in a *condensed phase* (solid or liquid) is more likely to be one to two molecular diameters, thereby explaining why the molar volume of a solid or liquid is so much smaller than the molar volume of a gas.

> We sometimes call a solid or a liquid a 'condensed phase'.

> 'Intermolecular' means 'between molecules'.

As a direct consequence of the large intermolecular separations, we can safely say no *interactions* form between the molecules in ammonia gas. The molecules are simply too far apart. We saw in the previous chapter how the property known as pressure is a *macroscopic* manifestation of the *microscopic* collisions occurring between gas particles and, say, a solid object such as a container's walls. But the gas particles can also strike each other on the same microscopic scale: we say the resultant interactions between molecules are *intermolecular*.

> A 'formal bond' involves the *permanent* involvement of electrons in covalent or ionic bonds; see p. 64. Interactions between molecules in a compressed gas are *temporary*.

Intermolecular interactions only operate over relatively short distances, so we assume that, under normal conditions, each molecule in a gas is wholly unaffected by all the others. By contrast, when the gas is compressed and the particles come to within two or three molecular diameters of each other, they start to 'notice' each other. We say the outer-shell electrons on an atom are *perturbed* by the charges of the electrons on adjacent atoms, causing an *interaction*. We call these interactions *bonds*, even though they may be too weak to be *formal* bonds such as those permanently connecting the atoms or ions in a molecule.

The intermolecular interactions between molecules of gas are generally *attractive*; so, by way of response, we find that, once atoms are close enough to interact, they prefer to remain close – indeed, once a tentative interaction forms, the atoms or molecules generally draw themselves closer, which itself makes the interaction stronger. We see a simple analogy with everyday magnets: once two magnets are brought close enough to induce an interaction, we feel the attractive force dragging them closer still.

As soon as the particles of a gas attract, the inertia of the aggregate species increases, thereby slowing down all translational motion. And slower particles, such as these aggregates, are an easier target for further collisions than fast-moving gas atoms and molecules. The same principle explains why it is impossible to catch someone who is running very fast during a playground game of 'tig'. Only when the runners tire and slow down can they be caught. In practice, as soon as an aggregate forms, we find that other gas particles soon adhere to it, causing eventual coalescence and the formation of a droplet of liquid. We say that *nucleation* occurs.

> Translational motion is movement through space, rather than a vibration about a mean point or a rotation about an axis.

> Formation of an aggregate facilitates further coalescence (eventually forming a condensed phase). We say 'nucleation' occurs.

With the same reasoning as that above, we can force the molecules of ammonia still closer together by applying a yet larger pressure, to form a denser state such as a solid.

Why does steam condense in a cold bathroom?

Elastic and inelastic collisions

In the previous example, we looked at the interactions induced when changing the external pressure, forcing the molecules into close proximity. We look here at the effects of changing the temperature.

A bathroom mirror is usually colder than the temperature of the steam rising from a hot bath. Each molecule of steam (gaseous water) has an enormous energy, which comes ultimately from the boiler that heats the water. The particles of steam would remain as liquid if they had less energy. In practice, particles evaporate from the bath to form energetic molecules of steam. We see this energy as *kinetic* energy, so the particles move fast (see p. 30). The typical speeds at which gas particles move make it inevitable that steam molecules will collide with the mirror. We say such a collision is *elastic* if no energy transfers during the collision between the gas particle and the mirror; but if energy does transfer – and it usually does – we say the collision is *inelastic*.

> No energy is exchanged during an 'elastic' collision, but energy *is* exchanged during an 'inelastic' collision.

The energy transferred during an inelastic collision passes *from* the hot molecule of steam *to* the cooler mirror. This energy flows in this direction because the steam initially possessed more energy per molecule than the mirror as a consequence of its higher temperature. It is merely a manifestation of the minus-oneth law of thermodynamics, as discussed in Chapter 1.

But there are consequences to the collisions being inelastic: the molecules of steam have less energy following the collision because some of their energy has transferred.

Energy is never lost or gained, only transferred or converted; see Chapter 3.

We generally assume that all particles in an ideal gas do not interact, meaning that the gas obeys the ideal-gas equation. This assumption is sometimes poor.

We perceive this lower energy as a cooler temperature, meaning that the water vapour in a steam-filled bathroom will cool down; conversely, the mirror (and walls) become warmer as they receive the energy that was previously possessed by the steam. These changes in the temperatures of gas and mirror occur in a complementary sense, so no energy is gained or lost.

These changes in temperature represent a macroscopic proof that microscopic processes do occur. Indeed, it is difficult to envisage a transfer of energy between the gas particles with the cold mirror *without* these microscopic interactions.

We spent quite a lot of time looking at the concept of an *ideal* gas in Chapter 1. The simplest definition of an ideal gas is that it obeys the ideal-gas equation (Equation (1.13)). Most gases can be considered as ideal most of the time. The most common cause of a gas disobeying the ideal-gas equation is the formation of interactions, and the results of intermolecular collisions.

How does a liquid-crystal display work?

Electronegativity and electropositivity

Liquid crystals are organic compounds that exhibit properties somewhere between those of a solid crystal and a liquid. Compounds **I** and **II** in Figure 2.1 both form liquid crystals at room temperature.

We observe that liquid crystals can flow like any other viscous liquid, but they also possess some of the properties of crystalline solids, such as physical order, rather than random chaos. Unlike most other liquids, liquid crystals have some properties

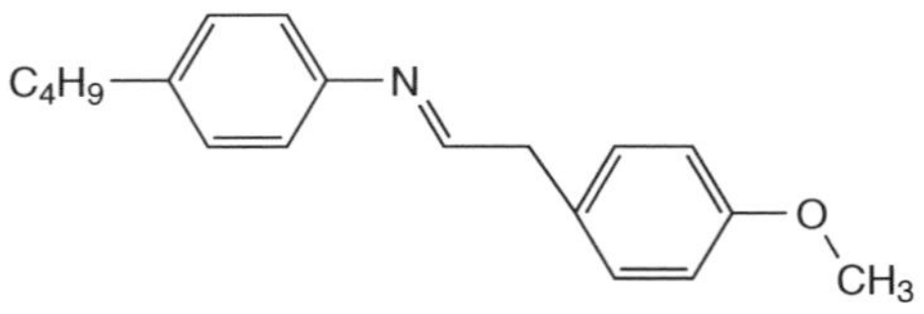

(I)

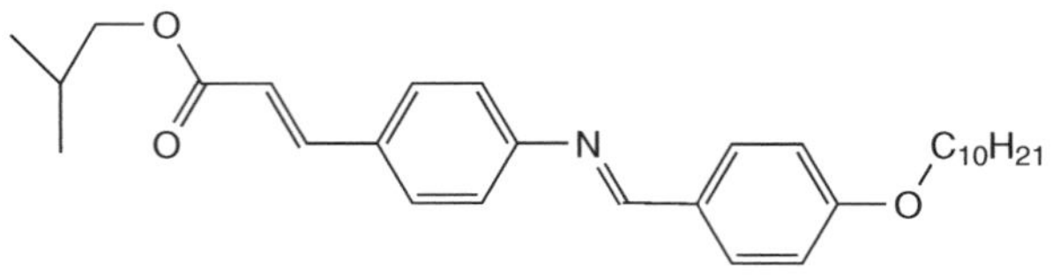

(II)

Figure 2.1   Compounds that form room-temperature liquid crystals

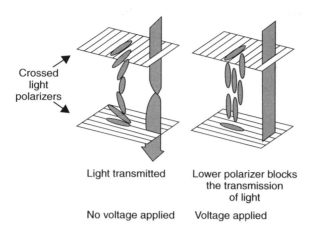

Crossed
light
polarizers

Light transmitted

Lower polarizer blocks
the transmission
of light

No voltage applied   Voltage applied

Figure 2.2   The transparent electrodes in an LCD are coated with crossed polarizers. The liquid crystals (depicted as slender lozenges) form helices, thereby 'guiding' polarized light from the upper electrode through the LCD, enabling transmission through to the lower polarizer. This is why the display has no colour. The helical structure is destroyed when a voltage is applied, because the polar liquid crystals align with the electrodes' field. No light can transmit, so the display looks black

that depend on the direction of measurement, because of the alignment of their long, rod-like structures.

In a liquid-crystal display (LCD) device, the two electrodes are parallel and separated by a thin layer of liquid crystal (see Figure 2.2). The liquid crystals in this layer naturally adopt a helical structure.

> A physicist would say the liquid crystal adopted a twisted *nematic structure*.

Light can be represented as a transverse electromagnetic wave made up of fluctuating electric and magnetic fields, moving in mutually perpendicular directions (see Chapter 9). Ordinary light is made up of waves that fluctuate at all possible angles, which normally cannot be separated. A *polarizer* is a material that allows only light with a specific angle of vibration to transmit. We place a light polarizer on one side of either transparent electrode in the LCD, each similar to one lens in a pair of polaroid sunglasses. The helix of the liquid crystal twists the polarized light as it transmits through the LCD, guiding it from the upper polarizer and allowing it unhindered passage through the 'sandwich' and lower polarizer. The transmitting state of an LCD (at zero voltage) is thus 'clear'.

Applying a voltage to a *pixel* within the cell causes the molecules to move, aligning themselves parallel with the electric field imparted by the electrodes. This realignment destroys the helical structure, precluding the unhindered transmission of light, and the display appears black.

> 'Pixel' is short for 'picture element'. An LCD image comprises many thousands of pixels.

Molecules of this type are influenced by an external electric field because they possess a *dipole*: one end of the molecule is electron withdrawing while the other is electron attracting, with the result that one end possesses a higher electron density than the other. As a result, the molecule behaves much like a miniature bar magnet. Applying a voltage between the two electrodes of the LCD causes the 'magnet' to reorientate in just the same way as a magnet moves when another magnet is brought close to it.

These dipoles form because of the way parts of the molecule attract electrons to differing extents. The power of an element (when part of a compound) to attract electrons is termed its 'electronegativity' χ. Highly electron-attracting atoms tend to exert control over the outer, valence electrons of adjacent atoms. The most electronegative elements are those placed near the right-hand side of the periodic table, such as oxygen and sulphur in Group VI(b) or the halogens in Group VII(b).

> Atoms or groups are 'electronegative' if they tend to acquire negative charge at the expense of juxtaposed atoms or groups. Groups acquiring a positive charge are 'electropositive'.

There have been a large number of attempts to quantify electronegativities χ, either theoretically or semi-empirically, but none has been wholly successful. All the better methods rely on bond strengths or the physical dimensions of atoms.

Similar to the concept of electronegativity is the *electropositivity* of an element, which is the power of its atoms (when part of a compound) to *lose* an electron. The most electropositive elements are the metals on the far-left of the periodic table, particularly Groups I(a) and II(a), which prefer to exist as cations. Being the opposite concept to electronegativity, electropositivity is not employed often. Rather, we tend to say that an atom such as sodium has a tiny electronegativity instead of being very electropositive.

Why does dew form on a cool morning?

Van der Waals forces

Many people love cool autumn mornings, with the scent of the cool air and a rich dew underfoot on the grass and paths. The dew forms when molecules of water from the air coalesce, because of the cool temperature, to form minute aggregates that subsequently nucleate to form visible drops of water. These water drops form a stable colloid (see Chapter 10).

Real gases are never wholly ideal: there will always be some extent of non-ideality. At one extreme are the monatomic rare gases such as argon and neon, which are non-polar. Hydrocarbons, like propane, are also relatively non-polar, thereby precluding stronger molecular interactions. Water, at the opposite extreme, is very polar because some parts of the molecule are more electron withdrawing than others. The central oxygen is relatively electronegative and the two hydrogen atoms are electropositive, with the result that the oxygen is more negative than either of the hydrogen atoms. We say it has a slight *excess charge*, which we write as δ^-. Similar reasoning shows how the hydrogen atoms are more positive than the oxygen, with excess charges of δ^+.

> The symbol δ means 'a small amount of . . .', so 'δ^-' is a small amount of negative charge.

These excess charges form in consequence of the molecule incorporating a variety of atoms. For example, the magnitude of δ^- on the chlorine of H–Cl is larger than the excess charges in the F–Cl molecule, because the difference in electronegativity

Table 2.1 Values of electronegativity χ for some main-group elements

H						
2.1						
Li	**Be**	**B**	**C**	**N**	**O**	**F**
1.0	1.5	2.0	2.5	3.0	3.5	4.0
Na	**Mg**	**Al**	**Si**	**P**	**S**	**Cl**
0.9	1.2	1.5	1.8	2.1	2.5	3.0
K	**Ca**	**Ga**	**Ge**	**As**	**Se**	**Br**
0.8	1.0	1.6	1.8	2.0	2.4	2.8
Rb	**Sr**					**I**
0.8	1.0					2.5

χ between H and Cl is greater than the difference between F and Cl. There will be no excess charge in the two molecules H–H or Cl–Cl because the atoms in both are the same – we say they are *homonuclear*. Table 2.1 contains a few electronegativities.

SAQ 2.2 By looking at the electronegativities in Table 2.1, suggest whether the bonds in the following molecules will be polar or non-polar: (a) hydrogen bromide, HBr; (b) silicon carbide, SiC; (c) sulphur dioxide, O=S=O; and (d) sodium iodide, NaI.

The actual magnitude of the excess charge is generally unknown, although we do know they are small. Whereas some calculations suggest that δ is perhaps as much as 0.1 of a full, formal charge, others suggest about 0.01 or even less.

While debate persists concerning the magnitudes of each excess charge within a molecule, it is certain that the *overall* charge on the molecule is zero, meaning that the two positive charges in water cancel out the central negative charge on the oxygen. We reason this by saying that water is a neutral molecule.

Figure 2.3 shows the 'V' shape of the water molecule. The top of the molecule (as drawn) has a negative excess charge and the bottom is positive. The δ^+ and δ^- charges are separated spatially, which we call a *dipole*. Such dipoles are crucial when explaining why water vapour so readily forms a liquid: those parts of the molecule bearing a slight positive charge (δ^+) attract those parts of adjacent molecules that bear a slight negative charge (δ^-). The interaction is *electrostatic*, and forms in much a similar manner to the north pole of a magnet attracting the south pole of another magnet.

Electrostatic interactions of this type are called 'dipole–dipole interactions', or 'van der Waals forces' after the Dutch physicist Johannes Diderik van der Waals (1837–1923) who first postulated their existence. A van der Waals force operates over a relatively

> Water is a *neutral* molecule, so the central negative charge in the water molecule counteracts the two positive charges.

> A 'dipole' forms when equal and opposite charges are separated by a short distance. 'Di' means two, and 'pole' indicates the two ends of a magnet.

> 'Van der Waals forces' are electrostatic interactions between dipoles. (Note how we pronounce 'Waals' as 'vahls'.)

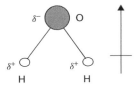

Figure 2.3   The water molecule has a 'V' shape. Experiments show that gaseous water has an O–H length of 0.957 18 Å; the H–O–H angle is 104.474°. Water is polar because the central oxygen is electronegative and the two hydrogen atoms are electropositive. The vertical arrow indicates the resultant dipole, with its head pointing toward the more negative end of the molecule

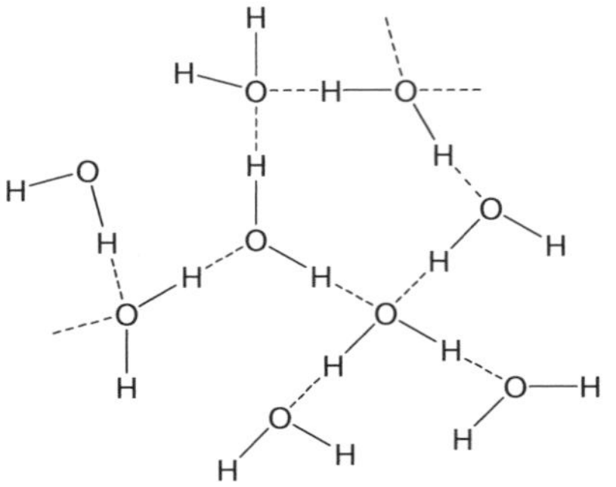

Figure 2.4   Water would be a gas rather than a liquid at room temperature if no van der Waals forces were present to 'glue' them together, as indicated with dotted lines in this two-dimensional representation. In fact, water coalesces as a direct consequence of this *three*-dimensional network of dipole–dipole interactions. Note how all the O–H $\cdots$ O bonds are linear

short distance because the influence of a dipole is not large. In practice, we find that the oxygen atoms can interact with hydrogen atoms on an adjacent molecule of water, but no further.

The interactions between the two molecules helps to 'glue' them together. It is a sobering thought that water would be a gas rather than a liquid if hydrogen bonds (which are merely a particularly strong form of van der Waals forces) did not promote the coalescence of water. The Earth would be uninhabitable without them. Figure 2.4 shows the way that liquid water possesses a three-dimensional network, held together with van der Waals interactions.

Each H_2O molecule in liquid water undergoes at least one interaction with another molecule of H_2O (sometimes two). Nevertheless, the interactions are not particularly strong – perhaps as much as 20 kJ mol^{-1}.

Whereas the dipoles themselves are permanent, van der Waals interactions are not. They are sufficiently weak that they continually break and re-form as part of a *dynamic* process.

How is the three-dimensional structure maintained within the DNA double helix?

Hydrogen bonds

DNA is a natural polymer. It was first isolated in 1869 by Meischer, but its role in determining heredity remained unrecognized until 1944, by which time it was appreciated that it is the *chromosomes* within a cell nucleus that dictate hereditary traits. And such chromosomes consist of DNA and protein. In 1944, the American bacteriologist Oswald Avery showed how it was the DNA that carried genetic information, not the protein.

The next breakthrough came in 1952, when Francis Crick and Donald Watson applied X-ray diffraction techniques to DNA and elucidated its structure, as shown schematically in Figure 2.5. They showed how its now famous 'double helix' is held together via a series of unusually strong dipole–dipole interactions between precisely positioned organic bases situated along the DNA polymer's backbone.

There are four bases in DNA: guanine, thymine, cytosine and adenine. Each has a ketone C=O group in which the oxygen is quite electronegative and bears an excess negative charge δ^-, and an amine in which the electropositive hydrogen atoms bear an excess

> The word 'theory' comes from the Greek *theoreō*, meaning 'I look at'. A theory is something we look at, pending acceptance or rejection.

> The rules of 'base pairing' (or nucleotide pairing) in DNA are: adenine (A) always pairs with thymine (T); cytosine (C) always pairs with guanine (G).

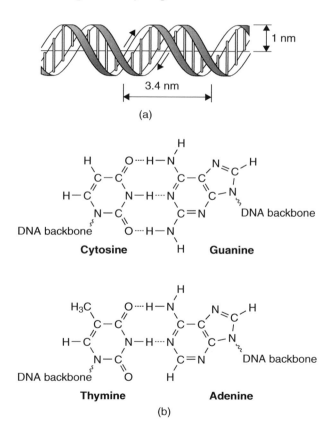

(a)

(b)

Figure 2.5   (a) The structure of the 'double helix' at the heart of DNA. The slender 'rods' represent the hydrogen bonds that form between the organic bases situated on opposing strands of the helix. (b) Hydrogen bonds (the dotted lines) link adenine with thymine, and guanine with cytosine

Table 2.2   The energies of hydrogen bonds

Atoms in H-bond	Typical energy/kJ mol^{-1}
H and N	20
H and O	25
H and F	40

positive charge δ^+. Since the hydrogen atom is so small and so electropositive, its excess charge leads to the formation of an unusually strong dipole, itself leading to a strong van der Waals bond. The bond is usually permanent (unlike a typical dipole–dipole interaction), thereby 'locking' the structure of DNA into its pair of parallel helices, much like the interleaving teeth of a zip binding together two pieces of cloth.

IUPAC is the International Union of Pure and Applied Chemistry. It defines terms, quantities and concepts in chemistry.

Strictly, the dipole–dipole interactions discussed on p. 42 are also hydrogen bonds, since we discussed the interactions arising between H–O-bonded species.

We call these extra-strong dipole–dipole bonds 'hydrogen bonds', and these are defined by the IUPAC as 'a form of association between an electronegative atom and a hydrogen atom attached to a second, relatively electronegative atom'. All hydrogen bonds involve two dipoles: one always comprises a bond ending with hydrogen; the other terminates with an unusually electronegative atom. It is best considered as an electrostatic interaction, heightened by the small size of hydrogen, which permits proximity of the interacting dipoles or charges. Table 2.2 contains typical energies for a few hydrogen bonds. Both electronegative atoms are usually (but not necessarily) from the first row of the periodic table, i.e. N, O or F. Hydrogen bonds may be intermolecular or intramolecular.

Finally, as a simple illustration of how weak these forces are, note how the energy required to break the hydrogen bonds in liquid hydrogen chloride (i.e. the energy required to vaporize it) is 16 kJ mol^{-1}, yet the energy needed break the chemical bond between atoms of hydrogen and chlorine in H–Cl is almost 30 times stronger, at 431 kJ mol^{-1}.

Aside

The ancient Greeks recognized that organisms often pass on traits to their offspring, but it was the experimental work of the Austrian monk Gregor Mendel (1822–1884) that led to a modern hereditary 'theory'. He entered the Augustinian monastery at Brünn (now Brno in the Czech Republic) and taught in its technical school.

He cultivated and tested the plants at the monastery garden for 7 years. Starting in 1856, painstakingly analysing seven pairs of seed and plant characteristics for at least 28 000 pea plants. These tedious experiments resulted in two generalizations, which were later called Mendel's laws of heredity. Mendel published his work in 1866, but it remained almost unnoticed until 1900 when the Dutch botanist Hugo Marie de Vries referred to it. The full significance of Mendel's work was only realized in the 1930s.

His observations also led him to coin two terms that still persist in present-day genetics: *dominance* for a trait that shows up in an offspring, and *recessiveness* for a trait masked by a dominant gene.

How do we make liquid nitrogen?

London dispersion forces

Liquid nitrogen is widely employed when freezing sperm or eggs when preparing for the *in vitro* fertilization (IVF). It is also essential for maintaining the cold temperature of the superconducting magnet at the heart of the NMR spectrometers used for structure elucidation or magnetic resonance imaging (MRI).

Nitrogen condenses to form a liquid at $-196\,^{\circ}\mathrm{C}$ (77 K), which is so much lower than the temperature of 373.15 K at which water condenses that we suspect a different physicochemical process is in evidence. Below $-196\,^{\circ}\mathrm{C}$, molecules of nitrogen interact, causing condensation. That there is any interaction at all should surprise us, because the dipoles above were a feature of *heteronuclear* bonds, but the di-nitrogen molecule ($N\equiv N$) is homonuclear, meaning both atoms are the same.

> A 'homonuclear' molecule comprises atoms from only one element; *homo* is Greek for 'same'. Most molecules are 'heteronuclear' and comprise atoms from several elements; *hetero* is Greek for 'other' or 'different'.

We need to invoke a new type of interaction. The triple bond between the two nitrogen atoms in the di-nitrogen molecule incorporates a huge amount of electron density. These electrons are never still, but move continually; so, at any instant in time, one end of a molecule might be slightly more negative than the other. A fraction of a second later and the imbalance departs. But a tiny dipole forms during the instant while the charges are unbalanced: we call this an *induced dipole*; see Figure 2.6.

> Dipoles are usually a feature of heteronuclear bonds, although a fuller treatment needs to consider the electronic environment of atoms and groups beyond the bond of interest.

The electron density changes continually, so induced dipoles never last more than about 10^{-11} s. Nevertheless, they last sufficiently long for an interaction to form with the induced dipole of another nitrogen molecule nearby. We call this new interaction the *London dispersion force* after Fritz London, who first postulated their existence in 1930.

London dispersion forces form between all molecules, whether polar or non-polar. In a large atom or molecule, the separation between the nucleus and valence electrons is quite large; conversely, the nucleus–electron separation in a lighter atom or molecule is smaller, implying that the electrons are more tightly held. The tighter binding precludes the ready formation of an induced dipole. For this reason, larger (and therefore heavier) atoms and molecules generally exhibit stronger dispersion forces than those that are smaller and lighter.

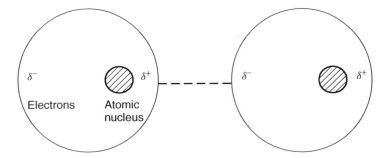

Figure 2.6   Schematic diagram to show how an induced dipole forms when polarizable electrons move within their orbitals and cause a localized imbalance of charge (an 'induced dipole' in which the negative electrons on one atom attract the positive nucleus on another). The dotted line represents the electrostatic dipole interaction

---

Aside

The existence of an attractive force between non-polar molecules was first recognized by van der Waals, who published his classic work in 1873. The origin of these forces was not understood until 1930 when Fritz London (1900–1954) published his quantum-mechanical discussion of the interaction between fluctuating dipoles. He showed how these temporary dipoles arose from the motions of the outer electrons on the two molecules.

We often use the term 'dispersion force' to describe these attractions. Some texts prefer the term 'London–van der Waals' forces.

Polarizability

> The ease with which the electron distribution around an atom or molecule can be distorted is called its 'polarizability'.

The electrons in a molecule's outer orbitals are relatively free to move. If we could compare 'snapshots' of the molecule at two different instants in time then we would see slight differences in the charge distributions, reflecting the changing positions of the electrons in their orbitals. The ease with which the electrons can move with time depends on the molecule's *polarizability*, which itself measures how easily the electrons can move within their orbitals.

> The weakest of all the intermolecular forces in nature are always London dispersion forces.

In general, polarizability increases as the orbital increases in size: negative electrons orbit the positive nucleus at a greater distance in such atoms, and consequently experience a weaker electrostatic interaction. For this reason, London dispersion forces tend to be stronger between molecules that are easily polarized, and weaker between molecules that are not easily polarized.

Why is petrol a liquid at room temperature but butane is a gas?

The magnitude of London dispersion forces

The major component of the petrol that fuels a car is octane, present as a mixture of various isomers. It is liquid at room temperature because its boiling point temperature $T_{(boil)}$ is about $125.7\,^{\circ}C$. The methane gas powering an oven has a $T_{(boil)} = -162.5\,^{\circ}C$ and the butane propellant in a can of air freshener has $T_{(boil)} = -0.5\,^{\circ}C$. Octadecane is a gel and paraffin wax is a solid. Figure 2.7 shows the trend in $T_{(boil)}$ for a series of straight-chain alkanes Each hydrocarbon experiences exactly the same intermolecular forces, so what causes the difference in $T_{(boil)}$?

Interactions always form between molecules because London forces cannot be eradicated. Bigger molecules experience greater intermolecular forces. These dispersion forces are weak, having a magnitude of between 0.001 and 0.1 per cent of the strength of a typical covalent bond binding the two atoms in diatomic molecule, H_2. These forces, therefore, are so small that they may be ignored within molecules held together by stronger forces, such as network covalent bonds or large permanent dipoles or ions.

The strength of the London dispersion forces becomes stronger with increased polarizability, so larger molecule (or atoms) form stronger bonds. This observation helps explain the trends in physical state of the Group VII(b) halogens: I_2 is a solid, Br_2 is a liquid, and Cl_2 and F_2 are gases.

But the overall dispersion force strength also depends on the total number of electrons in the atom or molecule. It is a cumulative effect. Butane contains 14 atoms and 58 electrons, whereas octane has 26 atoms and 114 electrons. The greater number of electrons increases the total number of interactions possible and, since both melting

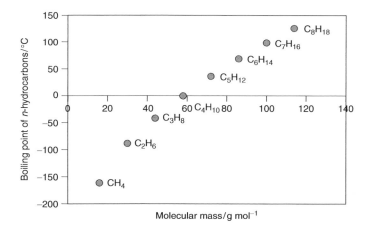

Figure 2.7 The boiling temperature of simple linear hydrocarbons increases as a function of molecular mass, as a consequence of a greater number of induced dipoles

and boiling points depend directly on the strength of intermolecular bonds, the overall strength of the London forces varies as the molecule becomes larger.

Aside

When is a dispersion force sufficiently strong that we can safely call it a hydrogen bond?

Hydrogen bonds are much stronger than London dispersion forces for two principal reasons:

(1)   The induced dipole is permanent, so the bond is permanent.

(2)   The molecule incorporates a formal H–X covalent bond in which X is a relatively electronegative element (see p. 42).

We call an interaction 'a hydrogen bond' when it fulfils both criteria.

2.2   Quantifying the interactions and their influence

How does mist form?

Condensation and the critical state

Why is it that no dew forms if the air pressure is low, however cool the air temperature?

To understand this question, we must first appreciate how molecules come closer together when applying a pressure. The Irish physical chemist Thomas Andrews (1813–1885) was one of the first to study the behaviour of gases as they liquefy: most of his data refer to CO_2. In his most famous experiments, he observed liquid CO_2 at constant pressure, while gradually raising its temperature. He readily discerned a clear meniscus between condensed and gaseous phases in his tube at low temperatures, but the boundary between the phases vanished at temperatures of about 31 °C. Above this temperature, no amount of pressure could bring about liquefaction of the gas.

The 'critical temperature' $T_{(critical)}$ is that temperature above which it is impossible to liquefy a gas.

Andrews suggested that each gas has a certain 'critical' temperature, above which condensation is impossible, implying that no liquid will form by changes in pressure alone. He called this temperature the 'critical temperature' $T_{(critical)}$.

Figure 2.8 shows a Boyle's-law plot of pressure p (as y) against volume V (as x) for carbon dioxide. The figure is drawn as a function of temperature. Each line on the graph represents data obtained at a single, constant temperature, and helps explain why we call each line an *isotherm*. The uppermost isotherm represents data collected at 31.5 °C. Its shape is essentially straightforward, although it clearly shows distortion. The middle trace (at

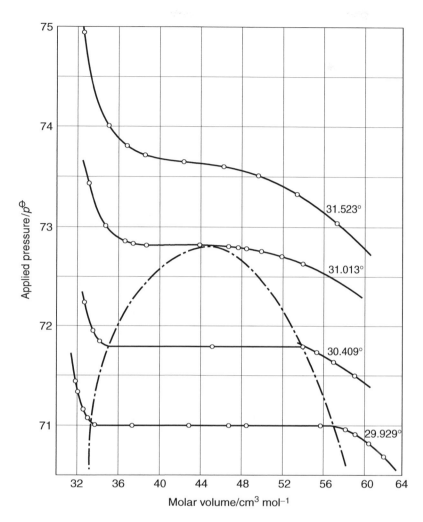

Figure 2.8 Isotherms of carbon dioxide near the critical point of 31.013 °C. The shaded parabolic region indicates those pressures and volumes at which it is possible to condense carbon dioxide

the cooler temperature of 30.4 °C) is more distorted – it even shows a small region where the pressure experienced is independent of temperature, which explains why the graph is linear and horizontal. And in the bottom trace the isotherm at 29.9 °C is even more distorted still.

The area enclosed by the parabola at the bottom of Figure 2.8 represents those values of pressure and temperature at which CO_2 *will* condense. If the temperature is higher than that at the apex of the parabola (i.e. warmer than 31.013 °C) then, whatever the pressure, CO_2 will *not* liquefy. For this reason, we call 31.013 °C the critical temperature $T_{(critical)}$ of CO_2. Similarly, the gas will not liquefy unless the pressure is above a minimum that we call the 'critical pressure' $p_{(critical)}$.

Critical fluids are discussed in Chapter 5, where values of $T_{(critical)}$ are listed.

As the temperature rises above the critical temperature and the pressure drops below the critical pressure, so the gas approximates increasingly to an ideal gas, i.e. one in which there are no interactions and which obeys the ideal-gas equation (Equation (1.13)).

How do we liquefy petroleum gas?

Quantifying the non-ideality

Liquefied petroleum gas (LPG) is increasingly employed as a fuel. We produce it by applying a huge pressure ($10-20 \times p^{\ominus}$) to the petroleum gas obtained from oil fields.

Above a certain, critical, pressure the hydrocarbon gas condenses; we say it reaches the *dew point*, when droplets of liquid first form. The proportion of the gas liquefying increases with increased pressure until, eventually, all of it has liquefied.

Increasing the pressure forces the molecules closer together, and the intermolecular interactions become more pronounced. Such interactions are not particularly strong because petroleum gas is a non-polar hydrocarbon, explaining why it is a gas at room temperature and pressure. We discuss other ramifications later.

> We often call a gas that is non-ideal, a *real* gas.

The particles of an ideal gas (whether atoms or molecules) do not interact, so the gas obeys the ideal-gas equation all the time. As soon as interactions form, the gas is said to be non-ideal, with the result that we lose ideality, and the ideal-gas equation (Equation (1.13)) breaks down. We find that $pV \neq nRT$.

> The physical chemistry underlying the liquefaction of a gas is surprisingly complicated, so we shall not return to the question until Chapter 5.

When steam (gaseous water) is cooled below a certain temperature, the molecules have insufficient energy to maintain their high-speed motion and they slow down. At these slower speeds, they attract one another, thereby decreasing the molar volume.

Figure 2.9 shows a graph of the quotient $pV \div nRT$ (as y) against pressure p (as x). We sometime call such a graph an *Andrews plot*. It is clear from the ideal-gas equation (Equation (1.13)) that if $pV = nRT$ then $pV \div nRT$ should always equal one: the horizontal line drawn through $y = 1$, therefore, indicates the behaviour of an ideal gas.

> We form a 'quotient' when dividing one thing by another. We meet the word frequently when discussing a person's IQ, their 'intelligence quotient', which we define as: (a person's score in an intelligence test $\div$ the average score) $\times$ 100.

But the plots in Figure 2.9 are *not* horizontal. The deviation of a trace from the $y = 1$ line quantifies the extent to which a gas deviates from the ideal-gas equation; the magnitude of the deviation depends on pressure. The deviations for ammonia and ethene are clearly greater than for nitrogen or methane: we say that ammonia *deviates* from ideality more than does nitrogen. Notice how the deviations are worse at high pressure, leading to the empirical observation that a real gas behaves more like an ideal gas at lower pressures.

Figure 2.10 shows a similar graph, and displays Andrews plots for methane as a function of temperature. The graph clearly

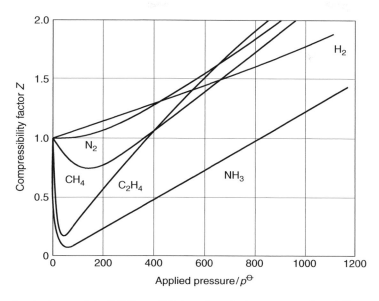

Figure 2.9   An Andrews plot of $PV \div nRT$ (as y) against pressure p (as x) for a series of real gases, showing ideal behaviour only at low pressures. The function on the y-axis is sometimes called the compressibility Z

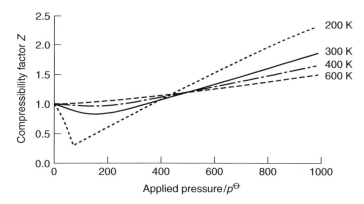

Figure 2.10   An Andrews plot of $PV \div nRT$ (as y) against pressure p (as x) for methane gas as a function of temperature. Methane behaves more like an ideal gas at elevated temperatures

demonstrates how deviations from ideality become less severe with increasing temperature. In fact, we should expect the deviations to *decrease* as the temperature *increases*, because a higher temperature tells us how the particles have more energy, decreasing the likelihood of interparticle interactions being permanent.

Drawing graphs such as Figure 2.10 for other gases suggests a second empirical law, that gases behave more like ideal gases as the temperature rises. The ideal-gas equation (Equation (1.13)) is so useful that we do not want to lose it. Accordingly, we adapt it

> Gases behave more like ideal gases at higher temperatures.

somewhat, writing it as

$$pV = Z \times (nRT) \tag{2.1}$$

> We sometimes call the function '$pV \div nRT$' the 'compressibility' or 'compressibility factor' Z.

where Z is the *compressibility* or *compressibility factor*. The value of Z will always be one for an ideal gas, but Z rarely has a value of one for a real gas, except at very low pressures. As soon as p increases, the gas molecules approach close enough to interact and $pV \neq nRT$. The value of Z tells us a lot about the interactions between gas particles.

Aside

The gas constant R is generally given the value $8.314 \, \mathrm{J \, K^{-1} mol^{-1}}$, but in fact this *numerical* value only holds if each unit is the SI standard, i.e. pressure expressed in pascals, temperature in kelvin and volume in cubic metres.

The value of R changes if we express the ideal-gas equation (Equation (1.13)) with different units. Table 2.3 gives values of R in various other units. We must note an important philosophical truth here: the value of the gas constant is truly constant, but the actual numerical value we cite will depend on the units with which we express it. We met a similar argument before on p. 19, when we saw how a standard prefix (such as deca, milli or mega) will change the *appearance* of a number, so $V = 1 \, \mathrm{dm^3} = 10^3 \, \mathrm{cm^3}$. In reality, the number remains unaltered.

We extend this concept here by showing how the units themselves alter the numerical value of a constant.

Table 2.3 Values of the gas constant R expressed with various units[a]

$8.3145 \, \mathrm{J \, K^{-1} \, mol^{-1}}$
$2 \, \mathrm{cal \, K^{-1} \, mol^{-1}}$
$0.083\,145 \, \mathrm{dm^3 \, bar \, mol^{-1} \, K^{-1}}$
$83.145 \, \mathrm{cm^3 \, bar \, mol^{-1} \, K^{-1}}$
$0.082\,058 \, \mathrm{dm^3 \, atm \, mol^{-1} \, K^{-1}}$
$82.058 \, \mathrm{cm^3 \, atm \, mol^{-1} \, K^{-1}}$

[a] $1 \, \mathrm{bar} = p^{\ominus} = 10^5 \, \mathrm{Pa}$. $1 \, \mathrm{atm} = 1.013\,25 \times 10^5$ Pa. The 'calorie' is a wholly non-SI unit of energy; $1 \, \mathrm{cal} = 4.157 \, \mathrm{J}$.

Why is the molar volume of a gas not zero at 0 K?

The van der Waals equation

In Chapter 1 we recalled how Lord Kelvin devised his temperature scale after cooling gases and observing their volumes. If the simplistic graph in Figure 1.5 was obeyed, then a gas would have a zero volume at $-273.15\,°C$. In fact, the molar volume of a gas is always significant, even at temperatures close to absolute zero. Why the deviation from Kelvin's concept?

Every gas consists of particles, whether as atoms (such as neon) or as molecules (such as methane). To a relatively good first approximation, any atom can be regarded as a small, incompressible sphere. The reason why we can compress a gas relates to the large separation between the gas particles. The first effect of compressing a gas is to decrease these interparticle distances.

Particles attract whenever they approach to within a minimum distance. Whatever the magnitude of the interparticle attraction, energetic molecules will separate and continue moving after their encounter; but, conversely, molecules of lower energy do not separate after the collision because the attraction force is enough to overwhelm the momentum that would cause the particles to bounce apart. The process of coalescence has begun.

Compressing a gas brings the particles into close proximity, thereby increasing the probability of interparticle collisions, and magnifying the number of interactions. At this point, we need to consider two physicochemical effects that operate in opposing directions. Firstly, interparticle interactions are usually *attractive*, encouraging the particles to get closer, with the result that the gas has a smaller molar volume than expected. Secondly, since the particles have their own intrinsic volume, the molar volume of a gas is described not only by the separations between particles but also by the particles themselves. We need to account for these two factors when we describe the physical properties of a real gas.

The Dutch scientist van der Waals was well aware that the ideal-gas equation was simplistic, and suggested an adaptation, which we now call the *van der Waals equation of state*:

> The *a* term reflects the strength of the interaction between gas particles, and the *b* term reflects the particle's size.

$$\left(p + \frac{n^2a}{V^2}\right)(V - nb) = nRT \qquad (2.2)$$

where the constants *a* and *b* are called the 'van der Waals constants', the values of which depend on the gas and which are best obtained experimentally. Table 2.4 contains a few sample values. The constant *a* reflects the strength of the interaction between gas molecules; so, a value of 18.9 for benzene suggests a strong interaction whereas 0.03 for helium represents a negligible interaction. Incidentally, this latter value reinforces the idea that inert gases *are* truly inert. The magnitude of the constant *b* reflects the physical size of the gas particles, and are again seen to follow a predictable trend. The magnitudes of *a* and *b* dictate the extent to which the gases deviate from ideality.

Note how Equation (2.2) simplifies to become the ideal-gas equation (Equation (1.13)) if the volume *V* is large. We expect this result, because a large volume not only implies a low pressure, but also yields the best conditions for minimizing all instances of interparticle collisions.

> Equation (2.2) simplifies to become the ideal-gas equation (Equation (1.13)) whenever the volume *V* is large.

Table 2.4   Van der Waals constants for various gases

Gas	$a/(\text{mol dm}^{-3})^{-2}$ bar	$b/(\text{mol dm}^{-3})^{-1}$
Monatomic gases		
Helium	0.034 589	0.023 733
Neon	0.216 66	0.017 383
Argon	1.3483	0.031 830
Krypton	2.2836	0.038 650
Diatomic gases		
Hydrogen	0.246 46	0.026 665
Nitrogen	1.3661	0.038 577
Oxygen	1.3820	0.031 860
Carbon monoxide	1.4734	0.039 523
Polyatomic gases		
Ammonia	4.3044	0.037 847
Methane	2.3026	0.043 067
Ethane	5.5818	0.065 144
Propane	9.3919	0.090 494
Butane	13.888	0.116 41
Benzene	18.876	0.119 74

> The value of p calculated with the ideal-gas equation is 3.63×10^5 Pa, or 3.63 bar.

Worked Example 2.1   0.04 mol of methane gas is enclosed within a flask of volume 0.25 dm^3. The temperature is 0 °C. From Table 2.4, $a = 2.3026$ dm^6 bar mol^{-2} and $b = 0.043\,067$ dm^3 mol^{-1}. What is the pressure p exerted?

We first rearrange Equation (2.2), starting by dividing both sides by the term $(V - nb)$, to yield

$$p + \frac{n^2 a}{V^2} = \frac{nRT}{(V - nb)}$$

We then subtract $(n^2 a) \div V^2$ from both sides:

$$p = \frac{nRT}{(V - nb)} - \left(\frac{n}{V}\right)^2 a$$

> Calculations with the van der Waals equation are complicated because of the need to convert the units to accommodate the SI system. The value of R comes from Table 2.3.

Next, we insert values and convert to SI units, i.e. 0 °C is expressed as 273.15 K.

$$p = \frac{0.04 \text{ mol} \times (8.314 \times 10^{-2} \text{ dm}^3 \text{ bar K}^{-1} \text{ mol}^{-1}) \times 273.15 \text{ K}}{(0.25 \text{ dm}^3 - 0.04 \text{ mol} \times 0.043\,067 \text{ dm}^3 \text{ mol}^{-1})}$$

$$- \left(\frac{0.04 \text{ mol}}{0.25 \text{ dm}^3}\right)^2 \times 2.3026 \text{ (dm}^3 \text{ mol}^{-1})^2 \text{ bar}$$

$$p = 3.65887 \text{ bar} - 0.05895 \text{ bar}$$

$$p = 3.59992 \text{ bar}$$

The pressure calculated with the ideal-gas equation (Equation (1.13)) is 3.63 bar, so the value we calculate with the van der Waals equation (Equation (2.2)) is 1 per cent *smaller*. The experimental value is 3.11 bar, so the result with the van der Waals equation is superior.

The lower pressure causes coalescence of gas particles, which decreases their kinetic energy. Accordingly, the impact between the aggregate particle and the container's walls is less violent, which lowers the observed pressure.

The virial equation

An alternative approach to quantifying the interactions and deviations from the ideal-gas equation is to write Equation (1.13) in terms of 'virial coefficients':

$$p \left(\frac{V}{n} \right) = RT \ (1 + B'p + C'p^2 + \ldots) \qquad (2.3)$$

where the $V \div n$ term is often rewritten as V_m and called the *molar* volume.

Equation (2.3) is clearly similar to the ideal-gas equation, Equation (1.13), except that we introduce additional terms, each expressed as powers of pressure. We call the constants, B', C' etc., 'virial coefficients', and we determine them experimentally. We call B' the second virial coefficient, C' the third, and so on.

> The word 'virial' comes from the Latin for force or powerful.

Equation (2.3) becomes the ideal-gas equation if both B' and C' are tiny. In fact, these successive terms are often regarded as effectively 'fine-tuning' the values of p or V_m. The C' coefficient is often so small that we can ignore it; and D' is so minuscule that it is extremely unlikely that we will ever include a fourth virial coefficient in any calculation. Unfortunately, we must exercise care, because B' constants are themselves a function of temperature.

Worked Example 2.2   What is the molar volume V_m of oxygen gas at 273 K and $p^{\ominus}$? Ignore the third and subsequent virial terms, and take $B' = -4.626 \times 10^{-2} \text{ bar}^{-1}$.

> *Care*: the odd-looking units of B' require us to cite the gas constant R in SI units with prefixes.

From Equation (2.3)

$$V_m = \frac{V}{n} = \frac{RT}{p} \times (1 + B'p)$$

Inserting numbers (and taking care how we cite the value of R) yields

$$V_m = \frac{83.145 \text{ cm}^3 \text{ bar mol}^{-1} \text{ K}^{-1} \times 273 \text{ K}}{1 \text{ bar}}$$

$$\times (1 - 4.626 \times 10^{-2} \text{ bar}^{-1} \times 1 \text{ bar})$$

$$V_m = 22\,697 \times 0.954 \text{ cm}^3$$

$$V_m = 21\,647 \text{ cm}^3$$

> The value of V_m calculated with the ideal-gas equation (Equation (1.13)) is 4.4 per cent higher.

1 cm^3 represents a volume of $1 \times 10^{-6} \text{ m}^3$, so expressing this value of V_m in SI units yields $21.6 \times 10^{-3} \text{ m}^3$.

SAQ 2.3 Calculate the temperature at which the molar volume of oxygen is 24 dm^3. [Hint: you will need some of the data from Worked Example 2.2. Assume that B' has not changed, and be careful with the units, i.e. $V_m = 24\,000$ cm^3.]

> The relationship between B and B' is $B' = (B \div RT)$.

An alternative form of the virial equation is expressed in terms of molar volume V_m rather than pressure:

> A positive virial coefficient indicates repulsive interactions between the particles. The magnitude of B indicates the strength of these interaction.

$$PV_m = RT \left(1 + \frac{B}{V_m} + \frac{C}{V_m} + \cdots \right) \tag{2.4}$$

Note that the constants in Equation (2.4) are distinguishable from those in Equation (2.3) because they lack the prime symbol. For both Equations (2.3) and (2.4), the terms in brackets represents the molar compressibility Z. Table 2.5 lists a few virial coefficients.

SAQ 2.4 Calculate the pressure of 1 mol of gaseous argon housed within 2.3 dm^3 at 600 K. Take $B = 11.9$ cm^3 mol^{-1}, and ignore the third virial term, C'. [Hint: take care with all units; e.g. remember to convert the volume to m^3.]

Table 2.5   Virial coefficients B for real gases as a function of temperature, and expressed in units of cm^3 mol^{-1}

Gas	100 K	273 K	373 K
Argon	−187.0	−21.7	−4.2
Hydrogen	−2.0	13.7	15.6
Helium	11.4	12.0	11.3
Nitrogen	−160.0	−10.5	6.2
Neon	−6.0	10.4	12.3
Oxygen	−197.5	−22.0	−3.7

2.3 Creating formal chemical bonds

Why is chlorine gas lethal yet sodium chloride is vital for life?

The interaction requires electrons

Chlorine gas is very reactive, and causes horrific burns to the eyes and throat; see p. 243. The two atoms are held together by means of a single, non-polar covalent bond. Cl_2 has a yellow–green colour and, for a gas, is relatively dense at s.t.p. Conversely, table salt (sodium chloride) is an ionic solid comprising Na^+ and Cl^- ions, held together in a three-dimensional array. What is the reason for their differences in behaviour?

> The word 'chlorine' derives from the Greek *chloros*, meaning 'green'.

The outer shell of each 'atom' in Cl_2 possesses a full octet of electrons: seven electrons of its own (which explains why it belongs to Group VII(b) of the periodic table) and an extra electron from covalent 'sharing' with the other atom in the Cl_2 molecule. The only other simple interactions in molecular chlorine are the inevitable induced dipolar forces, which are too weak at room temperature to allow for the liquefying of $Cl_{2(g)}$.

Each chloride *ion* in NaCl also has eight electrons: again, seven electrons come from the element prior to formation of a chloride ion, but the extra eighth electron comes from *ionizing* the sodium counter ion. This extra electron resides entirely on the chloride ion, so no electrons are shared. The interactions in solid NaCl are wholly ionic in nature. Induced dipoles will also exist within each ion, but their magnitude is utterly negligible when compared with the strength of the *formal* charges on the Na^+ and Cl^- ions. We are wise to treat them as absent.

So, in summary, the principal differences between $Cl_{2(g)}$ and $NaCl_{(s)}$ lie in the location and the interactions of electrons in the atoms' outer shells. We say these electrons reside in an atom's *frontier orbitals*, meaning that we can ignore the inner electrons, which are tightly bound to the nucleus.

Why does a bicycle tyre get hot when inflated?

Bonds and interactions involve energy changes

A bicycle tyre gets quite hot during its inflation. The work of inflating the tyre explains in part why the temperature increases, but careful calculations (e.g. see pp. 86 and 89) show that additional factors are responsible for the rise in temperature.

> We look on p. 86 at the effect of performing 'work' while inflating a bicycle tyre, and the way work impinges on the internal energy of the gas.

On a macroscopic level, we say we compress the gas into the confined space within the tyre; on a microscopic level, interparticle interactions form as soon as the gas particles come into close proximity.

All matter seeks to minimize its energy and entropy; see Chapter 4. This concept explains, for example, why a ball rolls down a hill, and only stops when it reaches its position of lowest potential energy. These interparticle interactions form for a similar reason.

When we say that two atoms interact, we mean that the outer electrons on the two atoms 'respond' to each other. The electrons within the inner orbitals are buried too deeply within the atom to be available for interactions or bonding. We indicate this situation by saying the electrons that interact reside within the 'frontier' orbitals.

And this interaction always occurs in such as way as to minimize the energy. We could describe the interaction schematically by

$$A + B \longrightarrow \text{product} + \text{energy} \qquad (2.5)$$

where A and B are particles of gas which interact when their frontier orbitals are sufficiently close to form a 'product' of some kind; the product is generally a molecule or association complex. (A less naïve view should also accommodate changes in entropy; see Chapter 4.)

Energy is liberated when bonds and interactions form.	We saw earlier (on p. 33) that measuring the temperature is the simplest macroscopic test for an increased energy content. Therefore, we understand that the tyre becomes warmer during inflation because interactions form between the particles with the concurrent release of energy (Equation (2.5)).

How does a fridge cooler work?

Introduction to the energetics of bond formation

At the heart of a fridge's cooling mechanism is a large flask containing volatile organic liquids, such as alkanes that have been partially fluorinated and/or chlorinated, which are often known as *halons* or chlorofluorocarbons (CFCs). We place this flask behind the fridge cabinet, and connect it to the fridge interior with a thin-walled pipe. The CFCs circulate continually between the fridge interior and the rear, through a heat exchanger.

Now imagine placing a chunk of cheese in the refrigerator. We need to cool the cheese from its original temperature to, say, 5 °C. Because the cheese is warmer than the fridge interior, energy in the form of heat transfers *from* the cheese *to* the fridge, as a consequence of the zeroth law of thermodynamics (see p. 8). This energy passes ultimately to the volatile CFCs in the cooling system.

Converting the liquid CFC to a gas (i.e. boiling) is analogous to putting energy into a kettle, and watching the water boil off as steam.	The CFC is initially a liquid because of intermolecular interactions (of the London dispersion type). Imagine that the interactions involves 4 kJ of energy but cooling the cheese to 5 °C we liberate about 6 kJ of energy: it should be clear that more energy is liberated than is needed to overcome the induced dipoles. We say that

absorption of the energy from the cheese 'overcomes' the interactions – i.e. breaks them – and enables the CFC to convert from its liquid form to form a gas:

$$CFC_{(l)} + energy \longrightarrow CFC_{(g)} \qquad (2.6)$$

The fridge pump circulates the CFC, so the hotter (gaseous) CFC is removed from the fridge interior and replaced with cooler CFC (liquid). We increase the pressure of the gaseous CFC with a pump. The higher pressure causes the CFC to condense back to a liquid. The heat is removed from the fridge through a so-called heat exchanger. Incidentally, this emission of heat also explains why the rear of a fridge is generally warm – the heat emitted is the energy liberated when the cheese cooled.

> Forming bonds and interactions liberates energy; breaking bonds and interactions requires the addition of energy.

In summary, interactions form with the liberation of energy, but adding an equal or greater amount of energy to the system can break the interactions. Stated another way, forming bonds and interactions liberates energy; breaking bonds and interactions requires the addition of energy.

A similar mechanism operates at the heart of an air-conditioning mechanism in a car or office.

Why does steam warm up a cappuccino coffee?

Forming a bond releases energy: introducing calorimetry

To make a cappuccino coffee, pass high-pressure steam through a cup of cold milk to make it hot, then pour coffee through the milk froth. The necessary steam comes from a kettle or boiler.

A kettle or boiler heats water to its boiling point to effect the process:

$$H_2O_{(l)} + heat\ energy \longrightarrow H_2O_{(g)} \qquad (2.7)$$

Water is a liquid at room temperature because cohesive forces bind together the molecules; the bonds in this case are hydrogen bonds – see p. 44. To effect the *phase transition*, liquid $\rightarrow$ gas, we overcome the hydrogen bonds, which explains why we must put energy *into* liquid water to generate gaseous steam. Stated another way, steam is a high-energy form of water.

Much of the steam condenses as it passes through the cool milk. This condensation occurs in tandem with forming the hydrogen bonds responsible for the water being a liquid. These bonds form concurrently with the liberation of energy. This energy transfers to the milk, explaining why its temperature increases.

> The word 'calorimetry' comes from the Latin *calor*, which means heat. We get the word 'calorie' from the same root.

Calorimetry is the measurement of energy changes accompanying chemical or physical changes. We usually want to know how much energy is liberated or consumed per unit mass or mole of

Strictly, this amount of energy is liberated only when the temperature remains at 100 °C during the condensation process. Any changes in temperature need to be considered separately.

substance undergoing the process. Most chemists prefer data to be presented in the form of energy *per mole*. In practice, we measure accurately the amount of heat energy liberated or consumed by a known amount of steam while it condenses.

A physical chemist reading from a data book learns that 40.7 kJ mol^{-1} of energy are liberated when 1 mol of water condenses and will 'translate' this information to say that when 1 mol (18 g) of steam condenses to form liquid water, bonds form concurrently with the liberation of 40 700 J of energy.

As 40.7 kJ mol^{-1} is the *molar* energy (the energy per mole), we can readily calculate the energy necessary, whatever the amount of water involved. In fact, every time the experiment is performed, the same amount of energy will be liberated when 18 g condense.

Worked Example 2.3 How much energy is liberated when 128 g of water condenses?

Note the way the units of 'g' *cancel*, to leave n expressed in the units of moles.

Firstly, we calculate the amount of material n involved using

$$\text{amount of material } n = \frac{\text{mass in grams}}{\text{molar mass in grams per mole}} \quad (2.8)$$

so, as 1 mol has a mass of 18 g mol^{-1}

$$n = \frac{128 \text{ g}}{18 \text{ g mol}^{-1}}$$

$$n = 7.11 \text{ mol}$$

Secondly, the energy liberated *per mole* is 40.7 kJ mol^{-1}, so the overall amount of energy given out is 40.7 kJ mol^{-1} × 7.11 mol = 289 kJ.

SAQ 2.5 How much energy will be liberated when 21 g of water condense?

Cappuccino coffee is named after Marco d'Aviano, a 'Capuchin' monk who was recently made a saint. He entered a looted Turkish army camp, and found sacks of roasted coffee beans. He mixed it with milk and honey to moderate its bitter flavour.

A physical chemist will go one stage further, and say that this energy of 40.7 kJ mol^{-1} relates directly to processes occurring *during the condensation process*. In this case, the energy relates to the formation of hydrogen bonds.

As each water *molecule* forms two hydrogen bonds, so 1 mol of water generates 2 mol of hydrogen bonds. The energy per hydrogen bond is therefore (40.7 kJ mol^{-1} ÷ 2); so the energy of forming a hydrogen bond is 20.35 kJ mol^{-1}.

In summary, the macroscopic changes in energy measured in an experiment such as this are a direct reflection of microscopic energy changes occurring on the molecular level. The milk of a cappuccino coffee is warmed when steam passes through it because the steam condenses to form liquid water; and the water is a liquid because of the formation of intermolecular forces in the form of hydrogen bonds.

Why does land become more fertile after a thunderstorm?

Breaking bonds requires an input of energy

A plant accumulates nutrients from the soil as it grows. Such accumulation depletes the amount of nutrient remaining in the soil; so, harvesting an arable crop, such as maize, barley or corn, removes nutrients from the field. A farmer needs to replenish the nutrients continually if the land is not to become 'exhausted' after a few seasons.

> The reaction of elemental nitrogen to form compounds that can be readily metabolized by a plant is termed 'fixing'. All the principal means of fixing nitrogen involve bacteria.

In the context here, 'nutrients' principally comprise compounds of nitrogen, most of which come from bacteria that employ naturally occurring catalysts (enzymes) which feed on elemental nitrogen – a process known as *fixing*. An example is the bacterium *Rhizobium* which lives on beans and peas. The bacteria convert atmospheric nitrogen into ammonia, which is subsequently available for important biological molecules such as amino acids, proteins, vitamins and nucleic acids.

Other than natural fixing, the principal sources of nutrients are the man-made fertilizers applied artificially by the farmer, the most common being inorganic ammonium nitrate (NH_4NO_3), which is unusually rich in nitrogen.

But *lightning* is also an efficient fertilizer. The mixture of gases we breathe comprises nitrogen (78 per cent), oxygen (21 per cent) and argon (1 per cent) as its principal components. The nitrogen atoms in the N_2 molecule are bound together tightly via a triple bond, which is so strong that most reactions occurring during plant growth (*photosynthesis*) cannot cleave it: N_2 is inert. But the incredible energies unleashed by atmospheric lightning *are* able to overcome the $N\equiv N$ bond.

> Notice the difference between the two words 'princiPAL' (meaning 'best', 'top' or 'most important') and 'princiPLE' (meaning 'idea', 'thought' or 'concept').

The actual mechanism by which the $N\equiv N$ molecule cleaves is very complicated, and is not fully understood yet. It is nevertheless clear that much nitrogen is oxidized to form nitrous oxide, NO. This NO dissolves in the water that inevitably accompanies lightning and forms water-soluble nitrous acid HNO_2, which further oxidizes during the storm to form nitric acid, HNO_3. Nitric acid functions as a high-quality fertilizer. It has been estimated that a thunderstorm can yield many tonnes of fertilizer per acre of land.

To summarize, the $N\equiv N$ bond in the nitrogen molecule is very strong and cannot be cleaved unless a large amount of energy is available to overcome it. Whereas bacteria can fix nitrogen, the biological processes within crops, such as corn and maize, cannot provide sufficient energy. But the energy unleashed during a

> We require energy to cleave bonds: bond energies are discussed on p. 114.

thunderstorm easily overcomes the N≡N *bond energy*, fixing the nitrogen without recourse to a catalyst.

Aside

In the high mountains of Pashawa in Pakistan, near the border with Afghanistan, thunderstorms are so common that the soil is saturated with nitrates deriving from the nitric acid formed by lightning. The soil is naturally rich in potassium compounds. Ion-exchange processes occur between the nitric acid and potassium ions to form large amounts of potassium nitrate, KNO_3, which forms a thick crust of white crystals on the ground, sometimes lending the appearance of fresh snow.

High concentrations of KNO_3 are relatively toxic to plant growth because the ratio of K^+ to Na^+ is too high, and so the soil is not fertile.

Why does a satellite need an inert coating?

Covalency and bond formation

A satellite, e.g. for radio or TV communication, needs to be robust to withstand its environment in space. In particular, it needs to be protected from the tremendous gravitational forces exerted during take off, from the deep vacuum of space, *and from atoms in space*.

> The hydrogen atoms in space form a 'hydride' with the materials on the surface of the satellite.

Being a deep vacuum, there is a negligible 'atmosphere' surrounding a satellite as it orbits in space. All matter will exist solely as unattached atoms (most of them are hydrogen). These atoms impinge on the satellite's outer surface as it orbits. On Earth, hydrogen atoms always seek to form a single bond. The hydrogen atoms in space interact similarly, but with the satellite's tough outer skin.

Such interactions are much stronger than the permanent hydrogen bonds or the weaker, temporary induced dipoles we met in Section 2.1. They form a stronger interaction, which we call a *covalent* bond.

The great American scientist G. N. Lewis coined the word covalent, early in the 20th century. He wanted to express the way that a bond formed by means of electron *sharing*. Each covalent bond comprises a pair of electrons. This pairing is permanent, so we sometimes say a covalent bond is a *formal bond*, to distinguish it from weak and temporary interactions such as induced dipoles.

The extreme strength of the covalent bond derives from the way electrons accumulate *between* the two atoms. The space occupied by the electrons as they accumulate is not random; rather, the two electrons occupy a *molecular orbital* that is orientated spatially in such a way that the highest probability of finding the electronic charges is directly between the two atomic nuclei.

As we learn about the distribution of electrons within a covalent bond, we start with a popular representation known as a *Lewis structure*. Figure 2.11 depicts the

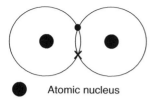

●   Atomic nucleus

● ✕   Electrons

Figure 2.11 Lewis structure of the covalent hydrogen molecule in which electrons are shared

Lewis structure of the hydrogen molecule, in which each atom of hydrogen (atomic number 1) provides a single electron. The resultant molecule may be defined as two atoms held together by means of sharing electrons. Incidentally, we note that the glue holding the two atoms together (the 'bond') involves two electrons. This result is common: each covalent bond requires two electrons.

> We often call these Lewis structures 'dot–cross diagrams'.

Aside

Why call it a 'molecule'?

The word 'molecule' has a long history. The word itself comes from the old French *molécule*, itself derived from the Latin *molecula*, the diminutive of *moles*, meaning 'mass'.

One of the earliest cited uses of the word dates from 1794, when Adams wrote, 'Fermentation disengages a great quantity of air, that is disseminated among the fluid molecules'; and in 1799, Kirwen said, 'The molecules of solid abraded and carried from some spots are often annually recruited by vegetation'. In modern parlance, both Kirwen and Adams meant 'very small particle'.

Later, by 1840, Kirwen's small particle meant '*microscopic* particle'. For example, the great Sir Michael Faraday described a colloidal suspension of gold, known then as *Purple of Cassius*, as comprising molecules which

> Colloids are discussed in Chapter 10.

were 'small particles'. The surgeon William Wilkinson said in 1851, 'Molecules are merely indistinct granules; but under a higher magnifying power, molecules become [distinct] granules'.

Only in 1859 did the modern definition come into being, when the Italian scientist Stanislao Cannizarro (1826–1910) defined a molecule as 'the smallest fundamental unit comprising a group of atoms of a chemical compound'. This statement arose while Cannizarro publicized the earlier work of his compatriot, the chemist and physicist Amedeo Avogadro (1776–1856).

This definition of a molecule soon gained popularity. Before modern theories of bonding were developed, Tyndall had clearly assimilated Cannizarro's definition of a molecule when he described the way atoms assemble, when he said, 'A molecule is a group of atoms drawn and held together by what chemists term affinity'.

Why does water have the formula H₂O?

Covalent bonds and valence

The water molecule always has a composition in which two hydrogen atoms combine with one oxygen. Why?

The Lewis structure in Figure 2.11 represents water, H_2O. Oxygen is element number eight in the periodic table, and each oxygen atom possesses six electrons in its outer shell. Being a member of the second row of the periodic table, each oxygen atom seeks to have an outer shell of eight electrons – we call this trend the 'octet rule'. Each oxygen atom, therefore, has a deficiency of two electrons. As we saw immediately above, an atom of hydrogen has a single electron and, being a row 1 element, requires just one more to complete its outer shell.

> The concept of the full outer shell is crucial if we wish to understand covalent bonds.

The Lewis structure in Figure 2.12 shows the simplest way in which nature satisfies the valence requirements of each element: each hydrogen shares its single electron with the oxygen, meaning that the oxygen atom now has eight electrons (six of its own – the crosses in the figure) and two from the hydrogen atoms. Looking now at each hydrogen atom (the two are identical), we see how each now has two electrons: its own original electron (the dot in the diagram) together with one extra electron each from the oxygen (depicted as crosses).

We have not increased the number of electrons at all. All we have done is shared them between the two elements, thereby enabling each atom to have a full outer shell. This approach is known as the *electron-pair* theory.

Valence bond theory

The *valence bond theory* was developed by Linus Pauling (1901–1992) and others in the 1930s to amalgamate the existing electron-pair bonding theory of G. N. Lewis and new data concerning molecular geometry. Pauling wanted a single, unifying theory.

He produced a conceptual framework to explain molecular bonding, but in practice it could not explain the shapes of many molecules.

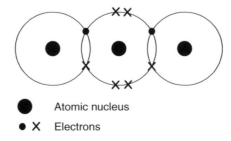

Figure 2.12   Lewis structure of the covalent water molecule. The inner shell of the oxygen atom has been omitted for clarity

Nevertheless, even today, we often discuss the bonding of organic compounds in terms of Lewis structures and valence bond theory.

Why is petroleum gel so soft?

Properties of covalent compounds

Clear petroleum gel is a common product, comprising a mixture of simple hydro-carbons, principally *n*-octadecane (**III**). It is not quite a solid at room temperature; neither is it really a liquid, because it is very viscous. We call it a *gel*. Its principal applications are to lubricate (in a car) or to act as a water-impermeable barrier (e.g. between a baby and its nappy, or on chapped hands).

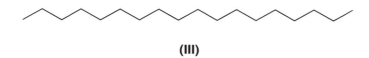

(III)

We saw on p. 52 how methane is a gas unless condensed by compression at high pressure or frozen to low temperatures. But octadecane is neither a solid nor a gas. Why?

There are several, separate types of interaction in **III**: both cova-lent bonds and dipoles. Induced dipoles involve a *partial* charge, which we called δ^+ or δ^-, but, by contrast, covalent bonds involve whole numbers of electrons. A normal covalent bond, such as that between a hydrogen atom and one of the carbon atoms in the back-bone of **III**, requires two electrons. A 'double bond' consists simply of two covalent bonds, so four electrons are shared. Six electrons are incorporated in each of the rare instances of a covalent 'triple bond'. A few quadruple bonds occur in organometallic chemistry, but we will ignore them here.

Most covalent bonds are relatively non-polar. Some are com-pletely non-polar: the diatomic hydrogen molecule is held together with two electrons located equidistantly from the two hydrogen nuclei. Each of the two atoms has an equal 'claim' on the elec-trons, with the consequence that there is no partial charge on the atoms: each is wholly neutral. Only homonuclear molecules such as H_2, F_2, O_2 or N_2 are wholly non-polar, implying that the major-ity of covalent bonds do possess a slight polarity, arising from an unequal sharing of the electrons bound up within the bond.

We see the possibility of a substance having several types of bond. Consider water for example. Formal covalent bonds hold together the hydrogen and oxygen atoms, but the individual water molecules cohere by means of hydrogen bonds. Conversely, paraf-fin wax (n-$C_{15}H_{32}$) is a solid. Each carbon is bonded covalently

> Molecules made of only one element are called 'homonuclear', since *homo* is Greek for 'same'. Examples of homonuclear molecules are H_2, N_2, S_8 and ful-lerene C_{60}.

> Even a covalent bond can possess a perma-nent induced dipole.

> Covalent compounds tend to be gases or liq-uids. Even when solid, they tend to be soft. But many covalent compounds are only solid at lower temper-atures and/or higher pressures, i.e. by max-imizing the incidence of induced dipoles.

Table 2.6 Typical properties covalent compounds

Property	Example
Low melting point	Ice melts in the mouth
Low boiling point	Molecular nitrogen is a gas at room temperature
Physically soft	We use petroleum jelly as a lubricant
Malleable, not brittle	Butter is easily spread on a piece of bread
Low electrical conductivity	We insulate electrical cables with plastic[a]
Dissolve in non-polar solvents	We remove grease with methylated spirit[b]
Insoluble in polar solvents	Polyurethane paint protects the window frame from rain

[a]The polythene coating on an electrical wire comprises a long-chain alkane.
[b]'Methylated spirit' is the industrial name for a mixture of ethanol and methanol.

Figure 2.13   Diamond has a giant macroscopic structure in which each atom is held in a rigid three-dimensional array. Other covalent solids include silica and other p-block oxides such as Al_2O_3

to one or two others to form a linear chain; the hydrogen atoms are bound to this backbone, again with covalent bonds. But the wax is a solid because dispersion forces 'glue' together the molecules. Table 2.6 lists some of the common properties of covalent compounds.

Finally, macromolecular *covalent solids* are unusual in comprising atoms held together in a gigantic three-dimensional array of bonds. Diamond and silica are the simplest examples; see Figure 2.13. Giant macroscopic structures are always solid.

Aside

The word covalent was coined in 1919 when the great American Chemist Irving Langmuir said, 'it is proposed to define valence as the number of pairs of electrons which a given atom shares with others. In view of the fact ... that 'valence' is very often used to express something quite different, it is recommended that the word *covalence* be used to denote valence defined as above.' He added, 'In [ionic] sodium chloride, the covalence of both sodium and chlorine is zero'.

The modern definition from IUPAC says, 'A covalent bond is a region of relatively high electron density between nuclei which arises (at least partially) from sharing of electrons, and gives rise to an attractive force and characteristic inter-nuclear distance'.

Why does salt form when sodium and chlorine react?

Bond formation with ions

Ionic interactions are electrostatic by nature, and occur between ions of opposite charge. The overwhelming majority of ionic compounds are solids, although a few biological exceptions do occur. Table 2.7 lists a few typical properties of ionic compounds.

It is generally unwise to think of ionic compounds as holding together with physical bonds; it is better to think of an array of point charges, held together by the balance of their mutual electrostatic interactions. (By 'mutual' here, we imply equal numbers of positive and negative ions, which therefore impart an overall charge of zero to the solid.)

Ionic compounds generally form following the reaction of metallic elements; non-metals rarely have sufficient energy to provide the necessary energy needed to form ions (see p. 123).

The structure in Figure 2.14 shows the result of an ionic reaction: sodium metal has reacted with chlorine gas to yield white crystalline sodium chloride, NaCl. Each Na atom has lost an electron to form an Na^+ cation and each chlorine atom has gained an electron and is hence a Cl^- anion. In practice, the new electron possessed by the chloride came from the sodium atom.

The electron has transferred and in no way is it shared. Sodium chloride is a compound held together with an ionic bond, the strength of the bond coming from an electrostatic interaction between the positive and negative charges on the ions.

> *Care*: chlorINE is an elemental gas; chlorIDE is a negatively charged anion.

> The chloride ion has a negative charge because, following ionization, it possesses more electrons than protons.

Why heat a neon lamp before it will generate light?

Ionization energy

Neon lamps generate a pleasant pink–red glow. Gaseous neon within the tube (at low pressure) is subjected to a strong electric discharge. One electron per neon atom

Table 2.7   Typical properties of ionic compounds

Property	Example
High melting point	We need a blast furnace to melt metals
High boiling point	A lightning strike is needed to volatilize some substances
Physically hard	Ceramics (e.g. plates) can bear heavy weights
Often physically brittle	Table salt can be crushed to form a powder
High electrical conductivity in solution	Using a hair dryer in the bath risks electrocution
Dissolve in polar solvents	Table salt dissolves in water
Insoluble in non-polar solvents	We dry an organic solvent by adding solid $CaCl_2$

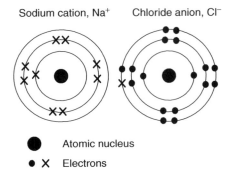

Sodium cation, Na$^+$ Chloride anion, Cl$^-$

● Atomic nucleus

● X Electrons

Figure 2.14 Lewis structure of ionic sodium chloride. Note how the outer shell of the sodium ion is empty, so the next (inner) shell is full

is lost, forming positively charged Ne$^+$ ions:

$$Ne_{(g)} \longrightarrow Ne^+_{(g)} + e^-_{(g)} \tag{2.9}$$

Generally, the flask holding the neon gas contains a small amount of sodium to catalyse ('kick start') the ionization process – see p. 481.

We 'ionized' the neon atoms to form Ne$^+$ *cations*, i.e. each bears a positive charge. (On p. 480 we discuss in detail the photochemical processes occurring at the heart of the neon lamp.)

We generally need quite a lot of energy to ionize an atom or molecule. For example, 2080 kJ of energy are required to ionize 1 mol of monatomic neon gas. This energy is large and explains the need to heat the neon strongly via a strong electric discharge. We call this energy the *ionization energy*, and give it the symbol I (some people symbolize it as I_e). Ionization energy is defined formally as the minimum energy required to ionize 1 mol of an element, generating 1 mol of electrons and 1 mol of positively charged cations.

'Monatomic' is an abbreviation for 'monoatomic', meaning the 'molecule' contains only one atom. The word generally applies to the Group VIII(a) rare gases.

The energy required will vary slightly depending on the conditions employed, so we need to systematize our terminology. While the definition of I is simple enough for neon gas, we need to be more careful for elements that are not normally gaseous. For example, consider the process of ionizing the sodium catalyst at the heart of the neon lamp. In fact, there are *two* energetically distinct processes:

(1) Vaporization of the sodium, to form a gas of sodium atoms: $Na_{(s)} \rightarrow Na_{(g)}$.

(2) Ionization of gaseous atoms to form ions: $Na_{(g)} \rightarrow Na^+_{(g)} + e^-_{(g)}$.

To remove any possible confusion, we further refine the definition of ionization energy, and say that I is the minimum energy required to ionize 1 mol of a *gaseous* element. The ionization energy I relates to process (2); process (1) is additional.

Table 2.8   Ionization energies $I_{(n)}$. For convenience, the figures in the table are given in MJ mol^{-1} rather than the more usual kJ mol^{-1} to emphasize their magnitudes

Element	$I_{(1)}$	$I_{(2)}$	$I_{(3)}$	$I_{(4)}$	$I_{(5)}$	$I_{(6)}$	$I_{(7)}$	$I_{(8)}$	$I_{(9)}$	$I_{(10)}$
Hydrogen	1.318									
Helium	2.379	5.257								
Lithium	0.526	7.305	11.822							
Beryllium	0.906	1.763	14.855	21.013						
Boron	0.807	2.433	3.666	25.033	32.834					
Carbon	1.093	2.359	4.627	6.229	37.838	47.285				
Nitrogen	1.407	2.862	4.585	7.482	9.452	53.274	64.368			
Oxygen	1.320	3.395	5.307	7.476	10.996	13.333	71.343	84.086		
Fluorine	1.687	3.381	6.057	8.414	11.029	15.171	17.874	92.047	106.443	
Neon	2.097	3.959	6.128	9.376	12.184	15.245	20.006	23.076	115.389	131.442

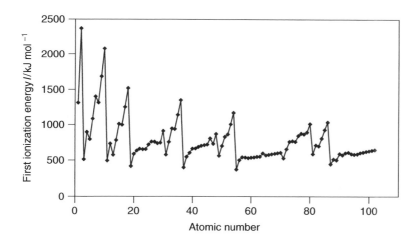

Figure 2.15   The first ionization energies I of the first 105 elements (as y) against atomic number (as x)

Table 2.8 lists several ionization energies: notice that all of them are positive. Figure 2.15 depicts the *first* ionization energies $I_{(1)}$ (as y) for the elements hydrogen to nobelium (elements 1–102) drawn as a function of atomic number (as x).

It is clear from Figure 2.15 that the rare gases in Group VIII(b) have the highest values of I, which is best accounted for by noting that they each have a full outer shell of electrons and, therefore, are unlikely to benefit energetically from being ionized. Similarly, the halogens in Group VII(b) have high values of I because their natural tendency is to *accept* electrons and become anions X$^-$, rather than to *lose* electrons.

The alkali metals in Group I(a) have the lowest ionization energies, which is again expected since they always form cations with a +1 valence. There is little variation in I across the d-block and f-block elements, with a slight increase in I as the atomic number increases.

Why does lightning conduct through air?

Electron affinity

Lightning is one of the more impressive manifestations of the power in nature: the sky lights up with a brilliant flash of light, as huge amounts of electrical energy pass through the air.

As an excellent generalization, gases may be thought of as electrical insulators, so why do we see the lightning travel through the air? How does it conduct? Applying a huge voltage across a sample of gas generates an *electric discharge*, which is apparent by the appearance of light. In fact, the colour of the light depends on the nature of the gas, so neon gives a red colour, krypton gives a green colour and helium is invisible to the eye, but emits ultraviolet light.

The source of the light seen with an electric discharge is the *plasma* formed by the electricity, which is a mixture of ions and electrons, and unionized atoms. If, for example, we look solely at nitrogen, which represents 78 per cent of the air, an electric discharge would form a plasma comprising N_2^+, N^+, electrons e^-, nitrogen radicals $N^\bullet$, as well as unreacted N_2. Incidentally, the formation of these ions explains how air may conduct electricity.

Very soon after the electric discharge, most of the electrons and nitrogen cations reassemble to form uncharged nitrogen, N_2. The recombination produces so much energy that we see it as visible light – lightning. Some of the electrons combine with nitrogen atoms to form nitrogen anions N^-, via the reaction

$$N_{(g)} + e^-_{(g)} \longrightarrow N^-_{(g)} \tag{2.10}$$

> *Care*: do not confuse the symbols for electron affinity $E_{(ea)}$ and activation energy E_a from kinetics (see Chapter 8).

and, finally, some, N_2^+ cations react with water or oxygen in the air to form ammonium or hydroxylamine species.

The energy exchanged during the reaction (in Equation (2.10)) is called the 'electron affinity' $E_{(ea)}$. This energy (also called the electron attachment energy) is defined as the change in the internal energy that occurs when 1 mol of atoms in the gas phase are converted by electron attachment to form 1 mol of gaseous ions.

> The negative value of $E_{(ea)}$ illustrates how the energy of the species $X^-_{(g)}$ is *lower* than that of its precursor, $X_{(g)}$.

The negative ions formed in Equation (2.10) are called *anions*. Most elements are sufficiently electronegative that their electron affinities are negative, implying that energy is given out during the electron attachment. For example, the first electron affinity of nitrogen is only $7\,\text{kJ mol}^{-1}$, but for chlorine $E_{(ea)} = -364\,\text{kJ mol}^{-1}$. Table 2.9 lists the electron affinities of gaseous halogens, and Figure 2.16 depicts the electron affinities for the first 20 elements (hydrogen–calcium).

The electron affinity measures the attractive force between the incoming electron and the nucleus: the stronger the attraction, the more energy there is released. The factors that affect this attraction are exactly the same as those relating to ionization

Table 2.9 The first electron affinities of the Group VII(a) elements

Gas	$E_{(ea)}$/kJ mol^{-1}
F_2	−348
Cl_2	−364
Br_2	−342
I_2	−314

Note: there is much disagreement in the literature about the exact values of electron affinity. These values are taken from the *Chemistry Data Book* by Stark and Wallace. If we use a different data source, we may find slightly different numbers. The trends will be the same, whichever source we consult.

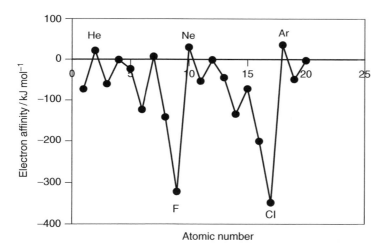

Figure 2.16 Graph of the electron affinities $E_{(ea)}$ of the first 30 elements (as y) against atomic number (as x)

energies – nuclear charge, the distance between the nucleus and the electron, as well as the number of electrons residing between the nucleus and the outer, valence electrons.

Aside

We must be careful with the definition above: in many older textbooks, the electron affinity is defined as the energy *released* when an electron attaches to a neutral atom. This different definition causes $E_{(ea)}$ to change its sign.

Why is argon gas inert?

First electron affinity and reactivity

Gases such as helium, neon and argon are so unreactive that we call them the *inert gases*. They form no chemical compounds, and their only interactions are of the London dispersion force type. They cannot form hydrogen bonds, since they are not able to bond with hydrogen and are not electronegative.

> Krypton, xenon and radon *will* form a very limited number of compounds, e.g. with fluorine, but only under quite exceptional conditions.

The outer shell of the helium atom is full and complete: the shell can only accept two electrons and, indeed, is occupied by two electrons. Similarly, argon has a complete octet of electrons in its outer shell. Further reaction would increase the number of electrons if argon were to undergo a covalent bond or become an anion, or would decrease the number of electrons below the 'perfect' eight if a cation were to form. There is no impetus for reaction because the monatomic argon is already at its position of lowest energy, and we recall that bonds form in order to decrease the energy.

Sodium atoms always seek to lose a single electron to form the Na^+ monocation, because the outer valence shell contains only one electron – that is *why* we assign sodium to Group I(a) of the periodic table. This single electron helps us explain why it is so favourable, energetically, to form the Na^+ cation: loss of the electron empties the outer shell, to reveal a complete inner shell, much like removing the partial skin of an onion to expose a perfectly formed inner layer. So, again, removal of sodium's single outer electron occurs in order to generate a full shell of electrons.

But if we look at an element like magnesium, there are several ionization processes possible:

(1) Formation of a monocation: $Mg_{(g)} \rightarrow Mg^+_{(g)} + e^-_{(g)}$.

(2) Formation of a dication: $Mg^+_{(g)} \rightarrow Mg^{2+}_{(g)} + e^-_{(g)}$.

> *Care*: do not confuse the symbols for molecular iodine I_2 and the second ionization energy $I_{(2)}$. Hint: note carefully the use of *italic* type.

The energy change in reaction (1) is called the *first ionization energy* and the energy associated with reaction (2) is the *second ionization energy*. We symbolize the two processes as $I_{(1)}$ and $I_{(2)}$ respectively.

The second ionization energy is always larger than the first, because we are removing a negative electron from a positively charged cation, so we need to overcome the attractive force between them. The value of $I_{(1)}$ for a magnesium atom is 734 kJ mol^{-1}, but $I_{(2)}$ for removing an electron from the Mg^+ monocation is 1451 kJ mol^{-1}. Both ionization energies are huge, but $I_{(2)}$ is clearly much the larger. Table 2.8 contains many other ionization energies for elements 1–10.

It is clear from Table 2.8 that each ionization energy is larger than the one before. Also note that the last two ionization energies of an element are always larger than the others. The sudden rise follows because the last two energies represent the removal of the two 1 s electrons: removal of electrons from the 2s and 2p orbitals is easier.

Why is silver iodide yellow?

Mixed bonding

Silver chloride is white; silver bromide is pale yellow; and silver iodide has a rich yellow colour. We might first think that the change in colour was due to AgI incorporating the iodide anion, yet NaI or HI are both colourless, so the colour does not come from the iodide ions on their own. We need to find a different explanation.

Silver iodide also has other anomalous properties: it is physically soft – it can even be beaten into a sheet, unlike the overwhelming majority of ionic compounds. More unusual still, it is slightly soluble in ethanol. Clearly, silver iodide is not a straightforward ionic compound. In fact, its properties appear to overlap between covalent (see Table 2.6) and ionic (see Table 2.7).

Silver iodide is neither wholly covalent nor wholly ionic; its bonding shows contributions from both. In fact, most formal chemical bonds comprise a contribution from both covalent *and* ionic forces. The only exceptions to this general rule are homonuclear molecules such as hydrogen or chlorine, in which the bonding is 100 per cent covalent. The extent of covalency in compounds we prefer to think of as ionic will usually be quite small: less than 0.1 per cent in NaCl. For example, each C–H bond in methane is about 4% ionic, but the bonding can be quite unusual in compounds comprising elements from the p- and d-blocks of the periodic table. For example, aluminium chloride, $AlCl_3$, has a high vapour pressure (see p. 221); tungsten trioxide will sublime under reduced pressures to form covalent W_3O_9 trimers; sulphur trioxide is a gas but will dissolve in water. Each, therefore, demonstrates a mixture of ionic and covalent bonding.

> A 'trimer' is a species comprising three components (the Latin *tri* means 'three'). The W_3O_9 trimer has a triangle structure, with a WO_3 unit at each vertex.

In other words, the valence bonds approach is suitable for compounds showing purely ionic or purely covalent behaviour; we require molecular orbitals for a more mature description of the bonding in such materials. So the yellow colour of silver iodide reflects the way the bonding is neither ionic nor covalent. We find, in fact, that the charge clouds of the silver and iodide ions overlap to some extent, allowing change to transfer between them. We will look at *charge transfer* in more detail on p. 459.

> We require 'molecular orbitals' for a more mature description of the bonding in such materials.

Oxidation numbers

Valency is the number of electrons lost, borrowed or shared in a chemical bond. Formal charges are indicated with Arabic numerals, so the formal charge on a copper cation is expressed as Cu^{2+}, meaning each copper cation has a deficiency of two electrons. In this system of thought, the charge on the central carbon of methane is zero.

> Numbers written as 1,2,3,..., etc. are called 'Arabic numerals'.

Table 2.10   Rule for assigning oxidation numbers

1. In a binary compound, the metal has a positive oxidation number and, if a non-metal, it has a negative oxidation number.

2. The oxidation number of a free ion equals the charge on the ion, e.g. in Na^+ the sodium has a $+I$ oxidation number and chlorine in the Cl^- ion has an oxidation number of $-I$. The oxidation number of the MnO_4^- ion is $-I$, oxide O^{2-} is $-II$ and the sulphate SO_4^{2-} ion is $-II$.

3. The sum of the oxidation numbers in a polyatomic ion equals the oxidation number of the ions incorporated: e.g. consider MnO_4^- ion. Overall, its oxidation number is $-I$ (because the ion's charge is -1). Each oxide contributes $-II$ to this sum, so the oxidation number of the central manganese must be $+VII$.

4. The oxidation number of a neutral compound is zero. The oxidation number of an uncombined element is zero.

5. Variable oxidation numbers:
   $H = +I$ (except in the case of hydrides)
   $Cl = -I$ (except in compounds and ions containing oxygen)
   $O = -II$ (except in peroxides and superoxides)

Unfortunately, many compounds contain bonds that are a mixture of ionic and covalent. In such a case, a formal charge as written is unlikely to represent the actual number of charges gained or lost. For example, the complex ferrocyanide anion $[Fe(CN)_6]^{4-}$ is prepared from aqueous Fe^{2+}, but the central iron atom in the complex definitely does not bear a $+2$ charge (in fact, the charge is likely to be nearer $+1.5$). Therefore, we employ the concept of *oxidation number*. Oxidation numbers are cited with Roman numbers, so the oxidation number of the iron atom in the ferrocyanide complex is $+II$. The IUPAC name for the complex requires the oxidation number: we call it hexacyanoferrate (II).

> Numbers written as I, II, III, … etc. are called 'Roman numerals'.

Considering the changes in oxidation number during a reaction can dramatically simplify the concept of oxidation and reduction: oxidation is an increase in oxidation number and reduction is a decrease in oxidation number (see Chapter 7). Be aware, though, oxidation numbers rarely correlate with the charge on an ion. For example, consider the sulphate anion SO_4^{2-} (**IV**).

$$O=S=O$$

with O above (double bond, with $-$) and O below (double bond)

(IV)

The central sulphur has eight bonds. The ion has an overall charge of -2. The oxidation number of the sulphur is therefore $8 - 2 = +6$. We generally indicate oxidation numbers with roman numerals, though, so we write S(VI). Table 2.10 lists the rules required to assign an oxidation number.

3

Energy and the first law of thermodynamics

Introduction

In this chapter we look at the way energy may be converted from one form to another, by breaking and forming bonds and interactions. We also look at ways of measuring these energy changes.

While the change in internal energy ΔU is relatively easy to visualize, chemists generally concentrate on the *net* energy ΔH, where H is the enthalpy. ΔH relates to changes in ΔU after adjusting for pressure–volume expansion work, e.g. against the atmosphere and after transfer of energy q into and out from the reaction environment.

Finally, we look at indirect ways of measuring these energies. Both internal energy and enthalpy are state functions, so energy cycles may be constructed according to Hess's law; we look also at Born–Haber cycles for systems in which ionization processes occur.

3.1 Introduction to thermodynamics: internal energy

Why does the mouth get cold when eating ice cream?

Energy

Eating ice cream soon causes the mouth to get cold, possibly to the extent of making it feel quite uncomfortable. The mouth of a normal, healthy adult has a temperature of about $37\,°C$, and the ice cream has a maximum temperature of $0\,°C$, although it is likely to be in the range -5 to $-10\,°C$ if it recently came from the freezer. A large difference in temperature exists, so energy transfers *from* the mouth *to* the ice cream, causing it to melt.

Ice cream melts as it warms in the mouth and surpasses its normal melting temperature; see Chapter 5.

The evidence for such a transfer of energy between the mouth and the ice cream is the change in temperature, itself a response to the minus-oneth law of thermodynamics (p. 7), which says heat travels from hot to cold. Furthermore, the zeroth law (p. 8) tells us energy will continue to transfer from the mouth (the hotter object) to the ice cream (the colder) until they are at the same temperature, i.e. when they are in *thermal equilibrium*.

Internal energy U

Absolutely everything possesses energy. We cannot 'see' this energy directly, nor do we experience it except under certain conditions. It appears to be invisible because it is effectively 'locked' within a species. We call the energy possessed by the object the '*internal energy*', and give it the symbol U.

> We cannot know how much energy a body or system has 'locked' within it. Experimentally, we can only study *changes* in the internal energy, ΔU.

The internal energy U is defined as the total energy of a body's components. Unfortunately, there is no way of telling how much energy is locked away. In consequence, the experimentalist can only look at *changes* in U.

The energy is 'locked up' within a body or species in three principal ways (or 'modes'). First, energy is locked within the atomic nuclei. The only way to release it is to split the nucleus, as happens in atomic weapons and nuclear power stations to yield *nuclear energy*. The changes in energy caused by splitting nuclei are massive. We will briefly mention nuclear energy in Chapter 8, but the topic will not be discussed otherwise. It is too rare for most physical chemists to consider further.

> The energy E locked into the atomic nucleus is related to its mass m and the speed of light c, according to the Einstein equation, $E = mc^2$.

This second way in which energy is locked away is within chemical bonds. We call this form of energy the *chemical energy*, which is the subject of this chapter. Chemical energies are smaller than nuclear energies.

And third, energy is possessed by virtue of the potential energy, and the translational, vibrational, rotational energy states of the atoms and bonds within the substance, be it atomic, molecular or ionic. The energy within each of these states is quantized, and will be discussed in greater detail in Chapter 9 within the subject of spectroscopy. These energies are normally much smaller than the energies of chemical bonds.

> Strictly, the bonds are held together with 'outer-shell' electrons.

As thermodynamicists, we generally study the second of these modes of energy change, following the breaking and formation of bonds (which are held together with *electrons*), although we occasionally consider potential energy. The magnitude of the chemical energy will change during a reaction, i.e. while altering the number and/or nature of the bonds in a chemical. We give the name *calorimetry* to the study of energy changes occurring during bond changes.

Chemists need to understand the physical chemistry underlying these changes in chemical energy. We generally prefer to write in shorthand, so we don't say 'changes in internal energy' nor the shorter phrase 'changes in U', but say instead 'ΔU'. But we need to be careful: the symbol Δ does not just mean 'change in'. We define it more precisely with Equation (3.1):

$$\Delta U_{(overall)} = U_{(final\ state)} - U_{(initial\ state)} \qquad (3.1)$$

where the phrases 'initial state' and 'final state' can refer to a single chemical or to a mixture of chemicals as they react. This way, ΔU has both a magnitude and a *sign*.

> Placing the Greek letter Δ (Delta) before the symbol for a parameter such as U indicates the change in U while passing *from* an initial *to* a final state. We define the change in a parameter X as $\Delta X = X_{(final\ state)} - X_{(initial\ state)}$.

Aside

In some texts, Equation (3.1) is assumed rather than defined, so we have to work out which are the final and initial states each time, and remember which comes first in expressions like Equation (3.1). In other texts, the final state is written as a subscript and the initial state as a superscript. The value of ΔU for melting ice cream would be written as $\Delta U_{(initial\ state)}^{(final\ state)}$, i.e. $\Delta U_{(ice\ cream\ before\ melting)}^{(melted\ ice\ cream)}$. It may even be abbreviated to ΔU_s^l, where s = solid and l = liquid.

To further complicate matters, other books employ yet another notation. They retain the sub- and super-scripts, but place them *before* the variable, so the last expression in the previous paragraph would be written as $\Delta_s^l U$.

We will not use any of these notation styles in this book.

Why is skin scalded by steam?

Exothermic reactions

Water in the form of a gas is called 'steam'. Two things happen concurrently when human skin comes into close contact with steam – it could happen, for example, when we get too close to a boiling kettle. Firstly, the flesh in contact with the steam gets burnt and hurts. Secondly, steam converts from its gaseous form to become liquid water. We say it *condenses*. We summarize the condensation reaction thus:

> The 'condensation' reaction is one of the simplest forms of a 'phase change', which we discuss in greater depth in Chapter 5.

$$H_2O_{(g)} \longrightarrow H_2O_{(l)} \qquad (3.2)$$

> We will use the word 'process' here to mean any physical chemistry requiring a change in energy.

From Equation (3.1), which defines the changes to internal energy, ΔU for the process in Equation (3.2) is $\Delta U_{(condensation)} = U_{(water, \, l)} - U_{(water, \, g)}$.

As we saw in Chapter 2, the simplest way of telling whether something gains energy is to ascertain whether its temperature goes up. The temperature of the skin does increase greatly (so it feels hot); its energy *increases* following the condensation reaction. Conversely, the temperature of the water decreases – indeed, its temperature decreases to below its boiling temperature, so it condenses. The water has *lost* energy. In summary, we see how energy is transferred, with energy passing *from* the steam *to* the skin.

> The word 'exothermic' comes from two Greek roots: *thermo*, meaning 'energy' or 'temperature', and *exo* meaning 'outside' or 'beyond'. An exothermic process therefore *gives out energy*.

When energy passes from one body to another, we say the process is *thermodynamic*. The condensation of water is a thermodynamic process, with the energy of the water being lower following condensation. Stated another way, the precursor steam had more energy than the liquid water product, so $U_{(final)}$ is lower than $U_{(initial)}$. Figure 3.1 represents this situation visually, and clearly shows how the change in internal energy ΔU during steam condensation is negative. We say the change in U is *exothermic*.

The energy lost by the steam passes to the skin, which therefore gains energy. We experience this excess energy as burning: with the skin being an insulator, the energy from the steam remains within the skin and causes damaging thermal processes. Nerve endings in the skin report the damage to the brain, which leads to the experience of pain.

But none of the energy is *lost* during condensation, so exactly the same amount of energy is given out by the steam as is given to the skin. (In saying this, we assume no other thermodynamic processes occur, such as warming of the surrounding air. Even if other thermodynamic processes *do* occur, we can still say confidently that no energy is lost. It's just more difficult to act as an 'energy auditor', and thereby follow where it goes.)

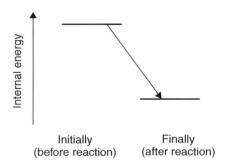

Figure 3.1 In an *exothermic* process, the final product has *less* energy than the initial starting materials. Energy has been given out

Worked Example 3.1   Use Equation (3.1) to demonstrate that ΔU is negative for the condensation of steam if, say, $U_{(\text{final})} = 12$ J and $U_{(\text{initial})} = 25$ J.

Inserting values into Equation (3.1):

$$\Delta U = U_{(\text{final})} - U_{(\text{initial})}$$

$$\Delta U = (12 - 25) \text{ J}$$

$$\Delta U = -13 \text{ J}$$

> *Important*: although we have assigned numerical values to $U_{(\text{final})}$ and $U_{(\text{initial})}$, it is, in fact, impossible to know their values. In reality, we only know the *difference* between them.

So we calculate the value of ΔU as -13 J. The change in U is negative and, therefore, exothermic, as expected.

We see that ΔU is negative. We could have *reasoned* this result by saying $U_{(\text{final})} < U_{(\text{initial})}$, and subtracting a larger energy from a smaller one generates a deficit.

> The symbol 'J' here means *joule*, which is the SI unit of energy.

Why do we sweat?

Endothermic reactions

We all sweat at some time or other, e.g. after running hard, living in a hot climate or perhaps during an illness when our temperature is raised due to an infection (which is why we sometimes say, we have 'got a temperature').

Producing sweat is one of the body's natural ways of cooling itself, and it operates as follows. Sweat is an aqueous solution of salt and natural oils, and is secreted by glands just below the surface of the skin. The glands generate this mixture whenever the body feels too hot. Every time air moves over a sweaty limb, from a mechanical fan or natural breeze, the skin feels cooler following evaporation of water from the sweat.

When we say the water evaporates when a breeze blows, we mean it undergoes a phase transition from liquid to vapour, i.e. a phase transition proceeding in the opposite direction to that in the previous example, so Equation (3.2) occurs backwards. When we consider the internal energy changes, we see $U_{(\text{final})} = U_{(\text{water, g})}$ and $U_{(\text{initial})} = U_{(\text{water, l})}$, so the final state of the water here is *more* energetic than was its initial state. Figure 3.2 shows a schematic representation of the energy change involved.

> We need the salt in sweat to decrease the water's surface tension in order to speed up the evaporation process (we feel cooler more quickly). The oils in sweat prevents the skin from drying out, which would make it susceptible to sunburn.

> Evaporation is also called 'vaporization'. It is a thermodynamic process, because energy is transferred.

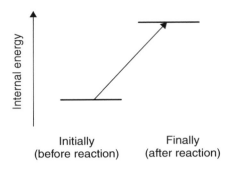

Figure 3.2   In an *endothermic* process, the final product has *more* energy than the initial starting materials. Energy has been taken in

A process is endo-thermic if the final state has more energy than does the initial state. The word derives from the Greek roots *thermo* (meaning 'energy' or 'temperature') and *endo* (meaning 'inside' or 'within'). An endother-mic process takes in energy.

Worked Example 3.2   What is the change in internal energy during sweating?

The definition of ΔU in Equation (3.1) is $\Delta U = U_{(\text{final})} - U_{(\text{initial})}$, so the value of $\Delta U_{(\text{evaporation})}$ is obtained as $U_{(\text{water, g})} - U_{(\text{water, l})}$. We already know that the final state of the water is more energetic than its initial state, so the value of ΔU is positive. We say such a process is *endothermic*.

We feel cooler when sweating because the skin loses energy by transferring it to the water on its surface, which then evapo-rates. This process of water evaporation (sweating) is endothermic because energy passes *from* the skin *to* the water, and a body containing less energy has a lower temperature, which is why we feel cooler.

Aside

Heat is absorbed from the surroundings while a liquid evaporates. This heat does not change the temperature of the liquid because the energy absorbed equates exactly to the energy needed to break intermolecular forces in the liquid (see Chapter 2). Without these forces the liquid would, in fact, be a gas.

At constant temperature, the heat absorbed during evaporation is often called the *latent heat* of evaporation. This choice of words arises from the way evaporation occurs without heating of the liquid; 'latent' is Latin for 'hidden', since the energy added to is not 'seen' as a temperature rise.

Why do we still feel hot while sweating on a humid beach?

State functions

Sometimes we feel hot even when sweating, particularly in a humid environment like a beach by the sea on a hot day. Two processes occur in tandem on the skin: evaporation (liquid water $\rightarrow$ gaseous water) and condensation (gaseous water $\rightarrow$ liquid water). It is quite possible that the same water condenses on our face as evaporated earlier. In effect, then, a cycle of 'liquid $\rightarrow$ gas $\rightarrow$ liquid' occurs. The two halves of this cycle operate in opposite senses, since both exo- and endo-thermic processes occur simultaneously. The *net* change in energy is, therefore, negligible, and we feel no cooler.

These two examples of energy change involve water. The only difference between them is the *direction* of change, and hence the sign of ΔU. But these two factors are related. If we were to *condense* exactly 1 mol of steam then the amount of energy released into the skin would be 40 700 J. The change in internal energy ΔU (ignoring volume changes) is negative because energy is given out during the condensation process, so $\Delta U = -40\,700$ J.

Conversely, if we were to *vaporize* exactly 1 mol of water from the skin of a sweaty body, the change in internal energy would be $+40\,700$ J. In other words, the magnitude of the change is identical, but the sign is different.

While the chemical substance involved dictates the *magnitude* of ΔU (i.e. the amount of it), its *sign* derives from the direction of the thermodynamic process. We can go further: if the same mass of substance is converted from state A to state B, then the change in internal energy is equal and opposite to the same process occurring in the reverse direction, from B to A. This essential truth is depicted schematically in Figure 3.3.

The value of ΔU when condensing exactly 1 mol of water is termed the *molar change* in internal energy. We will call it $\Delta U_{\text{m (condensation)}}$, where the small 'm' indicates that a mole is involved in the thermodynamic process. Similarly, the molar

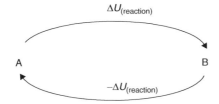

Figure 3.3   The change in internal energy when converting a material from state A to state B is equal and opposite to the change in U obtained when performing the same process in reverse, from B to A

> We often omit the small 'm'; so, from now on, we assume that changes in internal energy are *molar* quantities.

change in energy during vaporization can be symbolized as $\Delta U_{\text{m (vaporization)}}$. If we compare the molar energies for these two similar processes, we see the following relation:

$$\Delta U_{\text{m (condensation)}} = -\Delta U_{\text{m (vaporization)}} \qquad (3.3)$$

The two energies are equal and opposite because one process occurs in the opposite direction to the other, yet the same amount of material (and hence the same amount of energy) is involved in both.

> Internal energy U is a 'state function' because: (1) it is a thermodynamic property; and (2) its value depends only on the present state of the system, i.e. is independent of the previous history.

Following from Equation (3.3), we say that internal energy is a *state function*. A more formal definition of state function is, 'A thermodynamic property (such as internal energy) that depends only on the present state of the system, and is independent of its previous history'. In other words, a 'state function' depends only on those variables that define the *current* state of the system, such as how much material is present, whether it is a solid, liquid or gas, etc.

The concept of a state function can be quite difficult, so let us consider a simple example from outside chemistry. Geographical position has analogies to a thermodynamic state function, insofar as it does not matter whether we have travelled from London to New York via Athens or flew direct. The net difference in position is identical in either case. Figure 3.4 shows this truth diagrammatically. In a similar way, the value of ΔU for the process A → C is the same as the overall change for the process A → B → C. We shall look further at the consequences of U being a state function on p. 98.

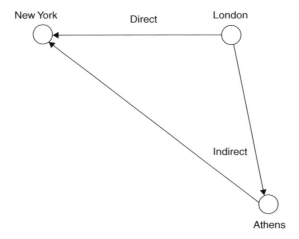

Figure 3.4   If geographical position were a thermodynamic variable, it would be a state function because it would not matter if we travelled from London to New York via Athens or simply flew direct. The *net* difference in position would be identical. Similarly, internal energy, enthalpy, entropy and the Gibbs function (see Chapter 4) are all state functions

Furthermore, because internal energy is a state function, the overall change in U is zero following a series of changes described by a *closed loop*. As an example, imagine three processes: a change from A → B, then B → C and finally from C → A. The only reason why the net value of ΔU for this cycle is zero is because we have neither lost nor picked up any energy *over the cycle*. We can summarize this aspect of physical chemistry by saying, 'energy cannot be created or destroyed, only converted' – a vital truth called the *first law of thermodynamics*.

If we measure ΔU over a thermodynamic *cycle* and obtain a non-zero value, straightaway we know the cycle is either incomplete (with one or more processes not accounted for) or we employed a sloppy technique while measuring ΔU.

> The 'first law of thermodynamics' says energy can neither be created nor destroyed, only converted from one form to another.

Aside

William Rankine was the first to propose the first law of thermodynamics explicitly, in 1853 (he was famous for his work on steam engines). The law was already implicit in the work of other, earlier, thermodynamicists, such as Kelvin, Helmholtz and Clausius. None of these scientists sought to prove their theories experimentally; only Joule published experimental proof of the first law.

Why is the water at the top of a waterfall cooler than the water at its base?

The mechanical equivalence of work and energy

Two of the architects of modern thermodynamics were William Thompson (better known as Lord Kelvin) and his friend James Prescott Joule – a scientist of great vision, and a master of accurate thermodynamic measurement, as well as being something of an English eccentric. For example, while on a holiday in Switzerland in 1847, Thompson met Joule. Let Thompson describe what he saw:

> I was walking down from Chamonix to commence a tour of Mont Blanc, and whom should I meet walking up but Joule, with a long thermometer in his hand and a carriage with a lady in, not far off. He told me that he had been married since we parted in Oxford [two weeks earlier] and that he was going to try for the elevation of temperature in waterfalls.

Despite it being his honeymoon, Joule possessed a gigantic thermometer fully 4 to 5 feet in length (the reports vary). He spent much of his spare time during his honeymoon in making painstaking measurements of the temperature at the top and bottom of elongated Swiss waterfalls. He determined the temperature difference between the water at the bottom and top of the waterfall, finding it to be about 1 °F warmer at the bottom. In Joule's own words, 'A [water]fall of 817 feet [249 m] will generate one degree [Fahrenheit] of temperature'. This result is not attributable to colder air at the top of the waterfall, nor due to friction or viscous drag, or other effects occurring *during* the water's descent, but is wholly due to a change in internal energy. The water was simply changing its altitude.

The potential energy of a raised object is given by the expression

$$\text{potential energy} = mgh \tag{3.4}$$

where *m* is the mass, *g* is the acceleration due to gravity and *h* the height by which it is raised. The potential energy of the water decreases during descent because its height decreases. This energy is liberated; and, as we have noted several times already, the simplest way to tell if the internal energy has increased is to determine its temperature. Joule showed the temperature of the water of the waterfalls had indeed increased.

We could summarize by saying that thermodynamic work *w* is energetically equivalent to the lowering or raising of a weight (like the water of the waterfall, above), as discussed below.

Why is it such hard work pumping up a bicycle tyre?

Thermodynamic work

No one who has pumped up a bicycle tyre says it's easy. Pumping a car tyre is harder still. It requires a lot of energy, and we really have to work at it.

> The pressure inside a party balloon is higher than the external, atmospheric pressure, as evidenced by the way it whizzes around a room when punctured.

We saw in Chapter 1 how increasing the amount of a gas causes its volume to increase. This increase in volume is needed to oppose any increases in pressure. It also explains why blowing into a party balloon causes it to get bigger. By contrast, a car tyre cannot expand greatly during pumping, so increasing the amount of gas it contains will increase its internal pressure. In a fully inflated car tyre, the internal pressure is about 10 times greater than 'standard pressure' $p^{\ominus}$, where $p^{\ominus}$ has a value of 10^5 Pa.

> Work is a form of energy. The word 'energy' comes from the Greek *en ergon*, meaning 'from work'.

The first law of thermodynamics states that energy may be converted between forms, but cannot be created or destroyed. Joule was a superb experimentalist, and performed various types of work, each time generating energy in the form of heat. In one set of experiments, for example, he rotated small paddles immersed in a water trough and noted the rise in temperature. This experiment was apparently performed publicly in St Anne's Square, Manchester.

Joule discerned a relationship between energy and *work* (symbol *w*). We have to perform thermodynamic work to increase the pressure within the tyre. Such work is performed every time a system alters its volume against an opposing pressure or force, or alters the pressure of a system housed within a constant volume.

Work and energy can be considered as interchangeable: we perform work whenever energy powers a physical process, e.g. to propel a car or raise a spoon to the mouth. The work done *on* a system increases its energy, so the value of U increases, itself causing ΔU to be positive). Work done *by* a system corresponds to a negative value of ΔU.

> Work done *on* a system increases its energy, so ΔU is positive. Work done *by* a system corresponds to a negative value of ΔU.

Why does a sausage become warm when placed in an oven?

Isothermal changes in heat and work

At first sight, the answer to our title question is obvious: from the minus-oneth law of thermodynamics, heat travels from the hot oven to the cold(er) item of food we place in it. Also, from the zeroth law, thermal equilibrium is attained only when the sausage and the oven are at the same temperature. So the simplest answer to why a sausage gets hot is to say the energy content of the sausage (in the form of heat) increases, causing its internal energy to rise. And, yet again, we see how the simplest test of an increasing internal energy is an increased temperature.

We can express this truth by saying the sausage gets warmer as the magnitude of its internal energy increases; so, from Equation (3.1), $\Delta U = U_{(final)} - U_{(initial)}$, hence $\Delta U = U_{(after\ heating)} - U_{(before\ heating)}$. We see how the value of ΔU is *positive* since $U_{(after\ heating)} > U_{(before\ heating)}$.

But we can now be more specific. The internal energy U changes in response to two variables, work w and heat energy q, as defined by

$$\Delta U = q + w \qquad (3.5)$$

> *Care*: in the past, Equation (3.5) was often written as $\Delta U = q - w$, where the minus sign is intended to show how the internal energy decreases following work done *by* a system. We will use Equation (3.5), which is the more usual form.

We have already met the first law of thermodynamics. Equation (3.5) here is *the* definitive statement of this law, and is expressed in terms of the transfer of energy between a system and its environment. In other words, the magnitude of ΔU is the sum of the changes in the heat q added (or extracted) from a system, and the work w performed by (or done to) it.

The internal energy can increase or decrease even if one or other of the two variables, q and w, remains fixed. Although the sausage does no work w in the oven, the magnitude of ΔU increases because the food receives heat energy q from the oven.

> *Care*: The symbol of the joule is J. A small 'j' does not mean joules; it represents another variable from a completely different branch of physical chemistry.

Worked Example 3.3   What is the energy of the sausage after heating, if its original energy is 4000 J, and 20 000 J is added to it?

No work is done, so $w = 0$.

The definition of ΔU is given by Equation (3.1):

$$\Delta U = U_{(final)} - U_{(initial)}$$

Rearranging to make $U_{(final)}$ the subject, we obtain

$$U_{(final)} = U_{(initial)} + \Delta U$$

Equation (3.5) is another expression for ΔU. Substituting for ΔU in Equation (3.1) allows us to say

$$U_{(final)} = U_{(initial)} + q + w$$

Inserting values into this equation, we obtain

$$U_{(final)} = 4000 \text{ J} + 20\,000 \text{ J} = 24\,000 \text{ J}$$

> *Care*: this is a highly artificial calculation and is intended for illustrative purposes only. In practice, we *never* know values of U, only changes in U, i.e. ΔU.

The example above illustrates how energy flows in response to the minus-oneth law of thermodynamics, to achieve thermal equilibrium. The impetus for energy flow is the equalization of temperature (via the zeroth law), so we say that the measurement is *isothermal*.

We often want to perform thermodynamic studies isothermally because, that way, we need no subsequent corrections for inequalities in temperature; isothermal measurements generally simplify our calculations.

> The word 'isothermal' can be understood by looking at its Greek roots. *Iso* means 'same' and *thermo* means 'energy' or 'temperature', so a measurement is *isothermal* when performed at a constant temperature.

Why, when letting down a bicycle tyre, is the expelled air so cold?

Thermodynamic work

When a fully inflated car tyre is allowed to deflate, the air streaming through the nozzle is cold to the touch. The pressure of the air within the tyre is fairly high, so opening the tyre valve allows it to leave the tyre rapidly – the air movement may even cause a breeze. We could feel a jet of cold air on our face if we were close enough.

As it leaves the tyre, this jet of air pushes away atmospheric air, which requires an effort. We say that *work* is performed. (It is a form of pressure–volume work, and will be discussed in more depth later, in Section 3.2.)

The internal energy of the gas must change if work is performed, because $\Delta U = q + w$. It is unlikely that any energy is exchanged so, in this simplistic example, we assume that $q = 0$.

Energy is consumed because work w is performed by the gas, causing the energy of the gas to *decrease*, and the change in internal

> Energy *added* to, or work done on, a system is *positive*. Energy *removed* from, or work done by, a system is *negative*.

energy is negative. If ΔU is negative and $q = 0$, then w is also negative. By corollary, the value of w in Equation (3.5) is negative whenever the gas performs work.

From Chapter 2, we remember again that the simplest way to tell whether the internal energy decreases is to check whether the temperature also decreases. We see that the gas coming from the tyre is cold because it performs work, which decreases its internal energy.

Why does a tyre get hot during inflation?

Adiabatic changes

Anyone inflating a tyre with a hand pump will agree that much hard work is needed. A car or bicycle tyre usually gets hot during inflation. In the previous example, the released gas did thermodynamic work and the value of w was negative. In this example, work is done *to* the gas in the tyre, so the value of w is positive. Again, we assume that no energy is transferred, which again allows us to take q as zero.

> The temperature of a tyre also increases when inflated, and is caused by interparticle interactions forming; see p. 59.

Looking again at Equation (3.5), $\Delta U = q + w$, we see that if $q = 0$ and w increases (w is positive), then ΔU increases. This increase in ΔU explains why the temperature of the gas in the tyre increases.

Let us return to the assertion that q is zero, which implies that the system is energetically closed, i.e. that no energy can enter or leave the tyre. This statement is not wholly true because the temperature of the gas within the tyre will equilibrate eventually with the rubber of the tyre, and hence with the outside air, so the tyre becomes cooler in accordance with the minus-oneth and zeroth laws of thermodynamics. But the rubber with which tyre is made is a fairly good thermal insulator, and equilibration is slow. We then make the good approximation that the system is closed, energetically. We say the change in energy is *adiabatic*.

Energetic changes are adiabatic if they can be envisaged to occur while contained within a boundary across which no energy can pass. In other words, the energy content within the system stays fixed. For this reason, there may be a steep temperature jump in going from inside the sealed system to its surroundings – the gas in the tyre is hot, but the surrounding air is cooler.

> A thermodynamic process is *adiabatic* if it occurs within a (conceptual) boundary across which no energy can flow.

In fact, a truly adiabatic system cannot be attained, since even the most insulatory materials will slowly conduct heat. The best approximations are devices such as a Dewar flask (sometimes called a 'vacuum flask').

Can a tyre be inflated without a rise in temperature?

Thermodynamic reversibility

A tyre can indeed be inflated without a rise in temperature, most simply by filling it with a pre-cooled gas, although some might regard this 'adaptation' as cheating!

Alternatively, we could consider inflating the tyre with a series of, say, 100 short steps – each separated by a short pause. The difference in pressure before and after each of these small steps would be so slight that the gas within the tyre would be allowed to reach equilibrium with its surroundings after adding each increment, and before the next. Stated a different way, the difference between the pressure of the gas in the hand pump and in the tyre will always be slight.

> 'Infinitesimal' is the reciprocal of infinite, i.e. incredibly small.

We can take this idea further. We need to realize that if there is no real difference in pressure between the hand pump and the gas within the tyre, then no work would be needed to inflate because there would never be the need to pump *against* a pressure. Alternatively, if the inflation were accomplished at a rate so slow that it was infinitesimally slow, then there would never be a difference in pressure, ensuring w was always zero. And if w was zero, then U would stay constant per increment. (We need to be aware that this argument requires us to perform the process *isothermally*.)

It should be clear that inflating a tyre under such conditions is never going to occur in practice, because we would not have the time, and the inflation would never be complete. But as a conceptual experiment, we see that working at an infinitesimally slow rate does not constitute work in the thermodynamic sense.

> A thermodynamic process is *reversible* if an infinitesimal change in an external variable (e.g. pressure) can change the direction in which the process occurs.

It is often useful to perform thought experiments of this type, changing a thermodynamic variable at an infinitely slow rate: we say we perform the change *reversibly*. (If we perform a process in a non-reversible manner then we say it is 'irreversible'.) As a simple definition, a process is said to be reversible if the change occurs at an infinitesimal rate, and if an infinitesimal change in an external variable (such as pressure) could change the direction of the thermodynamic process. It is seen that a change is only reversible if it occurs with the system and surroundings in equilibrium *at all times*. In practice this condition is never attained, but we can sometimes come quite close.

> The amount of work that can be performed during a thermodynamic process is maximized by performing it reversibly.

Reversibility can be a fairly difficult concept to grasp, but it is invaluable. In fact, the amount of work that can be performed during a thermodynamic process is maximized when performing it reversibly.

The discussion here has focused on work done by changes in pressure, but we could equally have discussed it in terms of volume changes, electrical work (see Chapter 7) or chemical changes (see Chapter 4).

How fast does the air in an oven warm up?

Absorbing energy

The air inside an oven begins to get warm as soon as we switch it on. We can regard the interior of the oven as a fixed system, so the internal energy U of the gases increases as soon as energy is added, since $\Delta U = q + w$ (Equation (3.5)). For simplicity, in this argument we ignore the expansion of the atmosphere inside the oven.

But what is the temperature inside the oven? And by how much does the temperature increase? To understand the relationship between the temperature and the amount of heat entering the system, we must first appreciate that all energies are quantized. The macroscopic phenomenon of temperature rise reflects the microscopic absorption of energy. During absorption, quanta of energy enter a substance at the lowest energy level possible, and only enter higher quantal states when the lower energy states are filled. We see the same principle at work when we fill a jar with marbles: the first marbles fall to the jar bottom (the position of lowest potential energy); and we only see marbles at the top of the jar when all the lower energy levels are filled. Continuing the analogy, a wide jar fills more slowly than does a narrower jar, even when we add marbles at a constant rate.

On a macroscopic level, the rate at which the quantal states are filled as a body absorbs energy is reflected by its heat capacity C. We can tell how quickly the quantum states are occupied because the temperature of a body is in direct proportion to the proportion of states filled. A body having a large number of quantum states requires a large number of energy quanta for the temperature to increase, whereas a body having fewer quantum states fills more quickly, and becomes hot faster.

Why does water boil more quickly in a kettle than in a pan on a stove?

Heat capacity

Most modern kettles contain a powerful element (the salesman's word for 'heater'), operating at a power of 1000 W or more. A heater emits 1 W if it gives out $1\ J\,s^{-1}$; so, a heater rated at 1000 W emits $1000\ J\,s^{-1}$. We may see this power expressed as 1 kW (remember that a small 'k' is shorthand for kilo, meaning 1000). By contrast, an electrical ring on the stove will probably operate between 600 and 800 W, so it emits a smaller amount of heat per second. Because the water absorbs less heat energy per unit time on a stove, its temperature rises more slowly.

> The SI unit of power is the watt (W). A heater rated at a power of 1 W emits $1\ J\,s^{-1}$.

The amount of energy a material or body must absorb for its temperature to increase is termed its 'heat capacity' C. A fixed amount of water will, therefore, get warmer at a slower rate if the amount of heat energy absorbed is smaller per unit time.

> The heat capacity C of a material or body relates the amount of energy absorbed when raising its temperature.

Equation (3.6) expresses the heat capacity C in a mathematical form:

$$C_V = \left(\frac{dU}{dT}\right)_V \qquad (3.6)$$

The expression in Equation (3.6) is really a *partial differential*: the value of U depends on both T and V, the values of which are connected via Equation (1.13). Accordingly, we need to keep one variable constant if we are unambiguously to attribute changes in C_V to the other. The two subscript 'V' terms tell us C is measured while maintaining the volume constant. When the derivative is a partial derivative, it is usual to write the 'd' as '∂'.

> We also call C_V the isochoric heat capacity.

We call C_V 'the heat capacity at constant volume'. With the volume constant, we measure C_V without performing any work (so $w = 0$), so we can write Equation (3.6) differently with dq rather than dU.

Unfortunately, the value of C_V changes slightly with temperature; so, in reality, a value of C_V is obtained as the *tangent* to the graph of internal energy (as y) against temperature (as x); see Figure 3.5.

> A tangent is a straight line that meets a curve at a point, but not does cross it. If the heat capacity changes slightly with temperature, then we obtain the value of C_V as the *gradient* of the tangent to a curve of ΔU (as y) against T (as x).

If the change in temperature is small, then we can usually assume that C_V has no temperature dependence, and write an approximate form of Equation (3.6), saying

$$C_V = \left(\frac{\Delta U}{\Delta T} \right) \qquad (3.7)$$

Analysing Equations (3.6) and (3.7) helps us remember how the SI unit of heat capacity C_V is J K^{-1}. Chemists usually cite a heat capacity after dividing it by the amount of material, calling it the *specific heat capacity*, either in terms of J K^{-1} mol^{-1} or J K^{-1}g^{-1}. As an example, the heat capacity of water is 4.18 J K^{-1}g^{-1}, which means that the temperature of 1 g of water increases by 1 K for every 4.18 J of energy absorbed.

SAQ 3.1 Show that the *molar* heat capacity of water is 75.24 J K^{-1} mol^{-1} if $C_V = 4.18$ J K^{-1}g^{-1}. [Hint: first calculate the molar mass of H$_2$O.]

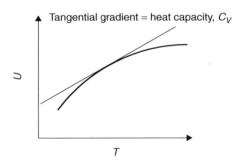

Figure 3.5 The value of the heat capacity at constant volume C_V changes slightly with temperature, so its value is best obtained as the *gradient* of a graph of internal energy (as y) against temperature (as x)

Worked Example 3.4   An electrical heater warms 12 g of water. Its initial temperature is 35.0 °C. The heater emits 15 W for 1 min. What is the new temperature of the water?

Answer strategy. Firstly, we will calculate the energy produced by the heater, in joules. Secondly, knowing the heat capacity of the water C, we divide this energy by C to obtain the temperature rise.

(1) *To calculate the energy produced by the heater.* Remember that $1\ W = 1\ J\,s^{-1}$, so a wattage of 15 W means $15\ J\,s^{-1}$. The heater operates for 1 min (i.e. 60 s), so the energy produced is $15\ J\,s^{-1} \times 60\,s = 900\ J$.

> The word 'strategy' comes from the Greek *stratos* meaning 'army'. Strategy originally concerned military manoeuvres.

This amount of energy is absorbed by 12 g of water, so the energy absorbed per gram is

$$\frac{900\ J}{12\ g} = 75\ J\ g^{-1}$$

(2) *To calculate the temperature rise.* The change in temperature ΔT is sufficiently small that we are justified in assuming that the value of C_V is independent of temperature. This assumption allows us to employ the approximate equation, Equation (3.7). We rearrange it to make ΔT the subject:

$$\Delta T = \frac{\Delta U}{C_V}$$

Inserting values:

$$\Delta T = \frac{75\ J\ g^{-1}}{4.18\ J\ K^{-1} g^{-1}}$$

yielding

$$\Delta T = 17.9\ K$$

As $1\,°C = 1\ K$, the final temperature of the water is $(25.0 + 17.9)\ °C = 42.9\ °C$.

SAQ 3.2   How much energy must be added to 1.35 kg of water in a pan if it is to be warmed from 20 °C to its boiling temperature of 100 °C? Assume C_V does not vary from $4.18\ J\,K^{-1}\,mol^{-1}$.

The heat capacity C_V is an *extensive* quantity, so its value depends on how much of a material we want to warm up. As chemists, we usually want a value of C_V expressed per *mole* of material. A *molar* heat capacity is an *intensive* quantity.

Aside

Another heat capacity is C_p, the heat capacity measured at constant *pressure* (which is also called the *isobaric* heat capacity). The values of C_p and C_V will differ, by perhaps as much as 5–10 per cent. We will look at C_p in more depth in the next section.

Why does a match emit heat when lit?

Reintroducing calorimetry

'Lighting' a match means initiating a simple combustion reaction. Carbohydrates in the wood combine chemically with oxygen in the air to form water and carbon dioxide. The amount of heat liberated is so great that it catches fire (causing the water to form as steam rather than liquid water).

Heat is evolved because the internal energy of the system changes during the combustion reaction. Previously, the oxygen was a gaseous element characterized by O=O bonds, and the wood was a solid characterized by C–C, C–H and C–O bonds. The burning reaction completely changes the number and type of bonds, so the internal energies of the oxygen and the wood alter. This explains the change in ΔU.

We know from Equation (3.5) that $\Delta U = q + w$. Because ΔU changes, one or both of q and w must change. It is certain that much energy is liberated because we feel the heat, so the value of q is negative. Perhaps work w is also performed because gases are produced by the combustion reaction, causing movement of the atmosphere around the match (i.e. w is positive).

The simplest way to measure the change in internal energy ΔU is to perform a reaction in a vessel of constant volume and to look at the amount of heat evolved. We perform a reaction in a sealed vessel of constant volume called a *calorimeter*. In practice, we perform the reaction and look at the rise in temperature. The calorimeter is completely immersed in a large reservoir of water (see Figure 3.6) and its temperature is monitored closely before, during, and after the reaction. If we know the heat

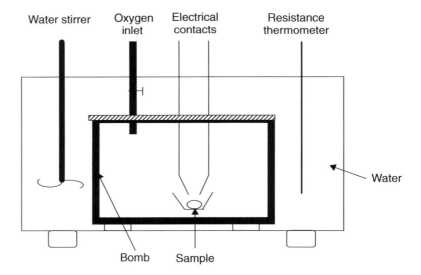

Figure 3.6 Schematic representation of the bomb calorimeter for measuring the changes in internal energy that occur during combustion. The whole apparatus approximates to an adiabatic chamber, so we enclose it within a vacuum jacket (like a Dewar flask)

capacity C of both the calorimeter itself and the surrounding water, then we can readily calculate the change in energy ΔU accompanying the reaction.

Why does it always take 4 min to boil an egg properly?

Thermochemistry

Most people prefer their eggs to be lightly boiled, with the yellow yolk still liquid and the *albumen* solid and white. We say the egg white has been 'denatured'. The variation in egg size is not great. An average egg contains essentially a constant amount of yolk and albumen, so the energy necessary to heat both the yolk and albumen (and to denature the albumen) is, more or less, the same for any egg.

If the energy required to cook an egg is the same per egg, then the simplest way to cook the egg perfectly every time is to ensure carefully that the same amount of energy is absorbed. Most people find that the simplest way to do this is to immerse an egg in boiling water (so the amount of energy entering the egg per unit time is constant), and then to say, 'total energy = energy per second × number of seconds'. In practice, it seems that most people prefer an egg immersed in boiling water for about 240 s, or 4 min.

This simple example introduces the topic of *thermochemistry*. In a physical chemist's laboratory, we generally perform a similar type of experiment but in reverse, placing a sample in the calorimeter and measuring the energy *released* rather than *absorbed*. The most commonly performed calorimetry experiment is combustion inside a *bomb calorimeter* (Figure 3.6). We place the sample in the calorimeter and surround it with oxygen gas at high pressure, then seal the calorimeter securely to prevent its internal contents leaking away, i.e. we maintain a constant volume. An electrical spark then

> *Thermochemistry* is the branch of thermodynamics concerned with the way energy is transferred, released or consumed during a chemical reaction.

ignites the sample, burning it completely. A fearsome amount of energy is liberated in consequence of the ignition, which is why we call this calorimeter a 'bomb'.

The overall heat capacity of the calorimeter is a simple function of the amount of steel the bomb comprises and the amount of water surrounding it. If the mass is m and the heat capacity is C, then the overall heat capacity is expressed by

$$C_{\text{(overall)}} = (m_{\text{(steel)}} \times C_{\text{m (steel)}}) + (m_{\text{(water)}} \times C_{\text{m (water)}}) \qquad (3.8)$$

If the amount of compound burnt in the calorimeter is n, and remembering that no work is done, then a combination of Equations (3.7) and (3.8) suggests that the change in internal energy occurring during combustion is given by

> $C_{\text{(overall)}}$ is the heat capacity of the reaction mixture and the calorimeter.

$$\Delta U_{\text{m (combustion)}} = -\frac{C_{\text{(overall)}} \Delta T}{n} \qquad (3.9)$$

where the 'm' means 'molar'. The negative sign arises from the conventions above, since heat is given out if the temperature goes up, as shown by ΔT being positive.

> $\Delta U_{(combustion)}$ for ben-
> zoic acid is -3.2231 MJ
> mol^{-1} at 298 K.

It is wise first to calibrate the calorimeter by determining an accurate value of $C_{(overall)}$. This is achieved by burning a compound for which the change in internal energy during combustion is known, and then accurately warming the bomb and its reservoir with an electrical heater. Benzoic acid (**I**) is the usual standard of choice when calibrating a bomb calorimeter.

(I)

The electrical energy passed is q, defined by

$$q = V \times I \times t \tag{3.10}$$

where V is the voltage and I the current of the heater, which operates for a time of t seconds.

SAQ 3.3 A voltage of 10 V produces a current of 1.2 A when applied across a heater coil. The heater is operated for 2 min and 40 s. Show that the energy produced by the heater is 1920 J.

> We can assume C_p is
> constant only if ΔT is
> small. For this reason,
> we immerse the 'bomb'
> in a large volume of
> water. This explains
> why we need to oper-
> ate the heater for a
> long time.

Worked Example 3.5 A sample of glucose (10.58 g) is burnt completely in a bomb calorimeter. What is the change in internal energy ΔU if the temperature rises by 1.224 K? The same heater as that in SAQ 3.3 is operated for 11 240 s to achieve a rise in temperature of 1.00 K.

Firstly, we calculate the energy evolved by the reaction. From Equation (3.10), the energy given out by the heater is $q = 10$ V $\times$ 1.2 A $\times$ 11 240 s $=$ 134 880 J.

Secondly, we determine the value of $C_{(overall)}$ for the calorimeter, saying from Equation (3.9)

$$C = \frac{\text{energy released}}{\text{change in temperature}} = \frac{134\,880 \text{ J}}{1.00 \text{ K}}$$

so

$$C = 134\,880 \text{ J K}^{-1}.$$

Thirdly, we determine the amount of glucose consumed n. We obtain the value of n as 'amount = mass $\div$ molar mass'. The molar mass of glucose is 180 g mol^{-1}, so the number of moles is 5.88×10^{-2} mol.

Finally, we calculate the value of ΔU from Equation (3.9). Inserting values:

$$\Delta U_{(combustion)} = -\frac{134\,880 \text{ J K}^{-1} \times 1.224 \text{ K}}{0.0588 \text{ mol}}$$

so

$$\Delta U_{(combustion)} = -2.808 \text{ MJ mol}^{-1}$$

Notice how this value of ΔU is negative. As a good generalization, the change in internal energy ΔU liberated during combustion is negative, which helps explain why so many fires are self-sustaining (although see Chapter 4).

The value of $\Delta U_{(combustion)}$ for glucose is huge, but most values of ΔU are smaller, and are expressed in kilo joules per mole.

> The minus sign is a consequence of the way Equation (3.9) is written.

> An energy change of MJ mol^{-1} is exceptional. Most changes in ΔU are smaller, of the order of kJ mol^{-1}.

> We often calculate a volume of ΔU but cite the answer after adjusting for pressure–volume work; see p. 102.

SAQ 3.4   A sample of anthracene ($C_{14}H_{10}$, **II**) was burnt in a bomb calorimeter. A voltage of 10 V and a current of 1.2 A were passed for exactly 15 min to achieve the same rise in temperature as that caused by the burning of 0.40 g. Calculate the molar energy liberated by the anthracene.

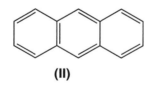

(II)

Aside

The large value of ΔU in Worked Example 3.5 helps explain why sweets, meals and drinks containing sugar are so fattening. If we say a single spoonful of sugar comprises 5 g of glucose, then the energy released by metabolizing it is the same as that needed to raise a 3.5 kg weight from the ground to waist level *1000 times*.

(We calculate the energy per lift with Equation (3.4), saying $E = m \times g \times h$, where m is the mass, g is the acceleration due to gravity and h is the height through which the weight is lifted.)

Why does a watched pot always take so long to boil?

Introduction to Hess's law

We sometimes say, 'A watched pot never boils'. This empirical observation – that we get bored waiting a long time for the pot to boil – follows because we need to put a lot of energy (heat) in order for the water to boil. The amount of energy we can put into the water per unit time was always low in the days of coal and wood fires. Accordingly, a long time was required to boil the water, hence the long wait.

> The popular saying 'A watched pot never boils' arose when most fires were wood or coal, neither of which generates heat as fast as, say, a modern 1 kW kettle.

Imagine we want to convert 1 mol of water starting at a room temperature of, say, 25 °C to steam. In fact we must consider *two* separate thermodynamic processes: we first consider the heat needed to warm the water from 25 °C to its boiling temperature of 100 °C. The water remains liquid during this heating process. Next, we convert 1 mol of the liquid water at 100 °C to gaseous water (i.e. we boil it), but without altering the temperature.

> This argument relies only on words. In reality, the situation is somewhat more complicated because water expands slightly on heating, and greatly on boiling.

We will at the moment ignore once more the problems caused by volume changes. The change in internal energy $\Delta U_{(overall)}$ for the overall process $H_2O_{(l)}$ at 25 °C $\rightarrow$ $H_2O_{(g)}$ at 100 °C can be separated into two components:

Energy ΔU_1 relates to the process

$$H_2O_{(l)} \text{ at } 25\,°C \longrightarrow H_2O_{(l)} \text{ at } 100\,°C$$

Energy ΔU_2 relates to the process

$$H_2O_{(l)} \text{ at } 100\,°C \longrightarrow H_2O_{(g)} \text{ at } 100\,°C$$

so ΔU_1 relates to warming the water until it reaches the boiling temperature, and ΔU_2 relates to the actual boiling process itself.

We can obtain $\Delta U_{(overall)}$ algebraically, according to

$$\Delta U_{(overall)} = \Delta U_1 + \Delta U_2 \tag{3.11}$$

> We can obtain the answer in several different ways because internal energy is a 'state function'.

In practice, we could have measured $\Delta U_{(overall)}$ directly in the laboratory. Alternatively, we could have measured ΔU_1 or ΔU_2 in the laboratory and found the ΔU values we did not know in a book of tables. Either way, we will get the same answer from these two calculation routes.

Equation (3.11) follows directly from ΔU being a state function, and is an expression of *Hess's law*. The great German thermodynamicist Hess observed in 1840 that, 'If a reaction is performed in more than one stage, the overall enthalpy change is a sum of the enthalpy changes involved in the separate stages'.

> *Hess's law* states that the value of an energy obtained is independent of the number of intermediate reaction steps taken.

We shall see shortly how the addition of energies in this way provides the physical chemist with an extremely powerful tool.

Hess's law is a restatement if the first law of thermodynamics. We do not need to measure an energy change directly but can, in practice, divide the reaction into several constituent parts. These parts need not be realizable, so we can actually calculate the energy change for a reaction that is impossible to perform in the laboratory. The only stipulation is for all chemical reactions to balance.

The importance of Hess's law lies in its ability to access information about a reaction that may be difficult (or impossible) to obtain experimentally, by looking at a series of other, related reactions.

.2   Enthalpy

How does a whistling kettle work?

Pressure–volume work

The word 'work' in the question above could confuse. In common parlance, we say a kettle works or does not work, meaning it either functions as a kettle or is useless. But following the example in the previous section, we now realize how the word 'work' has a carefully defined thermodynamic meaning. 'Operate' would be a better choice in this context. In fact, a kettle does not perform any work at all, since it has no moving parts and does not itself move.

In a modern, automatic kettle, an electric heater warms the water inside the kettle – we call it the 'element'. The electric circuit stops when the water reaches $100\,^\circ\mathrm{C}$ because a temperature-sensing bimetallic strip is triggered. But the energy for a more old-fashioned, whistling kettle comes from a gas or a coal hob. The water boils on heating and converts to form copious amounts of gas (steam), which passes through a small valve in the kettle lid to form a shrill note, much like in a football referee's whistle.

The whistle functions because boiling is accompanied by a change in volume, so the steam has to leave the kettle. And the volume change is large: the volume per mole of liquid water is $18\ \mathrm{cm}^3$ (about the size of a small plum) but the volume of a mole of gaseous water (steam) is huge.

SAQ 3.5 Assuming steam to be an ideal gas, use the ideal-gas equation (Equation (1.13)) to prove that 1 mol of steam at $100\,^\circ\mathrm{C}$ (373 K) and standard pressure ($p^\ominus = 10^5$ Pa) has a volume of is $0.031\ \mathrm{m}^3$.

A volume of $0.031\ \mathrm{m}^3$ corresponds to $31\ \mathrm{dm}^3$, so the water increases its volume by a factor of almost 2000 when boiled to generate steam. This staggering result helps us realize just how great the increase in *pressure* is inside the kettle when water boils.

The volume inside a typical kettle is no more than $2\ \mathrm{dm}^3$. To avoid a rapid build up of pressure within the kettle (which could cause an explosion), the steam seeks to leave the kettle, exiting through the small aperture in the whistle. All the vapour passes through this valve just like a referee blowing 'time' after a game. And the large volume ensures a rapid exit of steam, so the kettle produces an intense, shrill whistle sound.

The steam expelled from the kettle must exert a pressure against the air as it leaves the kettle, pushing it aside. Unless stated otherwise, the pressure of the air surrounding the kettle will be 10^5 Pa, which we call 'standard pressure' $p^\ominus$. The value of $p^\ominus$ is 10^5 Pa. The steam must push against this pressure when leaving the kettle. If it does not do so, then it will not move, and will remain trapped within the kettle. This pushing against the air represents *work*. Specifically, we call it *pressure–volume work*, because the volume can only increase by exerting work against an external pressure.

The magnitude of this pressure–volume work is w, and is expressed by

$$w = -\Delta(pV) \tag{3.12}$$

where $\Delta(pV)$ means a change in the product of $p \times V$. Work is done *to* the gas when it is compressed at constant pressure, i.e. the minus sign is needed to make w positive.

Equation (3.12) could have been written as $\Delta p \times \Delta V$ if both the pressure and the volume changed at the same time (an example would be the pushing of a piston in a car engine, to cause the volume to decrease at the same time as the pressure increases). In most of the physicochemical processes we will consider here, either p or V will be constant so, in practice, there is only one variable. And with one variable, Equation (3.12) becomes either $w = p \times \Delta V$ or $w = \Delta V \times p$, depending on whether we hold p or V constant.

| For most purposes, a chemist can say $w = p \times \Delta V$. |

Most chemists perform experiments in which the contents of our beaker, flask or apparatus are open to the air – obvious examples include titrations and refluxes, as well as the kinetic and electro-chemical systems we consider in later chapters. The pressure is the air pressure (usually $p^\ominus$), which does not change, so any pressure–volume work is the work necessary to push back the atmosphere. For most purposes, we can say $w = p\Delta V$.

It should be obvious that the variable held constant – whether p or V – cannot be negative, so the sign of w depends on which of the variables we *change*, so the sign of w in Equation (3.12) depends on the sign(s) of Δp or ΔV. The sign of w will be negative if we decrease the volume or pressure while performing work.

Worked Example 3.6   We generate 1 mol of water vapour in a kettle by boiling liquid water. What is the work w performed by expansion of the resultant steam?

We have already seen how 1 mol of water vapour occupies a volume of 0.031 m^3 (see SAQ 3.5). This volume of air must be pushed back if the steam is to leave the kettle. The external pressure is $p^\ominus$, i.e. 10^5 Pa.

The change in volume

$$\Delta V = V_{(\text{final})} - V_{(\text{initial})}$$

so

$$\Delta V = V_{(\text{water, g})} - V_{(\text{water, l})}$$

Inserting numbers yields

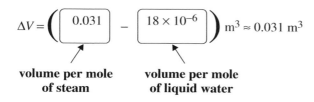

volume per mole
of steam   volume per mole
of liquid water

Each aliquot of 1 cm^3
represents a volume of
1×10^{-6} m^3.

Inserting values into Equation (3.12), $w = p\Delta V$:

$$w = 10^5 \text{ Pa} \times 0.031 \text{ m}^3$$

$$w = 3.1 \times 10^3 \text{ J}$$

so the pressure–volume work is 3100 J.

We often encounter energies of the order of thousands of joules. As a shorthand, we often want to abbreviate, so we rewrite the answer to Worked Example 3.6, and say $w = 3.1 \times 1000$ J. Next, we substitute for the factor of 1000 with an abbreviation, generally choosing a small letter 'k'. We rewrite, saying $w = 3.1$ kJ.

SAQ 3.6 What is the work done when the gas from a party balloon is released? Assume the inflated balloon has a volume of 2 dm^3, and a volume of 10 cm^3 when deflated. Assume there is no pressure change, so $p = p^\ominus$. [Hint: $1 \text{ l} = 1 \times 10^{-3}$ m^3].

An energy expressed with a letter 'k' in the answer means 'thousands of joules'. We say *kilojoules*. The choice of 'k' comes from the Greek word for thousand, which is *kilo*.

Worked Example 3.7 What is the work performed when inflating a car tyre from $p^\ominus$ to $6 \times p^\ominus$. Assume the volume inside the tyre stays constant at 0.3 m^3.

Firstly, we calculate the change in pressure, from an equation like Equation (3.1), $\Delta p = p_{(\text{final})} - p_{(\text{initial})}$, so $\Delta p = (6 - 1) \times p^\ominus$, i.e.

$$\Delta p = 5 \times p^\ominus$$

$$\Delta p = 5 \times 10^5 \text{ Pa}$$

Then, inserting values into Equation (3.5), $w = \Delta p \times V$:

$$w = 5 \times 10^5 \text{ Pa} \times 0.3 \text{ m}^3$$

yielding

$$w = 15\,000 \text{ J} = 15 \text{ kJ}$$

How much energy do we require during a distillation?

The effect of work on ΔU: introducing enthalpy H

Performing a simple distillation experiment is every chemist's delight. We gently warm a mixture of liquids, allowing each component to boil off at its own characteristic temperature (the 'boiling temperature' $T_{(boil)}$). Each gaseous component cools and condenses to allow collection. Purification and separation are thereby effected.

Although we have looked already at boiling and condensation, until now we have always assumed that no work was done. We now see how invalid this assumption was. A heater located within the distillation apparatus, such as an isomantle, supplies heat energy q to molecules of the liquid. Heating the flask increases the internal energy U of the liquids sufficiently for it to vaporize and thence become a gas.

But not all of the heater's energy q goes into raising U. We need some of it to perform pressure–volume work, since the vapour formed on boiling works to push back the external atmosphere. The difference between the internal energy U and the available energy (the enthalpy) is given by

$$\Delta H = \Delta U + p\Delta V \tag{3.13}$$

H is a state function since p, V and U are each state functions. As a state function, the enthalpy is convenient for dealing with systems in which the pressure is constant but the volume is free to change. This way, an enthalpy can be equated with the energy supplied as heat, so $q = \Delta H$.

Worked Example 3.8   A mole of water vaporizes. What is the change in enthalpy, ΔH? Take pressure as $p^{\ominus}$.

We have already seen in the previous section that $\Delta U = +40.7$ kJ per mole of water, and from SAQ 3.5 the volume of 1 mol of water vapour is 0.031 m³ per mole of water.
    Inserting values into Equation (3.13):

$$\Delta H = \boxed{40700 \text{ J mol}^{-1}}_{\Delta U} + \left(\boxed{10^5 \text{ Pa}}_{p} \times \boxed{0.031 \text{ m}^3 \text{ mol}^{-1}}_{\Delta V} \right)$$

so

$$\Delta H = 40\,700 \text{ J mol}^{-1} + (3100 \text{ J mol}^{-1})$$

and

$$\Delta H = 45.8 \text{ kJ mol}^{-1}$$

In this example, the difference between ΔU and ΔH is about 11 per cent. The magnitude of the difference will increase as the values of ΔH and ΔU get smaller.

Justification Box 3.1

We saw above how the work w performed by a gas is pV. Because performing work will decrease the internal energy, we say

$$w = -pV \tag{3.14}$$

Substitution of this simple relationship into the definition of internal energy in Equation (3.5) yields

$$U = q - pV \tag{3.15}$$

and rearranging Equation (3.15) yields $q = U + pV$. This combination of variables occurs so often in physical chemistry, that we give it a name: we call it the *enthalpy*, and give it the symbol H. Accordingly, we rewrite Equation (3.15) as:

> We often call a collection of variables a 'compound variable'.

$$H = U + pV \tag{3.16}$$

The *change* in enthalpy ΔH during a thermodynamic process is defined in terms of internal energy and pressure–volume work by

$$\Delta H = \Delta U + \Delta(pV) \tag{3.17}$$

Because it is usual to perform a chemical experiment with the top of the beaker open to the open air, the pressure p during most chemical reactions and thermodynamic processes is the atmospheric pressure $p^{\ominus}$. Furthermore, this pressure will not vary. In other words, we usually simplify $\Delta(pV)$ saying $p\Delta V$ because only the volume changes.

Accordingly, Equation (3.17) becomes

$$\Delta H = \Delta U + p\Delta V \tag{3.18}$$

The equation in the form of Equation (3.18) is the usual form we use. Changes in U are not equal to the energy supplied as heat (at constant pressure p) because the system employs some of its energy to push back the surroundings as they expand. The pV term for work is, therefore, a correction for the loss of energy as work. Because many, if not most, physicochemical measurements occur under conditions of constant pressure, changes in enthalpy are vitally important because it automatically corrects for the loss of energy to the surroundings.

Aside

One of the most common mistakes we make during calculations of this kind is forgetting the way 'k' stands for '1000'. Think of it this way: a job advertisement offers a salary of £14 k. We would be very upset if, at the end of the first year, we were given just £14 and the employer said he 'forgot' the 'k' in his advert!

Why does the enthalpy of melting ice decrease as the temperature decreases?

Temperature dependence of enthalpy

The enthalpy of freezing water is $-6.00 \text{ kJ mol}^{-1}$ at its normal freezing temperature of 273.15 K. The value is negative because energy is *liberated* during freezing. But the freezing temperature of water changes if the external pressure is altered; so, for example, water freezes at the lower temperature of 253 K when a pressure of about $100 \times p^{\ominus}$ is applied. This high pressure is the same as that along the leading edge of an aeroplane wing. At this lower temperature, $\Delta H_{(\text{melt})} = 5.2 \text{ kJ mol}^{-1}$.

> The word 'normal' in this context means 'at a pressure of $p^{\ominus}$'.

The principal cause of $\Delta H_{(\text{melt})}$ changing is the decreased temperature. The magnitude of an enthalpy depends on the temperature. For this reason, we need to cite the temperature at which an enthalpy is determined. If the conditions are not cited, we assume a temperature of 298 K and a pressure of $p^{\ominus}$. We recognize these conditions as s.t.p. Values of enthalpy are often written as $\Delta H^{\ominus}_{\text{r 298 K}}$ for this reason.

The temperature dependence of the standard enthalpy is related by *Kirchhoff's law*:

$$\Delta H^{\ominus}_{\text{r } T_2} = \Delta H^{\ominus}_{\text{r } T_1} + \int_{T_1}^{T_2} \Delta C_p(T) \, \mathrm{d}T \qquad (3.19)$$

> *Reminder*: The 'curly d' symbols ∂ tells us the bracketed term in the equation is a 'partial differential'.

where C_p is the molar heat capacity at constant *pressure* of the substance in its standard state at a temperature of T. We define C_p according to

$$C_p = \left(\frac{\partial H}{\partial T} \right)_p \qquad (3.20)$$

> $C_p(T)$ means C_p as a function of thermodynamic temperature.

The value of C_p is itself a function of temperature (see p. 140), which explains why we integrate $C_p(T)$ rather than C_p alone.

The Kirchhoff law is a direct consequence of the heat capacity at constant pressure being the derivative of enthalpy with respect to temperature. It is usually sufficient to assume that the heat capacity C_p is itself independent of temperature over the range of temperatures required, in which case Equation (3.19) simplifies to

$$\Delta H^{\ominus}_{\text{r } T_2} = \Delta H^{\ominus}_{\text{r } T_1} + \Delta C_p(T_2 - T_1) \qquad (3.21)$$

The experimental scientist should ensure the range of temperatures is slight if calculating with Equation (3.21).

Worked Example 3.9 The standard enthalpy of combustion $\Delta H^{\ominus}_{\text{c}}$ for benzoic acid (**I**) is $-3223.1 \text{ kJ mol}^{-1}$ at 20 °C. What is $\Delta H^{\ominus}_{\text{c 298 K}}$? The change in C_p during the reaction is $118.5 \, \mathrm{J \, K^{-1} \, mol^{-1}}$. Assume this value is temperature independent over this small temperature interval.

Inserting data into Equation (3.21):

$$\Delta H^{\ominus}_{r \, 298 \, K} = -3223.7 \, \mathrm{kJ \, mol^{-1}} + 118.5 \, \mathrm{J \, K^{-1} \, mol^{-1}} (298 - 293) \, K$$

$$\Delta H^{\ominus}_{r \, 298 \, K} = -3223.7 \, \mathrm{kJ \, mol^{-1}} + 592.5 \, \mathrm{J \, mol^{-1}}$$

so

$$\Delta H^{\ominus}_{r \, 298 \, K} = -3223.1 \, \mathrm{kJ \, mol^{-1}} \text{ or } -3.2231 \, \mathrm{MJ \, mol^{-1}}$$

SAQ 3.7 Ethane burns completely in oxygen to form carbon dioxide and water with an enthalpy of $\Delta H^{\ominus}_{c} = -1558.8 \, \mathrm{kJ \, mol^{-1}}$ at $25 \, ^{\circ}C$. What is $\Delta H^{\ominus}_{c}$ at $80 \, ^{\circ}C$? First calculate the change in heat capacity C_p from the data in the following table and Equation (3.22).

Substance C_p at $80 \, ^{\circ}C$/J K^{-1} mol^{-1}	$C_2H_{6(g)}$	$O_{2(g)}$	$CO_{2(g)}$	$H_2O_{(l)}$
	52.6	29.4	37.1	75.3

$$\Delta C_p = \sum_{\text{products}} v C_p - \sum_{\text{reactants}} v C_p \qquad (3.22)$$

where the upper-case Greek letter Sigma Σ means 'sum of', and the lower-case Greek letters v (nu) represent the *stoichiometric number* of each species, which are the numbers of each reagent in a fully balanced equation. In the convention we adopt here, the values of v are positive for products and negative for reactants.

Justification Box 3.2

Starting with the definition of heat capacity in Equation (3.20):

$$\left(\frac{\partial H}{\partial T} \right)_p = C_p$$

This equation represents C_p for a single, pure substance. Separating the variables yields

$$dH = C_p \, dT$$

Then we integrate between limits, saying the enthalpy is H_1 at T_1 and H_2 at T_2:

$$\int_{H_1}^{H_2} dH = \int_{T_1}^{T_2} C_p \, dT$$

Integrating yields

$$(H_2 - H_1) = C_p(T_2 - T_1) \qquad (3.23)$$

where the term on the left-hand side is ΔH. Equation (3.23) relates to a single, pure substance.

If we consider a chemical reaction in which several chemicals combine, we can write an expression like this for each chemical. Each chemical has a unique value of ΔH and C_p, but the temperature change $(T_2 - T_1)$ remains the same for each.

We combine each of the ΔH terms to yield $\Delta H_{r\ T_2}^{\ominus}$ (i.e. $\Delta H_r^{\ominus}$ at T_2) and $\Delta H_{r\ T_1}^{\ominus}$. Combining the C_p terms according to Equation (3.22) yields ΔC_p. Accordingly, Equation (3.23) then becomes Equation (3.21), i.e.:

$$\Delta H_{r\ T_2}^{\ominus} = \Delta H_{r\ T_1}^{\ominus} + \Delta C_p(T_2 - T_1)$$

Why does water take longer to heat in a pressure cooker than in an open pan?

The differences between C_V and C_p

A pressure cooker is a sealed cooking pan. Being sealed, as soon as boiling occurs, the pressure of steam within the pan increases dramatically, reaching a maximum pressure of about $6 \times p^{\ominus}$, causing the final boiling temperature to increase (see Fig. 5.12 on p. 200). Unlike other pans, the internal volume is fixed and the pressure can vary; the pressure in most pans is atmospheric pressure ($\sim p^{\ominus}$), but the volume of the steam increases continually.

See p. 199 to see why a pressure cooker can cook faster than a conventional, open pan.

The heat capacity of the contents in a pressure cooker is C_V because the internal volume is constant. By contrast, the heat capacity of the food or whatever inside a conventional pan is C_p. The water is a pressure cooker warms slower because the value of C_p is always smaller than C_V. And being smaller, the temperature increases faster per unit input of energy.

In fact, the relationship between C_V and C_p is given by

$$C_V - C_p = nR \qquad (3.24)$$

It is relatively rare that we need C_V values; most reactions are performed at constant pressure, e.g. refluxing a flask at atmospheric pressure.

where we have met all terms previously.

Worked Example 3.10 What is the heat capacity C_V of 1 mol of water? Take the value of C_p from SAQ 3.7.

Rearranging Equation (3.24) slightly yields

$$C_V = nR + C_p$$

Inserting values:

$$C_V = (1 \times 8.3 + 75.3) \text{ J K}^{-1}\text{mol}^{-1}$$

so $C_V = 83.6 \text{ J K}^{-1} \text{ mol}^{-1}$.

The value of C_V is 11 per cent higher than C_p, so the water in the pressure cooker will require 11 per cent more energy than if heated in an open pan.

Justification Box 3.3

Starting with the definition of enthalpy in Equation (3.16):

$$H = U + pV$$

The pV term can be replaced with 'nRT' via the ideal-gas equation (Equation (1.13)), giving

$$H = U + nRT$$

The differential for a small change in temperature is

$$dH = dU + nR\,dT$$

> The values of n and R are constants and do not change.

dividing throughout by dT yields

$$\left(\frac{\partial H}{\partial T}\right) = \left(\frac{\partial U}{\partial T}\right) + nR$$

The first bracket equals C_V and the second bracket equals C_p, so

$$C_V = C_p + nR$$

which is just Equation (3.24) rearranged. Dividing throughout by n yields the *molar* heat capacities:

$$C_V = C_p + R \qquad\qquad (3.25)$$

Why does the temperature change during a reaction?

Enthalpies and standard enthalpies of reaction: ΔH_r and $\Delta H_r^{\ominus}$

One of the simplest definitions of a chemical reaction is 'changes in the bonds'. All reactions proceed with some bonds cleaving concurrently with others forming. Each bond requires energy to form, and each bond liberates energy when breaking (see p. 63 ff). Typically, the amount of energy consumed or liberated is characteristic of the bond involved, so each C–H bond in methane releases about 220 kJ mol^{-1} of energy. And, as we have consistently reported, the best macroscopic indicator of a microscopic energy change is a change in temperature.

Like internal energy, we can never know the enthalpy of a reagent; only the *change* in enthalpy during a reaction or process is knowable. Nevertheless, we can think of changes in H. Consider the preparation of ammonia:

$$N_{2(g)} + 3H_{2(g)} \longrightarrow 2NH_{3(g)} \tag{3.26}$$

We obtain the standard enthalpy change on reaction $\Delta H_r^\ominus$ as a sum of the molar enthalpies of each chemical participating in the reaction:

$$\Delta H_r^\ominus = \sum_{\text{products}} \nu H_m^\ominus - \sum_{\text{reactants}} \nu H_m^\ominus \tag{3.27}$$

The values of ν for the reaction in Equation (3.26) are $\nu_{(NH_3)} = +2$, $\nu_{(H_2)} = -3$ and $\nu_{(N_2)} = -1$. We obtain the standard molar enthalpy of forming ammonia after inserting values into Equation (3.27), as

$$\Delta H_r^\ominus = 2H_m^\ominus(NH_3) - [H_m^\ominus(N_2) + 3H_m^\ominus(H_2)]$$

SAQ 3.8 Write out an expression for $\Delta H_r^\ominus$ for the reaction $2NO + O_2 \rightarrow 2NO_2$ in the style of Equation (3.27).

Unfortunately, we do not know the enthalpies of any reagent. All we can know is a *change* in enthalpy for a reaction or process. But what is the magnitude of this energy change? As a consequence of Hess's law (see p. 98), the overall change in enthalpy accompanying a reaction follows from the number and nature of the bonds involved. We call the overall enthalpy change during a reaction the 'reaction enthalpy' ΔH_r, and define it as 'the change in energy occurring when 1 mol of reaction occurs'. In consequence, its units are J mol^{-1}, although chemists will usually want to express ΔH in kJ mol^{-1}.

In practice, we generally prefer to tighten the definition of ΔH_r above, and look at reagents in their *standard states*. Furthermore, we maintain the temperature T at 298 K, and the pressure p at $p^\ominus$. We call these conditions standard temperature and pressure, or s.t.p. for short. We need to specify the conditions because temperature and pressure can so readily change the physical conditions of the reactants and products. As a simple example, elemental bromine is a liquid at s.t.p., so we say the standard state of bromine at s.t.p. is $Br_{2(l)}$. If a reaction required gaseous bromine $Br_{2(g)}$ then we would need to consider an additional energy – the energy of vaporization to effect the process $Br_{2(l)} \rightarrow Br_{2(g)}$. Because we restricted ourselves to s.t.p. conditions, we no longer talk of the reaction enthalpy, but the 'standard reaction enthalpies' $\Delta H_r^\ominus$, where we indicate the standard state with the plimsoll sign '$\ominus$'.

In summary, the temperature of a reaction mixture changes because energy is released or liberated. The temperature of the reaction mixture is only ever constant in the unlikely event of $\Delta H_r^\ominus$ being zero. (This argument requires an *adiabatic* reaction vessel; see p. 89.)

Some standard enthalpies have special names. We consider below some of the more important cases.

Are diamonds forever?

Enthalpies of formation

We often hear it said that 'diamonds are forever'. There was even a James Bond novel and film with this title. Under most conditions, a diamond will indeed last forever, or as near 'for ever' as makes no difference. But is it an *absolute* statement of fact?

Diamond is one of the naturally occurring allotropes of carbon, the other common allotrope being graphite. (Other, less common, allotropes include buckminster fullerine.) If we could observe a diamond over an extremely long time scale – in this case, several billions of years – we would observe a slow conversion from brilliant, clear diamond into grey, opaque graphite. The conversion occurs because diamond is slightly less stable, thermodynamically, than graphite.

> Some elements exist in several different crystallographic forms. The differing crystal forms are called *allotropes*.

Heating graphite at the same time as compressing it under enormous pressure will yield diamond. The energy needed to convert 1 mol of graphite to diamond is 2.4 kJ mol^{-1}. We say the 'enthalpy of formation' ΔH_f for the diamond is $+2.4$ kJ mol^{-1} because graphite is the standard state of carbon.

We define the 'standard enthalpy of formation' $\Delta H_f^\ominus$ as the enthalpy change involved in forming 1 mol of a compound from its elements, each element existing in its standard form. Both T and p need to be specified, because both variables influence the magnitude of ΔH. Most books and tables cite $\Delta H_f^\ominus$ at standard pressure $p^\ominus$ and at a temperature of 298 K. Table 3.1 cites a few representative values of $\Delta H_f^\ominus$.

> The 'standard enthalpy of formation' $\Delta H_f^\ominus$ is the enthalpy change involved in forming 1 mol of a compound or non-stable allotrope from its elements, each element being in its standard form, at s.t.p.

It will be immediately clear from Table 3.1 that several values of ΔH_f are zero. This value arises from the definition we chose, above: as ΔH_f relates to forming a compound *from its constituent elements*, it follows that the enthalpy of forming an element can only be zero, provided it exists in its standard state. Incidentally, it also explains why $\Delta H_f(Br_2, l) = 0$ but $\Delta H_f(Br_2, g) = 29.5$ kJ mol^{-1}, because the stable form of bromine is liquid at s.t.p.

For completeness, we stipulate that the elements must exist in their *standard states*. This sub-clause is necessary, because whereas most elements exist in a single form at s.t.p. (in which case their enthalpy of formation is zero), some elements, such as carbon

> We *define* the enthalpy of formation of an element (in its *normal* state) as zero.

Table 3.1   Standard enthalpies of formation $\Delta H_f^\ominus$ at 298 K

Compound	$\Delta H_f^\ominus /\text{kJ mol}^{-1}$
Organic	
Hydrocarbons	
methane (CH_4, g)	−74.8
ethane (CH_3CH_3, g)	−84.7
propane ($CH_3CH_2CH_3$, g)	−103.9
n-butane (C_4H_{10}, g)	−126.2
ethane ($CH_2{=}CH_2$, g)	54.3
ethyne ($CH{\equiv}CH$, g)	226.7
cis-2-butene (C_4H_8, g)	−7.00
$trans$-2-butene (C_4H_8, g)	−11.2
n-hexane (C_6H_{14}, l)	−198.7
cyclohexane (C_6H_{12}, l)	−156
Alcohols	
methanol (CH_3OH, l)	−238.7
ethanol (C_2H_5OH, l)	−277.7
Aromatics	
benzene (C_6H_6, l)	49.0
benzene (C_6H_6, g)	82.9
toluene ($CH_3C_6H_5$, l)	50.0
Sugars	
α-D-glucose ($C_6H_{12}O_6$, s)	−1274
β-D-glucose ($C_6H_{12}O_6$, s)	−1268
sucrose ($C_{12}H_{22}O_{11}$, s)	−2222
Elements	
bromine (Br_2, l)	0.00
bromine (Br_2, g)	30.9
chlorine (Cl_2, g)	0.00
chlorine (Cl, g)	121.7
copper (Cu, s)	0.00
copper (Cu, g)	338.3
fluorine (F_2, g)	0.00
fluorine (F, g)	78.99
iodine (I_2, s)	0.00
iodine (I_2, g)	62.4
iodine (I, g)	106.8
nitrogen (N, g)	472.7
phosphorus (P, white, s)	0.00
phosphorus (P, red, s)	15.9
sodium (Na, g)	107.3
sulphur (S, rhombic, s)	0.00
sulphur (S, monoclinic, s)	0.33
Inorganic	
carbon (diamond, s)	2.4
carbon monoxide (CO, g)	−110.5

Table 3.1   (*continued*)

Compound	$\Delta H_f^{\ominus}$/kJ mol^{-1}
carbon dioxide (CO$_2$, g)	−393.0
copper oxide (CuO, s)	−157.3
hydrogen oxide (H$_2$O, l)	−285.8
hydrogen oxide (H$_2$O, g)	−241.8
hydrogen fluoride (HF, g)	−271.1
hydrogen chloride (HCl, g)	−92.3
nitrogen hydride (NH$_3$, g)	−46.1
nitrogen hydride (NH$_3$, aq)	−80.3
nitrogen monoxide (NO, g)	90.3
nitrogen dioxide (NO$_2$, g)	33.2
phosphine (PH$_3$, g)	5.4
silicon dioxide (SiO$_2$, s)	−910.9
sodium hydroxide (NaOH, s)	−425.6
sulphur dioxide (SO$_2$, g)	−296.8
sulphur trioxide (SO$_3$, g)	−395.7
sulphuric acid (H$_2$SO$_4$, l)	−909.3

(above), sulphur or phosphorus, have *allotropes*. The enthalpy of formation for the *stable* allotrope is always zero, but the value of ΔH_f for the non-stable allotropes will not be. In fact, the value of ΔH_f for the non-stable allotrope is cited *with respect to* the stable allotrope. As an example, ΔH_f for white phosphorus is zero by definition (it is the stable allotrope at s.t.p.), but the value of ΔH_f for forming red phosphorus from white phosphorus is 15.9 kJ mol^{-1}.

If the value of ΔH_f is determined within these three constraints of standard T, standard p and standard allotropic form, we call the enthalpy a *standard* enthalpy, which we indicate using the plimsoll symbol '$\ominus$' as $\Delta H_f^{\ominus}$.

To conclude: are diamonds forever? No. They convert slowly into graphite, which is the stablest form of carbon. Graphite has the lowest energy for any of the allotropes of carbon, and will not convert to diamond without the addition of energy.

Why do we burn fuel when cold?

Enthalpies of combustion

A common picture in any book describing our Stone Age forebears shows short, hairy people crouched, warming themselves round a flickering fire. In fact, fire was one of the first chemical reactions discovered by our prehistoric ancestors. Primeval fire was needed for warmth. Cooking and warding off dangerous animals with fire was a later 'discovery'.

But why do they burn wood, say, when cold? The principal reactions occurring when natural materials burn involve chemical oxidation, with carbohydrates combining with elemental oxygen to yield water and carbon dioxide. Nitrogen compounds yield nitrogen oxide, and sulphur compounds yield sulphur dioxide, which itself oxidizes to form SO_3.

Let us simplify and look at the *combustion* of the simplest hydrocarbon, methane. CH_4 reacts with oxygen according to

$$CH_{4(g)} + 2O_{2(g)} \longrightarrow CO_{2(g)} + 2H_2O_{(g)} \qquad (3.28)$$

The reaction is very exothermic, which explains why much of the developed world employs methane as a heating fuel. We can measure the enthalpy change accompanying the reaction inside a calorimeter, or we can calculate a value with thermochemical data.

> Most authors abbreviate 'combustion' to just 'c', and symbolize the enthalpy change as $\Delta H_c^{\ominus}$. Others write $\Delta H_{(comb)}^{\ominus}$.

This enthalpy has a special name: we call it the *enthalpy of combustion*, and define it as the change in enthalpy accompanying the burning of methane, and symbolize it as $\Delta H_{(combustion)}$ or just ΔH_c. In fact, we rarely perform calculations with ΔH_c but with the *standard* enthalpy of combustion $\Delta H_c^{\ominus}$, where the plimsoll symbol '$\ominus$' implies s.t.p. conditions.

Table 3.2 contains values of $\Delta H_c^{\ominus}$ for a few selected organic compounds. The table shows how all value of $\Delta H_c^{\ominus}$ are negative, reminding us that energy is given out during a combustion reaction. We say combustion is *exothermic*, meaning energy is emitted. All exothermic reactions are characterized by a negative value of $\Delta H_c^{\ominus}$.

> We can use equations like Equation (3.27) for *any* form of enthalpy, not just combustion.

But we do not have to measure each value of $\Delta H_c^{\ominus}$: we can calculate them if we know the enthalpies of formation of each chemical, product and reactant, we can adapt the expression in eq. (3.27), saying:

$$\Delta H_c^{\ominus} = \sum_{products} \nu \Delta H_f^{\ominus} - \sum_{reactants} \nu \Delta H_f^{\ominus} \qquad (3.29)$$

> We could not perform cycles of this type unless enthalpy was a *state function*.

where each ΔH term on the right-hand side of the equation is a molar enthalpy of formation, which can be obtained from tables.

Worked Example 3.11 The wood mentioned in our title question is a complicated mixture of organic chemicals; so, for simplicity, we update the scene. Rather than prehistoric men sitting around a fire, we consider the calorific value of methane in a modern central-heating system. Calculate the value of ΔH_c for methane at 25 °C using molar enthalpies of formation $\Delta H_f^{\ominus}$.

> The word 'calorific' means heat containing, and comes from the Latin *calor*, meaning 'heat'.

The necessary values of $\Delta H_f^{\ominus}$ are:

Species (all as gases)	CH_4	O_2	CO_2	H_2O
$\Delta H_f^{\ominus}$/kJ mol^{-1}	−74.81	0	−393.51	−285.83

Table 3.2 Standard enthalpies of combustion $\Delta H_c^{\ominus}$ for a few organic compounds (all values are at 298 K)

Substance	$\Delta H_c^{\ominus}/\text{kJ mol}^{-1}$
Hydrocarbons	
methane (CH_4, g)	−890
ethane (CH_3CH_3, g)	−1560
propane ($CH_3CH_2CH_3$, g)	−2220
n-butane (C_4H_{10}, g)	−2878
cyclopropane (C_3H_6, g)	−2091
propene (C_3H_6, g)	−2058
1-butene (C_4H_8, g)	−2717
cis-2-butene (C_4H_8, g)	−2710
trans-2-butene (C_4H_8, g)	−2707
Alcohols	
methanol (CH_3OH, l)	−726
ethanol (C_2H_5OH, l)	−1368
Aromatics	
benzene (C_6H_6, l)	−3268
toluene ($CH_3C_6H_5$, l)	−3953
naphthalene ($C_{10}H_8$, s)	−5147
Acids	
methanoic (HCO_2H, l)	−255
ethanoic (CH_3CO_2H, l)	−875
oxalic ($HCO_2 \cdot CO_2H$, s)	−254
benzoic ($C_6H_5 \cdot CO_2H$, s)	−3227
Sugars	
α-D-glucose ($C_6H_{12}O_6$, s)	−2808
β-D-glucose ($C_6H_{12}O_6$, s)	−2810
sucrose ($C_{12}H_{22}O_{11}$, s)	−5645

O_2 is an element, so its value of $\Delta H_f^{\ominus}$ is zero. The other values of $\Delta H_f^{\ominus}$ are exothermic. Inserting values into Equation (3.29):

$$\Delta H_c^{\ominus} = [(-393.51) + (2 \times -285.83)] - [(1 \times -74.81) + (2 \times 0)] \text{ kJ mol}^{-1}$$

$$\Delta H_c^{\ominus} = -965.17 - (-74.81) \text{ kJ mol}^{-1}$$

$$\Delta H_c^{\ominus} = -886.36 \text{ kJ mol}^{-1}$$

which is very close to the experimental value of -890 kJ mol^{-1} in Table 3.2.

SAQ 3.9 Calculate the standard enthalpy of combustion $\Delta H_c^{\ominus}$ for burning β-D-glucose, $C_6H_{12}O_6$. The required values of $\Delta H_f^{\ominus}$ may be found in Table 3.1.

The massive value of $\Delta H_c^{\ominus}$ for glucose explains why athletes consume glucose tablets to provide them with energy.

We defined the value of $\Delta H_c^{\ominus}$ during combustion as $H_{(final)} - H_{(initial)}$, so a negative sign for $\Delta H_c^{\ominus}$ suggests the final enthalpy is more negative after combustion. In other words, energy is given out during the reaction. Our Stone Age forebears absorbed this energy by their fires in the night, which is another way of saying 'they warmed themselves'.

Why does butane burn with a hotter flame than methane?

Bond enthalpies

Methane is easily bottled for transportation because it is a gas. It burns with a clean flame, unlike coal or oil. It is a good fuel. The value of $\Delta H_c^{\ominus}$ for methane is -886 kJ mol^{-1}, but $\Delta H_c^{\ominus}$ for n-butane is $-2878 \text{ kJ mol}^{-1}$. Burning butane is clearly far more exothermic, explaining why it burns with a hotter flame. In other words, butane is a better fuel.

The overall enthalpy change during combustion is $\Delta H_c^{\ominus}$. An alternative way of calculating an enthalpy change during reaction dispenses with enthalpies of formation $\Delta H_f^{\ominus}$ and looks at the individual numbers of bonds formed and broken. We saw in Chapter 2 how we always need energy to break a bond, and release energy each time a bond forms. Its magnitude depends entirely on the enthalpy change for breaking or making the bonds, and on the respective numbers of each. For example, Equation (3.28) proceeds with six bonds cleaving (four C–H bonds and two O=O bonds) at the same time as six bonds form (two C=O bonds and four H–O bonds).

A quick glance at Worked Example 3.11 shows how the energy released during combustion is associated with forming the CO_2 and H_2O. If we could generate more CO_2 and H_2O, then the overall change in ΔH would be greater, and hence the fuel would be superior. In fact, many companies prefer butane to methane because it releases more energy per mole.

To simplify the calculation, we pretend the reaction proceeds with all bonds breaking at once; then, an instant later, different bonds form, again all at once. Such an idea is mechanistic nonsense but it simplifies the calculation.

We can calculate an enthalpy of reaction with bond enthalpies by assuming the reaction consists of two steps: first, bonds break, and then different bonds form. This approach can be simplified further if we consider the reaction consists only of reactive fragments, and the products form from these fragments. The majority of the molecule can remain completely unchanged, e.g. we only need to consider the hydroxyl of the alcohol and the carboxyl of the acid during a simple esterification reaction.

Worked Example 3.12   What fragments do we need to consider during the esterification of 1-butanol with ethanoic acid?

We first draw out the reaction in full:

(3.30)

> The butyl ethanoate produced by Equation (3.30) is an ester, and smells of pear drops.

Second, we look for those parts that change and those that remain unchanged. In this example, the bonds that cleave are the O–H bond on the acid and the C–O bond on the alcohol. Such cleavage will require energy. The bonds that form are an O–H bond (to yield water) and a C–O bond in the product ester. All bonds release energy as they form. In this example, the bonds outside the box do not change and hence do not change their energy content, and can be ignored.

The value of ΔH_r relates to bond *changes*. In this example, equal numbers of O–H and C–O bonds break as form, so we expect an equal amount of energy to be released as is consumed, leading to an enthalpy change of zero. In fact, the value of ΔH_r is tiny at -12 kJ mol^{-1}.

Worked Example 3.12 is somewhat artificial, because most reactions proceed with differing numbers of bonds breaking and forming. A more rigorous approach quantifies the energy per bond – the 'bond enthalpy' $\Delta H_{BE}^{\ominus}$ (also called the 'bond *dissociation* energy').

> In some texts, $\Delta H_{BE}^{\ominus}$ is written simply as 'BE'.

$\Delta H_{BE}^{\ominus}$ is the energy needed to cleave 1 mol of bonds. For this reason, values of $\Delta H_{BE}^{\ominus}$ are always *positive*, because energy is consumed.

The chemical environment of a given atom in a molecule will influence the magnitude of the bond enthalpy, so tabulated data such as that in Table 3.3 represent average values.

We can calculate a value of $\Delta H_r^{\ominus}$ with an adapted form of Equation (3.29):

$$\Delta H_r^{\ominus} = -\left(\overset{\text{products}}{\underset{i}{\sum}} \nu \Delta H_{BE}^{\ominus} - \overset{\text{reactants}}{\underset{i}{\sum}} \nu \Delta H_{BE}^{\ominus} \right) \qquad (3.31)$$

where the subscripted i means those bonds that cleave or form within each reactant or product species during the reaction. We need the minus sign because of the way we defined the bond *dissociation* enthalpy. All the values in Table 3.3 are positive because $\Delta H_{BE}^{\ominus}$ relates to bond dissociation.

The stoichiometric numbers ν here can be quite large unless the molecules are small. The combustion of butane, for example, proceeds with the loss of 10 C–H bonds. A

Table 3.3   Table of mean bond enthalpies $\Delta H_{BE}^{\ominus}$ as a function of bond order and atoms. All values cited in kJ mol^{-1} and relate to data obtained (or corrected) to 298 K

	C	N	S	O	I	Br	Cl	F	H
H–	414	389	368	464	297	368	431	569	435
F–	490	280	343	213	280	285	255	159	
Cl–	326	201	272	205	209	218	243		
Br–	272	163	209	–	176	192			
I–	218	–	–	–	151				
O–	326	230	423	142					
O=	803^a	590^b	523	498					
O≡	1075	–	–	–					
S–	289	–	247						
S=	582	–	–						
N–	285	159							
N=	515	473							
N≡	858	946							
C–	331								
C=	590^c								
C≡	812								

^{a}728 if –C=O.
^{b}406 if –NO$_2$; 368 if –NO$_3$.
^{c}506 if alternating – and =.

moment's thought suggests an alternative way of writing Equation (3.31), i.e.:

Values of $\Delta H_r^{\ominus}$ can vary markedly from experimental values if calculated in terms of $\Delta H_{BE}^{\ominus}$.

$$\Delta H_r^{\ominus} = -\left(\overbrace{\sum_i v \Delta H_{BE}^{\ominus}}^{\text{bonds formed}} - \overbrace{\sum_i v \Delta H_{BE}^{\ominus}}^{\text{bonds broken}} \right) \quad (3.32)$$

Note again the minus sign, which we retain for the same reason as for Equation (3.31).

Each of these bond enthalpies is an *average* enthalpy, measured from a series of similar molecules. Values of $\Delta H_{BE}^{\ominus}$ for, say, C–H bonds in hydrocarbons are likely to be fairly similar, as shown by the values in Table 3.3. The bond energies of C–H bonds will differ (sometimes quite markedly) in more exceptional molecules, such as those bearing ionic charges, e.g. carbocations. $\Delta H_{BE}^{\ominus}$ values differ for the OH bond in an alcohol, in a carboxylic acid and in a phenol.

These energies relate to bond rearrangement in *gaseous* molecules, but calculations are often performed for reactions of condensed phases, by combining the enthalpies of vaporization, sublimation, etc. We can calculate a value without further correction if a crude value of ΔH_r is sufficient, or we do not know the enthalpies of phase changes.

Worked Example 3.13   Use the bond enthalpies in Table 3.3 to calculate the enthalpy of burning methane (Equation (3.28)). Assume all processes occur in the gas phase.

Strategy. We start by writing a list of the bonds that break and form.

$$\text{Broken: } 4 \times \text{C--H and } 2 \times \text{O=O}$$

$$\text{Formed: } 2 \times \text{C=O and } 4 \times \text{O--H}$$

so

$$\Delta H_r^\ominus = -[2 \times \Delta H_{BE(C=O)}^\ominus + 4 \times \Delta H_{BE(O-H)}^\ominus] - [4 \times \Delta H_{BE(C-H)}^\ominus + 2 \times \Delta H_{BE(O=O)}^\ominus]$$

Inserting values of $\Delta H_{BE}^\ominus$ from Table 3.3 into Equation (3.32):

$$\Delta H_r^\ominus = -[(2 \times 803) + (4 \times 464)]$$
$$- [(4 \times 414) + (2 \times 498)] \text{ kJ mol}^{-1}$$
$$\Delta H_r^\ominus = -(3462 - 2652) \text{ kJ mol}^{-1}$$

> *Reminder*: all $\Delta H_{BE}^\ominus$ values are positive because they relate to *dissociation* of bonds.

so

$$\Delta H_r^\ominus = -810 \text{ kJ mol}^{-1}$$

which is similar to the value in Worked Example 3.11, but less exothermic.

Aside

Calculations with bond enthalpies $\Delta H_{BE}^\ominus$ tend to be relatively inaccurate because each energy is an *average*. As a simple example, consider the sequential dissociation of ammonia.

(1) NH_3 dissociates to form $NH_2^\bullet$ and $H^\bullet$, and requires an energy of 449 kJ mol^{-1}.

(2) $NH_2^\bullet$ dissociates to form $NH^\bullet$ and $H^\bullet$, and requires an energy of 384 kJ mol^{-1}.

> The symbol '•' means a radical species, i.e. with unpaired electron(s).

(3) $NH^\bullet$ dissociates to form $N^\bullet$ and $H^\bullet$, and requires an energy of 339 kJ mol^{-1}.

The variations in $\Delta H_{BE}^\ominus$ are clearly huge, so we usually work with an *average* bond enthalpy, which is sometimes written as $\overline{BE}$ or $\Delta H_{\overline{BE}}^\ominus$. The average bond enthalpy for the three processes above is 390.9 kJ mol^{-1}.

3.3 Indirect measurement of enthalpy

How do we make 'industrial alcohol'?

Enthalpy Cycles from Hess's Law

Industrial alcohol is an impure form of ethanol made by hydrolysing ethene, $CH_2=CH_2$:

$$(3.33)$$

We pass ethene and water (as a vapour) at high pressure over a suitable catalyst, causing water to add across the double bond of the ethene molecule. The industrial alcohol is somewhat impure because it contains trace quantities of ethylene glycol (1,2-dihydroxyethane, **III**), which is toxic to humans. It also contains unreacted water, and some dissolved ethene.

(III)

> We may rephrase Hess's law, saying 'The standard enthalpy of an overall reaction is the sum of the standard enthalpies of the individual reactions into which the reaction may be divided'.

But what is the enthalpy of the hydration reaction in Equation (3.33)? We first met Hess's law on p. 98. We now rephrase it by saying 'The standard enthalpy of an overall reaction is the sum of the standard enthalpies of the individual reactions into which the reaction may be divided.'

Accordingly, we can obtain the enthalpy of reaction by drawing a Hess cycle, or we can obtain it algebraically. In this example, we will use the cycle method.

Worked Example 3.14   What is the enthalpy change ΔH_r of the reaction in Equation (3.33)?

> Notice that each of these formation reactions is highly exothermic, explaining why energy is needed to obtain the pure elements.

We start by looking up the enthalpies of formation ΔH_f for ethene, ethanol and water. Values are readily found in books of data; Table 3.1 contains a suitable selection.

$$\Delta H_{f(1)}[CH_2=CH_2] = -52 \text{ kJ mol}^{-1}$$

$$\Delta H_{f(2)}[CH_3CH_2OH] = -235 \text{ kJ mol}^{-1}$$

$$\Delta H_{f(3)}[H_2O] = -286 \text{ kJ mol}^{-1}$$

(We have numbered these three (1) to (3) simply to avoid the necessity of rewriting the equations.)

To obtain the enthalpy of forming ethanol, we first draw a cycle. It is usual to start by writing the reaction of interest along the top, and the elements parallel, along the bottom. Remember, the value of ΔH_r is our ultimate goal.

Step 1

$$CH_2{=}CH_2 + H_2O \xrightarrow{\Delta H_r} CH_3CH_2OH \quad \text{Reaction of interest}$$

$$2C + 3H_2 + \tfrac{1}{2}O_2 \qquad \text{Elements}$$

The next three stages are inserting the three enthalpies ΔH_f (1) to (3). Starting on the left-hand side, we insert $\Delta H_{f(1)}$:

Step 2

$$CH_2{=}CH_2 + H_2O \xrightarrow{\Delta H_r} CH_3CH_2OH$$
$$\Delta H_{f(1)} \nwarrow$$
$$2C + 3H_2 + \tfrac{1}{2}O_2$$

We then put in the enthalpy of forming water, $\Delta H_{f(3)}$:

Step 3

$$CH_2{=}CH_2 + H_2O \xrightarrow{\Delta H_r} CH_3CH_2OH$$
$$\Delta H_{f(1)} \quad \Delta H_{f(3)}$$
$$2C + 3H_2 + \tfrac{1}{2}O_2$$

And finally we position the enthalpy of forming the product ethanol, $\Delta H_{f(2)}$:

Step 4

$$CH_2{=}CH_2 + H_2O \xrightarrow{\Delta H_r} CH_3CH_2OH$$
$$\Delta H_{f(1)} \quad \Delta H_{f(3)} \quad \Delta H_{f(2)}$$
$$2C + 3H_2 + \tfrac{1}{2}O_2$$

We can now determine the value of ΔH_r. Notice how we only draw *one* arrow per reaction. The rules are as follows:

> We are only allowed to make a choice of route like this because enthalpy is a state function.

(1)   We wish to go from the left-hand side of the reaction to the right-hand side. We can either follow the arrow labelled ΔH_r, or we pass via the elements (along the bottom line) and thence back up to the ethanol.

(2)   If we go along an arrow in the *same* direction as the arrow is pointing, then we use the value of ΔH as it is written.

(3)   If we have to go along an arrow, but in the *opposite* direction to the direction in which it points, then we multiply the value of ΔH by '-1'.

In the example here, to go from the left-hand side to the right-hand side via the elements, we need to go along two arrows $\Delta H_{f(1)}$ and $\Delta H_{f(3)}$ in the opposite directions to the arrows, so we multiply the respective values of ΔH and multiply each by -1. We then go along the arrow $\Delta H_{f(2)}$, but this time we move in the *same* direction as the arrow, so we leave the sign of the enthalpy unaltered.

And then we tie the threads together and say:

> Note there are three arrows, so there are three ΔH terms within ΔH_r.

$$\Delta H_r = (-1 \times \Delta H_{f(1)}) + (-1 \times \Delta H_{f(3)}) + \Delta H_{f(2)}$$

Inserting values into this equation:

$$\Delta H_r = (-1 \times -52 \text{ kJ mol}^{-1}) + (-1 \times -286 \text{ kJ mol}^{-1}) + (-235 \text{ kJ mol}^{-1})$$

$$\Delta H_r = 52 \text{ kJ mol}^{-1} + 286 \text{ kJ mol}^{-1} + (-235 \text{ kJ mol}^{-1})$$

so

$$\Delta H_r = 103 \text{ kJ mol}^{-1}$$

We obtained this value of ΔH_r knowing the other enthalpies in the cycle, and remembering that enthalpy is a state function. Experimentally, the value of $\Delta H_r = 99 \text{ kJ mol}^{-1}$, so this indirect measurement with Hess's law provides relatively good data.

Sometimes, these cycles are considerably harder than the example here. In such cases, it is usual to write out a cycle for each reaction, and then use the results from each cycle to compile another, bigger cycle.

How does an 'anti-smoking pipe' work?

Hess's Law Cycles with Enthalpies of Combustion

Smoking causes severe damage to the heart, lungs and respiratory system. The tobacco in a cigarette or cigar is a naturally occurring substance, and principally comprises the elements carbon, oxygen, hydrogen and nitrogen.

Unfortunately, because the tobacco is contained within the bowl of a pipe or a paper wrapper, complete combustion is rare, meaning that the oxidation is incomplete. One of the worse side effects of incomplete combustion during smoking is the formation of carbon monoxide (CO) in relatively large quantities. Gaseous CO is highly toxic, and forms an irreversible complex with haemoglobin in the blood. This complex helps explain why people who smoke are often breathless.

A simple way of overcoming the toxic effects of CO is to oxidize it before the smoker inhales the tobacco smoke. This is where the 'anti-smoking' pipe works. (In fact, the name is a misnomer: it does not stop someone smoking, but merely makes the smoke less toxic.) The cigarette is inserted into one end of a long, hollow tube (see Figure 3.7) and the smoker inhales from the other. Along the tube's length are a series of small holes. As the smoker inhales, oxygen enters the holes, mixes with the CO and combines chemically with it according to

$$CO_{(g)} + \tfrac{1}{2}O_2 \longrightarrow CO_{2(g)} \tag{3.34}$$

The $CO_{2(g)}$ produced is considerably less toxic than $CO_{(g)}$, thereby averting at least one aspect of tobacco poisoning.

We might wonder: What is the enthalpy change of forming the CO in Equation (3.34)? It is relatively easy to make CO in the laboratory (for example by dehydrating formic acid with concentrated sulphuric acid), so the enthalpy of oxidizing CO to CO_2 is readily determined. Similarly, it is easy to determine the enthalpy of formation of CO_2, by burning elemental carbon; but it is almost impossible to determined ΔH_c for the reaction $C + \tfrac{1}{2}O_2 \rightarrow CO$, because the pressure of oxygen in a bomb calorimeter is so high that all the carbon is oxidized directly to CO_2 rather than CO. Therefore, we will employ Hess's law once more, but this time employing enthalpies of combustion ΔH_c.

The enthalpies of combustion of carbon and CO are obtained readily from books of data. We can readily find out the following from such data books or Table 3.2:

> The enthalpy $\Delta H_{c(1)}$ is huge, and helps explain why we employ coke and coal to warm a house; this reaction occurs when a coal fire burns.

$$\Delta H_{c(1)}[C_{(s)} + O_{2(g)} \longrightarrow CO_{2(g)}] = -393.5 \text{ kJ mol}^{-1}$$
$$\Delta H_{c(2)}[CO_{(g)} + \tfrac{1}{2}O_{2(g)} \longrightarrow CO_{2(g)}] = -283.0 \text{ kJ mol}^{-1}$$

Again, we have numbered the enthalpies, to save time.

Once more, we start by drawing a Hess-law cycle with the elements at the bottom of the page. This time, it is not convenient to write the reaction of interest along the

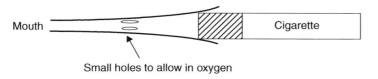

Small holes to allow in oxygen

Figure 3.7 An anti-smoking device: the cigarette is inserted into the wider end. Partially oxidized carbon monoxide combines chemically with oxygen inside the device after leaving the end of the cigarette but before entering the smoker's mouth; the oxygen necessary to effect this oxidation enters the device through the small circular holes positioned along its length

top, so we have drawn it on the left as ΔH_r.

Step 1

$$CO + \tfrac{1}{2}O_2 \qquad CO_2 \quad \text{Compounds}$$

ΔH_r

$$C + O_2 \qquad \text{Constituent elements}$$

Next we insert the enthalpies for the reactions of interest; we first insert $\Delta H_{c(1)}$:

Step 2

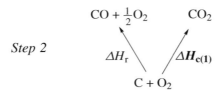

$$CO + \tfrac{1}{2}O_2 \qquad CO_2$$

$\Delta H_r \qquad \Delta H_{c(1)}$

$$C + O_2$$

and finally, we insert $\Delta H_{c(2)}$:

Step 3

$$\overset{\Delta H_{c(2)}}{CO + \tfrac{1}{2}O_2 \longrightarrow CO_2}$$

$\Delta H_r \qquad \Delta H_{c(1)}$

$$C + O_2$$

We want to calculate a value of ΔH_r. Employing the same laws as before, we see that we can either go along the arrow for ΔH_r directly, or along $\Delta H_{c(1)}$ in the same direction as the arrow (so we do not change its sign), then along the arrow $\Delta H_{c(2)}$. Concerning this last arrow, we go in the *opposite* direction to the arrow, so we multiply its value by -1.

The value of ΔH_r is given as

$$\Delta H_r = \Delta H_{c(1)} + (-1 \times \Delta H_{c(2)})$$

Inserting values:

$$\Delta H_r = -393.5 \text{ kJ mol}^{-1} + (+283 \text{ kJ mol}^{-1})$$

so

$$\Delta H_r = -110.5 \text{ kJ mol}^{-1}$$

SAQ 3.10 Calcite and aragonite are both forms of calcium carbonate, $CaCO_3$. Calcite converts to form aragonite. If $\Delta H^{\ominus}_{f \text{ (calcite)}} = -1206.92$ kJ

mol^{-1} and $\Delta H^{\ominus}_{f \text{ (aragonite)}} = -1207.13 \text{ kJ mol}^{-1}$, calculate the value of ΔH_r for the transition process:

$$CaCO_{3(s, \text{ calcite})} \longrightarrow CaCO_{3(s, \text{ aragonite})} \qquad (3.35)$$

Why does dissolving a salt in water liberate heat?

Hess's Law Applied to Ions: Constructing Born–Haber Cycles

Dissolving an ionic salt in water often liberates energy. For example, 32.8 kJ mol^{-1} of energy are released when 1 mol of potassium nitrate dissolves in water. Energy is released, as experienced by the test tube getting warmer.

Before we dissolved the salt in water, the ions within the crystal were held together by strong electrostatic interactions, which obeyed Coulomb's law (see p. 313). We call the energetic sum of these interactions the *lattice enthalpy* (see p. 124). We need to overcome the lattice enthalpy if the salt is to dissolve. Stated another way, salts like magnesium sulphate are effectively insoluble in water because water, as a solvent, is unable to overcome the lattice enthalpy.

> The 'lattice enthalpy' is defined as the standard change in enthalpy when a solid substance is converted from solid to form gaseous constituent ions. Accordingly, values of $\Delta H_{(\text{lattice})}$ are *always* positive.

But what is the magnitude of the lattice enthalpy? We cannot measure it directly experimentally, so we measure it indirectly, with a Hess's law energy cycle. The first scientists to determine lattice enthalpies this way were the German scientists Born and Haber: we construct a *Born–Haber cycle*, which is a form of Hess's-law cycle.

Before we start, we perform a thought experiment; and, for convenience, we will consider making 1 mol of sodium chloride at $25\,^{\circ}C$. There are two possible ways to generate 1 mol of gaseous Na^+ and Cl^- ions: we could start with 1 mol of solid NaCl and vaporize it: the energy needed is $\Delta H^{\ominus}_{(\text{lattice})}$. Alternatively, we could start with 1 mol of sodium chloride and convert it back to the ele-

> It is common to see values of $\Delta H_{(\text{lattice})}$ called 'lattice *energy*'. Strictly, this latter term is only correct when the temperature T is 0 K.

ments (1 mol of metallic sodium and 0.5 mol of elemental chlorine gas (for which the energy is $-\Delta H^{\ominus}_f$) and, then vaporize the elements one at a time, and ionize each in the gas phase. The energies needed to effect ionization are I for the sodium and $E_{(\text{ea})}$ for the chlorine.

In practice, we do not perform these two experiments because we can *calculate* a value of lattice enthalpy $\Delta H_{(\text{lattice})}$ with an energy cycle. Next, we appreciate how generating ions from metallic sodium and elemental chlorine involves several processes. If we first consider the sodium, we must: (i) convert it from its solid state to gaseous atoms (for which the energy is $\Delta H^{\ominus}_{(\text{sublimation})}$); (ii) convert the gaseous atoms to gaseous cations (for which the energy is the ionization energy I). We next consider the chlorine, which is already a gas, so we do not need to volatilize it. But: (i) we must cleave each diatomic molecule to form atoms (for which the energy is $\Delta H^{\ominus}_{\text{BE}}$);

(ii) ionize the gaseous atoms of chlorine to form anions (for which the energy is the electron affinity $E_{(ea)}$). Finally, we need to account for the way the sodium chloride forms from elemental sodium and chlorine, so the cycle must also include $\Delta H_f^{\ominus}$.

Worked Example 3.15   What is the lattice enthalpy $\Delta H_{(lattice)}$ of sodium chloride at 25 °C?

Strategy. (1) We start by compiling data from tables. (2) We construct an energy cycle. (3) Conceptually, we equate two energies: we say the lattice enthalpy is the same as the sum of a series of enthalpies that describe our converting solid NaCl first to the respective elements and thence the respective gas-phase ions.

(1) We compile the enthalpies:

for sodium chloride

These energies are huge. Much of this energy is incorporated into the lattice; other-wise, the value of $\Delta H_f^{\ominus}$ would be massive.

$$Na_{(s)} + \tfrac{1}{2}Cl_{2(g)} \longrightarrow NaCl_{(s)} \qquad \Delta H_f^{\ominus} = -411.15 \text{ kJ mol}^{-1}$$

for the sodium

$$Na_{(s)} \longrightarrow Na_{(g)} \qquad \Delta H_{(sublimation)}^{\ominus} = 107.32 \text{ kJ mol}^{-1}$$

$$Na_{(g)} \longrightarrow Na^+{}_{(g)} + e^- \qquad I_{(Na)} = 502.04 \text{ kJ mol}^{-1}$$

for the chlorine

$$\tfrac{1}{2}Cl_{(g)} \longrightarrow Cl^{\cdot}{}_{(g)} \qquad \tfrac{1}{2}\Delta H_{BE}^{\ominus} = 121.68 \text{ kJ mol}^{-1} \text{ (i.e. half of 243.36 kJ mol}^{-1})$$

$$Cl^{\cdot}{}_{(g)} + e^- \longrightarrow Cl^-{}_{(g)} \qquad E_{(ea)} = -354.81 \text{ kJ mol}^{-1}$$

(2) We construct the appropriate energy cycle; see Figure 3.8. For simplicity, it is usual to draw the cycle with positive enthalpies going up and negative enthalpies going down.

(3) We obtain a value of $\Delta H_{(lattice)}^{\ominus}$, equating it to the energy needed to convert solid NaCl to its elements and thence the gaseous ions. We construct the sum:

We multiply $\Delta H_f^{\ominus}$ by '−1' because we con-sider the reverse pro-cess to formation. (We travel in the *opposite* direction to the arrow representing $\Delta H_f^{\ominus}$ in Figure 3.8.)

$$\Delta H_{(lattice)}^{\ominus} = -\Delta H_f^{\ominus} + \Delta H_{(sublimation)}^{\ominus} + I_{(Na)} + \tfrac{1}{2}\Delta H_{BE}^{\ominus} + E_{(ea)}$$

Inserting values:

$$\Delta H_{(lattice)}^{\ominus} = -(-411.153) + 107.32 + 502.04$$
$$+ 121.676 + (-354.81) \text{ kJ mol}^{-1}$$

so

$$\Delta H_{(lattice)}^{\ominus} = 787.38 \text{ kJ mol}^{-1}$$

SAQ 3.11   Calculate the enthalpy of formation $\Delta H_f^{\ominus}$ for calcium fluoride. Take $\Delta H_{(lattice)}^{\ominus} = -2600 \text{ kJ mol}^{-1}$, $\Delta H_{(sublimation)}^{\ominus} = 178 \text{ kJ mol}^{-1}$; $I_{1(Ca \rightarrow Ca^+)}$

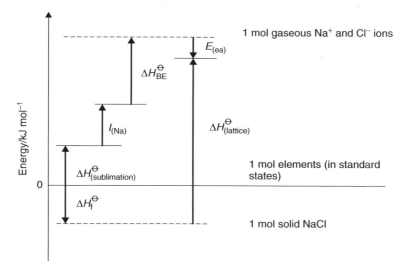

Figure 3.8   Born–Haber cycle constructed to obtain the lattice enthalpy $\Delta H^{\ominus}_{(lattice)}$ of sodium chloride. All arrows pointing up represent endothermic processes and arrows pointing down represent exothermic processes (the figure is not drawn to scale)

$= 596 \ kJ \, mol^{-1}$ and $I_{2(Ca^{+} \rightarrow Ca^{2+})} = 1152 \ kJ \, mol^{-1}$; $\Delta H^{\ominus}_{BE} = 157 \ kJ \, mol^{-1}$ and $E_{(ea)} = -334 \ kJ \, mol^{-1}$.

Why does our mouth feel cold after eating peppermint?

Enthalpy of Solution

Natural peppermint contains several components that, if ingested, lead to a cold sensation in the mouth. The best known and best understood is (−)-menthol (**IV**), which is the dominant component of the peppermint oil extracted from *Mentha piperiia* and *M. arvensia*.

The cause of the cooling sensation is the unusually positive enthalpy of solution. Most values of $\Delta H^{\ominus}_{(solution)}$ are positive, particularly for simple inorganic solutes.

Pure **IV** is a solid at s.t.p. Dissolving **IV** in the mouth disrupts its molecular structure, especially the breaking of the hydrogen bonds associated with the hydroxyl group. These bonds break concurrently with new hydrogen bonds forming with the water of the saliva. We require energy to break the existing bonds, and liberate energy as new bonds form. Energetically, dissolving (−)-menthol is seen to be endothermic, meaning we require energy. This energy comes from the mouth and, as we saw earlier, the macroscopic manifestation of a lower microscopic energy is a lower temperature. Our mouth feels cold.

The other substance sometimes added to foodstuffs to cause cooling of the mouth is xylitol (**V**). It is added as a solid to some sweets, chewing gum, toothpastes and mouth-wash solutions.

$$
\begin{array}{ccccc}
& \diagup CH & & \diagup CH & \\
CH_2 & | & \diagdown CH & | & \diagdown CH_2 \\
| & OH & | & OH & | \\
OH & & OH & & OH
\end{array}
$$

(V)

Measuring values of $\Delta H^{\ominus}_{(solution)}$

It is quite difficult to measure an accurate enthalpy of solution $\Delta H^{\ominus}_{(solution)}$ with a calorimeter, but we can measure it *indirectly*. Consider the example of sodium chloride, NaCl. The ions in solid NaCl are held together in a tight array by strong ionic bonds. While dissolving in water, the ionic bonds holding the constituent ions of Na^+ and Cl^- in place break, and new bonds form between the ions and molecules of water to yield *hydrated* species. Most simple ions are surrounded with six water molecules, like the $[Na(H_2O)_6]^+$ ion (**VI**). Exceptions include the proton with four water molecules (see p. 235) and lanthanide ions with eight.

$$
\left[
\begin{array}{ccc}
& OH_2 & \\
H_2O_{\prime\prime\prime\prime} & | & _{\prime\prime\prime}OH_2 \\
& Na & \\
H_2O & | & OH_2 \\
& OH_2 &
\end{array}
\right]^+
$$

(VI)

The positive charge does not reside on the central sodium alone. Some charge is distributed over the whole ion.

Each hydration bond is partially ionic and partially covalent. Each oxygen atom (from the water molecules) donates a small amount of charge to the central sodium; hence the ionicity. The orbitals also overlap to impart covalency to the bond.

Energy is needed to break the ionic bonds in the solid salt and energy is liberated forming hydration complexes like **VI**. We also break some of the natural hydrogen bonds in the water. The overall change in enthalpy is termed the enthalpy of solution, $\Delta H^{\ominus}_{(solution)}$. Typical values are -207 kJ mol^{-1} for nitric acid; 34 kJ mol^{-1} for potassium nitrate and -65.5 kJ mol^{-1} for silver chloride.

One of the most sensitive ways of determining a value of $\Delta H^{\ominus}_{(solution)}$ is to measure the temperature T at which a salt dissolves completely as a function of its solubility s. A plot of ln s (as y) against $1 \div T$ (as x) is usually linear. We obtain a value of $\Delta H^{\ominus}_{(solution)}$ by multiplying the gradient of the graph by $-R$, where R is the gas constant (as described in Chapter 5, p. 210).

How does a camper's 'emergency heat stick' work?

Enthalpies of Complexation

A camper is in great danger of exposure if alone on the moor or in the desert when night falls and the weather becomes very cold. If a camper has no additional heating, and knows that exposure is not far off, then he can employ an 'emergency heat stick'. The stick is long and thin. One of its ends contains a vial of water and, at the other, a salt such as anhydrous copper sulphate, $CuSO_4$. Both compartments are housed within a thin-walled glass tube, itself encased in plastic. Bending the stick breaks the glass, allowing the water to come into contact with the copper sulphate and effect the following hydration reaction:

'Exposure' is a condition of being exposed to the elements, leading to hypothermia, and can lead to death.

$$CuSO_{4(s)} + 5H_2O_{(l)} \longrightarrow CuSO_4 \cdot 5H_2O_{(s)} \qquad (3.36)$$

The reaction in Equation (3.36) is highly exothermic and releases 134 kJ mol^{-1} of energy. The camper is kept warm by this heat. The reaction in Equation (3.36) involves complexation. In this example, we could also call it 'hydration' or 'adding water of crystallization'. We will call the energy released the 'energy of complexation' $\Delta H_{(complexation)}$.

Heat is liberated when adding water to anhydrous copper sulphate because a new crystal lattice forms in response to strong, new bonds forming between the water and Cu^{2+} and SO_4^{2-} ions. As corroborative evidence of a change in the crystal structure, note how 'anhydrous' copper sulphate is off-white but the pentahydrate is blue.

4

Reaction spontaneity and the direction of thermodynamic change

Introduction

We start by introducing the concept of entropy S to explain why some reactions occur spontaneously, without needing additional energy, yet others do not. The sign of ΔS for a thermodynamic universe must be positive for spontaneity. We explore the temperature dependence of ΔS.

In the following sections, we introduce the concept of a thermodynamic universe (i.e. a system plus its surroundings). For a reaction to occur spontaneously in a system, we require the change in Gibbs function G to be negative. We then explore the thermodynamic behaviour of G as a function of pressure, temperature and reaction composition.

Finally, we investigate the relationship between ΔG and the respective equilibrium constant K, and outline the temperature interdependence of ΔG and K.

4.1 The direction of physicochemical change: entropy

Why does the colour spread when placing a drop of dye in a saucer of clean water?

Reaction spontaneity and the direction of change

However gently a drop of dye solution is added to a saucer of clean, pure water, the colour of the dye soon spreads into uncoloured regions of the water. This mixing occurs inevitably without warming or any kind of external agitation – the painter with watercolour would find his art impossible without this effect. Such mixing continues

> Mixing occurs spontaneously, but we never see the reverse process, with dye suddenly concentrating into a coloured blob surrounded by clear, uncoloured water.

> A reaction is 'spontaneous' if it occurs without any additional energy input.

> It used to be thought reactions were spontaneous if ΔH was negative. This simplistic idea is incorrect.

until the composition of the solution in the saucer is homogeneous, with the mixing complete. We never see the reverse process, with dye suddenly concentrating into a coloured blob surrounded by clear, uncoloured water.

In previous chapters we looked at the way heat travels from hot to cold, as described by the so called 'minus-oneth' law of thermodynamics, and the way net movements of heat cease at thermal equilibrium (as described by the zeroth law). Although this transfer of heat energy was quantified within the context of the first law, we have not so far been able to describe *why* such chemical systems occur. Thermodynamic changes only ever proceed spontaneously in one direction, but not the other. Why the difference?

In everyday life, we say the diffusion of a dye 'just happens' but, as scientists, we say the process is *spontaneous*. In years past, it was thought that all spontaneous reactions were exothermic, with non-spontaneous reactions being endothermic. There are now many exceptions to this overly simplistic rule; thus, we can confidently say that the sign of ΔH does not dictate whether the reaction is spontaneous or not, so we need a more sophisticated way of looking at the problem of spontaneity.

When we spill a bowl of sugar, why do the grains go everywhere and cause such a mess?

Changes in the extent of disorder

Surely everyone has dropped a bowl of sugar, flour or salt, and caused a mess! The powder from the container spreads everywhere, and seems to cover the maximum area possible. Spatial distribution of the sugar granules ensures a range of energies; so, for example, some particles reside on higher surfaces than others, thereby creating a range of potential energies. And some granules travel faster than others, ensuring a spread of kinetic energies.

> The granules of spilt sugar have a *range of energies*.

The mess caused by dropping sugar reflects the way nature always seeks to maximize disorder. Both examples so far, of dye diffusing in water and sugar causing a mess, demonstrate the achievement of greater disorder. But if we are specific, we should note how it is the *energetic* disorder that is maximized spontaneously.

It is easy to create disorder; it is difficult to create order. It requires effort to clean up the sugar when re-establishing order, showing in effect how reversing a spontaneous process requires an input of energy. This is why the converse situation – dropping a mess of sugar grains and creating a neat package of sugar – does not happen spontaneously in nature.

Why, when one end of the bath is hot and the other cold, do the temperatures equalize?

Entropy and the second law of thermodynamics

Quite often, when running a bath, the water is initially quite cold. After the hot water from the tank has had time to travel through the pipes, the water from the tap is hot. As a result, one end of the bath is hotter than the other. But a short time later, the temperature of the water is the same throughout the bath, with the hot end cooler and the cold end warmer. Temperature equilibration occurs even without stirring. Why?

We saw in Chapter 1 how the simplest way to gauge how much energy a molecule possesses is to look at its temperature. We deduce through a reasoning process such as this that molecules of water at the cold end of the bath have less energy than molecules at the hot end. Next, by combining the minus-oneth and zeroth laws of thermodynamics, we say that energy (in the form of heat) is transferred from molecules of water at the hot end of the bath to molecules at the cold end. Energy transfers until equilibrium is reached. All energy changes are *adiabatic* if the bath is lagged (to prevent energy loss), in accordance with the first law of thermodynamics.

As no chemical reactions occur, we note how these thermo-dynamic changes are purely physical. But since no bonds form or break, what is the impetus – the *cause* – of the transfer of energy? We have already seen the way processes occur with an attendant increase in disorder. We now introduce the concept of *entropy*. The extent of *energetic* disorder is given the name *entropy* (and has the symbol S). A bigger value of S corresponds to a greater extent of energetic disorder.

We now introduce the *second law of thermodynamics*: a physic-ochemical process only occurs spontaneously if accompanied by an increase in the entropy S. By corollary, a non-spontaneous process – one that we can force to occur by externally adding energy – would proceed concurrently with a decrease in the ener-getic disorder.

We can often think of entropy merely in terms of spatial disorder, like the example of the sugar grains above; but the entropy of a substance is properly the extent of *energetic* disorder. Molecules of hot and cold water in a bath exchange energy in order to maximize the randomness of their energies.

Figure 4.1 depicts a graph of the number of water molecules having the energy E (as y) against the energy of the water mole-cules E (as x). Trace (a) in Figure 4.1 shows the distribution of energies in a bath where half the molecules have one energy while the other half has a different energy, which explains why the graph contains two peaks. A *distribution* of energies soon forms as energy

The word 'entropy' comes from the Greek *en tropa*, meaning 'in change' or 'during transformation'.

The 'second law of ther-modynamics' says a process occurs sponta-neously only when the concomitant *energetic disorder* increases. We can usually approxi-mate, and talk in terms of 'disorder' alone.

The energy is trans-ferred via random, *inelastic* collisions bet-ween the molecules of water. Such molecular movement is some-times called *Brownian motion*; see p. 139.

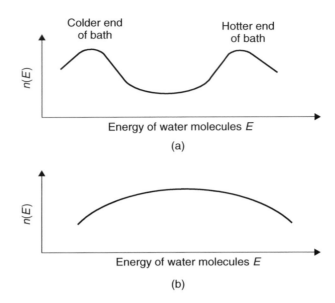

Figure 4.1   Graph of the number of water molecules of energy E against energy. (a) Soon after running the bath, so one end is hotter and the other cooler; and (b) after thermal equilibration. The (average) energy at the peak relates to the macroscopic temperature

is transferred from one set of water molecules to other. Trace (b) in Figure 4.1 shows the distribution of energies after equilibration. In other words, the energetic disorder S increases. The reading on a thermometer placed in the bath will represent an *average* energy.

The spread of energies in Figure 4.1 is a direct indication of entropy, with a wider spread indicating a greater entropy. Such energetic disorder is the consequence of having a *range* of energies. The spread widens spontaneously; an example of a non-spontaneous process would be the reverse process, with the molecules in a bath at, say, $50\,^\circ C$ suddenly reverting to one having a temperature of $30\,^\circ C$ at one end and a temperature of $70\,^\circ C$ at the other.

> In the thermodynamic sense, an 'engine' is a device or machine for converting energy into work. Clausius himself wanted to devise an efficient machine to convert heat energy (from a fuel) into mechanical work.

The German scientist Rudolf Clausius (1822–1888) was the first to understand the underlying physicochemical principles dictating reaction spontaneity. His early work aimed to understand the sky's blue colour, the red colours seen at sunrise and sunset, and the polarization of light. Like so many of the 'greats' of early thermodynamics, he was a mathematician. He was interested in *engines*, and was determined to improve the efficiency of steam-powered devices, such as pumping engines used to remove water from mines, and locomotives on the railways. Clausius was the first to introduce entropy as a new variable into physics and chemistry.

Why does a room containing oranges acquire their aroma?

Spontaneity and the sign of ΔS

When a bowl containing fresh oranges is placed on the dining room table, the room acquires their fragrance within a few hours. The organic substance we smell after its release from the oranges is the organic terpene (+)-limonene (**I**), each molecule of which is small and relatively non-polar. **I** readily evaporates at room temperature to form a vapour.

| We sometimes say these compounds *volatilize*. |

(**I**)

The process we detect when we note the intensifying smell of the oranges, is:

$$\text{limonene}_{(l)} \longrightarrow \text{limonene}_{(g)} \tag{4.1}$$

so the concentration of volatile limonene in the gas phase increases with time. But why does it evaporate in this way?

Liquids can flow (and hence transfer energy by inelastic collisions), so they will have a distribution of energies. Molecules in the liquid state possess a certain extent of energetic disorder and, therefore, have a certain extent of entropy S. By contrast, molecules in the gas phase have a greater freedom to move than do liquids, because there is a greater scope for physical movement: restrictions

| Gaseous materials have greater entropy than their respective liquids. |

arising from hydrogen bonds or other physicochemical interactions are absent, and the large distances between each molecule allow for wider variations in speed, and hence in energy. Gas molecules, therefore, have greater entropy than do the liquids from which they derive. We deduce the simple result $S_{(g)} > S_{(l)}$.

We could obtain this result more rigorously. We have met the symbol 'Δ' several times already, and recall its definition 'final state minus initial state', so the change in entropy ΔS for any process is given by the simple equation

$$\Delta S_{(process)} = S_{(final\ state)} - S_{(initial\ state)} \tag{4.2}$$

If the final disorder of a spontaneous process is greater than the initial disorder, then we appreciate from Equation (4.2) how a *spontaneous process* is accompanied by ΔS of *positive sign*. This will remain our working definition of spontaneity.

| A spontaneous process is accompanied by a *positive* value of ΔS. |

Ultimately, the sign of ΔS explains why the smell of the oranges increases with time.

Worked Example 4.1   Show mathematically how the entropy of a gas is higher than the entropy of its respective liquid.

If $S_{\text{(final state)}}$ is $S_{(g)}$ and $S_{\text{(initial state)}}$ is $S_{(l)}$, then $\Delta S = S_{(g)} - S_{(l)}$. Because the volatilizing of the compound is spontaneous, the sign of ΔS must be positive.

The only way to make ΔS positive is when $S_{(g)} > S_{(l)}$.

Why do damp clothes become dry when hung outside?

Reaction spontaneity by inspection

Everyone knows damp clothes become dry when hung outside on the washing line. Any residual water is lost by evaporation from the cloth. In fact, moisture evaporates even if the damp clothes hang limp in the absence of a breeze. The water *spontaneously* leaves the fabric to effect the physicochemical process $H_2O_{(l)} \rightarrow H_2O_{(g)}$.

> We obtain the energy for evaporating the water by lowering the internal energy of the garment fibres, so the clothes feel cool to the touch when dry.

The loss of water occurs during drying in order to increase the overall amount of entropy, because molecules of *gaseous* water have a greater energy than do molecules of *liquid*, merely as a result of being gas and liquid respectively. In summary, we could have employed our working definition of entropy (above), which leads to a *prediction* of the clothes becoming dry, given time, as a result of the requirement to increase the entropy. Inspection alone allows us a shrewd guess at whether a process will occur spontaneously or not.

Worked Example 4.2   By inspection alone, decide whether the sublimation of iodine (Equation (4.3)) will occur spontaneously or not:

$$I_{2(s)} \longrightarrow I_{2(g)} \tag{4.3}$$

Molecules in the gas phase have more entropy than molecules in the liquid phase; and molecules in the liquid phase have more entropy than molecules in the solid state. As an excellent generalization, the relative order of the entropies is given by

> This argument says nothing about the *rate* of sublimation. In fact, we do not see sublimation occurring significantly at room temperature because it is so slow.

$$S_{(g)} \gg S_{(l)} > S_{(s)} \tag{4.4}$$

The product of sublimation is a gas, and the precursor is a solid. Clearly, the product has greater entropy than the starting material, so ΔS increases during sublimation. The process is spontaneous because ΔS is positive.

In a bottle of iodine, the space above the solid I_2 always shows a slight purple hue, indicating the presence of iodine vapour.

SAQ 4.1 By inspection alone, decide whether the condensation of water, $H_2O_{(g)} \rightarrow H_2O_{(l)}$ is spontaneous or not.

Worked Example 4.3 Now consider the chemical process

$$SOCl_{2(l)} + H_2O_{(g)} \longrightarrow 2HCl_{(g)} + SO_{2(g)} \tag{4.5}$$

The reaction occurs spontaneously in the laboratory without recourse to heating or catalysis. The sight of 'smoke' above a beaker of $SOCl_{2(l)}$ is ample proof of reaction spontaneity.

> This increase in the entropy S of a gas explains why an open beaker of thionyl chloride $SOCl_2$ in the laboratory appears to be 'smoking'.

We see that the reaction in Equation (4.5) *consumes* 1 mol of *gas* (i.e. water vapour) and 1 mol of *liquid*, and generates 3 mol of gas. There is a small change in the number of moles: principally, the amount of gas increases. As was seen above, the entropy of a gas is greater than its respective liquid, so we see a net increase in the entropy of the reaction, making ΔS positive.

Worked Example 4.4 By inspection alone, decide whether the formation of ammonia by the Haber process (Equation (4.6)) is spontaneous or not.

$$N_{2(g)} + 3H_{2(g)} \longrightarrow 2NH_{3(g)} \tag{4.6}$$

All the species in Equation (4.6) are gases, so we cannot use the simple method of looking to see the respective phases of reactants and products (cf. Equation (4.4)).

But we notice the consumption of 4 mol of reactant to form 2 mol of product. As a crude generalization, then, we start by saying, '4 mol of energetic disorder are consumed during the process and 2 mol of energetic disorder are formed'. Next, with Equation (4.1) before us, we suggest the overall, crude entropy change ΔS is roughly -2 mol of disorder per mole of reaction, so the amount of disorder *decreases*. We suspect the process will *not* be spontaneous, because ΔS is negative.

In fact, we require heating to produce ammonia by the Haber process, so the reaction is definitely *not* spontaneous.

SAQ 4.2 By inspection alone, decide whether the oxidation of sulphur dioxide is thermodynamically spontaneous or not. The stoichiometry of the reaction is $\frac{1}{2}O_{2(g)} + SO_{2(g)} \rightarrow SO_{3(g)}$.

> The word 'stoichiometry' comes from the Greek *stoicheion*, meaning 'a part'.

Worked Example 4.5 By inspection alone, decide whether the reaction $Cl_{2(g)} + F_{2(g)} \rightarrow 2FCl_{(g)}$ should occur spontaneously.

Occasionally we need to be far subtler when we look at reaction spontaneity. The reaction here involves two molecules of diatomic gas reacting to form two molecules of a different diatomic gas. Also, there is no phase change during reaction, nor any change in the numbers of molecules, so any change in the overall entropy is likely to be slight.

Before we address this reaction, we need to emphasize how all these are *equilibrium* reactions: at completion, the reaction vessel contains product as well as unconsumed reactants. In consequence, there is a *mixture* at the completion of the reaction.

This change in ΔS arises from the *mixing* of the elements between the two reacting species: before reaction, all atoms of chlorine were bonded only to other chlorine atoms in elemental Cl_2. By contrast, after the reaction has commenced, a choice arises with some chlorine atoms bonded to other chlorine atoms (unreacted Cl_2) and others attached to fluorine in the product, FCl.

In fact, the experimental value of ΔS is very small and positive.

Aside

Entropy as 'the arrow of time'

The idea that entropy is continually increasing led many philosophers to call entropy 'the arrow of time'. The argument goes something like this. From the Clausius equality (see p. 142), entropy is the ratio of a body's energy to its temperature. Entropy is generally understood to signify an inherent tendency towards disorganization.

It has been claimed that the second law means that the universe as a whole must tend inexorably towards a state of maximum entropy. By an analogy with a closed system, the entire universe must eventually end up in a state of equilibrium, with the same temperature everywhere. The stars will run out of fuel. All life will cease. The universe will slowly peter out in a featureless expanse of nothingness. It will suffer a 'heat death'.

The idea was hugely influential. For example, it inspired the poet T. S. Eliot to write his poem *The Hollow Men* with perhaps his most famous lines

This the way the world ends
not with a bang but a whimper.

He wrote this in 1925. Eliot's poem, in turn, inspired others. In 1927, the astronomer Sir Arthur Eddington said that if entropy was always increasing, then we can know the direction in which time moves by looking at the direction in which it increases. The phrase 'entropy is the arrow of time' gripped the popular imagination, although it is rather meaningless.

In 1928, the English scientist and idealist Sir James Jean revived the old 'heat death' argument, augmented with elements from Einstein's relativity theory: since matter and energy are equivalents, he claimed, the universe must finally end up in the complete conversion of matter into energy:

The second law of thermodynamics compels materials in the universe to move ever in the same direction along the same road which ends only in death and annihilation.

Why does crystallization of a solute occur?

Thermodynamic systems and universes

Atoms or ions of solute leave solution during the process of crystallization to form a regular repeat lattice.

The extent of solute disorder is high *before* crystallization, because each ion or molecule resides in solution, and thereby experiences the same freedom as a molecule of liquid. Conversely, the extent of disorder *after* crystallization will inevitably be much smaller, since solute is incorporated within a solid comprising a regular repeat lattice.

> The extent of solute disorder *decreases* during crystallization.

The value of ΔS can only be negative because the symbol 'Δ' means 'final state minus initial state', and the extent of disorder during crystallization clearly follows the order 'solute disorder$_{(initial)}$ > solute disorder$_{(final)}$'. We see how the extent of disorder in the solute *decreases* during crystallization in consequence of forming a lattice and, therefore, do *not* expect crystallization to be a spontaneous process.

But crystallization *does* occur, causing us to ask, 'Why does crystallization occur even though ΔS for the process is negative?' To answer this question, we must consider *all* energetic considerations occurring during the process of crystallization, possibly including phenomena not directly related to the actual processes inside the beaker.

> A process is thermodynamically spontaneous only if the 'overall' value of ΔS is positive.

Before crystallization, each particle of solute is solvated. As a simple example, a chloride ion in water is attached to six water molecules, as $[Cl(H_2O)_6]^-$. Being bound to a solute species limits the freedom of solvent molecules, that is, when compared with *free*, unbound solvent.

Crystallization releases these six waters of solvation; see Figure 4.2:

$$[Cl(H_2O)_6]^-_{(aq)} \longrightarrow Cl^-_{(in\ solid\ lattice)} + 6H_2O_{(free,\ not\ solvating)} \qquad (4.7)$$

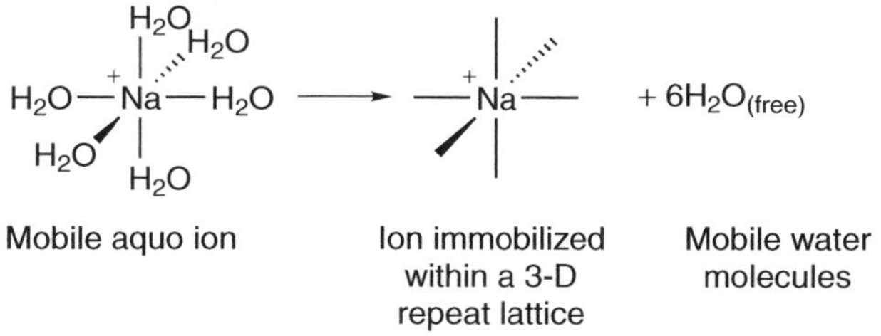

| Mobile aquo ion | Ion immobilized within a 3-D repeat lattice | Mobile water molecules |

Figure 4.2 Schematic representation of a crystallization process. Each solvated ion, here Na^+, releases six waters of solvation while incorporating into its crystal lattice. The *overall* entropy of the thermodynamic universe increases by this means

The word 'universe' in this context is completely different from a 'universe' in astronomy, so the two should not be confused. A thermodynamic 'universe' comprises both a 'system' and its 'surroundings'.

After their release from solvating this chloride ion, each water molecule has as much energetic disorder as did the whole chloride ion complex. Therefore, we expect a sizeable *increase* in the entropy of the solvent during crystallization because many water molecules are released.

When we look at the spontaneity of the crystallization process, we need to consider two entropy terms: (i) the solute (which decreases during crystallization) and (ii) the concurrent increase as solvent is freed. In summary, the entropy of the solute decreases while the entropy of the solvent increases.

The crystallization process involves a *system* (which we are interested in) and the *surroundings*. In terms of the component entropies in this example, we say $\Delta S_{(system)}$ is the entropy of the solute crystallizing and that $\Delta S_{(surroundings)}$ represents the entropy change of the solvent molecules released.

We define a thermodynamic universe as 'that volume large enough to enclose *all* the changes'; the size of the surroundings depends on the example.

We call the sum of the system and its surroundings the *thermodynamic universe* (see Figure 4.3). A thermodynamic universe is described as 'that volume large enough to enclose all the thermodynamic changes'. The entropy change of the thermodynamic universe during crystallization is $\Delta S_{(total)}$, which equates to

$$\Delta S_{(total)} = \Delta S_{(system)} + \Delta S_{(surroundings)} \tag{4.8}$$

The value of $\Delta S_{(system)}$ is negative in the example of crystallization. Accordingly, the value of $\Delta S_{(surroundings)}$ must be so much larger than $\Delta S_{(system)}$ that $\Delta S_{(total)}$ becomes positive. The crystallization is therefore spontaneous.

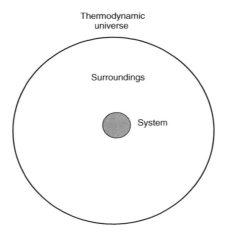

Figure 4.3 We call the sum of the system and its surroundings the 'thermodynamic universe'. Energy is exchanged between the system and its surroundings; no energy is exchanged beyond the surrounds, i.e. outside the boundaries of the thermodynamic universe. Hence, the definition 'a universe is that volume large enough to enclose *all* the thermodynamic changes'

This result of $\Delta S_{(total)}$ being positive helps explain how considering the entropy of a system's surroundings can obviate the apparent problems caused by only considering the processes occurring within a thermodynamic system. It also explains why crystallization is energetically feasible.

The concept of a thermodynamic system is essentially macroscopic, and assumes the participation of large numbers of molecules. Indeed, the word 'system' derives from the Greek *sustēma*, meaning to set up, to stay together.

> As a new criterion for reaction spontaneity, we say $\Delta S_{(total)}$ must be *positive*. We must consider the surroundings if we are to understand how the *overall* extent of energetic disorder increases during a process.

4.2 The temperature dependence of entropy

Why do dust particles move more quickly by Brownian motion in warm water?

Entropy is a function of temperature

Brownian motion is the random movement of small, solid particles sitting on the surface of water. They are held in position by the surface tension γ of the meniscus. When looking at the dust under a microscope, the dust particles appear to 'dance' and move randomly. But when the water is warmed, the particles, be they chalk or house dust, move faster than on cold water.

The cause of the Brownian motion is movement of water molecules, several hundred of which 'hold' on to the underside of each dust particle by surface tension. These water molecules move and jostle continually as a consequence of their own internal energy. Warming the water increases the internal energy, itself causing the molecules to move faster than if the water was cool.

> 'Brownian motion' is a *macroscopic* observation of entropy.

The faster molecules exhibit a greater randomness in their motion than do slower molecules, as witnessed by the dust particles, which we see 'dancing' more erratically. The Brownian motion is more extreme. The enhanced randomness is a consequence of the water molecules having higher entropy at the higher temperature. Entropy is a function of temperature.

> Entropy is a function of temperature.

Why does the jam of a jam tart burn more than does the pastry?

Relationship between entropy and heat capacity

When biting into a freshly baked jam tart, the jam burns the tongue but the pastry (which is at the same temperature) causes relatively little harm. The reason why the jam is more likely to burn is its higher 'heat capacity' C.

> The heat capacity $C_{p(ice)} = 39\ \mathrm{J\,K^{-1}\,mol^{-1}}$ tells us that adding 39 J of energy increases the temperature of 1 mol of water by 1 K.

Sections 3.1 and 3.2 describe heat capacity and explain how it may be determined at constant pressure C_p or at constant volume C_V. Most chemists need to make calculations with C_p, which represents the amount of energy (in the form of heat) that can be stored within a substance – the measurement having been performed at constant pressure p. For example, the heat capacity of solid water (ice) is 39 $\mathrm{J\,K^{-1}\,mol^{-1}}$. The value of C_p for liquid water is higher, at 75 $\mathrm{J\,K^{-1}\,mol^{-1}}$, so we store more energy in liquid water than when it is solid; stated another way, we need to add more energy to $H_2O_{(l)}$ if its temperature is to increase. C_p for steam ($H_2O_{(g)}$) is 34 $\mathrm{J\,K^{-1}\,mol^{-1}}$. C_p for solid sucrose (**II**) – a major component of any jam – is significantly higher at 425 $\mathrm{J\,K^{-1}\,mol^{-1}}$.

(II)

The heat capacity of a liquid is *always* greater than the heat capacity of the respective solid because the liquid, having a greater amount of energetic disorder, has a greater entropy according to

$$\Delta S = S_2 - S_1 = \int_{T_1}^{T_2} \frac{C_p}{T}\,\mathrm{d}T \tag{4.9}$$

More energy is 'stored' within a liquid than in its respective solid, as gauged by the relative values of C_p implied by the connection between the heat capacity and entropy S (of a pure material). This is to be expected from everyday experience: to continue with our simplistic example, when a freshly baked jam tart is removed from the oven, the jam burns the mouth and not the pastry, because the (liquid) jam holds much more energy, i.e. has a higher C_p than does the solid pastry, even though the two are at the same temperature. The jam, in cooling to the same temperature as the tongue, gives out more energy. The tongue cannot absorb all of this energy; the energy that is not absorbed causes other processes in the mouth, and hence the burn.

Aside

This argument is oversimplified because it is expressed in terms of jam:

(1) Jam, in comprising mainly water and sugar, will contain more moles per gram than does the pastry, which contains fats and polysaccharides, such as starch in the flour. Jam can, therefore, be considered to contain more energy per gram from a *molar* point of view, without even considering its liquid state.

(2) The jam is more likely to stick to the skin than does the pastry (because it is sticky liquid), thereby maximizing the possibility of heat transferring to the skin; the pastry is flaky and/or dusty, and will exhibit a lower efficiency in transferring energy.

Worked Example 4.6 Calculate the entropy change ΔS caused by heating 1 mol of sucrose from 360 K to 400 K, which is hot enough to badly burn the mouth. Take $C_p = 425 \text{ J K}^{-1} \text{ mol}^{-1}$.

Because the value of C_p has a constant value, we place it outside the integral, which allows us to rewrite Equation (4.9), saying

$$\Delta S = C_p \ln \left(\frac{T_2}{T_1} \right) \tag{4.10}$$

We insert data into Equation (4.10) to obtain

$$\Delta S = 425 \text{ J K}^{-1} \text{ mol}^{-1} \times \ln \left(\frac{400 \text{ K}}{360 \text{ K}} \right)$$

so

$$\Delta S = 425 \text{ J K}^{-1} \text{ mol}^{-1} \times \ln(1.11)$$

and

$$\Delta S = 44.8 \text{ J K}^{-1} \text{ mol}^{-1}$$

SAQ 4.3 We want to warm the ice in a freezer from a temperature of $-15\,^\circ$C to $0\,^\circ$C. Calculate the change in entropy caused by the warming (assuming no melting occurs). Take C_p for ice as $39 \text{ J K}^{-1} \text{ mol}^{-1}$. [Hint: remember to convert to K from $^\circ$C.]

Aside

C_p is not independent of temperature, but varies slightly. For this reason, the approach here is only valid for relatively narrow temperature ranges of, say, 30 K. When determining ΔS over wider temperature ranges, we can perform a calculation with Equation (4.9) provided that we know the way C_p varies with temperature, expressed as a mathematical power series in T. For example, C_p for liquid chloroform $CHCl_3$ is

$$C_p/\text{J K}^{-1}\,\text{mol}^{-1} = 91.47 + 7.5 \times 10^{-2}T$$

Alternatively, because Equation (4.9) has the form of an integral, we could plot a graph of $C_p \div T$ (as y) against T (as x) and determine the area beneath the curve. We would need to follow this approach if $C_p \div T$ was so complicated a function of T that we could not describe it mathematically.

Justification Box 4.1

Entropy is the ratio of a body's energy to its temperature according to the Clausius equality (as defined in the next section). For a *reversible* process, the change in entropy is defined by

$$dS = \frac{dq}{T} \qquad (4.11)$$

where q is the change in heat and T is the thermodynamic temperature. Multiplying the right-hand side of Equation (4.11) by dT/dT (which clearly equals one), yields

$$dS = \frac{dq}{T} \times \frac{dT}{dT} \qquad (4.12)$$

If no expansion work is done, we can safely assume that $q = H$. Substituting H for q, and rearranging slightly yields

$$dS = \left(\frac{dH}{dT}\right) \times \frac{1}{T}\,dT \qquad (4.13)$$

where the term in brackets is simply C_p. We write

$$dS = \frac{C_p}{T}\,dT \qquad (4.14)$$

Solution of Equation (4.14) takes two forms: (a) the case where C_p is considered not to depend on temperature (i.e. determining the value of ΔS over a limited range of temperatures) and (b) the more realistic case where C_p is recognized as having a finite temperature dependence.

(a) C_p is independent of temperature (over small temperature ranges).

$$\int_{S_1}^{S_2} dS = C_p \int_{T_1}^{T_2} \frac{1}{T}\,dT \qquad (4.15)$$

So

$$\Delta S = S_2 - S_1 = C_p[\ln T]_{T_1}^{T_2} \qquad (4.16)$$

and hence

$$\Delta S = C_p \ln\left(\frac{T_2}{T_1}\right) \qquad (4.17)$$

(b) C_p is not independent of temperature (over larger temperature ranges). We employ a similar approach to that above, except that C_p is incorporated *into* the integral, yielding

$$\int_{S_1}^{S_2} dS = \int_{T_1}^{T_2} \frac{C_p}{T} \, dT \qquad (4.18)$$

which, on integration, yields Equation (4.9).

Worked Example 4.7   What is the increase in entropy when warming 1 mol of chloroform (**III**) from 240 K to 330 K? Take the value of C_p for chloroform from the *Aside* box on p. 142.

(III)

We start with Equation (4.9), retaining the position of C_p within the integral; inserting values:

$$\Delta S = S_{330\ \text{K}} - S_{240\ \text{K}} = \int_{240\ \text{K}}^{330\ \text{K}} \frac{91.47 + 7.5 \times 10^{-2} T}{T} \, dT$$

> The 'T' on top and bottom cancel in the second term within the integral.

Rearranging:

$$\Delta S = \int_{240\ \text{K}}^{330\ \text{K}} \frac{91.47}{T} + 7.5 \times 10^{-2} \, dT$$

Performing the integration, we obtain

$$\Delta S = [91.47 \ln T]_{240\ \text{K}}^{330\ \text{K}} + 7.5 \times 10^{-2} \, [T]_{240\ \text{K}}^{330\ \text{K}}$$

Then, we insert the variables:

$$\Delta S = 91.47 \ln\left(\frac{330\ \text{K}}{240\ \text{K}}\right) + 7.5 \times 10^{-2}(330\ \text{K} - 240\ \text{K})$$

to yield

$$\Delta S = (29.13 + 6.75) \ \text{J K}^{-1} \, \text{mol}^{-1}$$

so

$$\Delta S = 35.9 \ \text{J K}^{-1} \, \text{mol}^{-1}$$

SAQ 4.4   1 mol of oxygen is warmed from 300 K to 350 K. Calculate the associated rise in entropy ΔS if $C_{p(O_2)}/\text{J K}^{-1}\,\text{mol}^{-1} = 25.8 + 1.2 \times 10^{-2}T/\text{K}$.

4.3   Introducing the Gibbs function

Why is burning hydrogen gas in air (to form liquid water) a spontaneous reaction?

Reaction spontaneity in a system

> The 'twin' subscript of 'l and g' arises because the reaction in Equation (4.19) is so exothermic that most of the water product will be steam.

Equation (4.19) describes the reaction occurring when hydrogen gas is burnt in air:

$$O_{2(g)} + 2H_{2(g)} \longrightarrow 2H_2O_{(l \text{ and } g)} \qquad (4.19)$$

We notice straightaway how the number of moles *decreases* from three to two during the reaction, so a consideration of the system alone suggests a non-spontaneous reaction. There may also be a concurrent phase change from gas to liquid during the reaction, which confirms our original diagnosis: we expect ΔS to be negative, and so we predict a non-spontaneous reaction.

But after a moment's reflection, we remember that one of the simplest tests for hydrogen gas generation in a test tube is to place a lighted splint nearby, and hear the 'pop' sound of an explosion, i.e. the reaction in Equation (4.19) occurs spontaneously.

The 'system' in this example comprises the volume within which chemicals combine. The 'surroundings' are the volume of air around the reaction vessel or flame; because of the explosive nature of reaction, we expect this volume to be huge. The surrounding air absorbs the energy liberated during the reaction; in this example, the energy is manifested as heat and sound. For example, the entropy of the air increases as it warms up. In fact, $\Delta S_{(\text{surroundings})}$ is sufficiently large and positive that the value of $\Delta S_{(\text{total})}$ is positive despite the value of $\Delta S_{(\text{system})}$ being negative. So we can now explain why reactions such as that in Equation (4.19) are spontaneous, although at first sight we might predict otherwise.

But, as chemists, we usually want to make *quantitative* predictions, which are clearly impossible here unless we can precisely determine the magnitude of $\Delta S_{(\text{surroundings})}$, i.e. quantify the influence of the surroundings on the reaction, which is usually not a trivial problem.

How does a reflux condenser work?

Quantifying the changes in a system: the Gibbs function

All preparative chemists are familiar with the familiar Liebig condenser, which we position on top of a refluxing flask to prevent the flask boiling dry. The evaporating solvent rises up the interior passage of the condenser from the flask, cools and thence condenses (Equation (4.20)) as it touches the inner surface of the condenser. Condensed liquid then trickles back into the flask beneath.

$$\text{solvent}_{(g)} \longrightarrow \text{solvent}_{(l)} + \text{energy} \qquad (4.20)$$

The energy is transferred to the glass inner surface of the condenser. We maintain a cool temperature inside the condenser by running a constant flow of water through the condenser's jacketed sleeve. The solvent releases a large amount of heat energy as it converts back to liquid, which passes to the water circulating within the jacket, and is then swept away.

Addition of heat energy to the flask causes several physicochemical changes. Firstly, energy allows the chemical reaction to proceed, but energy is also consumed in order to convert the liquid solvent into gas. An 'audit' of this energy is difficult, because so much of the energy is lost to the escaping solvent and thence to the surrounding water. It would be totally impossible to account for all the energy changes without also including the surroundings as well as the system.

So we see how the heater beneath the flask needs to provide energy to enable the reaction to proceed (which is what we want to happen) in addition to providing the energy to change the surroundings, causing the evaporation of the solvent, the extent of which we do not usually want to quantify, even if we could. In short, we need a simple means of taking account of all the surroundings without, for example, having to assess their spatial extent. From the second law of thermodynamics, we write

$$\Delta G_{(system)} = \Delta H_{(system)} - T \Delta S_{(system)} \qquad (4.21)$$

(see Justification Box 4.2) where the H, T and S terms have their usual definitions, as above, and G is the 'Gibbs function'. G is important because its value depends only on the system and not on the surroundings. By convention, a positive value of ΔH denotes an enthalpy absorbed by the system.

ΔH here is simply the energy given out by the system, i.e. by the reaction, or taken into it during endothermic reactions. This energy transfer affects the energy of the surroundings, which respectively absorb or receive energy from the reaction. And the change in the energy of the surroundings causes changes in the entropy of the surroundings. In effect, we can devise a 'words-only' definition of the Gibbs function, saying it represents 'The energy available for reaction (i.e. the *net* energy), after adjusting for the entropy changes of the surroundings'.

> The 'Gibbs function' G is named after Josiah Willard Gibbs (1839–1903), a humble American who contributed to most areas of physical chemistry. He also had a delightful sense of humour: 'A mathematician may say anything he pleases, but a physicist must be at least partially sane'.

> As well as calling G the Gibbs function, it is often called the 'Gibbs energy' or (incorrectly) 'free energy'.

> The Gibbs function is the energy available for reaction after adjusting for the entropy changes of the surroundings.

Justification Box 4.2

The total change in entropy is $\Delta S_{(total)}$, which must be positive for a spontaneous process. From Equation (4.8), we say

$$\Delta S_{(total)} = \Delta S_{(system)} + \Delta S_{(surroundings)} > 0$$

We usually know a value for $\Delta S_{(system)}$ from tables. Almost universally, we do not know a value for $\Delta S_{(surroundings)}$.

The *Clausius equality* says that a microscopic process is at equilibrium if $dS = dq/T$ where q is the heat change and T is the thermodynamic temperature (in kelvin). Similarly, for a macroscopic process, $\Delta S = \Delta q / T$. In a chemical reaction, the heat energy emitted is, in fact, the enthalpy change of reaction $\Delta H_{(system)}$, and the energy gained by the surroundings of the reaction vessel will therefore be $-\Delta H_{(system)}$. Accordingly, the value of $\Delta S_{(surroundings)}$ is $-\Delta H \div T$.

> This sign change occurring here follows since energy is *absorbed* by the surroundings if energy has been *emitted* by the reaction, and *vice versa*.

Rewriting Equation (4.8) by substituting for $\Delta S_{(surroundings)}$ gives

$$\Delta S_{(total)} = \Delta S_{(system)} - \frac{\Delta H}{T} \qquad (4.22a)$$

The right-hand side must be positive if the process is spontaneous, so

$$\Delta S_{(system)} - \frac{\Delta H_{(system)}}{T} > 0 \qquad (4.22b)$$

or

$$0 > \frac{\Delta H_{(system)}}{T} - \Delta S_{(system)}$$

Multiplying throughout by T gives

$$0 > \Delta H_{(system)} - T\Delta S_{(system)} \qquad (4.23)$$

So the compound variable $\Delta H_{(system)} - T\Delta S_{(system)}$ must be negative if a process is spontaneous.

This compound variable occurs so often in chemistry that we will give it a symbol of its own: G, which we call the Gibbs function. Accordingly, a spontaneous process in a system is characterized by saying,

$$0 > \Delta G_{(system)} \qquad (4.24)$$

In words, the Gibbs energy must be negative if a change occurs spontaneously.

The sign of ΔG

Equation (4.24) in Justification Box 4.2 shows clearly that a process only occurs spontaneously within a system if the change in Gibbs function is negative, even if the sign of $\Delta S_{(system)}$ is slightly negative or if $\Delta H_{(system)}$ is slightly positive. Analysing the reaction in terms of our new variable ΔG represents a great advance: previously, we could predict spontaneity if we knew that $\Delta S_{(total)}$ was positive – which we now realize is not necessarily a useful criterion, since we rarely know a value for $\Delta S_{(surroundings)}$. It is clear from Equation (4.21) that all three variables, G, H and S, each relate to the system alone, so we can calculate the value of ΔG by looking up values of ΔS and ΔH from tables, and without needing to consider the surroundings in a quantitative way.

A process occurring in a system is spontaneous if ΔG is negative, and it is not spontaneous if ΔG is positive, regardless of the sign of $\Delta S_{(system)}$. The size of ΔG (which is negative) is maximized for those processes and reactions for which ΔS is positive and which are exothermic, with a negative value of ΔH.

> The Gibbs energy must be *negative* if a change occurs spontaneously.

> We see the analytical power of ΔG when we realize how its value does not depend on the thermodynamic properties of the surroundings, but only on the system.

> A process occurring in a system is spontaneous if ΔG is negative, and is not spontaneous when ΔG is positive.

Worked Example 4.8 Methanol (**IV**) can be prepared in the gas phase by reacting carbon monoxide with hydrogen, according to Equation (4.25). Is the reaction feasible at 298 K if $\Delta H^{\ominus} = -90.7$ kJ mol^{-1} and $\Delta S^{\ominus} = -219$ J K^{-1} mol^{-1}?

$$CO_{(g)} + 2H_{2(g)} \longrightarrow CH_3OH_{(g)} \qquad (4.25)$$

(IV)

We shall use Equation (4.21), $\Delta G^{\ominus} = \Delta H^{\ominus} - T\Delta S^{\ominus}$. Inserting values (and remembering to convert from kJ to J):

$$\Delta G^{\ominus} = (-90\,700 \text{ J mol}^{-1}) - (298 \text{ K} \times -219 \text{ J K}^{-1} \text{ mol}^{-1})$$

$$\Delta G^{\ominus} = (-90\,700 + 65\,262) \text{ J mol}^{-1}$$

$$\Delta G^{\ominus} = -25.4 \text{ kJ mol}^{-1}$$

> The Gibbs function is a *function of state*, so values of $\Delta G^{\ominus}$ obtained with the van't Hoff isotherm (see p. 162) and routes such as Hess's law cycles are identical.

This value of $\Delta G^{\ominus}$ is negative, so the reaction will indeed be spontaneous in this example. This is an example of where the negative value of ΔH overcomes the unfavourable positive $-\Delta S$ term. In fact, although the reaction is thermodynamically feasible, the *rate* of reaction (see Chapter 8) is so small that we need to heat the reaction vessel strongly to about 550 K to generate significant quantities of product to make the reaction viable.

> Some zoologists believe this reaction inspired the myth of fire-breathing dragons: some large tropical lizards have glands beside their mouths to produce both H_2S and SO_2. Under advantageous conditions, the resultant sulphur reacts so vigorously with the air that flames form as a self-defence mechanism.

SAQ 4.5 Consider the reaction $2H_2S_{(g)} + SO_{2(g)} \rightarrow 2H_2O_{(g)} + 3S_{(s)}$, where all species are gaseous except the sulphur. Calculate $\Delta G_r^{\ominus}$ for this reaction at 298 K with the thermodynamic data below:

	H_2S	SO_2	H_2O	S
$\Delta H_f^{\ominus}/kJ\ mol^{-1}$	-22.2	-296.6	-285.8	0
$\Delta S_f^{\ominus}/J\ K^{-1}\ mol^{-1}$	205.6	247.9	70.1	31.9

[Hint: it is generally easier first to determine values of ΔH_r and ΔS_r by constructing separate Hess's-law-type cycles.]

SAQ 4.6 The thermodynamic quantities of charge-transfer complex formation for the reaction

methyl viologen + hydroquinone $\longrightarrow$ charge-transfer complex   (4.26)

are $\Delta H_r = -22.6$ kJ mol^{-1} and $\Delta S_r = -62.1$ J K^{-1} mol^{-1} at 298 K. Such values have been described as 'typical of weak charge-transfer complex interactions'. Calculate the value of ΔG_r.

4.4   The effect of pressure on thermodynamic variables

How much energy is needed?

The Gibbs–Duhem equation

'How much energy is needed?' is a pointless question. It is too imprecise to be useful to anyone. The amount of energy needed will depend on how much material we wish to investigate. It also depends on whether we wish to perform a chemical reaction or a physical change, such as compression. We cannot answer the question until we redefine it.

The total amount we need to pay when purchasing goods at a shop depends both on the identity of the items we buy and how many of each. When buying sweets and apples, the total price will depend on the price of each item, and the amounts of each that we purchase. We could write it as

$$d(money) = (\text{price of item } 1 \times \text{number of item } 1)$$
$$+ (\text{price of item } 2 \times \text{number of item } 2) \tag{4.27}$$

While more mathematical in form, we could have rewritten Equation (4.27)

$$d(money) = \frac{\partial(money)}{\partial(1)} \times N(1) + \frac{\partial(money)}{\partial(2)} \times N(2) + \cdots \tag{4.28}$$

where N is merely the number of item (1) or item (2), and each bracket represents the price of each item: it is the amount of money per item. An equation like Equation (4.28) is called a *total differential*.

In a similar way, we say that the value of the Gibbs function changes in response to changes in pressure and temperature. We write this as

$$G = f(p, T) \tag{4.29}$$

and say G is a function of pressure and temperature.

> We must use the symbol ∂ ('curly d') in a differential when several terms are changing. The term in the first bracket is the rate of change of one variable when all other variables are constant.

> The value of *G* for a single, pure material is a function of both its temperature and pressure.

So, what is the change in G for a single, pure substance as the temperature and pressure are altered? A mathematician would start answering this question by writing out the total differential of G:

$$dG = \left(\frac{\partial G}{\partial p}\right)_T dp + \left(\frac{\partial G}{\partial T}\right)_p dT \tag{4.30}$$

> The small subscripted *p* on the first bracket tells us the differential must be obtained at constant pressure. The subscripted *T* indicates constant temperature.

which should remind us of Equation (4.28). The first term on the right of Equation (4.30) is the change in G per unit change in pressure, and the subsequent dp term accounts for the actual change in pressure. The second bracket on the right-hand side is the change in G per unit change in temperature, and the final dT term accounts for the actual change in temperature.

Equation (4.30) certainly looks horrible, but in fact it's simply a statement of the obvious – and is directly analogous to the prices of apples and sweets we started by talking about, cf. Equation (4.27).

We derived Equation (4.30) from first principles, using pure mathematics. An alternative approach is to prepare a similar equation algebraically. The result of the algebraic derivation is the *Gibbs–Duhem equation*:

$$dG = V\,dp - S\,dT \tag{4.31}$$

Justification Box 4.3

We start with Equation (4.21) for a single phase, and write $G = H - TS$. Its differential is

$$dG = dH - T\,dS - S\,dT \qquad (4.32)$$

From Chapter 2, we recall that $H = U + pV$, the differential of which is

$$dH = dU + p\,dV + V\,dp \qquad (4.33)$$

> Equation (4.35) combines the first and second laws of thermodynamics: it derives from Equation (3.5) and says, in effect, $dU = dq + dw$. The $p\,dV$ term relates to expansion work and the $T\,dS$ term relates to the adiabatic transfer of heat energy.

Substituting for dH in Equation (4.32) with the expression for dH in Equation (4.33), we obtain

$$dG = (dU + p\,dV + V\,dp) - T\,dS - S\,dT \qquad (4.34)$$

For a closed system (i.e. one in which no expansion work is possible)

$$dU = T\,dS - p\,dV \qquad (4.35)$$

Substituting for the dU term in Equation (4.34) with the expression for dU in Equation (4.35) yields

> The *Gibbs–Duhem equation* is also commonly (mis-)spelt 'Gibbs–Duheme'.

$$dG = (T\,dS - p\,dV) + p\,dV + V\,dp - T\,dS - S\,dT \qquad (4.36)$$

The $T\,dS$ and $p\,dV$ terms will cancel, leaving the *Gibbs–Duhem equation*, Equation (4.31).

We now come to the exciting part. By comparing the total differential of Equation (4.30) with the Gibbs–Duhem equation in Equation (4.31) we can see a pattern emerge:

$$dG = \left(\frac{\partial G}{\partial p}\right)dp + \left(\frac{\partial G}{\partial T}\right)dT$$

$$dG = V\,dp - S\,dT$$

So, by direct analogy, comparing one equation with the other, we can say

$$V = \frac{\partial G}{\partial p} \qquad (4.37)$$

and

$$-S = \frac{\partial G}{\partial T} \qquad (4.38)$$

> These two equations are known as the *Maxwell relations*.

Equations (4.37) and (4.38) are known as the *Maxwell relations*. The second Maxwell relation (Equation (4.38)) may remind us of the form of the Clausius equality (see p. 142). Although the first Maxwell relation (Equation (4.37)) is not intuitively obvious, it will be of enormous help later when we look at the changes in G as a function of pressure.

Why does a vacuum 'suck'?

The value of G as a function of pressure

Consider two flasks of gas connected by a small tube. Imagine also that a tap separates them, as seen by the schematic illustration in Figure 4.4. One flask contains hydrogen gas at high pressure p, for example at 2 atm. The other has such a low pressure of hydrogen that it will be called a vacuum.

As soon as the tap is opened, molecules of hydrogen move spontaneously from the high-pressure flask to the vacuum flask. The movement of gas is usually so rapid that it makes a 'slurp' sound, which is why we often say the vacuum 'sucks'.

> The old dictum, 'nature abhors a vacuum' is not just an old wives tale, it is also a manifestation of the second law of thermodynamics.

Redistributing the hydrogen gas between the two flasks is essentially the same phenomenon as a dye diffusing, as we discussed at the start of this chapter: the redistribution is thermodynamically favourable because it increases the entropy, so ΔS is positive.

We see how the spontaneous movement of gas always occurs from high pressure to low pressure, and also explains why a balloon will deflate or pop on its own, but work is needed to blow up the balloon or inflate a bicycle tyre (i.e. inflating a tyre is not spontaneous).

> Gases move spontaneously from high pressure to low.

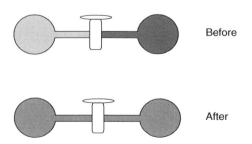

Before

After

Figure 4.4 Two flasks are connected by a tap. One contains gas at high pressure. As soon as the tap separating the two flasks is opened, molecules of gas move spontaneously from the flask under higher pressure to the flask at lower pressure. (The intensity of the shading represents the pressure of the gas)

Why do we sneeze?

The change in Gibbs function during gas movement: gas molecules move from high pressure to low

When we sneeze, the gases contained in the lungs are ejected through the throat so violently that they can move at extraordinary speeds of well over a hundred miles an hour. One of the obvious reasons for a sneeze is to expel germs, dust, etc. in the nose – which is why a sneeze can be so messy.

But a sneeze is much more sophisticated than mere germ removal: the pressure of the expelled gas is quite high because of its speed. Having left the mouth, a partial vacuum is left near the back of the throat as a result of the *Bernoulli effect* described below. Having a partial vacuum at the back of the throat is thermodynamically unstable, since the pressure in the nose is sure to be higher. As in the example above, gas from a region of high pressure (the nose) will be sucked into the region of low pressure (the back of the mouth) to equalize them. The nose is unblocked during this process of pressure equilibration, so one of the major reasons why we sneeze is to unblock the nose.

It is relatively easy to unblock a nose by blowing, but a sneeze is a superb means for unblocking the nose from the opposite direction.

Aside

The Bernoulli effect

Hold two corners of a piece of file paper along its narrow side. It will droop under its own weight because of the Earth's gravitational pull acting on it. But the paper will rise and stand out almost horizontally when we blow gently over its upper surface, as if by magic (see Figure 4.5). The paper droops before blowing. Blowing induces an additional force on the paper to counteract the force of gravity.

The air pressure above the upper side of the paper decreases because the air moves over its surface faster than the air stream running past the paper's underside. The

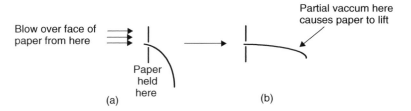

Figure 4.5   The Bernoulli effect occurs when the flow of fluid over one face of a body is greater than over another, leading to pressure inequalities. Try it: (a) hold a pace of paper in both hands, and feel it sag under its own weight. (b) Blow over the paper's upper surface, and see it lift

disparity in air speed leads to a *difference* in pressure. In effect, a partial vacuum forms above the paper, which 'sucks' the paper upwards. This is known as the *Bernoulli* effect.

A similar effect enables an aeroplane to fly: the curve on a plane's wing is carefully designed such that the pressure above the wing is less than that below. The air flows over the upper face of the wing with an increased speed, leading to a decrease in pressure. Because the upward thrust on the underside of the wing is great (because of the induced vacuum), it counterbalances the downward force due to gravity, allowing the plane to stay airborne.

How does a laboratory water pump work?

Gibbs function of pressure change

The water pump is another example of the Bernoulli effect, and is an everyday piece of equipment in most laboratories, for example being used during Büchner filtration. It comprises a piece of rubber tubing to connect the flask to be evacuated to a pump. Inside the pump, a rapid flow of water past one end of a small aperture inside the head decreases the pressure of the adjacent gas, so the pressure inside the pump soon decreases.

Gas passes from the flask to the pump where the pressure is lower. The change in Gibbs function associated with these pressure changes is given by

$$\Delta G = RT \, \ln \left(\frac{p_{(\text{final})}}{p_{(\text{initial})}} \right) \qquad (4.39)$$

where the ΔG term represents the change in G *per mole* of gas. We will say here that gas enters the pump at pressure $p_{(\text{final})}$ from a flask initially at pressure $p_{(\text{initial})}$. Accordingly, since $p_{(\text{final})} < p_{(\text{initial})}$ and the term in brackets is clearly less than one, the logarithm term is negative. ΔG is thus negative, showing that gas movement from a higher pressure $p_{(\text{initial})}$ to a lower pressure $p_{(\text{final})}$ is spontaneous.

It should be clear from Equation (4.39) that gas movement in the opposite direction, from low pressure ($p_{(\text{final})}$) to high ($p_{(\text{initial})}$) would cause ΔG to be positive, thereby explaining why the process of gas going from low pressure to high never occurs naturally. Stated another way, compression can only occur if energy is put *into* the system; so, *compression involves work*, which explains why pumping up a car tyre is difficult, yet the tyre will deflate of its own accord if punctured.

These highly oversimplified explanations ignore the effects of turbulent flow, and the formation of vortices.

The minimum pressure achievable with a water pump equals the vapour pressure of water, and has a value of about 28 mmHg.

ΔG is negative for the physical process of gas moving from higher to lower pressure.

Since Equation (4.39) relies on a *ratio* of pressures, we say that a gas moves from 'higher' to 'lower', rather than 'high' to 'low'.

Compression involves work.

Equation (4.39) involves a *ratio* of pressures, so, although mmHg (millimetres of mercury) is not an SI unit of pressure, we are permitted to use it here.

The pressure of vapour above a boiling liquid is the same as the atmospheric pressure.

Worked Example 4.9 The pressure inside a water pump is the same as the vapour pressure of water (28 mmHg). The pressure of gas inside a flask is the same as atmospheric pressure (760 mmHg). What is the change in Gibbs function per mole of gas that moves? Take $T = 298$ K.

Inserting values into Equation (4.39) yields

$$\Delta G = 8.414 \text{ J K}^{-1} \text{ mol}^{-1} \times 298 \text{ K} \times \ln\left(\frac{28 \text{ mmHg}}{760 \text{ mmHg}}\right)$$

$$\Delta G = 2477 \text{ J mol}^{-1} \times \ln(3.68 \times 10^{-2})$$

$$\Delta G = 2477 \text{ J mol}^{-1} \times (-3.301)$$

$$\Delta G = -8.2 \text{ kJ mol}^{-1}$$

SAQ 4.7 A flask of methyl-ethyl ether (**V**) is being evaporated. Its boiling temperature is 298 K (the same as room temperature) so the vapour pressure of ether above the liquid is the same as atmospheric pressure, i.e. at 100 kPa. The source of the vacuum is a water pump, so the pressure is the vapour pressure of water, 28 mmHg.

$$CH_3 \diagup \overset{\displaystyle CH_2}{} \diagdown \underset{\displaystyle O}{} \diagup CH_3$$

(V)

(1) Convert the vacuum pressure $p_{(vacuum)}$ into an SI pressure, remembering that 1 atm = 101 325 kPa = 760 mmHg.

(2) What is the molar change in Gibbs function that occurs when ether vapour is removed, i.e. when ether vapour goes from the flask at $p^{\ominus}$ into the water pump at $p_{(vacuum)}$?

Justification Box 4.4

We have already obtained the first Maxwell relation (Equation (4.37)) by comparing the Gibbs–Duhem equation with the total differential:

$$\frac{\partial G}{\partial p} = V$$

We obtain the molar volume V_m as $V \div n$.

The ideal-gas equation says $pV = nRT$, or, using a *molar* volume for the gas (Equation (1.13)):

$$pV_m = RT$$

Substituting for V_m from Equation (4.37) into Equation (1.13) gives

$$p\left(\frac{\mathrm{d}G}{\mathrm{d}p}\right) = RT \tag{4.40}$$

And separation of the variables gives

$$\frac{\mathrm{d}G}{\mathrm{d}p} = RT\frac{1}{p} \tag{4.41}$$

so

$$\mathrm{d}G = RT\frac{1}{p}\,\mathrm{d}p \tag{4.42}$$

Integration, taking G_1 at $p_{(\text{initial})}$ and G_2 at $p_{(\text{final})}$, yields

$$G_2 - G_1 = RT\ \ln\left(\frac{p_{(\text{final})}}{p_{(\text{initial})}}\right) \tag{4.43}$$

Or, more conveniently, if G_2 is the final value of G and G_1 the initial value of G, then the change in Gibbs function is

$$\Delta G = RT\ \ln\left(\frac{p_2}{p_1}\right)$$

i.e. Equation (4.39).

Aside

In practice, it is often found that compressing or decompressing a gas does not follow closely to the ideal-gas equation, particularly at high p or low T, as exemplified by the need for equations such as the van der Waals equation or a virial expression. The equation above is a good approximation, though.

A more thorough treatment takes one of two courses:

(1)   Utilize the concept of *virial* coefficients; see p. 57.
(2)   Use *fugacity* instead of pressure.

Fugacity f is defined as

$$f = p \times \gamma \tag{4.44}$$

The fugacity f can be regarded as an 'effective' pressure. The 'fugacity coefficient' γ represents the deviation from ideality. The value of γ tends to one as p tends to zero.

The word 'fugacity' comes from the Latin *fugere*, which means 'elusive' or 'difficult to capture'. The modern word 'fugitive' comes from the same source.

> where p is conventional pressure and γ is the *fugacity coefficient* representing the deviation from ideality. Values of γ can be measured or calculated.
>
> We employ both concepts to compensate for gas non-ideality.

4.5 Thermodynamics and the extent of reaction

Why is a 'weak' acid weak?

Incomplete reactions and extent of reaction

As chemists, we should perhaps re-cast the question 'Why is a 'weak' acid weak?' by asking 'How does the change in Gibbs function relate to the proportion of reactants that convert during a reaction to form products?'

An acid is defined as a proton donor within the Lowry–Brønsted theory (see Chapter 6). Molecules of acid ionize in aqueous solution to form an anion and a proton, both of which are solvated. An acid such as ethanoic acid (**VI**) is said to be 'weak' if the extent to which it dissociates is incomplete; we call it 'strong' if ionization is complete (see Section 6.2).

(VI)

Ionization is, in fact, a chemical reaction because bonds break and form. Consider the following general ionization reaction:

$$HA + H_2O \longrightarrow H_3O^+ + A^- \tag{4.45}$$

We give the Greek symbol ξ ('xi') to the *extent of reaction*. ξ is commonly mispronounced as 'ex-eye'.

We give the extent of the reaction in Equation (4.45) the Greek symbol ξ. It should be clear that ξ has a value of zero before the reaction commences. By convention, we say that $\xi = 1$ mol if the reaction goes to completion. The value of ξ can take any value between these two extremes, its value increasing as the reaction proceeds. A reaction going to completion only stops when no reactant remains, which we define as ξ having a value of 1 mol, although such a situation is comparatively rare except in inorganic redox reactions. In fact, to an excellent approximation, *all* preparative organic reactions fail to reach completion, so $0 < \xi < 1$.

The value of ξ only stops changing when the reaction stops, although the rate at which ξ changes belongs properly to the topic of kinetics (see Chapter 8). We say it reaches its *position of equilibrium*, for which the value of ξ has its equilibrium value $\xi_{(eq)}$. We propose perhaps the simplest of the many possible definitions of equilibrium: 'after an initial period of reaction, no further *net* changes in reaction composition occur'.

So when we say that a carboxylic acid is weak, we mean that $\xi_{(eq)}$ is small. Note how, by saying that $\xi_{(eq)}$ is small at equilibrium, we effectively imply that the extent of ionization is small because $[H_3O^+]$ and $[A^-]$ are both small.

But we need to be careful when talking about the magnitudes of ξ. Consider the case of sodium ethanoate dissolved in dilute mineral acid: the reaction occurring is, in fact, the *reverse* of that in Equation (4.45), with a proton and carboxylate anion associating to form undissociated acid. In this case, $\xi = 1$ mol before the reaction occurs, and its value decreases as the reaction proceeds. In other words, we need to define our reaction before we can speak knowledgeably about it. We can now rewrite our question, asking 'Why is $\xi \ll 1$ for a weak acid?'

The standard Gibbs function change for reaction is $\Delta G^\ominus$, and represents the energy available for reaction if 1 mol of reactants react until reaching equilibrium. Figure 4.6 relates ΔG and ξ, and clearly shows how the amount of energy available for reaction ΔG decreases during reaction (i.e. in going from left to right as ξ increases). Stated another way, the gradient of the curve is always negative before the position of equilibrium, so any increases in ξ cause the value of ΔG to become more negative.

> We assume such an equilibrium is fully *reversible* in the sense of being *dynamic* – the rate at which products form is equal and opposite to the rate at which reactants regenerate via a back reaction.

> *Reminder*: the energy released during reaction originates from the making and breaking of bonds, and the rearrangement of solvent. The *full* amount of energy given out is $\Delta H^\ominus$, but the *net* energy available is less that $\Delta H^\ominus$, being $\Delta H^\ominus - T\Delta S^\ominus$.

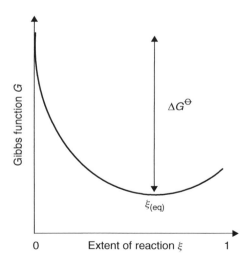

Figure 4.6 The value of the Gibbs function ΔG decreases as the extent of reaction ξ until, at $\xi_{(eq)}$, there is no longer any energy available for reaction, and $\Delta G = 0$. $\xi = 0$ represents no reaction and $\xi = 1$ mol represents complete reaction

The amount of energy liberated per incremental increase in reaction is quite large at the start of reaction, but decreases until, at equilibrium, a tiny increase in the extent of reaction would not change $\Delta G_{\text{(total)}}$. The graph has reached a minimum, so the gradient at the bottom of the trough is zero.

The minimum in the graph of ΔG against ξ is the reaction's position of equilibrium – we call it $\xi_{\text{(eq)}}$. The maximum amount of energy has already been expended at equilibrium, so ΔG is zero.

Any further reaction beyond $\xi_{\text{(eq)}}$ would not only fail to liberate any further energy, but also would in fact *consume* energy (we would start to go 'uphill' on the right-hand side of the figure). Any further increment of reaction would be character-ized by $\Delta G > 0$, implying a non-spontaneous process, which is why the reaction stops at $\xi_{\text{(eq)}}$.

Why does the pH of the weak acid remain constant?

The law of mass action and equilibrium constants

The amount of ethanoic acid existing as ionized ethanoate anion and solvated proton is always small (see p. 253). For that reason, the pH of a solution of weak acid is always higher than a solution of the same concentration of a strong acid. A naïve view suggests that, given time, all the undissociated acid will manage to dissociate, with the dual effect of making the acid strong, and hence lowering the pH.

We return to the graph in Figure 4.6 of Gibbs function (as y) against extent of reaction ξ (as x). At the position of the minimum, the amounts of free acid and ionized products remain constant because there is no longer any energy available for reaction, as explained in the example above.

The fundamental law of chemical equilibrium is the *law of mass action*, formulated in 1864 by Cato Maximilian Guldberg and Peter Waage. It has since been redefined several times. Consider the equilibrium between the four chemical species A, B, C and D:

$$a\text{A} + b\text{B} = c\text{C} + d\text{D} \tag{4.46}$$

where the respective stoichiometric numbers are $-a$, $-b$, c and d. The law of mass action states that, at equilibrium, the mathematical ratio of the concentrations of the two reactants $[\text{A}]^a \times [\text{B}]^b$ and the product of the two product concentrations $[\text{C}]^c \times [\text{D}]^d$, is equal. We could, therefore, define one of two possible fractions:

$$\frac{[\text{A}]^a[\text{B}]^b}{[\text{C}]^c[\text{D}]^d} \quad \text{or} \quad \frac{[\text{C}]^c[D]^d}{[\text{A}]^a[\text{B}]^b} \tag{4.47}$$

This ratio of concentrations is called an *equilibrium constant*, and is symbolized as K.

The two ratios above are clearly related, with one being the reciprocal of the other. Ultimately, the choice of which of these two we prefer is arbitrary, and usually relates to the way we write Equation (4.46). In consequence, the way we write this ratio is dictated by the sub-discipline of chemistry we practice. For example, in acid–base chemistry (see Chapter 6) we write a *dissociation* constant, but in complexation equilibria we write a *formation* constant.

In fact, an equilibrium constant is only ever useful when we have carefully defined the chemical process to which it refers.

The reaction quotient

It is well known that few reactions (other than inorganic redox reactions) ever reach completion. The value of $\xi_{(eq)}$ is always less than one.

The quotient of products to reactants during a reaction is

$$Q = \frac{\prod[\text{products}]^{\nu}}{\prod[\text{reactants}]^{\nu}} \qquad (4.48)$$

which is sometimes called the *reaction quotient*. The values of ν are the respective stoichiometric numbers. The mathematical value of Q *increases* continually during the course of reaction because of the way it relates to concentrations *during* reaction. Initially, the reaction commences with a value of $Q = 0$, because there is no product (so the numerator is zero).

> The mathematical symbol $\prod$ means the 'pi product', meaning the terms are multiplied together, so $\prod(2, 3, 4)$ is $2 \times 3 \times 4 = 24$.

Aside

A statement such as '$K = 0.4 \text{ mol dm}^{-3}$' is wrong, although we find examples in a great number of references and textbooks. We ought, rather, to say $K = 0.4$ when the equilibrium constant is formulated (i) in terms of concentrations, and (ii) where each concentration is expressed in the reference units of mol dm^{-3}. Equilibrium constants such as K_c or K_p are mere *numbers*.

Why does the voltage of a battery decrease to zero?

Relationships between K and $\Delta G^{\ominus}$

The voltage of a new torch battery (AA type) is about 1.5 V. After the battery has powered the torch for some time, its voltage drops, which we see in practice as the light beam becoming dimmer. If further power is withdrawn indefinitely then the voltage from the battery eventually drops to zero, at which point we say the battery is 'dead' and throw it away.

A battery is a device for converting chemical energy into electrical energy (see p. 344), so the discharging occurs as a consequence of chemical reactions inside the battery. The reaction is complete

> The battery produces 'power' W (energy per unit time) by passing a current through a resistor. The resister in a torch is the bulb filament.

The relationship between ΔG and voltage is discussed in Section 7.1.

when the battery is 'dead', i.e. the reaction has reached its equilibrium extent of reaction $\xi_{(eq)}$.

The battery voltage is proportional to the change in the Gibbs function associated with the battery reaction, call it $\Delta G_{(battery)}$. Therefore, we deduce that $\Delta G_{(battery)}$ must decrease to zero because the battery voltage drops to zero. Figure 4.7 shows a graph of battery voltage (as y) against time of battery discharge (as x); the time of discharge is directly analogous to extent of reaction ξ. Figure 4.7 is remarkably similar to the graph of ΔG against ξ in Figure 4.6.

The relationship between the energy available for reaction ΔG_r and the extent of reaction (expressed in terms of the reaction quotient Q) is given by

$$\Delta G_r = \Delta G_r^{\ominus} + RT \ln Q \qquad (4.49)$$

where ΔG is the energy available for reaction *during chemical changes*, and $\Delta G_r^{\ominus}$ is the standard change of Gibbs function $\Delta G^{\ominus}$, representing the change in Gibbs function from $\xi = 0$ to $\xi = \xi_{(eq)}$.

Equation (4.49) describes the shape of the graph in Figure 4.6. Before we look at Equation (4.46) in any quantitative sense, we note that if $RT \ln Q$ is smaller than $\Delta G_r^{\ominus}$, then ΔG_r is positive. The value of ΔG_r only reaches zero when $\Delta G_r^{\ominus}$ is exactly the same as $RT \ln Q$. In other words, there is no energy available for reaction when $\Delta G_r = 0$: we say the system has 'reached equilibrium'. In fact, $\Delta G_r = 0$ is one of the best definitions of equilibrium.

$\Delta G_r = 0$ is one of the best definitions of equilibrium.

In summary, the voltage of the battery drops to zero because the value of ΔG_r is zero, which happened at $\xi = \xi_{(eq)}$.

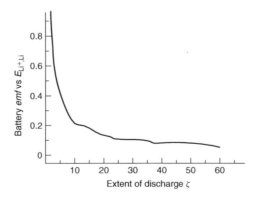

Figure 4.7 Graph of battery *emf* (as y) against extent of discharge (as x). Note the remarkable similarity between this figure and the left-hand side of Figure 4.6, which is not coincidental because *emf* $\propto \Delta G$, and extent of discharge is proportional to ξ. The trace represents the ninth discharge of a rechargeable lithium–graphite battery, constructed with a solid-state electrolyte of polyethylene glycol containing $LiClO_4$. The shakiness of the trace reflects the difficulty in obtaining a *reversible* measurement. Reprinted from S. Lemont and D. Billaud, *Journal of Power Sources* 1995; **54**: 338. Copyright © 1995, with permission from Elsevier

Justification Box 4.5

Consider again the simple reaction of Equation (4.46):

$$aA + bB = cC + dD$$

We ascertain the Gibbs energy change for this reaction. We start by saying

$$\Delta G = \sum \nu G_{\text{(products)}} - \sum \nu G_{\text{(reactants)}}$$

where ν is the respective stoichiometric number; so

$$\Delta G = cG_C + dG_D - aG_A - bG_B \tag{4.50}$$

From an equation like Equation (4.43), $G = G^{\ominus} + RT \ln(p/p^{\ominus})$, so each G term in Equation (4.50) may be converted to a *standard* Gibbs function by inserting a term like Equation (4.43):

$$\Delta G = \left[cG_C^{\ominus} + cRT \ln\left(\frac{p_C}{p^{\ominus}}\right) + dG_D^{\ominus} + dRT \ln\left(\frac{p_D}{p^{\ominus}}\right) - aG_A^{\ominus} \right.$$
$$\left. - aRT \ln\left(\frac{p_A}{p^{\ominus}}\right) - bG_B^{\ominus} - bRT \ln\left(\frac{p_B}{p^{\ominus}}\right) \right] \tag{4.51}$$

We can combine the $G^{\ominus}$ terms as $\Delta G^{\ominus}$ by saying

$$\Delta G^{\ominus} = cG_C^{\ominus} + dG_D^{\ominus} - aG_A^{\ominus} - bG_B^{\ominus} \tag{4.52}$$

So Equation (4.51) simplifies to become:

$$\Delta G = \Delta G^{\ominus} + cRT \ln\left(\frac{p_C}{p^{\ominus}}\right) + dRT \ln\left(\frac{p_D}{p^{\ominus}}\right) - aRT \ln\left(\frac{p_A}{p^{\ominus}}\right)$$
$$- bRT \ln\left(\frac{p_B}{p^{\ominus}}\right) \tag{4.53}$$

Then, using the laws of logarithms, we can simplify further:

$$\Delta G = \Delta G^{\ominus} + RT \ln\left[\frac{(p_C/p^{\ominus})^c (p_D/p^{\ominus})^d}{(p_A/p^{\ominus})^a (p_B/p^{\ominus})^b} \right] \tag{4.54}$$

The bracketed term is the reaction quotient, expressed in terms of pressures, allowing us to rewrite the equation in a less intimidating form of Equation (4.49):

$$\Delta G_r = \Delta G_r^{\ominus} + RT \ln Q$$

A similar proof may be used to derive an expression relating to $\Delta G^{\ominus}$ and K_c.

> We changed the positioned of each stoichiometric number via the laws of logarithms, saying $b \times \ln a = \ln a^b$.

Aside

A further complication arises from the ΔG term in Equation (4.49). The diagram above is clearer than the derivation: in reality, the differential quantity $\partial G/\partial \xi$ only corresponds to the change in Gibbs function ΔG under certain, well defined, and precisely controlled experimental conditions.

This partial differential is called the *reaction affinity* in older texts and in newer texts is called the *reaction free energy*.

Why does the concentration of product stop changing?

The van't Hoff isotherm

Because we only ever write K (rather than Q) at equilibrium, it is tautologous but very common to see K written as $K_{(eq)}$ or K_e.

The descriptor 'isotherm' derives from the Greek *iso* meaning 'same' and *thermos* meaning 'temperature'.

Jacobus van't Hoff was a Dutch scientist (1852–1911). Notice the peculiar arrangement of the apostrophe, and small and capital letters in his surname.

It would be beneficial if we could increase the yield of a chemical reaction by just leaving it to react longer. Unfortunately, the concentrations of reactant and product remain constant at the end of a reaction. In other words, the reaction quotient has reached a constant value.

At equilibrium, when the reaction stops, we give the reaction quotient the special name of *equilibrium constant*, and re-symbolize it with the letter K. The values of K and Q are exactly the same at equilibrium when the reaction stops. The value of Q is always smaller than K before equilibrium is reached, because some product has yet to form. In other words, before equilibrium, the top line of Equation (4.48) is artificially small and the bottom is artificially big.

Q and K only have the same value when the reaction has reached equilibrium, i.e. when $\Delta G_r = 0$. At this extent of reaction, the relationship between ξ and $\Delta G^{\ominus}$ is given by the *van't Hoff isotherm*:

$$\Delta G^{\ominus} = -RT \ln K \qquad (4.55)$$

where R and T have their usual thermodynamics meanings. The equation shows the relationship between $\Delta G^{\ominus}$ and K, indicating that these two parameters are interconvertible when the temperature is held constant.

SAQ 4.8 Show that the van't Hoff isotherm is dimensionally self-consistent.

Worked Example 4.10 Consider the dissociation of ethanoic (acetic) acid in water to form a solvated proton and a solvated ethanoate anion, $CH_3COOH + H_2O \rightarrow CH_3COO^-$

+ H_3O^+. This reaction has an equilibrium constant K of about 2×10^{-5} at room temperature (298 K) when formulated in the usual units of concentration (mol dm^{-3}). What is the associated change in Gibbs function of this reaction?

> The correct use of the van't Hoff isotherm necessitates using the *thermodynamic* temperature (expressed in kelvin).

Inserting values into the van't Hoff isotherm (Equation (4.55)):

$$\Delta G^\ominus = -8.314 \text{ J K}^{-1} \text{ mol}^{-1} \times 298 \text{ K} \times \ln(2 \times 10^{-5})$$

$$\Delta G^\ominus = -2478 \text{ J mol}^{-1} \times -10.8$$

$$\Delta G^\ominus = +26\,811 \text{ J mol}^{-1}$$

so

$$\Delta G^\ominus = +26.8 \text{ kJ mol}^{-1}$$

Note how $\Delta G^\ominus$ is positive here. We say it is *endogenic*.

> A process occurring with a *negative* value of ΔG is said to be *exogenic*. A process occurring with a *positive* value of ΔG is said to be *endogenic*.

Justification Box 4.6

We start with Equation (4.49):

$$\Delta G = \Delta G^\ominus + RT \ln Q$$

At equilibrium, the value of ΔG is zero. Also, the value of Q is called K:

$$0 = \Delta G^\ominus + RT \ln K \qquad (4.56)$$

Subtracting the '$-RT \ln K$' term from both sides yields the van't Hoff isotherm (Equation (4.55)):

$$\Delta G^\ominus = -RT \ln K$$

This derivation proves that equilibrium constants do exist. The value of $\Delta G^\ominus$ depends on T, so the value of K should be independent of the total pressure.

We sometimes want to know the value of K from a value of $\Delta G^\ominus$, in which case we employ a rearranged form of the isotherm:

$$K = \exp\left(\frac{-\Delta G^\ominus}{RT}\right) \qquad (4.57)$$

so a small change in the Gibbs function means a small value of K. Therefore, a weak acid is weak simply because $\Delta G^\ominus$ is small.

Care: we must always convert from kJ to J before calculating with Equation (4.57).

Care: if we calculate a value of K that is extremely close to one, almost certainly we forgot to convert from kJ to J, making the fraction in the bracket a thousand times too small.

Worked Example 4.11   Consider the reaction between ethanol and ethanoic acid to form a sweet-smelling ester and water:

$$CH_2CH_2OH + CH_3COOH \longrightarrow CH_2CH_2CO_2CH_3 + H_2O \quad (4.58)$$

What is the equilibrium constant K at room temperature (298 K) if the associated change in Gibbs function is *exogenic* at -3.4 kJ mol^{-1}?

Inserting values into eq. (4.57):

$$K = \exp\left[\frac{-(-3400 \text{ J mol}^{-1})}{8.314 \text{ J K}^{-1} \text{ mol}^{-1} \times 298 \text{ K}}\right]$$

$$K = \exp(+1.372)$$

$$K = 3.95$$

A value of K greater than one corresponds to a negative value of $\Delta G^{\ominus}$, so the esterification reaction is spontaneous and does occur to some extent without adding addition energy, e.g. by heating.

A few values of $\Delta G^{\ominus}$ are summarized as a function of K in Table 4.1 and values of K as a function of $\Delta G^{\ominus}$ are listed in Table 4.2. Clearly, K becomes larger as $\Delta G^{\ominus}$ becomes more negative. Conversely, $\Delta G^{\ominus}$ is positive if K is less than one.

Justification Box 4.7

We start with Equation (4.55):

$$\Delta G^{\ominus} = -RT \ln K$$

Both sides are divided by $-RT$, yielding

$$\left(\frac{-\Delta G^{\ominus}}{RT}\right) = \ln K \quad (4.59)$$

Then we take the exponential of both sides to generate Equation (4.57).

Table 4.1   The relationship between $\Delta G^{\ominus}$ and equilibrium constant K: values of $\Delta G^{\ominus}$ as a function of K

We see from Table 4.1 that every decade increase in K causes $\Delta G^{\ominus}$ to become more negative by 5.7 kJ mol^{-1} per tenfold increase in K.

K	$\Delta G^{\ominus}/\text{kJ mol}^{-1}$
1	0
10	-5.7
10^2	-11.4
10^3	-17.1
10^4	-22.8
10^{-1}	$+5.7$
10^{-2}	$+11.4$
10^{-3}	$+17.1$

Table 4.2 The relationship between $\Delta G^{\ominus}$ and equilibrium constant K: values of K as a function of $\Delta G^{\ominus}$

$\Delta G^{\ominus}/\text{kJ mol}^{-1}$	K
0	1
-1	1.50
-10	56.6
-10^2	3.38×10^{17}
-10^3	∞
$+1$	0.667
$+10$	0.0177
$+10^2$	2.96×10^{-18}

SAQ 4.9 What is the value of K corresponding to $\Delta G^{\ominus}_{298 \text{ K}} = -12 \text{ kJ mol}^{-1}$?

Why do chicken eggs have thinner shells in the summer?

The effect of altering the concentration on ξ

Egg shells are made of calcium carbonate, $CaCO_3$. The chicken ingeniously makes shells for its eggs by a process involving carbon dioxide dissolved in its blood, yielding carbonate ions which combine chemically with calcium ions. An equilibrium is soon established between these ions and solid chalk, according to

$$Ca^{2+}_{(aq)} + CO^{2-}_{3(aq)} = CaCO_{3(s, \text{ shell})} \tag{4.60}$$

Unfortunately, chickens have no sweat glands, so they cannot perspire. To dissipate any excess body heat during the warm summer months, they must pant just like a dog. Panting increases the amount of carbon dioxide exhaled, itself decreasing the concentration of CO_2 in a chicken's blood. The smaller concentration $[CO^{2-}_{3(aq)}]$ during the warm summer causes the reaction in Equation (4.60) to shift further toward the left-hand side than in the cooler winter, i.e. the amount of chalk formed decreases. The end result is a thinner eggshell.

Chicken farmers solve the problem of thin shells by carbonating the chickens' drinking water in the summer. We may never know what inspired the first farmer to follow this route, but any physical chemist could have solved this problem by first writing the equilibrium constant K for Equation (4.60):

$$K_{(\text{shell formation})} = \frac{[CaCO_{3(s)}]}{[Ca^{2+}_{(aq)}][CO^{2-}_{3(aq)}]} \tag{4.61}$$

The value of $K_{(\text{shell formation})}$ will not change provided the temperature is fixed. Therefore, we see that if the concentration of carbonate ions (see the bottom line) falls then the amount of chalk on the top line must also fall. These changes must occur in tandem if K is to remain constant. In other words, decreasing the amount of CO_2 in a chicken's blood means less chalk is available for shell production. Conversely, the same reasoning suggests that increasing the concentration of carbonate – by adding carbonated water to the chicken's drink – will increases the bottom line of Equation (4.61), and the chalk term on the top increases to maintain a constant value of K.

Le Chatelier's principle

Arguments of this type illustrate Le Chatelier's principle, which was formulated in 1888. It says:

> Le Chatelier's principle is named after Henri Louis le Chatelier (1850–1937). He also spelt his first name the English way, as 'Henry'.

Any system in stable chemical equilibrium, subjected to the influence of an external cause which tends to change either its temperature or its condensation (pressure, concentration, number of molecules in unit volume), either as a whole or in some of its parts, can only undergo such internal modifications as would, if produced alone, bring about a change of temperature or of condensation of opposite sign to that resulting from the external cause.

The principle represents a kind of 'chemical inertia', seeking to minimize the changes of the system. It has been summarized as, 'if a constraint is applied to a system in equilibrium, then the change that occurs is such that it tends to annul the constraint'. It is most readily seen in practice when:

(1) The pressure in a closed system is increased (at fixed temperature) and shifts the equilibrium in the direction that decreases the system's volume, i.e. to decrease the change in pressure.

(2) The temperature in a closed system is altered (at fixed pressure), and the equilibrium shifts in such a direction that the system absorbs heat from its surroundings to minimize the change in energy.

4.6 The effect of temperature on thermodynamic variables

Why does egg white denature when cooked but remain liquid at room temperature?

Effects of temperature on $\Delta G^{\ominus}$: the Gibbs–Helmholtz equation

Boiling an egg causes the transparent and gelatinous *albumen* ('egg white') to modify chemically, causing it to become a white, opaque solid. Like all chemical reactions, denaturing involves the rearrangement of *bonds* – in this case, of hydrogen bonds. For convenience, in the discussion below we say that bonds change from a spatial arrangement termed '1' to a different spatial arrangement '2'.

From everyday experience, we know that an egg will not denature at room temperature, however long it is left. We are not saying here that the egg denatures at an almost infinitesimal rate, so the lack of reaction at room temperature is *not* a kinetic phenomenon; rather, we see that denaturation is energetically non-spontaneous at one temperature (25 °C), and only becomes spontaneous as the temperature is raised above a certain threshold temperature, which we will call $T_{(critical)}$ (about 70 °C for an egg).

The sign of the Gibbs function determines reaction spontaneity, so a reaction *will* occur if $\Delta G^{\ominus}$ is negative and will *not* occur if $\Delta G^{\ominus}$ is positive. When the reaction is 'poised' at $T_{(critical)}$ between spontaneity and non-spontaneity, the value of $\Delta G^{\ominus} = 0$.

The changes to $\Delta G^{\ominus}$ with temperature may be quantified with the *Gibbs–Helmholtz* equation:

$$\frac{\Delta G_2^{\ominus}}{T_2} - \frac{\Delta G_1^{\ominus}}{T_1} = \Delta H^{\ominus}\left(\frac{1}{T_2} - \frac{1}{T_1}\right) \qquad (4.62)$$

where $\Delta G_2^{\ominus}$ is the change in Gibbs function at the temperature T_2 and $\Delta G_1^{\ominus}$ is the change in Gibbs function at temperature T_1. Note how values of T must be expressed in terms of *thermodynamic* temperatures. ΔH is the standard enthalpy of the chemical process or reaction, as determined experimentally by calorimetry or calculated via a Hess's-law-type cycle.

The value of ΔH_r for denaturing egg white is likely to be quite small, since it merely involves changes in hydrogen bonds. For the purposes of this calculation, we say ΔH_r has a value of 35 kJ mol^{-1}.

Additionally, we can propose an equilibrium constant of reaction, although we must call it a *pseudo* constant $K_{(pseudo)}$ because we cannot in reality determine its value. We need a value of $K_{(pseudo)}$ in order to describe the way hydrogen bonds change position during denaturing. We say that $K_{(pseudo,\ 1)}$ relates to hydrogen bonds in the pre-reaction position '1' (i.e. prior to denaturing) and $K_{(pseudo,\ 2)}$ relates to the number of hydrogen bonds reoriented in the post-reaction position '2' (i.e. after denaturing). We will say here that $K_{(pseudo)}$ is '1/10' before the denaturing reaction, i.e. before boiling the egg at 298 K. From the van't Hoff isotherm, $K_{(pseudo)}$ equates to a Gibbs function change of +5.7 kJ mol^{-1}.

This argument here has been oversimplified because the reaction is thermodynamically *irreversible* – after all, you cannot '*un*boil an egg'!

Some reactions are spontaneous at one temperature but not at others.

The temperature dependence of the Gibbs function change is described quantitatively by the *Gibbs–Helmholtz* equation.

The word 'pseudo' derives from the Greek stem *pseudes* meaning 'falsehood', which is often taken to mean having an appearance that belies the actual nature of a thing.

Denaturing *albumen* is an 'irreversible' process, yet the derivations below assume thermodynamic *reversibility*. In fact, *complete* reversibility is rarely essential; try to avoid making calculations if a significant extent of irreversibility is apparent.

Worked Example 4.12   The white of an egg denatures while immersed in water boiling at its normal boiling temperature of 373 K. What is the value of ΔG at this higher temperature? Take $\Delta H = 35$ kJ mol^{-1}.

> It does not matter which temperature is chosen as T_1 and which as T_2 so long as T_1 relates to ΔG_1 and T_2 relates to ΔG_2.

The value of $\Delta G^{\ominus}$ at 298 K is $+5.7$ kJ mol^{-1}, the positive sign explaining the lack of a spontaneous reaction. Inserting values into the Gibbs–Helmholtz equation, Equation (4.62), yields

$$\frac{\Delta G_{373\ \text{K}}}{373\ \text{K}} - \frac{5700\ \text{J mol}^{-1}}{298\ \text{K}} = +35\,000\ \text{J mol}^{-1}\left(\frac{1}{373\ \text{K}} - \frac{1}{298\ \text{K}}\right)$$

Note that T is a thermodynamic temperature, and is cited in kelvin. All energies have been converted from kJ mol^{-1} to J mol^{-1}.

$$\frac{\Delta G^{\ominus}_{373\ \text{K}}}{373\ \text{K}} = (19.13\ \text{J K}^{-1}\ \text{mol}^{-1}) + (35\,000\ \text{J mol}^{-1}) \times (-6.75 \times 10^{-4}\ \text{K}^{-1})$$

$$\frac{\Delta G^{\ominus}_{373\ \text{K}}}{373\ \text{K}} = (19.13\ \text{J K}^{-1}\ \text{mol}^{-1}) - (23.62\ \text{J K}^{-1}\ \text{mol}^{-1})$$

$$\frac{\Delta G^{\ominus}_{373\ \text{K}}}{373\ \text{K}} = -4.4\ \text{J K}^{-1}\ \text{mol}^{-1}$$

so

$$\Delta G^{\ominus}_{373\ \text{K}} = -4.4\ \text{J K}^{-1}\ \text{mol}^{-1} \times 373\ \text{K}$$

and

$$\Delta G^{\ominus}_{373\ \text{K}} = -1.67\ \text{kJ mol}^{-1}$$

$\Delta G^{\ominus}_{373\ \text{K}}$ has a negative value, implying that the reaction at this new, elevated temperature is now spontaneous. In summary, the Gibbs–Helmholtz equation quantifies a qualitative observation: the reaction to denature egg white is not spontaneous at room temperature, but it *is* spontaneous at elevated temperatures, e.g. when the egg is boiled in water.

SAQ 4.10   Consider the reaction in Equation (4.63), which occurs wholly in the gas phase:

$$NH_3 + \tfrac{5}{4}O_2 \longrightarrow NO + \tfrac{3}{2}H_2O \tag{4.63}$$

The value of $\Delta G^{\ominus}_r$ for this reaction is -239.9 kJ mol^{-1} at 298 K. If the enthalpy change of reaction $\Delta H_r = -406.9$ kJ mol^{-1}, then

(1) Calculate the associated entropy change for the reaction in Equation (4.63), and comment on its sign.

(2) What is the value of the Gibbs function for this reaction when the temperature is increased by a further 34 K?

Justification Box 4.8

We will use the quantity '$G \div T$' for the purposes of this derivation. Its differential is obtained by use of the product rule. In general terms, for a compound function ab, i.e. a function of the type $y = f(a, b)$:

$$\frac{dy}{dx} = a \times \frac{db}{dx} + b \times \frac{da}{dx} \qquad (4.64)$$

so here

$$\frac{d(G \div T)}{dT} = \frac{1}{T} \times \frac{dG}{dT} + G \times \left(-\frac{1}{T^2}\right) \qquad (4.65)$$

The function $G \div T$ occurs so often in thermodynamics that we call it the *Planck function*.

All standard signs $\ominus$ have been omitted for clarity

Note that the term $\dfrac{dG}{dT}$ is $-S$, so the equation becomes

$$\frac{d(G \div T)}{dT} = -\frac{S}{T} - \frac{G}{T^2} \qquad (4.66)$$

Recalling the now-familiar relationship $G = H - TS$, we may substitute for the $-S$ term by saying

$$-S = \frac{G - H}{T} \qquad (4.67)$$

$$\frac{d(G \div T)}{dT} = \left(\frac{G - H}{T}\right) \times \frac{1}{T} - \frac{G}{T^2} \qquad (4.68)$$

The term in brackets on the right-hand side is then split up; so

$$\frac{d(G \div T)}{dT} = \frac{G}{T^2} - \frac{H}{T^2} - \frac{G}{T^2} \qquad (4.69)$$

On the right-hand side, the first and third terms cancel, yielding

$$\frac{d(G \div T)}{dT} = -\frac{H}{T^2} \qquad (4.70)$$

Writing the equation in this way tells us that if we know the enthalpy of the system, we also know the temperature dependence of $G \div T$. Separating the variables and defining G_1 as the Gibbs function change at T_1 and similarly as the value of G_2 at T_2, yields

$$\int_{G_1/T_1}^{G_2/T_2} d(G/T) = H \int_{T_1}^{T_2} -\frac{1}{T^2} \, dT \qquad (4.71)$$

This derivation assumes that both H and S are temperature invariant – a safe assumption if the variation between T_1 and T_2 is small (say, 40 K or less).

so

$$\left[\frac{G}{T}\right]_{G_1/T_1}^{G_2/T_2} = H\left[\frac{1}{T}\right]_{T_1}^{T_2} \tag{4.72}$$

So, for a single chemical:

$$\frac{G_2}{T_2} - \frac{G_1}{T_1} = H\left(\frac{1}{T_2} - \frac{1}{T_1}\right) \tag{4.73}$$

And, for a chemical reaction we have Equation (4.62):

$$\frac{\Delta G_2}{T_2} - \frac{\Delta G_1}{T_1} = \Delta H\left(\frac{1}{T_2} - \frac{1}{T_1}\right)$$

We call this final equation the *Gibbs–Helmholtz* equation.

At what temperature will the egg start to denature?

Reactions 'poised' at the critical temperature

Care: the nomenclature $T_{(critical)}$ is employed in many other areas of physical chemistry (e.g. see pp. 50 and 189).

If $\Delta G^{\ominus}$ goes from positive to negative as the temperature alters, then clearly the value of $\Delta G^{\ominus}$ will transiently be zero at one unique temperature. At this 'point of reaction spontaneity', the value of $\Delta G^{\ominus} = 0$. We often call this the 'critical temperature' $T_{(critical)}$.

The value of $T_{(critical)}$, i.e. the temperature when the reaction first becomes thermodynamically feasible, can be determined approximately from

The reaction is 'poised' at the *critical temperature* with $\Delta G = 0$.

$$T_{(critical)} = \frac{\Delta H}{\Delta S} \tag{4.74}$$

Worked Example 4.13   At what temperature is the denaturation of egg albumen 'poised'?

We will employ the thermodynamic data from Worked Example 4.12. Inserting values into Equation (4.74):

This method yields only an *approximate* value of $T_{(critical)}$ because ΔS and ΔH are themselves functions of temperature.

$$T_{(critical)} = \frac{\Delta H}{\Delta S} = \frac{35\,000 \text{ J mol}^{-1}}{98.3 \text{ J K}^{-1} \text{ mol}^{-1}} = 356 \text{ K or } 83\,^{\circ}\text{C}$$

We deduce that an egg will start denaturing above about $T = 83\,^{\circ}\text{C}$, confirming what every cook knows, that an egg cooks in boiling water but not in water that is merely 'hot'.

SAQ 4.11   Water and carbon do not react at room temperature. Above what temperature is it feasible to prepare *synthesis gas* (a mixture of CO and H_2)? The reaction is

$$C_{(s)} + 2H_2O_{(g)} \longrightarrow CO_{(g)} + H_{2(g)} \qquad (4.75)$$

Take $\Delta H^\ominus = 132 \text{ kJ mol}^{-1}$ and $\Delta S^\ominus = 134 \text{ J K}^{-1} \text{ mol}^{-1}$.

Justification Box 4.9

We start with Equation (4.21):

$$\Delta G^\ominus = 0 = \Delta H - T\Delta S$$

When just 'poised', the value of $\Delta G^\ominus$ is equal to zero. Accordingly, $0 = \Delta H = T\Delta S$. Rearranging slightly, we obtain

$$\Delta H = T\Delta S \qquad (4.76)$$

which, after dividing both sides by 'ΔS', yields Equation (4.74).

Why does recrystallization work?

The effect of temperature on K: the van't Hoff isochore

To purify a freshly prepared sample, the preparative chemist will crystallize then recrystallize the compound until convinced it is pure. To recrystallize, we first dissolve the compound in hot solvent. The solubility s of the compound depends on the temperature T. The value of s is high at high temperature, but it decreases at lower temperatures until the solubility limit is first reached and then surpassed, and solute precipitates from solution (hopefully) to yield crystals.

The solubility s relates to a special equilibrium constant we call the 'solubility product' K_s, defined by

$$K_s = [\text{solute}]_{(\text{solution})} \qquad (4.77)$$

> We say the value of $[\text{solute}]_{(s)} = 1$ because its *activity* is unity; see Section 7.3.

The [solute] term may, in fact, comprise several component parts if the solute is ionic, or precipitation involves agglomeration. This equilibrium constant is not written as a fraction because the 'effective concentration' of the undissolved solute $[\text{solute}]_{(s)}$ can be taken to be unity.

Like all equilibrium constants, the magnitude of the equilibrium constant K_s depends quite strongly on temperature, according to

> The word 'isochore' implies constant pressure, since *iso* is Greek for 'same' and the root *chore* means pressure.

the *van't Hoff isochore*:

$$\ln\left(\frac{K_{s(2)}}{K_{s(1)}}\right) = \frac{-\Delta H_{(cryst)}^{\ominus}}{R}\left(\frac{1}{T_2} - \frac{1}{T_1}\right) \tag{4.78}$$

where R is the gas constant, $\Delta H_{(cryst)}^{\ominus}$ is the change in enthalpy associated with crystallization, and the two temperatures are expressed in kelvin, i.e. *thermodynamic* temperatures.

Worked Example 4.14   The solubility s of potassium nitrate is 140 g per 100 g of water at 70.9 °C, which decreases to 63.6 g per 100 g of water at 39.9 °C. Calculate the enthalpy of crystallization, $\Delta H_{(cryst)}^{\ominus}$.

Strategy. For convenience, we will call the higher temperature T_2 and the lower temperature T_1. (1) The van't Hoff isochore, Equation (4.78), is written in terms of a *ratio*, so we do not need the absolute values. In other words, in this example, we can employ the solubilities s without further manipulation. We can dispense with the units of s for the same reason. (2) We convert the two temperatures to kelvin, for the van't Hoff isochore requires thermodynamic temperatures, so $T_2 = 343.9$ K and $T_1 = 312.0$ K. (3) We insert values into the van't Hoff isochore (Equation (4.78)):

> Note how the two minus signs on the right will cancel.

$$\ln\left(\frac{140}{63.6}\right) = -\frac{-\Delta H_{(cryst)}^{\ominus}}{8.314 \text{ J K}^{-1}\text{ mol}^{-1}} \times \left(\frac{1}{343.9 \text{ K}} - \frac{1}{312.0 \text{ K}}\right)$$

$$\ln(2.20) = \frac{-\Delta H_{(cryst)}^{\ominus}}{8.314 \text{ J K}^{-1}\text{ mol}^{-1}} \times (-2.973 \times 10^{-4} \text{ K}^{-1})$$

$$\ln 2.20 = 0.7889$$

We then divide both sides by '2.973×10^{-4} K^{-1}', so:

$$\frac{0.7889}{2.973 \times 10^{-4} \text{ K}^{-1}} = \frac{\Delta H_{(cryst)}^{\ominus}}{8.314 \text{ J K}^{-1}\text{ mol}^{-1}}$$

> Only when the difference between T_2 and T_1 is less than ca. 40 K can we assume the reaction enthalpy $\Delta H^{\ominus}$ is independent of temperature. We otherwise correct for the temperature dependence of $\Delta H^{\ominus}$ with the *Kirchhoff equation* (Equation (3.19)).

The term on the left equals 2.654×10^3 K. Multiplying both sides by R then yields:

$$\Delta H_{(cryst)}^{\ominus} = 2.654 \times 10^3 \text{ K} \times 8.314 \text{ J K}^{-1}\text{ mol}^{-1}$$

so

$$\Delta H_{(cryst)}^{\ominus} = 22.1 \text{ kJ mol}^{-1}$$

SAQ 4.12   The simple aldehyde ethanal (**VII**) reacts with the di-alcohol ethylene glycol (**VIII**) to form a cyclic acetal (**IX**):

Calculate the enthalpy change $\Delta H^{\ominus}$ for the reaction if the equilibrium constant for the reaction halves when the temperature is raised from 300 K to 340 K.

Justification Box 4.10

We start with the Gibbs–Helmholtz equation (Equation (4.62)):

$$\frac{\Delta G_2^{\ominus}}{T_2} - \frac{\Delta G_1^{\ominus}}{T_1} = \Delta H^{\ominus}\left(\frac{1}{T_2} - \frac{1}{T_1}\right)$$

Each value of $\Delta G^{\ominus}$ can be converted to an equilibrium constant via the van't Hoff isotherm $\Delta G^{\ominus} = -RT \ln K$, Equation (4.55). We say, $\Delta G_1^{\ominus} = -RT_1 \ln K_1$ at T_1; and $\Delta G_2^{\ominus} = -RT_2 \ln K_2$ at T_2.

Substituting for each value of $\Delta G^{\ominus}$ yields

$$\frac{-RT_2 \ln K_2}{T_2} - \frac{-RT_1 \ln K_1}{T_1} = \Delta H^{\ominus}\left(\frac{1}{T_2} - \frac{1}{T_1}\right) \qquad (4.79)$$

which, after cancelling the T_1 and T_2 terms on the right-hand side, simplifies to

$$(-R \ln K_2) - (-R \ln K_1) = \Delta H^{\ominus}\left(\frac{1}{T_2} - \frac{1}{T_1}\right) \qquad (4.80)$$

Next, we divide throughout by '$-R$' to yield

$$\ln K_2 - \ln K_1 = -\frac{\Delta H^{\ominus}}{R}\left(\frac{1}{T_2} - \frac{1}{T_1}\right) \qquad (4.81)$$

and, by use of the laws of logarithms, $\ln a - \ln b = \ln(a \div b)$, so the left-hand side of the equation may be grouped, to generate the *van't Hoff isochore*. Note: it is common (but incorrect) to see $\Delta H^{\ominus}$ written without its plimsoll sign '$\ominus$'.

The isochore, Equation (4.81), was derived from the *integrated* form of the Gibbs–Helmholtz equation. It is readily shown that the van't Hoff isochore can be rewritten in a slightly different form, as:

> We often talk about 'the isochore' when we mean the 'van't Hoff isochore'.

$$\ln K = -\frac{\Delta H^{\ominus}}{R} \times \frac{1}{T} + \text{constant}$$

$$\quad\quad y \quad\quad\quad m \quad\quad\quad x \quad\quad\quad c$$

(4.82)

which is known as either the *linear* or *graphical* form of the equation. By analogy with the equation for a straight line ($y = mx + c$), a plot of $\ln K$ (as y) against $1 \div T$ (as x) should be linear, with a gradient of $-\Delta H^{\ominus}/R$.

Worked Example 4.15   The isomerization of 1-butene (**X**) to form *trans*-2-butene (**XI**). The equilibrium constants of reaction are given below. Determine the enthalpy of reaction $\Delta H^{\ominus}$ using a suitable graphical method.

(X)   **(XI)**

(4.83)

T/K	686	702	733	779	826
K	1.72	1.63	1.49	1.36	1.20

Strategy. To obtain $\Delta H^{\ominus}$: (1) we plot a graph of $\ln K$ (as y) against values of $1 \div T$ (as x) according to Equation (4.82); (2) we determine the gradient; and (3) multiply the gradient by $-R$.

(1) The graph is depicted in Figure 4.8. (2) The best gradient is 1415 K. (3) $\Delta H^{\ominus} =$ 'gradient $\times -R$', so

$$\Delta H^{\ominus} = 1415 \text{ K} \times (-8.314 \text{ J K}^{-1} \text{ mol}^{-1})$$

$$\Delta H^{\ominus} = -11.8 \text{ kJ mol}^{-1}$$

SAQ 4.13   The following data refer to the chemical reaction between ethanoic acid and glucose. Obtain $\Delta H^{\ominus}$ from the data using a suitable graphical method. (Hint: remember to convert all temperatures to kelvin.)

$T/^{\circ}$C	K
25	21.2×10^4
36	15.1×10^4
45	8.56×10^4
55	5.46×10^4

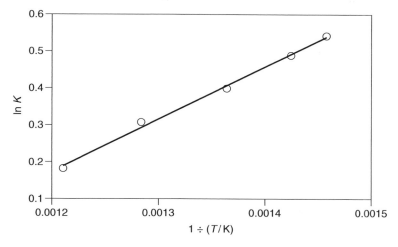

Figure 4.8   The equilibrium constant K for the isomerization of 1-butene depends on the temperature: van't Hoff isochore plot of $\ln K$ (as y) against $1 \div T$ (as x) from which a value of $\Delta H^{\ominus}$ can be calculated as 'gradient $\times -R$'

Aside

Writing table headers

Each term in the table for SAQ 4.13 has been multiplied by 10^4, which is repetitious, takes up extra space, and makes the table look messy and cumbersome.

Most of the time, we try to avoid writing tables in this way, by incorporating the common factor of '$\times 10^4$' into the header. We accomplish this by making use of the *quantity calculus* concept (see p. 13). Consider the second value of K. The table says that, at $36\,^{\circ}$C, $K = 15.1 \times 10^4$. If we divide both sides of this little equation by 10^4, we obtain, $K \div 10^4 = 15.1$. This equation is completely correct, but is more usually written as $10^{-4} K = 15.1$.

According, we might rewrite the first two lines of the table as:

$T/^{\circ}$C	$10^{-4} K$
25	21.2

The style of this latter version is wholly correct and probably more popular than the style we started with. Don't be fooled: a common mistake is to look at the table heading and say, 'we need to multiply K by 10^{-4}'. It has been already!

5

Phase equilibria

Introduction

Phase equilibrium describes the way phases (such as solid, liquid and/or gas) co-exist at some temperatures and pressure, but interchange at others.

First, the criteria for phase equilibria are discussed in terms of single-component systems. Then, when the ground rules are in place, multi-component systems are discussed in terms of partition, distillation and mixing.

The chapter also outlines the criteria for equilibrium in terms of the Gibbs function and chemical potential, together with the criteria for spontaneity.

5.1 Energetic introduction to phase equilibria

Why does an ice cube melt in the mouth?

Introduction to phase equilibria

The temperature of the mouth is about $37\,^\circ$C, so an overly simple explanation of why ice melts in the mouth is to say that the mouth is warmer than the transition temperature $T_{(melt)}$. And, being warmer, the mouth supplies energy to the immobilized water molecules, thereby allowing them to break free from those bonds that hold them rigid. In this process, solid H_2O turns to liquid H_2O – the ice melts.

Incidentally, this argument also explains why the mouth feels cold after the ice has melted, since the energy necessary to melt the ice comes entirely *from* the mouth. In consequence, the mouth has less energy after the melting than before; this statement is wholly in accord with the zeroth law of thermodynamics, since heat energy travels from the hot mouth to the cold ice. Furthermore, if the mouth is considered as an *adiabatic* chamber (see p. 89), then the only way for the energy to be found for melting is for the temperature of the mouth to fall.

Further thermodynamic background: terminology

In the thermodynamic sense, a *phase* is defined as part of a chemical system in which all the material has the same composition and state. Appropriately, the word comes from the Greek *phasis*, meaning 'appearance'. Ice, water and steam are the three simple phases of H_2O. Indeed, for almost all matter, the three simple phases are solid, liquid and gas, although we must note that there may be many different solid phases possible since $H_2O_{(s)}$ can adopt several different crystallographic forms. As a related example, the two stable phases of solid sulphur are its monoclinic and orthorhombic crystal forms.

> A *phase* is a component within a system, existing in a precisely defined physical state, e.g. gas, liquid, or a solid that has a single crystallographic form.

Ice is a solid form of water, and is its only stable form below $0\,°C$. The liquid form of H_2O is the only stable form in the temperature range $0 < T < 100\,°C$. Above $100\,°C$, the normal, stable phase is gaseous water, 'steam'. Water's *normal* melting temperature $T_{(melt)}$ is $0\,°C$ (273.15 K). The word 'normal' in this context implies 'at standard pressure $p^{\ominus}$'. The pressure $p^{\ominus}$ has a value of 10^5 Pa. This temperature $T_{(melt)}$ is often called the melting *point* because water and ice coexist indefinitely at this temperature and pressure, but at no other temperature can they coexist. We say they reside together at *equilibrium*.

> Concerning transitions between the two phases '1' and '2', Hess's Law states that $\Delta H_{(1 \to 2)} = -1 \times \Delta H_{(2 \to 1)}$.

To melt the ice, an amount of energy equal to $\Delta H^{\ominus}_{(melt)}$ must be added to overcome those forces that promote the water adopting a solid-state structure. Such forces will include hydrogen bonds. Re-cooling the melted water to re-solidify it back to ice involves the same amount of energy, but this time energy is *liberated*, so $\Delta H^{\ominus}_{(melt)} = -\Delta H^{\ominus}_{(freeze)}$. The freezing process is often called *fusion*. (Strictly, we ought to define the energy by saying that no pressure–volume work is performed during the melting and freezing processes, and that the melting and freezing processes occur without any changes in temperature.)

Table 5.1 gives a few everyday examples of phase changes, together with some useful vocabulary.

Two or more phases can coexist indefinitely provided that we maintain certain conditions of temperature T and pressure p. The normal boiling temperature of water is $100\,°C$, because this is the only temperature (at $p = p^{\ominus}$) at which both liquid and

Table 5.1   Summary of terms used to describe phase changes

Phase transition	Name of transition	Everyday examples
Solid $\to$ gas	Sublimation	'Smoke' formed from dry ice
Solid $\to$ liquid	Melting	Melting of snow or ice
Liquid $\to$ gas	Boiling or vaporization	Steam formed by a kettle
Liquid $\to$ solid	Freezing, solidification or fusion	Ice cubes formed in a fridge; hail
Gas $\to$ liquid	Condensation or liquification	Formation of dew or rain
Gas $\to$ solid	Condensation	Formation of frost

gaseous H_2O coexist at *equilibrium*. Note that this equilibrium is *dynamic*, because as liquid is converted to gas an equal amount of gas is also converted back to liquid.

However, the values of pressure and temperature at equilibrium depend on each other; so, if we change the pressure, then the temperature of equilibrium shifts accordingly (as discussed further in Section 5.2). If we plotted all the experimental values of pressure and temperature at which equilibrium exists, to see the way they affect the equilibrium changes, then we obtain a graph called a *phase diagram*, which looks something like the schematic graph in Figure 5.1.

> A *phase diagram* is a graph showing values of applied pressure and temperature at which equilibrium exists.

We call each solid line in this graph a *phase boundary*. If the values of p and T lie on a phase boundary, then equilibrium between two phases is guaranteed. There are three common phase boundaries: liquid–solid, liquid–gas and solid–gas. The line separating the regions labelled 'solid' and 'liquid', for example, represents values of pressure and temperature at which these two phases coexist – a line sometimes called the 'melting-point phase boundary'.

> A *phase boundary* is a line on a phase diagram representing values of applied pressure and temperature at which equilibrium exists.

The point where the three lines join is called the *triple point*, because *three* phases coexist at this single value of p and T. The triple point for water occurs at $T = 273.16$ K (i.e. at $0.01\,°C$) and $p = 610$ Pa ($0.006p^{\ominus}$). We will discuss the *critical point* later.

Only a single phase is stable if the applied pressure and temperature do not lie on a phase boundary, i.e. in one of the areas *between* the phase boundaries. For example, common sense tells is that on a warm and sunny summer's day, and at normal pressure, the only stable phase of H_2O is liquid water. These conditions of p and T are indicated on the figure as point '**D**'.

> The *triple point* on a phase diagram represents the value of pressure and temperature at which *three* phases coexist at equilibrium.

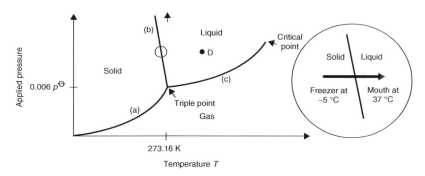

Figure 5.1 Schematic phase diagram showing pressures and temperatures at which two phases are at equilibrium. Phase boundary (a) represents the equilibrium between steam and ice; boundary (b) represents equilibrium between water and ice; and boundary (c) represents equilibrium between water and steam. The point **D** represents p and T on a warm, sunny day. *Inset*: warming an ice cube from $-5\,°C$ to the mouth at $37\,°C$ at constant pressure causes the stable phase to convert from solid to liquid. The phase change occurs at $0\,°C$ at $p^{\ominus}$

When labelling a phase diagram, recall how the only stable phase at high pressure and low temperature is a solid; a gas is most stable at low pressure and high temperature. The phase within the crook of the 'Y' is therefore a liquid.

We can predict whether an ice cube will melt just by looking carefully at the phase diagram. As an example, suppose we take an ice cube from a freezer at $-5\,°C$ and put it straightaway in our mouth at a temperature of $37\,°C$ (see the inset to Figure 5.1). The temperature of the ice cube is initially cooler than that of the mouth. The ice cube, therefore, will warm up as a consequence of the zeroth law of thermodynamics (see p. 8) until it reaches the temperature of the mouth. Only then will it attain equilibrium. But, as the temperature of the ice cube rises, it crosses the phase boundary, as represented by the bold horizontal arrow, and undergoes a *phase transition* from solid to liquid.

We know from Hess's law (see p. 98) that it is often useful to consider (mentally) a physical or chemical change by dissecting it into its component parts. Accordingly, we will consider the melting of the ice cube as comprising two processes: warming from $-5\,°C$ to $37\,°C$, and subsequent melting at $37\,°C$. During warming, the water crosses the phase boundary, implying that it changes from being a *stable* solid (when below $0\,°C$) to being an *unstable* solid (above $0\,°C$). Having reached the temperature of the mouth at $37\,°C$, the solid ice converts to its stable phase (water) in order to regain stability, i.e. the ice cube melts in the mouth. (It would be more realistic to consider three processes: warming to $0\,°C$, melting at constant temperature, then warming from 0 to $37\,°C$.)

The Greek root *meta* means 'adjacent to' or 'near to'. Something *metastable* is almost stable ... but not quite.

Although the situation with melting in two stages appears a little artificial, we ought to remind ourselves that the phase diagram is made up of *thermodynamic* data alone. In other words, it is possible to see liquid water at $105\,°C$, but it would be a *metastable* phase, i.e. it would not last long!

Aside

The arguments in this example are somewhat simplified.

Remember that the phase diagram's y-axis is the *applied* pressure. At room temperature and pressure, liquid water evaporates as a consequence of entropy (e.g. see p. 134). For this reason, both liquid and vapour are apparent even at s.t.p. The pressure of the vapour is known as the *saturated vapour pressure* (s.v.p.), and can be quite high.

The s.v.p. is not an applied pressure, so its magnitude is generally quite low. The s.v.p. of water will certainly be lower than atmospheric pressure. The s.v.p. increases with temperature until, at the boiling temperature, it equals the atmospheric pressure. One definition of boiling says that the s.v.p. equals the applied pressure.

The arguments in this section ignore the saturated vapour pressure.

Why does water placed in a freezer become ice?

Spontaneity of phase changes

It will be useful to concentrate on the diagram in Figure 5.2 when considering why a 'phase change' occurs *spontaneously*. We recall from Chapter 4 that one of the simplest tests of whether a thermodynamic event can occur is to ascertain whether the value of ΔG is negative (in which case the change is indeed spontaneous) or positive (when the change is not spontaneous).

The graph in Figure 5.2 shows the molar Gibbs function G_m as a function of temperature. (G_m decreases with temperature because of increasing entropy.) The value of G_m for ice follows the line on the left-hand side of the graph; the line in the centre of the graph gives values of G_m for liquid water; and the line on the right represents G_m for gaseous water, i.e. steam. We now consider the process of an ice cube being warmed from below $T_{(melt)}$ to above it. The molar Gibbs functions of water and ice become comparable when the temperature reaches $T_{(melt)}$. At $T_{(melt)}$ itself, the two values of G_m are the same – which is one definition of *equilibrium*. The two values diverge once more above $T_{(melt)}$.

Below $T_{(melt)}$, the two values of G_m are different, implying that the two forms of water are energetically different. It should be clear that if one energy is lower than the other, then the lower energy form is the stablest; in this case, the liquid water has a higher value of G_m and is less stable than solid ice (see the heavy vertical arrow, inset to Figure 5.2). Liquid water, therefore, is energetically unfavourable, and for that reason it is unstable. To attain stability, the liquid water must release energy and, in the process, undergo a phase change from liquid to solid, i.e. it freezes.

> Remember how the symbol Δ means 'final state minus initial state', so $\Delta G_m =$ $G_{m\ (final\ state)} -$ $G_{m\ (initial\ state)}$.

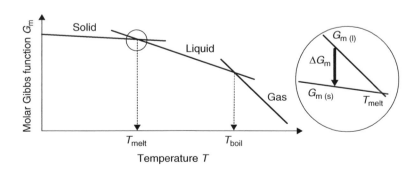

Figure 5.2   Graph of molar Gibbs function G_m as a function of temperature. *Inset*: at temperatures below $T_{(melt)}$ the phase transition from liquid to solid involves a negative change in Gibbs function, so it is spontaneous

These arguments represent a simple example of *phase equilibria*. This branch of thermodynamics tells us about the *direction* of change, but says nothing about the *rate* at which such changes occur.

It should be clear from the graph in Figure 5.2 that ΔG_m is negative (as required for a *spontaneous* change) only if the final state is *solid* ice and the initial state is *liquid* water. This sign of ΔG_m is all that is needed to explain why liquid water freezes at temperatures below $T_{(melt)}$.

Conversely, if an ice cube is warmed beyond $T_{(melt)}$ to the temperature of the mouth at $37\,°C$, now it is the *solid* water that has excess energy; to stabilize it relative to liquid water at $37\,°C$ requires a different phase change to occur, this time *from* ice *to* liquid water. This argument again relies on the relative magnitudes of the molar Gibbs function, so ΔG_m is only negative at this higher temperature if the final state is liquid and the initial state is solid.

Why was Napoleon's Russian campaign such a disaster?

Solid-state phase transitions

A large number of French soldiers froze to death in the winter of 1812 within a matter of weeks of their emperor Napoleon Bonaparte leading them into Russia. The loss of manpower was one of the principal reasons why Napoleon withdrew from the outskirts of Moscow, and hence lost his Russian campaign.

But why was so ruthless a general and so obsessively careful a tactician as Napoleon foolhardy enough to lead an unprepared army into the frozen wastes of Russia? In fact, he thought he *was* prepared, and his troops were originally well clothed with thick winter coats. The only problem was that, so the story goes, he chose at the last moment to replace the brass of the soldiers' buttons with tin, to save money.

Metallic tin has many allotropic forms: rhombic white tin (also called β-tin) is stable at temperatures above $13\,°C$, whereas the stable form at lower temperatures is cubic grey tin (also called α-tin). A transition such as $tin_{(white)} \rightarrow tin_{(grey)}$ is called a *solid-state phase transition*.

Figure 5.3 shows the phase diagram of tin, and clearly shows the transition from $tin_{(white)}$ to $tin_{(grey)}$. Unfortunately, the tin allotropes have very different densities ρ, so $\rho_{(tin,\ grey)} = 5.8\ \mathrm{g\,cm^{-3}}$ but $\rho_{(tin,\ white)} = 7.3\ \mathrm{g\,cm^{-3}}$. The difference in ρ during the transition from white to grey tin causes such an unbearable mechanical stress that the metal often cracks and turns to dust – a phenomenon sometimes called 'tin disease' or 'tin pest'.

The transition from white tin to grey was first noted in Europe during the Middle Ages, e.g. as the pipes of cathedral organs disintegrated, but the process was thought to be the work of the devil.

The air temperature when Napoleon entered Russia was apparently as low as $-35\,°C$, so the soldiers' tin buttons converted from white to grey tin and, concurrently, disintegrated into powder. So, if this story is true, then Napoleon's troops froze to death because they lacked effective coat fastenings. Other common metals, such

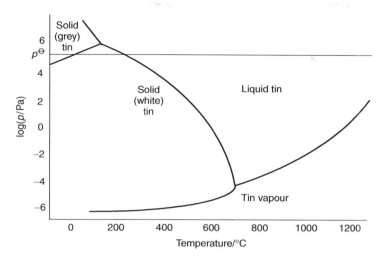

Figure 5.3 Phase diagram of tin computed from thermodynamic data, showing the transition from grey tin from white tin at temperatures below 13 °C. Note the logarithmic y-axis. At $p^{\ominus}$, $T_{(white \rightarrow grey)} = 13\,°C$, and $T_{(melt)} = 231.9\,°C$. (Figure constructed from data published in *Tin and its Alloys and Compounds*, B. T. K. Barry and C. J. Thwaits, Ellis Horwood, Chichester, 1983)

as copper or zinc, and alloys such as brass, do not undergo phase changes of this sort, implying that the troops could have survived but for Napoleon's last-minute change of button material.

The kinetics of phase changes

Like all spontaneous changes, the rate at which the two forms of tin interconvert is a function of temperature. Napoleon's troops would have survived if they had entered Russia in the summer or autumn, when the air temperature is similar to the phase-transition temperature. The rate of conversion would have been slower in the autumn, even if the air temperature had been slightly less than $T_{(transition)}$ – after all, the tin coating of a can of beans does not disintegrate while sitting in a cool cupboard! The conversion is only rapid enough to noticeably destroy the integrity of the buttons when the air temperature is much lower than $T_{(transition)}$, i.e. when the difference between $T_{(air)}$ and $T_{(transition)}$ is large.

Phase changes involving liquids and gases are generally fast, owing to the high mobility of the molecules. Conversely, while phase changes such as $tin_{(white)} \rightarrow tin_{(grey)}$ can and do occur in the solid state, the reaction is usually very much slower because it must occur wholly in the solid state, often causing any thermodynamic instabilities to remain 'locked in'; as an example, it is clear from the phase diagram of carbon in Figure 5.4 that graphite is the stable form of carbon (cf. p. 109), yet the phase change $carbon_{(diamond)} \rightarrow carbon_{(graphite)}$ is so slow that a significant extent of conversion requires millions of years.

We consider chemical kinetics further in Chapter 8.

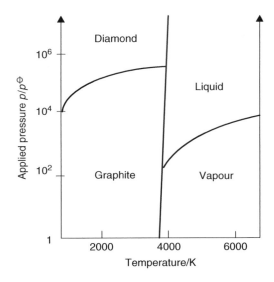

Figure 5.4   The phase diagram of carbon showing the two solid-state extremes of diamond and graphite. Graphite is the thermodynamically stable form of carbon at room temperature and pressure, but the rate of the transition $C_{(diamond)} \rightarrow C_{(graphite)}$ is virtually infinitesimal

5.2   Pressure and temperature changes with a single-component system: qualitative discussion

How is the 'Smoke' in horror films made?

Effect of temperature on a phase change: sublimation

Horror films commonly show scenes depicting smoke or fog billowing about the screen during the 'spooky' bits. Similarly, smoke is also popular during pop concerts, perhaps to distract the fans from something occurring on or off stage. In both cases, it is the adding of dry ice to water that produces the 'smoke'.

> *Dry ice* is solid carbon dioxide.

Dry ice is carbon dioxide (CO_2) in its solid phase. We call it 'dry' because it is wholly liquid-free at $p^{\ominus}$: such solid CO_2 looks similar to normal ice (solid water), but it 'melts' without leaving a puddle. We say it *sublimes*, i.e. undergoes a phase change involving direct conversion from solid to gas, without liquid forming as an intermediate phase. $CO_{2(l)}$ can only be formed at extreme pressures.

Solid CO_2 is slightly denser than water, so it sinks when placed in a bucket of water. The water is likely to have a temperature of $20\,°C$ or so at room temperature, while typically the dry ice has a maximum temperature of ca $-78\,°C$ (195 K). The stable phase at the temperature of the water is therefore gaseous CO_2. We should understand that the $CO_{2(s)}$ is thermodynamically unstable, causing the phase transition $CO_{2(s)} \rightarrow CO_{2(g)}$ on immersion in the water.

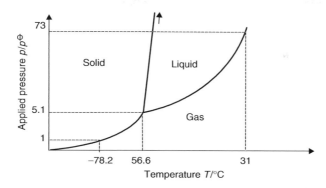

Figure 5.5   Phase diagram of a system that sublimes at room temperature: phase diagram of carbon dioxide. (Note that the y-axis here is logarithmic)

Incidentally, the water in the bucket is essential for generating the effect of theatrical 'smoke' because the large volumes of $CO_{2(g)}$ entrap minute particles of water (which forms a *colloid*; see Chapter 10.2). This colloidal water is visible because it creates the same atmospheric condition known as *fog*, which is opaque.

Look at the phase diagram of CO_2 in Figure 5.5, which is clearly similar in general form to the schematic phase diagram in Figure 5.1. A closer inspection shows that some features are different. Firstly, notice that the phase boundary between solid and liquid now has a positive gradient; in fact, water is almost unique in having a negative gradient for this line (see Section 5.1). Secondly, the conditions of room temperature ($T = 298$ K and $p = p^{\ominus}$) relate to conditions of the solid–gas phase boundary rather than the liquid–gas phase boundary.

By drawing a horizontal line across the figure at $p = p^{\ominus}$, we see how the line cuts the solid–gas phase boundary at $-78.2\,°C$. Below this temperature, the stable form of CO_2 is solid dry ice, and $CO_{2(g)}$ is the stable form above it. Liquid CO_2 is never the stable form at $p^{\ominus}$; in fact, Figure 5.5 shows that $CO_{2(l)}$ will not form at pressures below $5.1 \times p^{\ominus}$. In other words, liquid CO_2 is *never* seen naturally on Earth; which explains why dry ice sublimes rather than melts under s.t.p. conditions.

How does freeze-drying work?

Effect of pressure change on a phase change

Packets of instant coffee proudly proclaim that the product has been 'freeze-dried'. In practice, beans of coffee are ground, boiled in water and filtered to remove the depleted grounds. This process yields conventional 'fresh' coffee, as characterized by its usual colour and attractive smell. Finally, water is removed from the coffee solution to prepare granules of 'instant' coffee.

In principle, we could remove the water from the coffee by just boiling it off, to leave a solid residue as a form of 'instant coffee'. In fact, some early varieties of instant coffee were made in just this way, but the flavour was generally unpleasant as a result of charring during prolonged heating. Clearly, a better method of removing the water was required.

We now look at the phase diagram of water in Figure 5.6, which will help us follow the modern method of removing the water from coffee to yield anhydrous granules. A low temperature is desirable to avoid charring the coffee. Water vapour can be removed from the coffee solution at any temperature, because liquids are always surrounded by their respective vapour. The pressure of the vapour is the saturated vapour pressure, s.v.p. The water is removed faster when the applied pressure decreases. Again, a higher temperature increases the rate at which the vapour is removed. The fastest possible rate occurs when the solution boils at a temperature we call $T_{(boil)}$.

> Freeze-drying is a layman's description, and acknowledges that external conditions may alter the conditions of a phase change, i.e. the drying process (removal of water) occurs at a temperature lower than 100 °C.

Figure 5.6 shows the way in which the boiling temperature alters, with boiling becoming easier as the applied pressure decreases or the temperature increases, and suggests that the coffee solution will boil at a lower temperature when warmed in a partial vacuum. At a pressure of about $\frac{1}{100} \times p^{\ominus}$, water is removed from the coffee by warming it to temperatures of about 30 °C, when it boils. We see that the coffee is dried and yet is never subjected to a high temperature for long periods of time.

It is clear that decreasing the external pressure makes boiling easier. It is quite possible to remove the water from coffee at or near its freezing temperature – which explains the original name of freeze-drying.

In many laboratories, a *nomograph* (see Figure 5.7) is pinned to the wall behind a rotary evaporator. A nomograph allows for a simple *estimate* of the boiling temperature as a function of pressure. Typically, pressure is expressed in the old-fashioned units of atmospheres (atm) or millimetres of mercury (mmHg). 1 atm = 760 mmHg. (The curvature of the nomograph is a consequence of the mathematical nature of the way pressure and temperature are related; see Section 5.2).

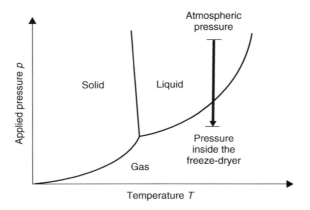

Figure 5.6   Freeze-drying works by decreasing the pressure, and causing a phase change; at higher pressure, the stable form of water is liquid, but the stable form at lower pressures is vapour. Consequently, water (as vapour) leaves a sample when placed in a vacuum or low-pressure chamber: we say the sample is 'freeze-dried'

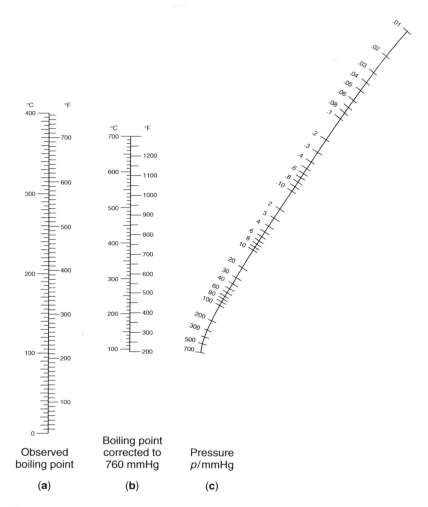

Figure 5.7   A typical nomograph for estimating the temperature at which a pure liquid boils when the pressure is decreased

   This is how a boiling temperature at reduced pressure is estimated with a nomograph: place a straight ruler against the applied pressure as indicated on the curved right-hand scale (**c**). The ruler must also pass through the 'normal' boiling temperature on the middle scale (**b**). The reduced-pressure boiling temperature is then read off the left-hand scale (**a**). As an example, if the normal boiling temperature is 200 °C, then the reduced boiling temperature may be halved to 100 °C if the applied pressure is approximately 20 mmHg.

SAQ 5.1   A liquid has a normal boiling temperature of 140 °C. Use the nomograph to *estimate* the applied pressure needed to decrease the boiling temperature to 90 °C.

How does a rotary evaporator work?

Thermodynamics of phase changes

Rotary evaporators are a common feature in most undergraduate laboratories. Their primary purpose is to remove solvent following a reflux, perhaps before crystallization of a reaction product.

To operate the evaporator, we place the reaction solution in a round-bottomed flask while the pressure inside the evaporator is decreased to about $\frac{1}{30} \times p^{\ominus}$. The flask is then rotated. The solvent evaporates more easily at this low pressure than at $p^{\ominus}$. The solvent removed under vacuum is trapped by a condenser and collected for easy re-use, or disposal in an environmentally sensitive way.

But molecules need energy if they are to leave the solution during boiling. The energy comes from the solution. The temperature of the solution would decrease rapidly if no external supply of energy was available, as a reflection of its depleted energy content (see p. 33). In fact, the solution would freeze during evaporation, so the rotating bulb is typically immersed in a bath of warm water.

> Strictly, the term s.v.p. applies to *pure* liquids. By using the term s.v.p., we are implying that all other components are wholly involatile, and the s.v.p. relates only to the solvent.

An atmosphere of vapour always resides above a liquid, whether the liquid is pure, part of a mixture, or has solute dissolved within it. We saw on p. 180 how the pressure of this gaseous phase is called its saturation vapour pressure, s.v.p. The s.v.p. increases with increased temperature until, at the boiling point $T_{(boil)}$, it equals the external pressure above the liquid. Evaporation occurs at temperatures below $T_{(boil)}$, and only above this temperature will the s.v.p. exceed $p^{\ominus}$. The applied pressure in a rotary evaporator is less than $p^{\ominus}$, so the s.v.p. of the solvent can exceed the applied pressure (and allow the liquid to boil) at pressures lower than $p^{\ominus}$.

> *Normal* in the context of phase equilibria means 'performed at a pressure of 1 bar, $p^{\ominus}$'.

We see this phenomenon in a different way when we look back at the phase diagram in Figure 5.6. The stable phase is liquid before applying a vacuum. After turning on the water pump, to decrease the applied pressure, the s.v.p. exceeds $p_{(applied)}$, and the solvent boils at a lower pressure. The bold arrow again indicates how a phase change occurs during a depression of the external pressure.

> The rotary evaporator is a simple example of a *vacuum distillation*.

We see how decreasing the pressure causes boiling of the solvent at a lower temperature than at its *normal* boiling temperature, i.e. if the external pressure were $p^{\ominus}$. Such a *vacuum distillation* is desirable for a preparative organic chemist, because a lower boiling temperature decreases the extent to which the compounds degrade.

Coffee, for example, itself does not evaporate even at low pressure, since it is a solid. Solids are generally much less volatile than liquids, owing to the stronger interactions between the particles. In consequence, the vapour pressure of a solid is several orders of magnitude smaller than that above a liquid.

How is coffee decaffeinated?

Critical and supercritical fluids

We continue our theme of 'coffee'. Most coffees contain a large amount of the heterocyclic stimulant *caffeine* (**I**). Some people prefer to decrease the amounts of caffeine they ingest for health reasons, or they simply do not like to consume it at all, and they ask for *decaffeinated* coffee instead.

(I)

The modern method of removing **I** from coffee resembles the operation of a coffee percolator, in which the water-soluble chemicals giving flavour, colour and aroma are leached from the ground-up coffee during constant irrigation with a stream of boiling water.

Figure 5.8 shows such a system: we call it a *Soxhlet* apparatus. Solvent is passed continually through a porous cup holding the ground coffee. The solvent removes the caffeine and trickles through the holes at the bottom of the cup, i.e. as a solution of caffeine. The solvent is then recycled: solvent at the bottom of the flask evaporates to form a gas, which condenses at the top of the column. This pure, clean solvent then irrigates the coffee a second time, and a third time, etc., until all the caffeine has been removed.

Water is a good choice of solvent in a standard kitchen percolator because it removes all the water-soluble components from the coffee – hence the flavour. Clearly, however, a different solvent is required if only the caffeine is to be removed. Such a solvent must be cheap, have a low boiling point to prevent charring of the coffee and, most importantly, should leave no toxic residues. The presence of any residue would be unsatisfactory to a customer, since it would almost certainly leave a taste; and there are also health and safety implications when residues persist.

The preferred solvent is *supercritical* CO_2. The reasons for this choice are many and various. Firstly, the CO_2 is not hot (CO_2 first becomes critical at 31 °C and 73 atm pressure; see Figure 5.5), so no charring of the coffee occurs during decaffeination. Furthermore, at such a low temperature, all the components within the coffee that impart the flavour and aroma remain within the solid coffee – try soaking coffee beans in cold water and see how the water tastes afterwards! Caffeine is removed while retaining a full flavour.

> CO_2 is *super*critical at temperatures and pressures *above* the critical point.

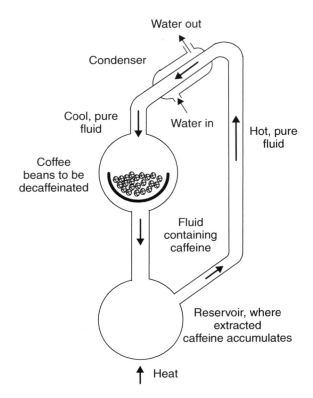

Figure 5.8   Coffee is decaffeinated by constantly irrigating the ground beans with supercritical carbon dioxide: schematic representation of a Soxhlet apparatus for removing caffeine from coffee

Secondly, solid CO_2 is relatively cheap. Finally, after caffeine removal, any occluded CO_2 will vaporize from the coffee without the need to heat it or employ expensive vacuum technology. Again, we retain the volatile essential oils of the coffee. Even if some CO_2 were to persist within the coffee granules, it is chemically inert, has no taste and would be released rapidly as soon as boiling water was added to the solid, decaffeinated coffee.

What is a critical or supercritical fluid?

We look once more at the phase diagram of CO_2 in Figure 5.5. The simplest way of obtaining the data needed to construct such a figure would be to take a sample of CO_2 and determine those temperatures and pressures at which the liquid, solid and gaseous phases coexist *at equilibrium*. (An appropriate apparatus involves a robust container having an observation window to allow us to observe the meniscus.) We then plot these values of p (as 'y') against T (as 'x').

We first looked at critical fluids on p. 50.

Let us consider more closely what happens as the conditions become more extreme inside the observation can. As heating proceeds, so the amount of $CO_{2(l)}$ converting to form gas increases. Accordingly, the amount of CO_2 within the gaseous phase increases, which will cause the density ρ of the vapour to increase. Conversely, if we consider the liquid, at no time does its *density* alter appreciably, even though its volume decreases as a result of liquid forming vapour.

From a consideration of the relative densities, we expect the liquid phase to reside at the bottom of the container, with the less-dense gaseous phase 'floating' above it. The 'critical' point is reached when the density of the gas has increased until it becomes the *same* as that of the liquid. In consequence, there is now no longer a lighter and a heavier phase, because $\rho_{(liquid)} = \rho_{(vapour)}$. Accordingly, we no longer see a meniscus separating liquid at the bottom of the container and vapour above it: it is impossible to see a clear distinction between the liquid and gas components. We say that the CO_2 is *critical*.

Further heating or additional increases in pressure generate *supercritical* CO_2. The pressure and temperature at which the fluid first becomes critical are respectively termed $T_{(critical)}$ and $p_{(critical)}$. Table 5.2 contains a few examples of $T_{(critical)}$ and $p_{(critical)}$.

The inability to distinguish liquid from gaseous CO_2 explains why we describe critical and supercritical systems as *fluids* – they are neither liquid nor gas.

It is impossible to distinguish between the liquid and gaseous phases of CO_2 at and above the critical point, which explains why a phase diagram has no phase boundary at temperatures and pressures above $T_{(critical)}$. The formation of a critical fluid has an unusual corollary: at temperatures above $T_{(critical)}$, we cannot cause the liquid and gaseous phases to separate by decreasing or increasing the pressure alone. The critical temperature, therefore, represents the maximum values of p and T at which liquification

> It is impossible to distinguish between the liquid and gaseous phases of CO_2 at temperatures and pressures at and above the critical point.

> The intensive properties of the liquid and gas (density, heat capacity, etc.) become equal at the *critical point*, which is the highest temperature and pressure at which both the liquid and gaseous phases of a given compound can coexist.

> IUPAC defines *supercritical chromatography* as a separation technique in which the mobile phase is kept above (or relatively close to) its critical temperature and pressure.

Table 5.2 Critical constants $T_{(critical)}$ and $p_{(critical)}$ for some common elements and bi-element compounds

Substance	$T_{(critical)}/K$	$p_{(critical)}/p^{\ominus}$
H_2	33.2	12.97
He	5.3	2.29
O_2	154.3	50.4
Cl_2	417	77.1
CO_2	304.16	73.9
SO_2	430	78.7
H_2O	647.1	220.6
NH_3	405.5	113.0

of the gas is possible. We say that there cannot be any $CO_{2(l)}$ at temperatures above $T_{(critical)}$.

Furthermore, supercritical CO_2 does not behave as merely a *mixture* of liquid and gaseous CO_2, but often exhibits an exceptional ability to solvate molecules in a specific way. The removal of caffeine from coffee relies on the chromatographic separation of caffeine and the other organic substances in a coffee bean; supercritical fluid chromatography is a growing and exciting branch of chemistry.

5.3   Quantitative effects of pressure and temperature change for a single-component system

Why is ice so slippery?

Effect of p and T on the position of a solid–liquid equilibrium

> The coefficient of friction μ (also called 'friction factor') is the quotient of the frictional force and the normal force. In other words, when we apply a force, is there a resistance to movement or not?

We say something is 'as slippery as an ice rink' if it is has a tiny coefficient of friction, and we cannot get a grip underfoot. This is odd because the coefficient of friction μ for ice is quite high – try dragging a fingernail along the surface of some ice fresh from the ice box. It requires quite a lot of effort (and hence work) for a body to move over the surface of ice.

At first sight, these facts appear to represent a contradiction in terms. In fact, the reason why it is so easy to slip on ice is that ice usually has a thin layer of liquid water covering its surface: it is this water–ice combination that is treacherous and slippery.

But why does any water form on the ice if the weather is sufficiently cold for water to have frozen to form ice? Consider the ice directly beneath the blade on a skater's ice-shoe in Figure 5.9: the edge of the blade is so sharp that an enormous pressure is exerted on the ice, as indicated by the grey tints.

> The sign of dp/dT for the liquid–solid line on a phase diagram is almost always positive. Water is the only common exception.

We now look at the phase diagram for water in Figure 5.10. Ice melts at $0\,^\circ\text{C}$ if the pressure is $p^{\ominus}$ (as represented by T_1 and p_1 respectively on the figure). If the pressure exerted on the ice increases to p_2, then the freezing temperature decreases to T_2. (The freezing temperature *decreases* in response to the *negative* slope of the liquid–solid phase boundary (see the inset to Figure 5.10), which is most unusual; virtually all other substances show a positive slope of dp/dT.)

If the temperature T_2 is lower than the freezing temperature of the water – and it usually is – then some of the ice converts to form liquid water; squeezing decreases the freezing temperature of the water. The water-on-ice beneath the skater's blade is slippery enough to allow effortless skating.

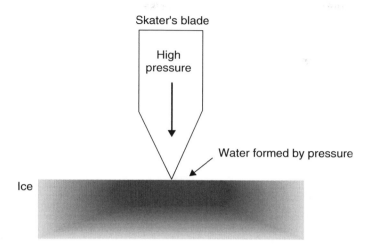

Figure 5.9   Skaters apply an enormous pressure beneath the blades of their skates. This pressure causes solid ice to melt and form liquid water

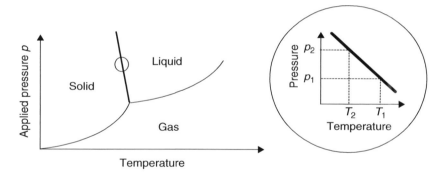

Figure 5.10   Phase diagram of water. *Inset*: applying a high pressure from p_1 (here $p^{\ominus}$) to p_2 causes the melting temperature of the ice to decrease from temperature T_1 (here $0\,^{\circ}\mathrm{C}$) to T_2

What is 'black ice'?

The Clapeyron equation

We give the name 'black ice' to the phenomenon of invisible ice on a road. In practice, anything applying a pressure to solid ice will cause a similar depression of the freezing temperature to that of the skater, so a car or heavy vehicle travelling over ice will also cause a momentary melting of the ice beneath its wheels. This water-on-ice causes the car to skid – often uncontrollably – and leads to many deaths every year. Such ice is particularly dangerous: whereas an ice skater wants the ice to be slippery, a driver does not.

We move from the *qualitative* argument that $T_{(melt)}$ decreases as p increases, and next look for a *quantitative* measure of the changes in melt temperature with pressure. We will employ the *Clapeyron* equation:

$$\frac{dp}{dT} = \frac{\Delta H^{\ominus}}{T \Delta V} \tag{5.1}$$

In fact, it does not matter whether ΔH relates to the direction of change of solid → liquid or of liquid → solid, provided that ΔV relates to the same direction of change.

where T is the normal melting temperature, dT is the change in the melting temperature caused by changing the applied pressure by an amount of dp (in SI units of pascals), where 1 Pa is the pressure exerted by a force of 1 N over an area of 1 m^2. ΔH is the enthalpy change associated with the melting of water and ΔV is the change in volume on melting. Strictly, both ΔH and ΔV are molar quantities, and are often written as ΔH_m and ΔV_m, although the 'm' is frequently omitted.

The molar change in volume ΔV_m has units of m^3 mol^{-1}. Values typically lie in the range 10^{-5}–10^{-6} m^3 mol^{-1}.

The molar change in volume ΔV_m has SI units of m^3 mol^{-1}. We should note how these volumes are *molar* volumes, so they refer to 1 mol of material, explaining why ΔV is always very small. The value of ΔV_m is usually about 10^{-6} to 10^{-5} m^3 mol^{-1} in magnitude, equating to 1 to 10 cm^3 mol^{-1} respectively.

We recall from Chapter 1 how the symbol Δ means 'final state minus initial state', so a positive value of ΔV_m during melting (which is $V_{m\,(liquid)} - V_{m\,(solid)}$) tells us that the liquid has a slightly larger volume than the solid from which it came. $\Delta V_{m\,(melt)}$ is positive in the overwhelming majority of cases, but for water $\Delta V_{m\,(melt)} = -1.6 \times 10^{-6}$ m^3 mol^{-1}. This minus sign is extremely unusual: it means that ice is less dense than water. This explains why an iceberg floats in water, yet most solids sink when immersed in their respective liquid phases.

The minus sign of ΔV_m reflects the way water *expands* on freezing. This expansion explains why a car radiator cracks in cold weather (if it contains no 'de-icer'): the water freezes and, in expanding, exerts a huge a pressure on the metal.

The enthalpy $\Delta H^{\ominus}_{(melt)}$ is the energy required to melt 1 mol of material at constant pressure. We need to be careful when obtaining data from tables, because many books cite the enthalpy of *fusion*, which is the energy released during the opposite process of solidification. We do not need to worry, though, because we know from Hess's law that $\Delta H^{\ominus}_{(melt)} = -\Delta H^{\ominus}_{(fusion)}$. The molar enthalpy of melting water is $+6.0$ kJ mol^{-1}.

Care: following Hess's law, we say:
$\Delta H^{\ominus}_{(melt)} = -\Delta H^{\ominus}_{(fusion)}$.

Worked Example 5.1 Consider a car weighing 1000 kg (about 2200 lbs) parked on a sheet of ice at 273.15 K. Take the area under wheels in contact with the ice as 100 cm^2 i.e. 10^{-2} m^2. What is the new melting temperature of the ice – call it $T_{(final)}$? Take $\Delta H^{\ominus}_{(melt)} = 6.0$ kJ mol^{-1} and water $\Delta V_{m\,(melt)} = -1.6 \times 10^{-6}$ m^3 mol^{-1}.

Strategy. (1) Calculate the pressure exerted and hence the pressure *change*. (2) Insert values into the Clapeyron equation (Equation (5.1)).

(1)   The pressure exerted by the car is given by the equation 'force ÷ area'. The force is simply the car weight expressed in newtons (N): force = 10 000 N, so we calculate the pressure exerted by the wheels as 10 000 N ÷ 10^{-2} m^2, which is 10^6 Pa. We see how a car exerts the astonishing pressure beneath its wheels of 10^6 Pa (about 10 bar).

> At sea level, a mass of 1 kg has a weight (i.e. exerts a force) of approximately 10 N.

(2)   Before inserting values into the Clapeyron equation, we rearrange it slightly, first by multiplying both sides by $T \Delta V$, then dividing both sides by $\Delta H^{\ominus}$, to give

$$dT = \frac{dp \, T \Delta V}{\Delta H^{\ominus}}$$

$$dT = \frac{10^6 \text{ Pa} \times 273.15 \text{ K} \times (-1.6 \times 10^{-6}) \text{ m}^3 \text{ mol}^{-1}}{6.0 \times 10^3 \text{ J mol}^{-1}}$$

$$dT = -0.07 \text{ K}$$

> Notice how the freezing temperature of water *decreases* when a pressure is applied. This decrease is directly attributable to the minus sign of ΔV.

Next, we recall that the symbol 'Δ' means 'final state − initial state'. Accordingly, we say

$$\Delta T = T_{(\text{final})} - T_{(\text{initial})}$$

where the temperatures relate to the melting of ice. The normal melting temperature of ice $T_{(\text{initial})}$ is 273.15 K. The final temperature $T_{(\text{final})}$ of the ice with the car resting on it is obtained by rearranging, and saying

$$\Delta T + T_{(\text{initial})} = -0.07 \text{ K} + 273.15 \text{ K}$$

$$T_{(\text{final})} = 273.08 \text{ K}$$

The new melting temperature of the ice $T_{(\text{final})}$ is 273.08 K. Note how we performed this calculation with the car parked and immobile on the ice. When driving rather than parked, the pressure exerted beneath its wheels is actually considerably greater. Since Equation (5.1) suggests that $dp \propto dT$, the change in freezing temperature dT will be proportionately larger (perhaps as much as −3 K), so there will be a layer of water on the surface of the ice even if the ambient temperature is −3 °C. Drive with care!

> *Care*: the word 'normal' here is code: it means 'at $p = p^{\ominus}$'.

SAQ 5.2 Paraffin wax has a normal melting temperature $T_{(\text{melt})}$ of 320 K. The temperature of equilibrium is raised by 1.2 K if the pressure is increased fivefold. Calculate ΔV_m for the wax as it melts. Take $\Delta H_{(melt)} = 8.064 \text{ kJ mol}^{-1}$.

Justification Box 5.1

Consider two phases (call them 1 and 2) that reside together in thermodynamic equilibrium. We can apply the *Gibbs–Duhem* equation (Equation 4.31) for each of the two phases, 1 and 2.

$$\text{For phase 1}: \qquad dG_{(1)} = (V_{m(1)} \ dp) - (S_{m(1)} \ dT) \qquad (5.2)$$

$$\text{For phase 2}: \qquad dG_{(2)} = (V_{m(2)} \ dp) - (S_{m(2)} \ dT) \qquad (5.3)$$

where the subscripts 'm' imply molar quantities, i.e. per mole of substance in each phase.

Now, because equilibrium exists between the two phases 1 and 2, the dG term in each equation must be the same. If they were different, then the change from phase 1 to phase 2 $(G_{(2)} - G_{(1)})$ would not be zero at all points; but at equilibrium, the value of ΔG will be zero, which occurs when $G_{(2)} = G_{(1)}$. In fact, along the line of the phase boundary we say $dG_{(1)} = dG_{(2)}$.

In consequence, we may equate the two equations, saying:

$$(V_{m(1)} \times dp) - (S_{m(1)} \times dT) = (V_{m(2)} \times dp) - (S_{m(2)} \times dT)$$

Factorizing will group together the two S and V terms to yield

$$(S_{m(2)} - S_{m(1)}) \ dT = (V_{m(2)} - V_{m(1)}) \ dp$$

which, after a little rearranging, becomes

$$\frac{dp}{dT} = \frac{\Delta S_{m(1 \to 2)}}{\Delta V_{m(1 \to 2)}} \qquad (5.4)$$

Finally, since $\Delta G^{\ominus} = \Delta H^{\ominus} - T \Delta S^{\ominus}$ (Equation (4.21)), and since $\Delta G^{\ominus} = 0$ at equilibrium:

$$\frac{\Delta H_m^{\ominus}}{T} = \Delta S_{m(1 \to 2)}$$

Inserting this relationship into Equation (5.4) yields the *Clapeyron* equation in its familiar form.

$$\frac{dp}{dT} = \frac{\Delta H_{m(1 \to 2)}^{\ominus}}{T \Delta V_{m(1 \to 2)}}$$

In fact, Equation (5.4) is also called the Clapeyron equation. This equation holds for phase changes between any two phases and, at heart, quantitatively defines the *phase boundaries* of a phase diagram. For example:

for the liquid → solid line $$\frac{dp}{dT} = \frac{\Delta H^{\ominus}_{m\ (l\to s)}}{T\Delta V_{m\ (l\to s)}}$$

for the vapour → solid line $$\frac{dp}{dT} = \frac{\Delta H^{\ominus}_{m\ (v\to s)}}{T\Delta^{s}_{v} V_{m\ (v\to s)}}$$

for the vapour → liquid line $$\frac{dp}{dT} = \frac{\Delta H^{\ominus}_{m\ (v\to s)}}{T\Delta^{l}_{v} V_{m\ (v\to s)}}$$

Approximations to the Clapeyron equation

We need to exercise a little caution with our terminology: we performed the calculation in Worked Example 5.1 with Equation (5.1) as written, but we should have written Δp rather than dp because 10^6 Pa is a very large change in pressure. Similarly, the resultant change in temperature should have been written as ΔT rather than dT, although 0.07 K is not large. To accommodate these larger changes in p and T, we ought to be rewrite Equation (5.1) in the related form:

> The 'd' in Equation (5.1) means an *infinitesimal* change, whereas the 'Δ' symbol here means a large, macroscopic, change.

$$\frac{\Delta p}{\Delta T} \approx \frac{\Delta H^{\ominus}_{m}}{T\Delta V_{m}}$$

We are permitted to assume that dp is directly proportional to dT because ΔH and ΔV are regarded as constants, although even a casual inspection of a phase diagram shows how curved the solid–gas and liquid–gas phase boundaries are. Such curvature clearly indicates that the Clapeyron equation fails to work except over extremely limited ranges of p and T. Why?

We assumed in Justification Box 5.1 that $\Delta H^{\ominus}_{(melt)}$ is independent of temperature and pressure, which is not quite true, although the dependence is usually sufficiently slight that we can legitimately ignore it. For accurate work, we need to recall the Kirchhoff equation (Equation (3.19)) to correct for changes in ΔH.

> The Clapeyron equation fails to work for phase changes involving gases, except over *extremely* limited ranges of p and T.

Also, we saw on p. 23 how Boyle's Law relates the volume of a gas to changes in the applied pressure. Similar expressions apply for liquids and solids (although such phases are usually much less compressible than gases). Furthermore, we assumed in the derivation of Equation (5.1) that ΔV_{m} does not depend on the pressure changes, which implies that the volumes of liquid and solid phases each change by an identical amount during compression. This approximation is only good when (1) the pressure change is not extreme, and (2) we are considering equilibria for the solid–liquid

> It is preferable to analyse the equilibria of gases in terms of the related Clausius–Clapeyron equation; see Equation (5.5).

phase boundary, which describes melting and solidification. For these reasons, the Clapeyron equation is most effective when dp is relatively small, i.e. 2–10 atm at most.

The worst deviations from the Clapeyron equation occur when one of the phases is a gas. This occurs because the volume of a gas depends strongly on temperature, whereas the volume of a liquid or solid does not. Accordingly, the value of ΔV_m is *not* independent of temperature when the equilibrium involves a gas.

Why does deflating the tyres on a car improve its road-holding on ice?

The Clapeyron equation, continued

We saw from the Clapeyron equation, Equation (5.1), how the decrease in freezing temperature dT is proportional to the applied pressure dp, so one of the easiest ways of avoiding the lethal conversion of solid ice forming liquid water is to apply a smaller pressure – which will decrease dT in direct proportion.

> The pressure beneath the blades of an ice-skater's shoe is enormous – maybe as much as 100 atm when the skater twists and turns at speed.

The pressure change dp is caused by the additional weight of, for example, a car, lorry or ice skater, travelling over the surface of the ice. We recall our definition of 'pressure' as 'force ÷ area'. There is rarely a straightforward way of decreasing the weight of a person or car exerting the force, so the best way to decrease the pressure is to apply the same force but over a larger area. An elementary example will suffice: cutting with a sharp knife is easier than with a blunt one, because the active area along the knife-edge is greater when the knife is blunt, thus causing p to decrease.

In a similar way, if we deflate slightly the tyres on a car, we see the tyre bulge a little, causing it to 'sag', with more of the tyre in contact with the road surface. So, although the weight of the car does not alter, deflating the tyre increases the area over which its weight (i.e. force) is exerted, with the result that we proportionately decrease the pressure.

In summary, we see that letting out some air from a car tyre decreases the value of dp, with the result that the change in melting temperature dT of the ice, as calculated with the Clapeyron equation (Equation (5.1)), also decreases, thereby making driving on ice much safer.

SAQ 5.3 A man is determined not to slip on the ice, so instead of wearing skates of area 10 cm^2 he now wears snow shoes, with the underside of each sole having an extremely large area to spread his 100 kg mass (equating to 1000 N). If the area of each snow shoe is 0.5 m^2, what is the depression of the freezing temperature of the ice caused by his walking over it?
Use the thermodynamic data for water given in Worked Example 5.1.

Aside

If water behaved in a similar fashion to most other materials and possessed a *positive* value of ΔV_m, then water would spontaneously freeze when pressure was applied, rather than solid ice melting under pressure. Furthermore, a positive value of ΔV_m would instantly remove the problems discussed above, caused by vehicles travelling over 'black' ice, because the ice would remain solid under pressure; and remember that the slipperiness occurs because liquid water forms on top of solid ice.

Unfortunately, a different problem would present itself if ΔV_m was positive! If ΔV_m was positive, then Equation (5.1) shows that applying a pressure to liquid water would convert it to ice, even at temperatures slightly higher than $0\,°C$, which provides a *different* source of black ice.

Why does a pressure cooker work?

The Clausius–Clapeyron equation

A pressure cooker is a sealed saucepan in which food cooks faster than it does in a simple saucepan – where 'simple', in this context means a saucepan that is open to the air. A pressure cooker is heated on top of a cooker or hob in the conventional way but, as the water inside it boils, the formation of steam rapidly causes the internal pressure to increase within its sealed cavity; see Figure 5.11. The internal pressure inside a good-quality pressure cooker can be as high 6 atm.

The phase diagram in Figure 5.12 highlights the pressure–temperature behaviour of the boiling (gas–liquid) equilibrium. The *normal* boiling temperature $T_{(boil)}$ of water is $100\,°C$, but $T_{(boil)}$ increases at higher pressures and decreases if the pressure decreases. As a simple example, a glass of water would boil instantly at the cold temperature of 3 K in the hard vacuum of deep space. The inset to Figure 5.12

> Remember that all equilibria are *dynamic*.

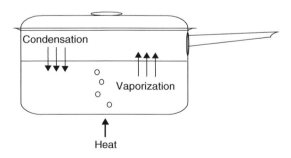

Figure 5.11 A pressure cooker enables food to cook fast because its internal pressure is high, which elevates the temperature at which food cooks

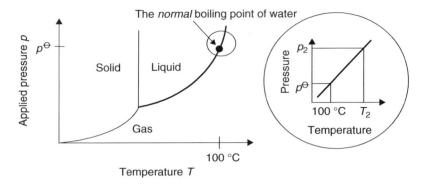

Figure 5.12   Phase diagram to show how a pressure cooker works. *Inset*: applying a high pressure from $p^{\ominus}$ to p_2 causes the boiling temperature of the water to increase from temperature $100\,^{\circ}\text{C}$ to T_2

shows why the water inside the pressure cooker boils at a higher temperature as a consequence of the pan's large internal pressure.

Having qualitatively discussed the way a pressure cooker facilitates rapid cooking, we now turn to a quantitative discussion. The Clapeyron equation, Equation (5.1), would lead us to suppose that $\mathrm{d}p \propto \mathrm{d}T$, but the liquid–gas phase boundary in Figure 5.12 is clearly curved, implying deviations from the equation. Therefore, we require a new version of the Clapeyron equation, adapted to cope with the large volume change of a gas. To this end, we introduce the *Clausius–Clapeyron equation*:

> The *Clausius–Clapeyron equation* quantifies the way a boiling temperature changes as a function of the applied pressure. At the boiling points of T_1 and T_2, the external pressures p_1 and p_2 are the same as the respective vapour pressures.

$$\ln\left(\frac{p_2 \text{ at } T_2}{p_1 \text{ at } T_1}\right) = -\frac{\Delta H^{\ominus}_{(\text{boil})}}{R}\left(\frac{1}{T_2} - \frac{1}{T_1}\right) \tag{5.5}$$

where R is the familiar gas constant, and $\Delta H^{\ominus}_{(\text{boil})}$ is the *enthalpy of vaporization*. $\Delta H^{\ominus}_{(\text{boil})}$ is always *positive* because energy must be put *in* to a liquid if it is to boil. T_2 here is the boiling temperature when the applied pressure is p_2, whereas changing the pressure to p_1 will cause the liquid to boil at a different temperature, T_1.

We need to understand that the Clausius–Clapeyron equation is really just a special case of the Clapeyron equation, and relates to phase changes in which *one of the phases is a gas*.

> It does not matter which of the values we choose as '1' and '2' provided that T_1 relates to p_1 and T_2 relates to p_2. It is permissible to swap T_1 for T_2 and p_1 for p_2 simultaneously, which amounts to multiplying both sides of the equation by '−1'.

Worked Example 5.2   What is the boiling temperature of pure water inside a pressure cooker? Let T_1 be the normal boiling temperature $T_{(\text{boil})}$ of water (i.e. $100\,^{\circ}\text{C}$, 373 K, at $p^{\ominus}$) and let p_2 of $6 \times p^{\ominus}$ be the pressure inside the pan. The enthalpy of boiling water is $50.0\ \text{kJ mol}^{-1}$.

In this example, it is simpler to insert values into Equation (5.5) and to rearrange later. Inserting values gives

Notice how the *ratio* within the bracket on the left-hand side of the Clausius–Clapeyron equation permits us to dispense with *absolute* pressures.

$$\ln\left(\frac{6 \times p^{\ominus}}{p^{\ominus}}\right) = \frac{-50\,000 \text{ J mol}^{-1}}{8.314 \text{ J K}^{-1} \text{ mol}^{-1}} \times \left(\frac{1}{T_2} - \frac{1}{373 \text{ K}}\right)$$

We can omit the units of the two pressures on the left-hand side because Equation (5.5) is written as a ratio, so the units cancel: we require only a *relative* change in pressure.

$$\ln 6.0 = -6104 \text{ K} \times \left(\frac{1}{T_2} - \frac{1}{373 \text{ K}}\right)$$

where ln 6.0 has a value of -1.79. Next, we rearrange slightly by dividing both sides by 6104 K, to yield:

$$\frac{-1.79}{6104 \text{ K}} = \frac{1}{T_2} - \frac{1}{373 \text{ K}}$$

so

$$\frac{1}{T_2} = -2.98 \times 10^{-4} \text{ K}^{-1} + \frac{1}{373 \text{ K}}$$

and

$$\frac{1}{T_2} = 2.38 \times 10^{-3} \text{ K}^{-1}$$

We obtain the temperature at which water boils by taking the reciprocal of both side. T_2, is 420 K, or 147 °C at a pressure of $6 \times p^{\ominus}$, which is much higher than the normal boiling temperature of 100 °C.

SAQ 5.4 A mountaineer climbs Mount Everest and wishes to make a strong cup of tea. He boils his kettle, but the final drink tastes lousy because the water boiled at too low a temperature, itself because the pressure at the top of the mountain is only $0.4 \times p^{\ominus}$. Again taking the enthalpy of boiling the water to be 50 kJ mol^{-1} and the normal boiling temperature of water to be 373 K, calculate the temperature of the water as it boils at the top of the mountain.

The form of the Clausius–Clapeyron equation in Equation (5.5) is called the *integrated* form. If pressures are known for more than two temperatures, an alternative form may be employed:

$$\ln p = -\frac{\Delta H^{\ominus}_{(\text{boil})}}{R} \times \frac{1}{T} + \text{constant} \qquad (5.6)$$

so a graph of the form '$y = mx + c$' is obtained by plotting $\ln p$ (as 'y') against $1/T$ (as 'x'). The gradient of this *Clapeyron graph* is '$-\Delta H_{(boil)}^{\ominus} \div R$', so we obtain $\Delta H_{(boil)}^{\ominus}$ as 'gradient $\times -1 \times R$'.

The intercept of a Clapeyron graph is not useful; its value may best be thought of as the pressure exerted by water boiling at infinite temperature. This alternative of the Clausius–Clapeyron equation is sometimes referred to as the *linear* (or *graphical*) form.

> We employ the *integrated* form of the Clausius–Clapeyron equation when we know two temperatures and pressures, and the *graphical* form for three or more.

Worked Example 5.3   The Clausius–Clapeyron equation need not apply merely to boiling (liquid–gas) equilibria, it also describes sublimation equilibria (gas–solid).

Consider the following thermodynamic data, which concern the sublimation of iodine:

$T_{(sublimation)}/K$	270	280	290	300	310	320	330	340
$p_{(I_2)}/Pa$	50	133	334	787	1755	3722	7542	14 659

> We obtain the value of ΔH as 'gradient$\times -1 \times R$'.

Figure 5.13 shows a plot of $\ln p_{(I_2)}$ (as 'y') against $1/T_{(sublimation)}$ (as 'x'). The enthalpy $\Delta H_{(sublimation)}^{\ominus}$ is obtained via the gradient of the graph 62 kJ mol^{-1} (note the positive sign).

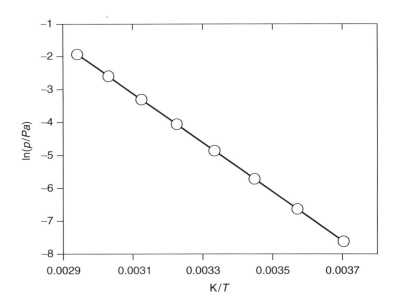

Figure 5.13   The linear form of the Clausius–Clapeyron equation: a graph of $\ln p$ (as 'y') against $1/T$ (as 'x') should be linear with a slope of $-\Delta H_{(vap)}^{\ominus} \div R$

Aside

Why does food cook faster at higher pressures?

The process of cooking involves a complicated series of chemical reactions, each of which proceeds with a rate constant of k. When boiling an egg, for example, the rate-limiting process is denaturation of the proteins from which albumen is made. Such denaturation has an activation energy E_a of about 40 kJ mol^{-1}.

The rate constant of reaction varies with temperature, with k increasing as the temperature increases. k is a function of T according to the well-known Arrhenius equation:

> We consider the *Arrhenius equation* in appropriate detail in Chapter 8.

$$\ln \left(\frac{k \text{ at } T_2}{k \text{ at } T_1} \right) = -\frac{E_a}{R} \left(\frac{1}{T_2} - \frac{1}{T_1} \right) \qquad (5.7)$$

We saw in Worked Example 5.2 how the temperature of the boiling water increases from 100 °C to 147 °C in a pressure cooker. A simple calculation with the Arrhenius equation (Equation (5.7)) shows that the rate constant of cooking increases by a little over fourfold at the higher temperature inside a pressure cooker.

Boiling an egg takes about 4 min at 100 °C, so boiling an egg in a pressure cooker takes about 1 min.

Justification Box 5.2

The Clapeyron equation, Equation (5.1), yields a quantitative description of a phase boundary on a phase diagram. Equation (5.1) works quite well for the liquid–solid phase boundary, but if the equilibrium is boiling or sublimation – both of which involve a gaseous phase – then the Clapeyron equation is a poor predictor.

For simplicity, we will suppose the phase change is the boiling of a liquid: liquid $\rightarrow$ gas. We must make three assumptions if we are to derive a variant that can accommodate the large changes in the volume of a gas:

Assumption 1: we assume the enthalpy of the phase change is independent of temperature and pressure. This assumption is good over limited ranges of both p and T, although note how the Kirchhoff equation (Equation (3.19)) quantifies changes in ΔH.

Assumption 2: we assume the gas is perfect, i.e. it obeys the *ideal-gas* equation, Equation (1.13), so

$$pV = nRT \quad \text{or} \quad pV_m = RT$$

where V_m is the molar volume of the gas.

Assumption 3: ΔV_m is the molar change in volume during the phase change. The value of $\Delta V_m = V_{m(g)} - V_{m(l)}$, where $V_{m(l)}$ is typically $20 \text{ cm}^3 \text{ mol}^{-1}$ and $V_{m(g)}$ is $22.4 \text{ dm}^3 \text{ mol}^{-1}$ (at s.t.p.), i.e. $22\,400 \text{ cm}^3 \text{ mol}^{-1}$. In response to the vast discrepancy between $V_{m(g)}$ and $V_{m(l)}$, we assume that $\Delta V_m \approx V_{m(g)}$, i.e. that $V_{m(l)}$ is negligible by comparison. This third approximation is generally good, and will only break down at very low temperatures.

First, we rewrite the Clapeyron equation in response to approximation 2:

$$\frac{dp}{dT} = \frac{\Delta H_m^\ominus}{T V_{m(g)}}$$

Next, since we assume the gas is ideal, we can substitute for the V_m term via the ideal-gas equation, and say $V_m = RT \div p$:

$$\frac{dp}{dT} = \frac{\Delta H_m^\ominus}{T} \times \frac{p}{RT}$$

Next, we multiply together the two T terms, rearrange and separate the variables, to give:

$$\frac{1}{p} dp = \frac{\Delta H_m^\ominus}{R} \times \frac{1}{T^2} dT$$

> We place the '$\Delta H_m^\ominus \div R$' term outside the right-hand integral because its value is constant.

Integrating with the limits p_2 at T_2 and p_1 at T_1 gives

$$\int_{p_1}^{p_2} \frac{1}{p} dp = \frac{\Delta H_m^\ominus}{R} \int_{T_1}^{T_2} \frac{1}{T^2} dT$$

Subsequent integration yields

$$[\ln p]_{p_1}^{p_2} = -\frac{\Delta H_m^\ominus}{R} \times \left[\frac{1}{T}\right]_{T_1}^{T_2}$$

Next, we insert limits:

$$\ln p_2 - \ln p_1 = -\frac{\Delta H_m^\ominus}{R} \left(\frac{1}{T_2} - \frac{1}{T_1}\right)$$

And, finally, we group together the two logarithmic terms to yield the *Clausius–Clapeyron* equation:

$$\ln\left(\frac{p_2}{p_1}\right) = -\frac{\Delta H^\ominus}{R} \left(\frac{1}{T_2} - \frac{1}{T_1}\right)$$

.4 Phase equilibria involving two-component systems: partition

Why does a fizzy drink lose its fizz and go flat?

Equilibrium constants of partition

Drinks such as lemonade, orangeade or coke contain dissolved CO_2 gas. As soon as the drink enters the warm interior of the mouth, CO_2 comes out of solution, imparting a sensation we say is 'fizzy'.

The CO_2 is pumped into the drink at the relatively high pressure of about 3 bar. After sealing the bottle, equilibrium soon forms between the gaseous CO_2 in the space above the drink and the CO_2 dissolved in the liquid drink (Figure 5.14). We say the CO_2 is *partitioned* between the gas and liquid phases.

The proportions of CO_2 in the space *above* the liquid and *in* the liquid are fixed according to an equilibrium constant, which we call the *partition constant*:

$$K_{(partition)} = \frac{\text{amount of } CO_2 \text{ in phase 1}}{\text{amount of } CO_2 \text{ in phase 2}} \qquad (5.8)$$

We need to note how the identities of phases 1 and 2 must be defined before K can be cited. We need to be aware that $K_{(partition)}$ is only ever useful if the identities of phases 1 and 2 are defined.

On opening the drink bottle we hear a hissing sound, which occurs because the pressure of the escaping CO_2 gas above the liquid is greater than the atmospheric pressure. We saw in Chapter 4 that the molar change in Gibbs function for movement of a gas is given by

> This equilibrium constant is often *incorrectly* called a 'partition function' – which is in fact a term from statistical mechanics.

$$\Delta G = RT \ln \left(\frac{p_{(final)}}{p_{(initial)}} \right) \qquad (5.9)$$

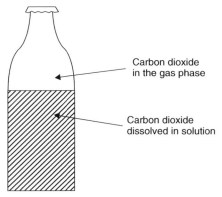

Figure 5.14 In a bottle of fizzy drink, carbon dioxide is partitioned between the gas and the solution phases

The value of ΔG is only ever negative, as required by a thermodynamically spontaneous process, if the initial pressure $p_{(initial)}$ is greater than the final pressure $p_{(final)}$, i.e. the fraction is less than one. In other words, Equation (5.9) shows why ΔG is negative only if the pressure of the CO_2 in the space above the liquid has a pressure that is greater than $p^{\ominus}$.

We disrupted the equilibrium in the bottle when we allowed out much of the CO_2 gas that formerly resided within the space above the liquid; conversely, the CO_2 dissolved in the liquid remains in solution.

After drinking a mouthful of the drink, we screw on the bottle top to stop any more CO_2 being lost, and come back to the bottle later when a thirst returns. The CO_2 re-equilibrates rapidly, with some of the CO_2 *in* the liquid phase passing *to* the gaseous phase. Movement of CO_2 occurs in order to maintain the constant value of $K_{(partition)}$: we call it 're-partitioning'.

A fizzy drink goes 'flat' after opening it several times because the water is depleted of CO_2.

Although the value of $K_{(partition)}$ does not alter, the amount of CO_2 in each of the phases has decreased because some of the CO_2 was lost on opening the bottle. The liquid, therefore, contains less CO_2 than before, which is why it is perceived to be less fizzy. And after opening the bottle several times, and losing gaseous CO_2 each time, the overall amount of CO_2 in the liquid is so depleted that the drink no longer sparkles, which is when we say it has 'gone flat'.

Worked Example 5.4   A bottle of fizzy pop contains CO_2. What are the relative amounts of CO_2 in the water and air if $K_{(partition)} = 4$?

Firstly, we need to note that stating a value of $K_{(partition)}$ is useless unless we know how the equilibrium constant $K_{(partition)}$ was written, i.e. which of the phases '1' and '2' in Equation (5.8) is the air and which is the liquid?

In fact, most of the CO_2 resides in the liquid, so Equation (5.8) would be written as

$$K_{(partition)} = \frac{\text{concentration of } CO_2 \text{ in the drink}}{\text{concentration of } CO_2 \text{ in the air above the liquid}}$$

A bottle of fizzy drink going flat is a fairly trivial example of partition, but the principle is vital to processes such as reactions in two-phase media or the operation of a high-performance liquid chromatography column.

This partition constant has a value of 4, which means that four times as much CO_2 resides in the drink as in the liquid of the space above the drink. Stated another way, four-fifths of the CO_2 is in the gas phase and one-fifth is in solution (in the drink).

SAQ 5.5   An aqueous solution of sucrose is prepared. It is shaken with an equal volume of pure chloroform. The two solutions do not mix. The sucrose partitions between the two solutions, and is more soluble in the water. The value of $K_{(partition)}$ for this water–chloroform system is 5.3. What percentage of the sucrose resides in the chloroform?

How does a separating funnel work?

Partition as a function of solvent

The operation of a *separating funnel* depends on partition. A solvent contains some solute. A different solvent, which is immiscible with the first, contains no compound.

Because the two solvents are *immiscible* – which means they do not mix – the separating funnel will show two distinct layers (see Figure 5.15). After shaking the funnel vigorously, and allowing its contents to settle, some of the solute will have partitioned between the two solvents, with some sample passing from the solution into the previously pure solvent 1.

> *Immiscible* solutions do not mix. The words 'miscible' and its converse 'immiscible' derive from the Latin word *miscere*, meaning 'to mix'.

We usually repeat this procedure two or three times during the practice of *solvent extraction*, and separate the two layers after each vigorous shake (we call this procedure 'running off' the heavier, lower layer of liquid). Several extractions are needed because $K_{(\text{partition})}$ is usually quite small, which implies that only a fraction of the solute is removed from the solution during each partition cycle.

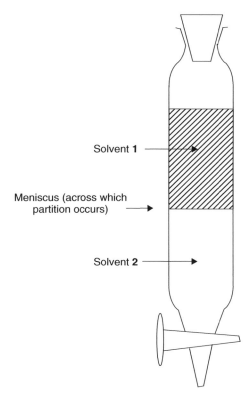

Solvent **1**

Meniscus (across which partition occurs)

Solvent **2**

Figure 5.15   A separating funnel is a good example of partition: solute is partitioned between two immiscible liquids

---

Aside

The reason we need to *shake* the two solutions together when partitioning is because the solute only passes from one solvent to the other across the *interface* between them, i.e. across the *meniscus*.

The meniscus is quite small if the funnel is kept still, and partitioning is slow. Conversely, shaking the funnel generates a large number of small globules of solvent, which greatly increases the 'active' surface area of the meniscus. Therefore, we shake the funnel to increase the *rate* of partitioning.

Why is an ice cube only misty at its centre?

The temperature dependence of partition

Most ice cubes look misty at their centre, but are otherwise quite clear. The ice from which the ice cubes are made is usually obtained from the tap, so it contains dissolved impurities such as chlorine (to ensure its sterility) and gases from the atmosphere. The mist at the centre of the ice cube comprises millions of minute air bubbles containing these gases, principally nitrogen and oxygen.

> A fish would not be able to 'breathe' in water if it contained no oxygen gas.

Gaseous oxygen readily partitions with oxygen dissolved in solution, in much the same way as the partitioning of CO_2 in the fizzy-drink example above. The exact amount of oxygen in solution depends on the value of $K_{(partition)}$, which itself depends on the temperature.

> Like all other equilibrium constants, the value of $K_{(partition)}$ depends strongly on temperature.

Tap water is always saturated with oxygen, the amount depending on the temperature. The maximum concentration of oxygen in water – about $0.02 \ mol \ dm^{-3}$ – occurs at a temperature of $3 \, °C$. The amount of oxygen dissolved in water will decrease below this temperature, since $K_{(partition)}$ decreases. Accordingly, much dissolved oxygen is expelled from solution as the water freezes, merely to keep track of the constant decreasing value of $K_{(partition)}$.

The tap water in the ice tray of our fridge undergoes some interesting phase changes during freezing. Even cold water straight from a tap is warmer than the air within a freezer. Water is a poor thermal conductor and does not freeze evenly, i.e. all at once; rather, it freezes progressively. The first part of the water to freeze is that adjacent to the freezer atmosphere; this outer layer of ice gradually becomes thicker with time, causing the amount of *liquid* water at the cube's core to decrease during freezing.

But ice cannot contain much dissolved oxygen, so air is expelled from solution each time an increment of water freezes. This oxygen enters any liquid water nearby, which clearly resides near the centre of the cube. We see how the oxygen from the water concentrates progressively near the cube centre during freezing.

Eventually, all the oxygen formerly in the water resides in a small volume of water near the cube centre. Finally, as the freezing process nears its completion and even this last portion solidifies, the amount of oxygen in solution exceeds $K_{(partition)}$ and leaves solution as gaseous oxygen. It is this expelled oxygen we see as tiny bubbles of gas.

Aside

Zone refining is a technique for decreasing the level of impurities in some metals, alloys, semiconductors, and other materials; this is particularly so for doped semiconductors, in which the amount of an impurity must be known and carefully controlled. The technique relies on the impurities being more soluble in a molten sample (like oxygen in water, as noted above) than in the solid state.

To exploit this observation, a cylindrical bar of material is passed slowly through an induction heater and a narrow molten 'zone' is moved along its length. This causes the impurities to segregate at one end of the bar and super-pure material at the other. In general, the impurities move in the same direction as the molten zone moves if the impurities lower the melting point of the material (see p. 212).

How does recrystallization work?

Partition and the solubility product

We say the solution is *saturated* if solute is partitioned between a liquid-phase solution and undissolved, solid material (Figure 5.16). In other words, the solution contains as much solute as is feasible, thermodynamically, while the remainder remains as solid. The best way to tell whether a solution is saturated, therefore, is to look for undissolved solid. If $K_{(partition)}$ is small then we say that not much of the solute resides in solution, so most of the salt remains as solid – we say the salt is not very soluble. Conversely, most, if not all, of the salt enters solution if $K_{(partition)}$ is large.

Like all equilibrium constants, the value of $K_{(partition)}$ depends on temperature, sometimes strongly so. It also depends on the solvent polarity. For example, $K_{(partition)}$

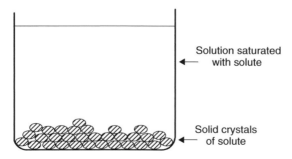

Solution saturated with solute

Solid crystals of solute

Figure 5.16   In a saturated solution, the solute is partitioned between the solid state and solute in solution

of sodium chloride (NaCl) in *water* is large, so a saturated solution has a concentration of about 4 mol dm^{-3}; a saturated solution of NaCl in *ethanol* contains less than 0.01 mol dm^{-3} of solute.

> Strictly, we should speak in terms of ionic *activities* rather than concentrations; see p. 312 ff.

An alternative way of expressing the partition constant of a sparingly soluble salt is to define its 'solubility product' K_{sp} (also called the 'solubility constant' K_s). K_s is defined as the product of the ion activities of an ionic solute in its saturated solution, each raised to its stoichiometric number v_i. K_s is expressed with due reference to the dissociation equilibria involved and the ions present.

We saw above how the extent of partition is temperature dependent; in that example, excess air was expelled from solution during freezing, since the solubility of air was exceeded in a cold freezer box, and the gas left solution in order for the value of $K_{(partition)}$ to be maintained.

> The improved purity of precipitated solute implies that $K_{(partition)}$ for the impurities is different from that for the major solute.

Like all equilibrium constants, $K_{(partition)}$ is a function of temperature, thereby allowing the preparative chemist to *recrystallize* a freshly made compound. In practice, we dissolve the compound in a solvent that is sufficiently hot so that $K_{(partition)}$ is large, as shown by the high solubility. Conversely, $K_{(partition)}$ decreases so much on cooling that much of the solute undergoes a phase change from the solution phase to solid in order to maintain the new, lower value of $K_{(partition)}$. The preparative chemist delights in the way that the precipitated solid retrieved is generally purer than that initially added to the hot solvent.

The energy necessary to dissolve 1 mol of solute is called the 'enthalpy of solution' $\Delta H^{\ominus}_{(solution)}$ (cf. p. 125). A value of ΔH can be estimated by analysing the solubility s of a solute (which is clearly a function of $K_{(partition)}$) with temperature T.

The value of $K_{(partition)}$ changes with temperature; the temperature dependence of an equilibrium constant is given by the van't Hoff isochore:

$$\ln\left(\frac{K_{(partition)2}}{K_{(partition)1}}\right) = -\frac{\Delta H^{\ominus}_{(solution)}}{R}\left(\frac{1}{T_2} - \frac{1}{T_1}\right) \tag{5.10}$$

> $\Delta H^{\ominus}_{(solution)}$ is sometimes called 'heat of solution', particularly in older books. The word 'heat' here can mislead, and tempts us to ignore the possibility of pressure–volume work.

so an approximate value of $\Delta H^{\ominus}_{(solution)}$ may be obtained from the gradient of a graph of an isochore plot of ln s (as 'y') against $1/T$ (as 'x'). Since s increases with increased T, we predict that $\Delta H^{\ominus}_{(solution)}$ will be positive.

Worked Example 5.5   Calculate the enthalpy of solution $\Delta H^{\ominus}_{(solution)}$ from the following solubilities s of potassium nitrate as a function of temperature T. Values of s were obtained from solubility experiments.

T/K	354	347.6	342	334	329	322	319	317
s/g per 100 g of water	140.0	117.0	100.0	79.8	68.7	54.6	49.4	46.1

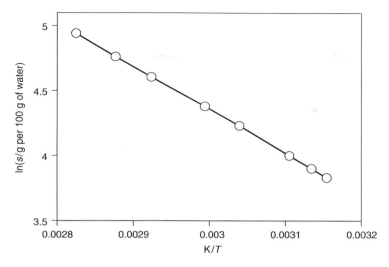

Figure 5.17 The solubility s of a partially soluble salt is related to the equilibrium constant $K_{(partition)}$ and obeys the van't Hoff isochore, so a plot of $\ln s$ (as 'y') against $1/T$ (as 'x') should be linear, with a slope of '$-\Delta H^{\ominus}_{(solution)} \div R$'. Note how the temperature is expressed in kelvin; a graph drawn with temperatures expressed in Celsius would have produced a curved plot. The label K/T on the x-axis comes from $1/T \div 1/K$

The solubility s is a function of $K_{(partition)}$; so, from Equation (5.10), a plot of $\ln s$ (as 'y') against $1/T$ (as 'x') yields the straight-line graph in Figure 5.17. A value of $\Delta H^{\ominus}_{(solution)} = 34 \text{ kJ mol}^{-1}$ is obtained by multiplying the gradient '$-1 \times R$'.

Why are some eggshells brown and some white?

Partition between two solid solutes

The major component within an eggshell is calcium carbonate (chalk). Binders and pigments make up the remainder of the eggshell mass, accounting for about 2–5%.

Before a hen lays its egg, the shell forms inside its body via a complicated series of precipitation reactions, the precursor for each being water soluble. Sometimes the hen's diet includes highly coloured compounds, such as corn husk. The chemicals forming the colour co-precipitate with the calcium carbonate of the shell during shell formation, which we see as the egg shell's colour. Any substantial change in the hen's diet causes a different combination of chemicals to precipitate during shell formation, explaining why we see differently coloured eggs.

In summary, this simple example illustrates the partition of solutes during precipitation: the colour of an egg shell results from the partitioning of chemicals, some coloured, between the growing shell and the gut of the hen during shell growth.

5.5   Phase equilibria and colligative properties

Why does a mixed-melting-point determination work?

Effects of impurity on phase equilibria in a two-component system

The best 'fail-safe' way of telling whether a freshly prepared compound is identical to a sample prepared previously is to perform a *mixed-melting-point* experiment.

> A 'mixed melting point' is the *only* absolutely fail-safe way of determining the purity of a sample.

In practice, we take two samples: the first comprises material whose origin and purity we know is good. The second is fresh from the laboratory bench: it may be pure and identical to the first sample, pure but a different compound, or impure, i.e. a mixture. We take the melting point of each separately, and call them respectively $T_{(melt, pure)}$ and $T_{(melt, unknown)}$. We know for sure that the samples are different if these two melting temperatures differ.

Ambiguity remains, though. What if the melting temperatures are the same but, by some strange coincidence, the new sample is different from the pure sample but has the same melting temperature? We therefore determine the melting temperature of a *mixture*. We mix some of the material known to be pure into the sample of unknown compound. If the two melting points are still the same then the two materials are indeed identical. But any decrease in $T_{(melt, impure)}$ means they are not the same. The value of $T_{(melt, mixture)}$ will *always* be lower than $T_{(melt, pure)}$ if the two samples are different, as evidenced by the decrease in $T_{(melt)}$. We call it a *depression of melting point* (or *depression of freezing point*).

Introduction to colligative properties: chemical potential

The depression of a melting point is one of the simplest manifestations of a *colligative* property. Other everyday examples include pressure, osmotic pressure, vapour pressure and elevation of boiling point.

> 'Colligative properties' depend on the *number*, rather than the *nature*, of the chemical particles (atoms or molecules) under study.

For simplicity, we will start by thinking of one compound as the 'host' with the other is a 'contaminant'. We find experimentally that the magnitude of the depression ΔT depends only on the *amount* of contaminant added to the host and not on the *identity* of the compounds involved – this is a general finding when working with colligative properties. A simple example will demonstrate how this finding can occur: consider a gas at room temperature. The ideal-gas equation (Equation (1.13)) says $pV = nRT$, and holds reasonably well under s.t.p. conditions. The equation makes it clear that the pressure p depends only on n, V and T, where V and T are thermodynamic variables, and n relates to the *number* of the particles but does not depend on the chemical *nature* of the compounds from which the gas is made. Therefore, we see how pressure is a colligative property within the above definition.

Earlier, on p. 181, we looked at the phase changes of a single-component system (our examples included the melting of an ice cube) in terms of changes in the molar Gibbs function ΔG_m. In a similar manner, we now look at changes in the Gibbs function for *each* component within the mixture; and because several components participate, we need to consider more variables, to describe both the host and the contaminant.

We are now in a position to understand why the melting point of a mixture is lower than that of the pure host. Previously, when we considered the melting of a simple single-component system, we framed our thinking in terms of the molar Gibbs function G_m. In a similar way, we now look at the molar Gibbs function of each component i within a mixture. Component i could be a contami-

> For a *pure* substance, the chemical potential μ is merely another name for the molar Gibbs function.

nant. But because i is only one part of a system, we call the value of G_m for material i the *partial* molar Gibbs function. The partial molar Gibbs function is also called the *chemical potential*, and is symbolized with the Greek letter mu, μ.

We define the 'mole fraction' x_i as the number of moles of component i expressed as a proportion of the total number of moles present:

$$x_i = \frac{\text{number of moles of component } i}{\text{total number of moles}} \qquad (5.11)$$

The value of μ_i – the molar Gibbs function of the contaminant – decreases as x_i decreases. In fact, the chemical potential μ_i of the contaminant is a function of its mole fraction within the host, according to Equation (5.11):

> The mole fraction x of the host DEcreases as the amount of contaminant INcreases. The sum of all the mole fractions must always equal one; and the mole fraction of a pure material is also one.

$$\mu_i = \mu_i^{\ominus} + RT \ln x_i \qquad (5.12)$$

where x_i is the mole fraction of the species i, and $\mu_i^{\ominus}$ is its *standard chemical potential*. Equation (5.12) should remind us of Equation (4.49), which relates ΔG and $\Delta G^{\ominus}$.

Notice that the mole fraction x has a maximum value of unity. The value of x decreases as the proportion of contaminant increases. Since the logarithm of a number less than one is always negative, we see how the $RT \ln x_i$ term on the right-hand side of

> Strictly, Equation (5.12) relates to an ideal mixture at constant p and T.

Equation (5.12) is zero for a pure material (implying $\mu_i = \mu_i^{\ominus}$). At all other times, $x_i < 1$, causing the term $RT \ln x_i$ to be negative. In other words, the value of μ will always *de*crease from a maximum value of $\mu_i^{\ominus}$ as the amount of contaminant *in*creases.

Figure 5.18 depicts graphically the relationship in Equation (5.12), and shows the partial molar Gibbs function of the host material as a function of temperature. We first consider the heavy bold lines, which relate to a pure host material, i.e. before contamination. The figure clearly shows two bold lines, one each for the material when solid and another at higher temperatures for the

> *Remember*: in this type of graph, the lines for solid and liquid intersect at the *melting temperature*.

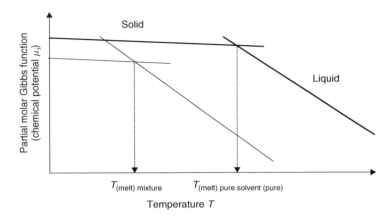

Figure 5.18 Adding a chemical to a host (mixing) causes its chemical potential μ to decrease, thereby explaining why a melting-point temperature is a good test of purity. The heavy solid lines represent the chemical potential of the pure material and the thin lines are those of the host containing impurities

respective liquid. In fact, when we remember that the chemical potential for a pure material is the same as the molar Gibbs function, we see how this graph (the bold line for the pure host material) is identical to Figure 5.2. And we recall from the start of this chapter how the lines representing G_m for solid and G_m for liquid intersect at the melting temperature, because liquid and solid are in equilibrium at $T_{(melt)}$, i.e. $G_{m(liquid)} = G_{m(solid)}$ at $T_{(melt)}$.

We look once more at Figure 5.18, but this time we concentrate on the thinner lines. These lines are seen to be parallel to the bold lines, but have been displaced down the page. These thin lines represent the values of G_m of the host within the mixture (i.e. the once pure material following contamination). The line for the solid mixture has been displaced to a lesser extent than the line for the liquid, simply because the Gibbs function for liquid phases is more sensitive to contamination.

> As the mole fraction of contaminant increases (as x_i gets larger), so we are forced to draw the line progressively lower down the figure.

The vertical difference between the upper bold line (representing $\mu^{\ominus}$) and the lower thin line (which is μ) arises from Eq. (5.12): it is a direct consequence of mixing. In fact, the mathematical composition of Eq. (5.12) dictates that we draw the line for an impure material (when $x_i < 1$) lower on the page than the line for the pure material.

It is now time to draw all the threads together, and look at the temperature at which the thin lines intersect. It is clear from Figure 5.18 that the intersection temperature for the mixture occurs at a cooler temperature than that for the pure material, showing why the melting point temperature for a mixture is depressed relative to a pure compound. The depression of freezing point is a direct consequence of chemical potentials as defined in Equation (5.12).

> A mixed-melting-point experiment is an ideal test of a material's purity since $T_{(melt)}$ never drops unless the compound is impure.

We now see why the melting-point temperature decreases following contamination, when its mole fraction deviates from unity. Conversely, the mole fraction does not change at all if the two components within the mixed-melting-point experiment are the same, in which $T_{(melt)}$ remains the same.

Justification Box 5.3

When we formulated the *total differential* of G (Equation (4.30)) in Chapter 4, we only considered the case of a pure substance, saying

$$dG = \left(\frac{\partial G}{\partial p}\right) dp + \left(\frac{\partial G}{\partial T}\right) dT$$

We assumed then the only variables were temperature and pressure. We must now rewrite Equation (4.30), but we add another variable, the amount of substance n_i in a mixture:

$$dG = \left(\frac{\partial G_i}{\partial p}\right) dp + \left(\frac{\partial G_i}{\partial T}\right) dT + \left(\frac{\partial G_i}{\partial n_i}\right) dn_i \qquad (5.13)$$

We append an additional subscript to this expression for dG to emphasize that we refer to the material i within a mixture. As written, Equation (5.13) could refer to either the host or the contaminant – so long as we define which is i.

The term $\partial G_i / \partial n_i$ occurs so often in second law of thermodynamics that it has its own name: the 'chemical potential' μ, which is defined more formally as

$$\mu_i = \left(\frac{\partial G_i}{\partial n_i}\right)_{p,T,n_j} \qquad (5.14)$$

where the subscripts to the bracket indicate that the variables p, T, and the amounts of all other components n_j in the mixture, each remain constant. The chemical potential is therefore seen to be the slope on a graph of Gibbs function G (as 'y') against the amount of substance n_i (as 'x'); see Figure 5.19. In general, the chemical potential varies with composition, according to Equation (5.12).

The chemical potential μ can be thought of as the constant of proportionality between a change in the amount of a species and the resultant change in the Gibbs function of a system.

The way we wrote ∂G in Equation (5.13) suggests the chemical potential μ is the Gibbs function of 1 mol of species i mixed into an infinite amount of host material. For example, if we dissolve 1 mol of sugar in a roomful of tea then the increase in Gibbs function is $\mu_{(sugar)}$. An alternative way to think of the chemical potential μ is to consider dissolving an infinitesimal amount of chemical i in 1 mol of host.

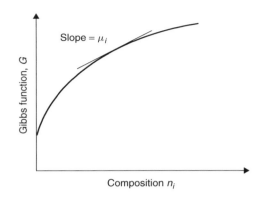

Figure 5.19   The chemical potential μ_i (the partial molar Gibbs function) of a species in a mixture is obtained as the slope of a graph of Gibbs function G as a function of composition

We need to employ 'mental acrobatics' of this type merely to ensure that our definition of μ is watertight – the overall composition of the mixture cannot be allowed to change significantly.

How did the Victorians make ice cream?

Cryoscopy and the depression of freezing point

The people of London and Paris in Victorian times (the second half of the nineteenth century) were always keen to experience the latest fad or novelty, just like many rich and prosperous people today. And one of their favourite 'new inventions' was ice cream and sorbets made of frozen fruit.

The ice cream was made this way: the fruit and/or cream to be frozen is packed into a small tub and suspended in an ice bath. Rock salt is then added to the ice, which depresses its freezing temperature (in effect causing the ice to melt). Energy is needed to melt the ice. $\Delta H_{(melt)} = 6.0 \text{ kJ mol}^{-1}$ for pure water. This energy comes from the fruit and cream in the tub. As energy from the cream and fruit passes through the tub wall to the ice, it freezes. Again, we see how a body's temperature is a good gauge of its internal energy (see p. 34).

The first satisfactory theory to explain how this cooling process works was that of François-Marie Raoult, in 1878. Though forgotten now, Raoult already knew 'Blag-

> Dissolving a solute in a solvent causes a depression of freezing point, in the same way as mixing solids.

den's law': a dissolved substance lowers the freezing point of a solvent in direct proportion to the concentration of the solute. In practice, this law was interpreted by saying that an ice–brine mixture (made with five cups of ice to one of rock salt) had a freezing point at about $-2.7\,°C$. Adding too much salt caused the temperature to fall too far and too fast, causing the outside of the ice cream to freeze prematurely while the core remained liquid. Adding too little salt meant that the ice did not melt, or remained at a temperature close to $0\,^\circ$C, so the cream and fruit juices remained liquid.

This depression of the freezing point occurs in just the same way as the lower melting point of an impure sample, as discussed previously. This determination of the depression of the freezing point is termed *crysoscopy*.

> The word 'cryoscopy' comes from the Greek *kryos*, which literally means 'frost'.

Why boil vegetables in salted water?

Ebullioscopy and the elevation of boiling point

We often boil vegetables in salted water (the concentration of table salt is usually in the range $0.01-0.05 \; \mathrm{mol\,dm^{-3}}$). The salt makes the food taste nicer, although we should wash off any excess salt water if we wish to maintain a healthy blood pressure.

But salted water boils at a higher temperature than does pure water, so the food cooks more quickly. (We saw on p. 203 how a hotter temperature promotes faster cooking.) The salt causes an *elevation of boiling point*, which is another colligative property. We call the determination of such an elevation *ebullioscopy*.

> The word 'ebullioscopy' comes from the Latin *(e)bulirre*, meaning 'bubbles' or 'bubbly'. In a related way, we say that someone is 'ebullient' if they have a 'bubbly' personality.

Look at Figure 5.20, the left-hand side of which should remind us of Figure 5.18; it has *two* intersection points. At the low-temperature end of the graph, we see again why the French ice-cream makers added salt to the ice, to depress its freezing point. But, when we look at the right-hand side of the figure, we see a second intersection, this time between the lines for liquid and gas: the temperature at which the lines intersect gives us the boiling point $T_{(\mathrm{boil})}$.

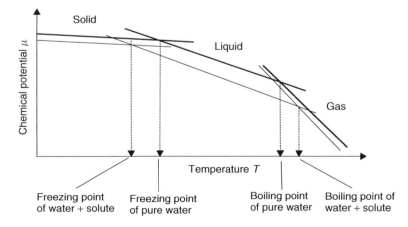

Figure 5.20   Salt in water causes the water to boil at a higher temperature and freeze at a lower temperature; adding a solute to a solvent decreases the chemical potential μ of the solvent. The bold lines represent pure water and the thinner lines represent water-containing solute

The figure shows how adding salt to the water has caused both the lines for liquid and for gas to drop down the page, thus causing the intersection temperature to change. Therefore, a second consequence of adding salt to water, in addition to changing its chemical potential, is to change the temperature at which boiling occurs. Note that the boiling temperature is raised, relative to that of pure water.

Why does the ice on a path melt when sprinkled with salt?

Quantitative cryoscopy

The ice on a path or road is slippery and dangerous, as we saw when considering black ice and ice skaters. One of the simplest ways to make a road or path safer is to sprinkle salt on it, which causes the ice to melt. In practice, rock salt is preferred to table salt, because it is cheap (it does not need to be purified) and because its coarse grains lend additional grip underfoot, even before the salt has dissolved fully.

The depression of freezing temperature occurs because ions from the salt enter the lattice of the solid ice. The contaminated ice melts at a lower temperature than does pure ice, and so the freezing point decreases. Even at temperatures below the normal melting temperatures of pure ice, salted water remains a liquid – which explains why the path or road is safer.

> The 'molaLity' m is the number of moles of solute dissolved per unit *mass* of solvent; 'molaRity' (note the different spelling) is the number of moles of solute dissolved per unit *volume*.

We must appreciate, however, that no *chemical* reaction occurs between the salt and the water; more or less, any ionic salt, when put on ice, will therefore cause it to melt. The chemical identity of the salt is irrelevant – it need not be sodium chloride at all. What matters is the *amount* of the salt added to the ice, which relates eventually to the mole fraction of salt. So, what is the magnitude of the freezing-point depression?

Let the depression of the freezing point be ΔT, the magnitude of which depends entirely on the amount of solute in the solvent. Re-interpreting Blagden's law gives

$$\Delta T \propto \text{molality} \tag{5.15}$$

> We prefer 'molaLity' m to 'molaRity' (i.e. concentration c) because the volume of a liquid or solution changes with temperature, whereas that of a mass does not. Accordingly, molality is temperature independent whereas concentration is not.

The amount is measured in terms of the *molality* of the solute. Molality (note the spelling) is defined as the amount of solute dissolved per unit *mass* of solvent:

$$\text{molality}, m = \frac{\text{moles of solute}}{\text{mass of solvent}} \tag{5.16}$$

where the number of moles of solute is equal to 'mass of solute ÷ molar mass of solute'. The proportionality constant in Equation (5.15) is the *cryoscopic constant* $K_{(\text{cryoscopic})}$. Table 5.3 contains a few typical values of $K_{(\text{cryoscopic})}$, from which it can be seen that

Table 5.3 Sample values of boiling and freezing points, and cryoscopic and ebullioscopic constants

Substance	Boiling point/°C	Freezing point/°C	$K_{(ebullioscopic)}$ /K kg mol^{-1}	$K_{(cryoscopic)}$ /K kg mol^{-1}
Acetic acid	118.5	16.60	3.08	3.59
Acetone	56.1	−94.7	1.71	−
Benzene	80.2	5.455	2.61	5.065
Camphor	208.0	179.5	5.95	40
Carbon disulphide	46.3	−111.5	2.40	3.83
Carbon tetrachloride	76.5	−22.99	5.03	29.8
Chloroform	61.2	−65.5	3.63	4.70
Cyclohexane	80.74	6.55	2.79	20.0
Ethanol	78.3	−114.6	1.07	1.99
Ethyl acetate	77.1	−83.6	2.77	−
Ethyl ether	34.5	−116.2	2.02	1.79
Methanol	64.7	−97.7	0.83	−
Methyl acetate	57	−98.1	2.15	−
n-Hexane	68.7	−95.3	2.75	−
n-Octane	125.7	−56.8	4.02	−
Naphthalene	217.9	80.3	6.94	5.80
Nitrobenzene	210.8	5.7	5.24	8.1
Phenol	181.8	40.9	3.56	7.27
Toluene	110.6	−95.0	3.33	−
Water	100	0	0.512	1.858

camphor as a solvent causes the largest depression. Note that K has the units of K kg mol^{-1}, whereas mass and molar mass are both expressed with the units in units of grammes, so any combination of Equations (5.15) and (5.16) requires a correction term of 1000 g kg^{-1}. Accordingly, Equation (5.15) becomes

$$\Delta T = K_{(cryoscopic)} \times 1000 \times \left(\frac{\text{mass of solute}}{\text{molar mass of solute}} \right) \times \frac{1}{\text{mass of solvent}} \quad (5.17)$$

where the term in parentheses is n, the number of moles of solute.

Worked Example 5.6 10 g of pure sodium chloride is dissolved in 1000 g of water. By how much is the freezing temperature depressed from its normal melting temperature of $T = 273.15$ K? Take $K_{(cryoscopic)}$ from Table 5.3 as 1.86 K kg mol^{-1}.

Inserting values into Equation (5.17) yields

$$\Delta T = 1.86 \text{ K kg mol}^{-1} \times 1000 \text{ g kg}^{-1} \times \frac{10 \text{ g}}{58.5 \text{ g mol}^{-1}} \times \frac{1}{1000 \text{ g}}$$

so $\Delta T = 0.32$ K

This value of ΔT represents the depression of the freezing temperature, so it is negative

showing that the water will freeze at the lower temperature of $(273.16 - 0.32)$ K.

SAQ 5.6 Pure water has a normal freezing point of 273.15 K. What will be the new normal freezing point of water if 11 g of KCl is dissolved in 0.9 dm^3 of water? The cryoscopic constant of water is 1.86 K kg^{-1} mol^{-1}; assume the density of water is 1 g cm^{-3}, i.e. molality and molarity are the same.

An almost identical equation relates the *elevation of boiling point* to the molality:

$$\Delta T_{\text{(elevation)}} = K_{\text{(ebullioscopic)}} \times 1000 \times \text{molality of the salt} \qquad (5.18)$$

where $K_{\text{(ebullioscopic)}}$ relates to the elevation of boiling temperature. Table 5.3 contains a few sample values of $K_{\text{(ebullioscopic)}}$. It can be seen from the relative values of $K_{\text{(ebullioscopic)}}$ and $K_{\text{(cryoscopic)}}$ in Table 5.3 that dissolving a solute in a solvent has a more pronounced effect on the freezing temperature than on the boiling temperature.

Aside

The ice on a car windscreen will also melt when squirted with *de-icer*. Similarly, we add *anti-freeze* to the water circulating in a car radiator to prevent it freezing; the radiator would probably crack on freezing without it; see the note on p. 194.

Windscreen de-icer and engine anti-freeze both depress the freezing point of water via the same principle as rock salt depressing the temperature at which ice freezes on a road. The active ingredient in these cryoscopic products is ethylene glycol (**II**), which is more environmentally friendly than rock salt. It has two physicochemical advantages over rock salt: (1) being liquid, it can more readily enter between the microscopic crystals of solid ice, thereby speeding up the process of cryoscopic melting; (2) rock salt is impure, whereas **II** is pure, so we need less **II** to effect the same depression of freezing point.

$$CH_2 - CH_2$$
$$\diagup \qquad \diagdown$$
$$OH \qquad\qquad OH$$
(II)

Ethylene glycol is also less destructive to the paintwork of a car than rock salt is, but it is toxic to humans.

.6   Phase equilibria involving vapour pressure

Why does petrol sometimes have a strong smell and sometimes not?

Dalton's law

The acrid smell of petrol on a station forecourt is sometimes overpoweringly strong, yet at other times it is so weak as to be almost absent. The smell is usually stronger on a still day with no wind, and inspection shows that someone has spilled some petrol on the ground nearby. At the other extreme, the smell is weaker when there is a breeze, which either blows away the spilt liquid or merely dilutes the petrol in the air.

The subjective experience of how strong a smell is relates to the amount of petrol in the air; and the amount is directly proportional to the pressure of gaseous petrol. We call this pressure of petrol the 'partial pressure' $p_{(\text{petrol})}$.

And if several gases exist together, which is the case for petrol in air, then the total pressure equals the sum of the partial pressures according to *Dalton's law*:

$$p_{(\text{total})} = \sum p_i \qquad (5.19)$$

In the case of a petrol smell near a station forecourt, the smell is strong when the partial pressure of the petrol vapour is large, and it is slight when $p_{(\text{petrol})}$ is small.

These differences in $p_{(\text{petrol})}$ need not mean any difference in the overall pressure $p_{(\text{total})}$, merely that the composition of the gaseous mixture we breathe is variable.

SAQ 5.7   What is the total pressure of 10 g of nitrogen gas and 15 g of methane at 298 K, and what is the partial pressure of nitrogen in the mixture? [Hint: you must first calculate the number of moles involved.]

Justification Box 5.4

The total number of moles equals the sum of its constituents, so

$$n_{(\text{total})} = n_A + n_B + \ldots$$

The ideal-gas equation (Equation (1.13)) says $pV = nRT$; thus $p_{(\text{petrol})}V = n_{(\text{petrol})}RT$, so $n_{(\text{petrol})} = p_{(\text{petrol})}V \div RT$.

Accordingly, in a mixture of gases such as petrol, oxygen and nitrogen:

$$\frac{p_{(\text{total})}V}{RT} = \frac{p_{(\text{petrol})}V}{RT} + \frac{p_{(\text{oxygen})}V}{RT} + \frac{p_{(\text{nitrogen})}V}{RT}$$

We can cancel the gas constant R, the volume and temperature, which are all constant, to yield

$$p_{(total)} = p_{(petrol)} + p_{(oxygen)} + p_{(nitrogen)}$$

which is *Dalton's law*, Equation (5.19).

How do anaesthetics work?

Gases dissolving in liquids: Henry's law

'Anaesthesia' is the science of making someone unconscious. The word comes from the Greek *aesthēsis*, meaning sensation (from which we get the modern English word 'aesthetic', i.e. to please the sensations). The initial *'ana'* makes the word negative, i.e. without sensation.

An anaesthetist administers chemicals such as halothane (**III**) to a patient before and during an operation to promote unconsciousness. Medical procedures such as operations would be impossible for the surgeon if the patient were awake and could move; and they would also be traumatic for a patient who was aware of what the surgery entailed.

(III)

A really deep, chemically induced sleep is termed 'narcosis', from the Greek *narke*, meaning 'numbness'. Similarly, we similarly call a class-A drug a 'narcotic'.

Although the topic of anaesthesia is hugely complicated, it is clear that the physiological effect of the compounds depends on their entrapment in the blood. Once dissolved, the compounds pass to the brain where they promote their narcotic effects. It is now clear that the best anaesthetics dissolve in the lipids from which cell membranes are generally made. The anaesthetic probably alters the properties of the cell membranes, altering the rates at which neurotransmitters enter and leave the cell.

A really deep 'sleep' requires a large amount of anaesthetic and a shallower sleep requires less material. A trained anaesthetist knows just how much anaesthetic to administer to induce the correct depth of sleep, and achieves this by varying the relative pressures of the gases breathed by the patient.

Henry's law is named after William Henry (1775–1836), and says that the amount of gas dissolved in a liquid or solid is in direct proportion to the partial pressure of the gas.

In effect, the anaesthetist relies on *Henry's law*, which states that the equilibrium amount of gas that dissolves in a liquid is proportional to the mole fraction of the gas above the liquid. Henry published his studies in 1803, and showed how the amount of gas dissolved in a liquid is directly proportional to the pressure (or

Table 5.4 Henry's law constants k_H for gases in water at 25 °C

Gas	k_H/mol dm^{-3} bar^{-1}
CO_2	3.38×10^{-2}
O_2	1.28×10^{-3}
CH_4	1.34×10^{-3}
N_2	6.48×10^{-4}

partial pressure) of the gas above it. Stated in another form, Henry's law says:

$$[i_{(\text{soln})}] = k_H p_i \qquad (5.20)$$

where p_i is the partial pressure of the gas i, and $[i_{(\text{soln})}]$ is the concentration of the material i in solution. The constant of proportionality k_H is the respective value of Henry's constant for the gas, which relates to the solubility of the gas in the medium of choice. Table 5.4 lists a few Henry's law constants, which relate to the solubility of gases in water.

> One of the simplest ways of removing gaseous oxygen from water is to bubble nitrogen gas through it (a process called 'sparging').

Worked Example 5.7   What is the concentration of molecular oxygen in water at 25 °C? The atmosphere above the water has a pressure of 10^5 Pa and contains 21 per cent of oxygen.

Strategy. (1) We calculate the partial pressure of oxygen $p_{(O_2)}$. (2) We calculate the concentration $[O_{2(\text{aq})}]$ using Henry's law, Equation (5.20), $[O_{2(\text{aq})}] = p_{(O_2)} \times k_{H(O_2)}$.

> Strictly, Henry's law only holds for *dilute* systems, typically in the mole-fraction range 0–2 per cent. The law tends to break down as the mole fraction x increases.

(1)   From the partial of oxygen $p_{(O_2)} = x_{(O_2)} \times$ the total pressure $p_{(\text{total})}$, where x is the mole fraction:

$$p_{(O_2)} = 0.21 \times 10^5 \text{ Pa}$$

$$p_{(O_2)} = 2.1 \times 10^4 \text{ Pa or } 0.21 \text{ bar}$$

(2)   To obtain the concentration of oxygen, we insert values into Henry's law, Equation (5.20):

$$[O_{2(\text{aq})}] = 0.21 \times p^{\ominus} \times 1.28 \times 10^{-3} \text{ mol dm}^{-3} \text{ bar}^{-1}$$

$$[O_{2(\text{aq})}] = 2.69 \times 10^{-4} \text{ mol dm}^{-3}$$

We need to be aware that k_H is an equilibrium constant, so its value depends strongly on temperature. For example, at 35 °C, water only accommodates 7.03 mg of oxygen per litre, which explains why fish in warm water sometimes die from oxygen starvation.

> This relatively high concentration of oxygen helps explain why fish can survive in water.

How do carbon monoxide sensors work?

Henry's law and solid-state systems

Small, portable sensors are now available to monitor the air we breathe for such toxins as carbon monoxide, CO. As soon as the air contains more than a critical concentration of CO, the sensor alerts the householder, who then opens a window or identifies the source of the gas.

At the 'heart' of the sensor is a slab of doped transition-metal oxide. Its mode of operation is to detect the concentration of CO *within* the oxide slab, which is in direct proportion to the concentration of CO gas in the air surrounding it, according to Henry's law.

A small voltage is applied across the metal oxide. When it contains no CO, the electrical conductivity of the oxide is quite poor, so the current through the sensor is minute (we argue this corollary from Ohm's law). But increasing the concentration of CO in the air causes a proportionate increase in the amount of CO incorporating into the solid oxide, which has a profound influence on electrical conductivity through the slab, causing the current through the slab to increase dramatically. A microchip within the sensor continually monitors the current. As soon as the current increases above its minimum permissible level, the alarm sounds.

So, in summary, CO gas partitions between the air and carefully formulated solid oxides. Henry's law dictates the amount of CO in the oxide.

> In general, Henry's law only applies over relatively small ranges of gas pressure.

Why does green petrol smell different from leaded petrol?

Effects of amount of material on vapour pressure

> Petrol is only useful in a car engine because it is *volatile*.

A car engine requires petrol as its source of fuel. Such petrol has a low boiling temperature of about $60\,^\circ$C. Being so volatile, the liquid petrol is always surrounded with petrol vapour. We say it has a high *vapour pressure* (also called 'saturated vapour pressure'), which explains why we smell it so readily.

Once started, the engine carburettor squirts a mixture of air and volatile petrol into a hot engine cylinder, where the mixture is ignited with a spark. The resultant explosion (we call it 'firing') provides the ultimate source of kinetic energy to propel the car.

A car engine typically requires four cylinders, which fire in a carefully synchronized manner. Unfortunately, these explosions sometimes occur prematurely, before the spark has been applied, so the explosions cease to be synchronized. It is clearly undesirable for a cylinder to fire out of sequence, since the kinetic energy is supplied in a jerky, irreproducible manner. The engine sounds dreadful, hence the word 'knock'.

Modern petrol contains small amounts of additives to inhibit this knocking. 'Leaded' petrol, for example, contains the organometallic compound lead tetraethyl, $PbEt_4$. Although $PbEt_4$ is excellent at stopping knocking, the lead by-products are toxic. In fact, most EU countries now ban $PbEt_4$.

So-called 'green' petrol is a preferred alternative to leaded petrol: it contains about 3 per cent of the aromatic hydrocarbon benzene (C_6H_6, **IV**) as an additive, the benzene acting as a lead-free alternative to $PbEt_4$ as an 'anti-knocking' compound.

> We experience *knocking* (which we colloquially call 'pinking') when explosions within a car engine are not synchronized.

> Lead tetraethyl is the most widely made *organometallic* compound in the world. It is toxic, and killed over 40 chemical workers during its early development.

(IV)

The $PbEt_4$ in petrol does not smell much because it is not volatile. By contrast, benzene is much more volatile – almost as volatile as petrol. The vapour above 'green' petrol, therefore, contains quite a high proportion of benzene (as detected by its cloying, sweet smell) as well as gaseous petrol. That is why green petrol has a sweeter smell than petrol on its own.

Why do some brands of 'green' petrol smell different from others?

Raoult's law

The 'petrol' we buy comprises a mixture of naturally occurring hydrocarbons, a principal component of which is octane; but the mixture also contains a small amount of benzene. Some brands of petrol contain more benzene than others, both because of variations in the conditions with which the crude oil is distilled into fractions, and also variations in the reservoir from which the crude oil is obtained. The proportion varies quite widely: the average is presently about 3 per cent.

Petrol containing a lot of benzene smells more strongly of benzene than petrol containing less of it. In fact, the intensity of the smell is in direct proportion to the amount of benzene in the petrol: at equilibrium, the pressure of vapour above a liquid mixture

> In the countries of North America, petrol is often called 'gas', which is short for gasoline'.

> *Raoult's law* is merely a special form of Henry's law.

depends on the liquid's composition, according to *Raoult's law*:

$$p_{(benzene)} = p^{\ominus}_{(benzene)} x_{(benzene)} \tag{5.21}$$

> Raoult's law states that (at constant temperature) the partial pressure of component *i* in the vapour residing at equilibrium above a liquid is proportional to the mole fraction x_i of component in the liquid.

where $x_{(benzene)}$ is the mole fraction of the benzene in the liquid. If we assume that liquid benzene and petrol have the same densities (which is entirely reasonable), then petrol containing 3 per cent of benzene represents a mole fraction $x_{(benzene)} = 0.03$; the mole fraction of the petrol in the liquid mixture is therefore 0.97 (or 97 per cent). The vapour above the petrol mixture will also be a mixture, containing some of each hydrocarbon in the petrol. We call the pressure due to the benzene component its partial pressure $p_{(benzene)}$. The constant of proportionality in Equation (5.21) is $p^{\ominus}_{(benzene)}$, which represents the pressure of gaseous benzene above *pure* (i.e. unmixed) liquid benzene.

Calculations with Raoult's law

> If a two-component system of A and B forms an ideal mixture, then we can calculate x_A if we know x_B because $x_A + x_B = 1$, so $x_B = (1 - x_A)$.

If we know the mole fraction of a liquid *i* (via Equation (5.11)) and the vapour pressures of the pure liquids $p^{\ominus}_i$, then we can ascertain the total vapour pressure of the gaseous mixture hovering at equilibrium above the liquid.

The intensity of the benzene smell is proportional to the amount of benzene in the vapour, $p_{(benzene)}$. According to Equation (5.21), $p_{(benzene)}$ is a simple function of how much benzene resides within the liquid petrol mixture. Figure 5.21 shows a graph of the partial pressures of benzene and octane above a mixture of the two liquids. (For convenience, we assume here that the mixture comprises only these two components.)

The extreme mole fractions, 0 and 1, at either end of the graph relate to pure petrol ($x = 0$) and pure benzene ($x = 1$) respectively. The mole fractions between these values represent mixtures of the two. The solid, bold line represents the total mole fraction while the dashed lines represent the vapour pressures of the two constituent vapours. It is clear that the sum of the two dashed lines equals the bold line, and represents another way of saying Dalton's law: the total vapour pressure above a mixture of liquids is the sum of the individual vapour pressures.

> Benzene is more volatile than bromobenzene because its vapour pressure is higher.

Worked Example 5.8 The two liquids benzene and bromobenzene are mixed intimately at 298 K. At equilibrium, the pressures of the gases above beakers of the *pure* liquids are 100.1 kPa and 60.4 kPa respectively. What is the vapour pressure above the mixture if 3 mol of benzene are mixed with 4 mol of bromobenzene?

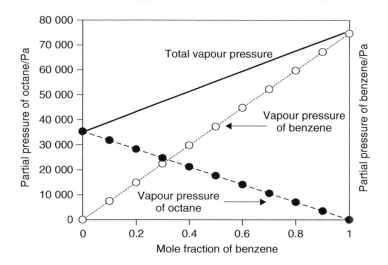

Figure 5.21   Petrol ('gasoline') is a mixture of liquid hydrocarbons. The partial pressure of benzene is nearly twice that of octane, making it much more volatile. The bold line represents the *total* pressure of vapour above a basin of petrol, and comprises the sum of two *partial* pressures: benzene (open circles) and octane (filled circles). Each partial pressure is proportional to the mole fraction of the respective liquid in the petrol mixture

From Dalton's law, the total vapour pressure is simply the sum of the individual vapour pressures:

$$p_{(total)} = p_{(benzene)} + p_{(bromobenzene)}$$

so, from Raoult's law, these partial pressures may be obtained by substituting each p term with $p_i^{\ominus} \times x_i$:

$$p_{(total)} = (p_{(benzene)}^{\ominus} \times x_{(benzene)}) + (p_{(bromobenzene)}^{\ominus} \times x_{(bromobenzene)})$$
(5.22)

> *Care*: do not confuse $p^{\ominus}$ (the standard pressure of 10^5 Pa) with $p_i^{\ominus}$, the vapour pressure of pure i.

We know from the question that there are 7 mol of liquid. We obtain the respective mole fractions x from Equation (5.11): the mole fraction of benzene is $\frac{3}{7}$ and the mole fraction of bromobenzene is $\frac{4}{7}$.

Substituting values of x_i and $p_i^{\ominus}$ into Equation (5.22) yields the total pressure $p_{(total)}$ as

$$p_{(total)} = \left(100.1 \text{ kPa} \times \tfrac{3}{7}\right) + \left(60.4 \text{ kPa} \times \tfrac{4}{7}\right)$$

$$p_{(total)} = (42.9 \text{ kPa}) + (34.5 \text{ kPa})$$

so

$$p_{(total)} = 77.4 \text{ kPa}$$

An *ideal mixture* comprises a pair (or more) of liquids that obey Raoult's law.

Because these two liquids, when mixed, obey Raoult's law, we say they form an *ideal mixture*. In fact, relatively few pairs of liquids form ideal mixtures: a few examples include benzene and bromobenzene, benzene and toluene, bromobenzene and chlorobenzene, *n*-pentane and i-pentane. Note how each set represents a pair of liquids showing a significant extent of similarity.

SAQ 5.8 Benzene and toluene form an ideal mixture, i.e. they obey Raoult's law. At 20 °C, the pressure $p^\ominus$ of benzene and toluene are $0.747 \times p^\ominus$ and $0.223 \times p^\ominus$ respectively. What is the pressure above a mixture of these two liquids that contains 12 mol% of benzene?

Worked Example 5.9   (Continuing from Worked Example 5.8.) What are the mole fractions of benzene and bromobenzene in the *vapour*?

From the definition of mole fraction x in Equation (5.11) above, we say

$$x_{(\text{benzene, vapour})} = \frac{\text{moles of benzene in the vapour}}{\text{total number of moles in the vapour phase}}$$

The numbers of moles n_i are directly proportional to the partial pressures p_i if we assume that each vapour behaves as an ideal gas (we assume here that T, R and V are constant). Accordingly, we can say

$$x_{(\text{benzene})} = \frac{\text{pressure of benzene}}{\text{total pressure}}$$

Note how the units cancel to yield a *dimensionless* mole fraction.

Substituting numbers from Worked Example 5.8:

$$x_{(\text{benzene})} = \frac{42.9 \text{ kPa}}{77.4 \text{ kPa}}$$

$$x_{(\text{benzene})} = 0.554$$

We need *four* mole fractions to define this two-component system – two for the liquid phases and two for the vapour phases.

The mole fraction of benzene in the vapour is 0.554, so it contains 55.4 per cent benzene. The remainder of the vapour comprises the second component bromobenzene, so the vapour contains $(100 - 55.4)\% = 44.6\%$ of bromobenzene.

Note how the liquid comprises 43 per cent benzene and 57 per cent bromobenzene, but the vapour contains proportionately more of the volatile benzene. We should expect the vapour to be richer in the more volatile component.

SAQ 5.9   Continuing with the system in SAQ 5.8, what is the mole fraction of toluene in the vapour above the mixture?

In fact, most liquid mixtures do not obey Raoult's law particularly well, owing to molecular interactions.

Why does a cup of hot coffee yield more steam than above a cup of boiling water at the same temperature?

The effects of poor mixing (immiscibility)

Prepare two cups: put boiling water into one and boiling coffee in the other. The temperature of each is the same because the water comes from the same kettle, yet the amount of steam coming from the coffee is seen to be greater. (We obtain a better view of the steam by placing both cups on a sunny window sill, and looking at the shadows cast on the opposing wall as the light passes through the vapour as it rises from the cups.)

When performing this little experiment, we will probably notice how the steam above the coffee has an extremely strong smell of coffee, although the smell dissipates rapidly as the rate of steam production decreases.

> The rate of steam production decreases with time as the water cools down because energy is lost from the cup as water molecules enter the gas phase.

This experiment is a simple example of *steam distillation*. Adding steam promotes the volatilization of otherwise non-volatile components, simplifying their extraction. For simplicity, we will say that the smell derives from a single sweet-smelling chemical 'coffee'. Coffee and water are not wholly miscible, with some of the essential oils from the coffee existing as tiny globules – we call the mixture a *colloid* (see Chapter 10). We have generated a two-phase system. Both phases, the water and the coffee, are saturated with each other. In fact, these globules would cause strong coffee to appear slightly misty, but for its strong colour blocking all light. We never see *phase separation* in the coffee cup, with a layer of oil floating above a layer of water, because the coffee's concentration is never high enough.

We say a pure liquid boils when its vapour pressure equals the external, atmospheric pressure (see p. 188). Similarly, when boiling a mixture, boiling occurs when the sum of the partial pressures ($p_{(water)} + p_{(coffee)}$) equals $p^{\ominus}$. It is for this reason that the steam above the coffee cup smells strongly of coffee, because the vapour contains the essential oils (e.g. esters) that impart the smell. But the water generates steam at a pressure of $p^{\ominus}$ when the water added to the cup is boiling, so the partial pressure of the coffee $p_{(coffee)}$ is additional. For this reason, we produce more steam than above the cup containing only water.

> The boiling of such a mixture requires the *sum* of the pressures, not just the pressure of one component, to equal $p^{\ominus}$.

How are essential oils for aromatherapy extracted from plants?

Steam distillation

The 'essential oils' of a plant or crop usually comprise a mixture of esters. At its simplest, the oils are extracted from a plant by distillation, as employed in a standard undergraduate laboratory. Since plants contain such a small amount of this precious oil, a ton of plant may be needed to produce a single fluid ounce. Some flowers, such as jasmine or tuberose, contain very small amounts of essential oil, and the petals are very temperature sensitive, so heating them would destroy the blossoms before releasing the essential oils.

To add to the cost further, many of these compounds are rather sensitive to temperature and would decompose before vaporizing. For example, oil of cloves (from *Eugenia caryophyllata*) is rich in the phenol eugenol (**V**), which has a boiling point of 250 °C). We cannot extract the oils via a conventional distillation apparatus.

(V)

Heat-sensitive or water-immiscible compounds are purified by *steam distillation* at temperatures considerably lower than their usual boiling temperatures.

The most common method of extracting essential oils is *steam distillation*. The plant is first crushed mechanically, to ensure a high surface area, and placed in a closed still. High-pressure steam is forced through the still, with the plant pulp becoming hot as the steam yields its heat of vaporization (see p. 79). The steam forces the microscopic pockets holding the essential oils to open and to release their contents. Tiny droplets of essential oil evaporate and mix in the gas-phase mixture with the steam. The mixture is then swept through the still before condensing in a similar manner to a conventional distillation.

Such 'steam heating' is even, and avoids the risk of overheating and decomposition that can occur in hot spots when external heating is used. The steam condenses back into water and the droplets coagulate to form liquid oil. Esters and essential oils do not mix with water, so phase separation occurs on cooling, and we see a layer of oil forming above a layer of condensed water. The oil is decanted or skimmed off the surface of the water, dried, and packaged.

Solvent extraction of essential oils tends to generate material that is contaminated with solvent (and cannot be sold); and *mechanical pressing* of a plant usually generates too poor a yield to be economically viable.

The only practical problem encountered when collecting organic compounds by steam distillation is that liquids of low volatility will usually distil slowly, since the proportion of compound in the vapour is proportional to the vapour pressure, according to

$$\frac{p_{(\text{oil})}}{p_{(\text{water})}} = \frac{n_{(\text{oil})}}{n_{(\text{water})}} \tag{5.23}$$

In practice, we force water vapour (steam) at high pressure through the clove pulp to obtain a significant partial pressure of eugenol (**V**).

Justification Box 5.5

When considering the theory behind steam distillation, we start with the ideal-gas equation (Equation (1.13)), $pV = nRT$. We will consider two components: oil and water. For the oil, we say $p_{(oil)}V = n_{(oil)}RT$, and for the water $p_{(water)}V = n_{(water)}RT$. Dividing the two equations by R and V (which are both constant) yields

$$p_{(oil)} = n_{(oil)} \times T \qquad \text{for the oil}$$

$$p_{(water)} = n_{(water)} \times T \qquad \text{for the water}$$

We then divide each pressure by the respective number of moles n_i, to obtain

$$p_{(oil)} \div n_{(oil)} = T \qquad \text{for the oil}$$

$$p_{(water)} \div n_{(water)} = T \qquad \text{for the water}$$

The temperature of the two materials will be T, which is the same for each as they are in thermal equilibrium. We therefore equate the two expressions, saying

$$p_{(oil)} \div n_{(oil)} = p_{(water)} \div n_{(water)}$$

Dividing both sides by $p_{(water)}$ and multiplying both sides by $n_{(oil)}$ yields Equation (5.23):

$$\frac{p_{(oil)}}{p_{(water)}} = \frac{n_{(oil)}}{n_{(water)}}$$

so we see how the percentage of each constituent in the vapour depends only on its vapour pressure at the distillation temperature.

To extract a relatively involatile oil such as eugenol (**V**) without charring requires a high pressure of steam, although the steam will not be hotter than 100 °C, so we generate a mixture of vapours at a temperature lower than that of the less volatile component.

6

Acids and Bases

Introduction

Equilibria involving acids and bases are discussed from within the Lowry–Brønsted theory, which defines an acid as a proton donor and a base as a proton acceptor (or 'abstracter'). The additional concept of pH is then introduced. 'Strong' and 'weak' acids are discussed in terms of the acidity constant K_a, and then conjugate acids and bases are identified.

Acid–base buffers comprise both a weak acid or base and its respective salt. Calculations with buffers employing the Henderson–Hasselbach equation are introduced and evaluated, thereby allowing the calculation of the pH of a buffer. Next, titrations and pH indicators are discussed, and their modes of action placed into context.

6.1 Properties of Lowry–Brønsted acids and bases

Why does vinegar taste sour?

The Lowry–Brønsted theory of acids

We instantly experience a sour, bitter taste when consuming anything containing vinegar. The component within the vinegar causing the sensation is ethanoic acid, CH_3COOH (**I**) (also called 'acetic acid' in industry). Vinegar contains between 10 and 15% by volume of ethanoic acid, the remainder being water (85–90%) and small amounts of other components such as caramel, which are added to impart extra flavour.

> The Latin word for 'sour' is *acidus*.

(I)

The German chemist Liebig, in 1838, was the first to suggest mobile, replaceable, hydrogen atoms being responsible for acidic properties. Arrhenius extended the idea in 1887, when he said the hydrogen existed as a proton.

The ethanoic acid molecule is essentially covalent, explaining why it is liquid when pure at room temperature. Nevertheless, the molecule is charged, with the O–H bond characterized by a high percentage of ionic character. Because water is so polar a solvent, it strongly solvates any solute dissolved within it. In aqueous solutions, water molecules strongly solvate the oxygen- and proton-containing ends of the O–H bond, causing the bond to break in a significant proportion of the ethanoic acid molecules, according to the following simplistic reaction:

$$CH_3COOH_{(aq)} \longrightarrow CH_3COO^-_{(aq)} + H^+_{(aq)} \qquad (6.1)$$

The O–H bond in an acid is sometimes said to be 'labile', since it is so easily broken. The word derives from the Latin *labi*, to lapse (i.e. to change).

We say the acid *dissociates*. The bare proton is very small, and has a large charge density, causing it to attract the negative end of the water dipole. The proton produced by Equation (6.1) is, therefore, hydrated in aqueous solutions, and is more accurately represented by saying $H^+_{(aq)}$.

We see how solvated protons impart the subjective impression of a sour, bitter flavour to the ethanoic acid in vinegar. In fact, not only the sour flavour, but also the majority of the properties we typically associate with an acid (see Table 6.1) can be attributed to an acidic material forming one or more solvated protons $H^+_{(aq)}$ in solution.

The 'Lowry–Brønsted theory' says an acid is a proton donor.

This classification of an acid is called the *Lowry–Brønsted theory* after the two scientists who (independently) proposed this definition of an acid in 1923. More succinctly, their theory says an acid is

Table 6.1 Typical properties of Lowry–Brønsted acids

Acid property	Example from everyday life
Acids dissolve a metal to form a salt plus hydrogen	Metallic sodium reacts with water, and 'fizzes' as hydrogen gas evolves
Acids dissolve a metal oxide to form a salt and water	The ability of vinegar to clean tarnished silver by dissolving away the coloured coating of Ag_2O
Acids react with metal carbonates to form a salt and carbon dioxide	The fizzing sensation in the mouth when eating sherbet (saliva is acidic, with a pH of 6.5); sherbet generally contains an organic acid, such as malic or ascorbic acids
Acids are corrosive	Teeth decay after eating sugar, and one of the first metabolites from sugar is lactic acid
Acids react with a base to form a salt and water ('neutralization')	Rubbing a dock leaf (which contains an organic base) on the site of a nettle sting (which contains acid) will neutralize the acid and relieve the pain

a substance capable of donating a proton. Therefore, we describe ethanoic acid as a 'Lowry–Brønsted acid'.

Why is it dangerous to allow water near an electrical appliance, if water is an insulator?

The solvated proton

We all need to know that electricity and water are an extremely dangerous combination, and explains why we are taught never to sit in the bath at the same time as shaving with a plug-in razor or drying our hair. Electrocution is almost inevitable, and is often fatal.

The quantity we call 'electricity' is a manifestation of charge Q passing through a suitable conductor. The electrical conductivity κ of water (be it dishwater, rainwater, bathwater, etc.) must be relatively high because electricity can readily conduct through water. Nevertheless, the value of κ for water is so low that we class water as an insulator. Surely there is a contradiction here?

> *Care*: the symbol of conductivity is not K but the Greek letter kappa, κ.

'Super-pure water' has been distilled several times, and is indeed an insulator: its conductivity κ is low at 6.2×10^{-8} S cm^{-1} at 298 K, and lies midway between classic insulators such as Teflon, with a conductivity of about 10^{-15} S cm^{-1}, and semiconductors such as doped silicon, for which $\kappa = 10^{-2}$ S cm^{-1}. The conductivity of metallic copper is as high as 10^6 S cm^{-1}.

The value of κ cited above was for *super-pure* water, i.e. the product of multiple distillations, but 'normal' water from a tap will inevitably contain solutes (hence the 'furring' inside a kettle or pipe). Inorganic solutes are generally ionic salts; most organic solutes are not ionic. The conductivity of super-pure water is low because the molecules of water are almost exclusively covalent, with the extent of ionicity being very slight. But as soon as a salt dissolves, the extent of ionicity in the water increases dramatically, causing more extensive water *dissociation*.

> Pure water is a mixture of three components: H_2O, and its two dissociation products, the solvated proton (H_3O^+) and the hydroxide ion (OH^-).

Ignoring for the moment the solute in solution, the water dissociation involves the splitting of water *itself* in a process called *autoprotolysis*. The reaction is usually represented as

$$2H_2O \longrightarrow H_3O^+_{(aq)} + OH^-_{(aq)} \qquad (6.2)$$

where the $H_3O^+_{(aq)}$ species is often called a *solvated proton*. It also has the names *hydroxonium ion* and *hydronium ion*. The complex ion $H_3O^+_{(aq)}$ is a more accurate representation of the proton responsible for acidic behaviour than the simplistic '$H^+_{(aq)}$' we wrote in Equation (6.1). Note how the left-hand side of Equation (6.2) is covalent and the right-hand side is ionic.

> It is safer in many instances to assume the solvated proton has the formula unit $[H(H_2O)_4]^+$, with four water molecules arranged tetrahedrally around a central proton, the proton being stabilized by a lone pair from each oxygen atom.

Dissolving a solute generally shifts the reaction in Equation (6.2) from left to right, thereby increasing the concentration of *ionic* species in solution. This increased number of ions causes the conductivity κ of water to increase, thereby making it a fatally efficient conductor of charge.

Why is bottled water 'neutral'?

Autoprotolysis

The word 'criterion' is used of a principle or thing we choose to use as a standard when judging a situation. The plural or criterion is 'criteria', not 'criterions'.

'Autoprotolysis' comes from *proto*–indicating the proton, and *lysis*, which is a Greek root meaning 'to cleave or split'. The prefix *auto* means 'by self' or 'without external assistance'.

The labels of many cosmetic products, as well as those on most bottles of drinking water, emphasize how the product is 'neutral', implying how it is neither acidic nor alkaline. This stipulation is deemed to show how healthy the water is. But how do they know? And, furthermore, what is their criterion for testing?

A better way of defining 'neutral' is to say equal numbers of protons and hydroxide ions reside in solution (both types of ion being solvated). How does this situation arise? *Autoprotolysis*, as mentioned above, represents the *self*-production of protons, which is achieved by the splitting of water according to Equation (6.2). It is clear from Equation (6.2) how the consequence of such auto-protolytic splitting is a solution with equal numbers of protons and hydroxide ions.

When water contains no dissolved solutes, the concentrations of the solvated protons and the hydroxide ions are equal. Accordingly, from our definition of 'neutral' above, we see why pure water should always be neutral, since $[H_3O^+_{(aq)}] = [OH^-_{(aq)}]$.

As with all physicochemical processes, the extent of Equation (6.2) may be quantified by an equilibrium constant K. We call it the *autoprotolysis constant*, as defined by

$$K = \frac{[H_3O^+_{(aq)}][OH^-_{(aq)}]}{[H_2O]^2} \qquad (6.3)$$

The water term in the denominator of Equation (6.3) is always large when compared with the other two concentrations on the top, so we say it remains constant. This assumption explains why it is rare to see the autoprotolysis constant written as Equation (6.3). Rather, we usually rewrite it as

$$K_w = [H_3O^+_{(aq)}][OH^-_{(aq)}] \qquad (6.4)$$

where K in Equation (6.3) $= K_w \times [H_2O]^2$ in Equation (6.4). We will only employ Equation (6.4) from now on. We call K_w the *autoprotolysis constant* or *ionic product* of water.

Table 6.2   Values of the autoprotolysis constant K_w as a function of temperature

Temperature $T / {}^\circ C$	0	18	25	34	50
$K_w \times 10^{14}$	0.12	0.61	1.04	2.05	5.66

Source: *Physical Chemistry*, W. J. Moore (4th Edn), Longmans, London, 1962, p. 365.

The value of K_w is 1.04×10^{-14} at 298 K when expressed in concentration units of mol dm^{-3}. Like all equilibrium constants, its value depends on the temperature. Table 6.2 lists a few values of K_w as a function of temperature. Note how K_w increases slightly as the temperature increases.

It should now be clear from Equation (6.4) how water splits (dissociates) to form *equal* number of protons and hydroxide ions, hence its neutrality, allowing us to calculate the numbers of each from the value of K_w.

> Note how K_w has units.

> K_w is often re-expressed as pK_w, where the 'p' is a mathematical operator (see p. 246). pK_w has a value of 14 at 298 K.

Worked Example 6.1   What is the concentration of the solvated proton in super-pure water?

Since $[H_3O^+{}_{(aq)}] = [OH^-{}_{(aq)}]$, we could rewrite Equation (6.4) as

$$K_w = [H_3O^+{}_{(aq)}]^2$$

Taking the square root of both sides of this expression, we obtain

$$[H_3O^+{}_{(aq)}] / \text{mol dm}^{-3} = \sqrt{K_w} = [10^{-14}]^{1/2}$$

so

$$[H_3O^+{}_{(aq)}] = 10^{-7} \text{ mol dm}^{-3}$$

The concentration of solvated protons in super-pure water is clearly very small.

What is 'acid rain'?

Hydrolysis

Acid rain is one of the worst manifestations of the damage we, as humans, inflict on our planet. Chemicals combine with elemental oxygen during the burning of fossil fuels, trees and rubbish to generate large amounts of 'acidic oxides' such as nitrogen monoxide (NO), carbon dioxide (CO_2) and sulphur dioxide (SO_2).

Natural coal and oil contain many compounds of nitrogen. One of the worst products of their combustion is the acidic oxide of nitrogen, NO. At once, we are startled by this terminology, because the Lowry–Brønsted definition of an acid involves the release of a proton, yet nitrogen monoxide NO has no proton to give.

As long ago as the 18th century, French chemists appreciated how burning elemental carbon, nitrogen or sulphur generated compounds which, when dissolved in water, yielded an acidic solution.

To understand the acidity of pollutants such as NO and CO_2, we need to appreciate how the gas does not so much dissolve in water as *react* with it, according to

$$CO_{2(g)} + 2H_2O_{(l)} \longrightarrow HCO^-_{3(aq)} + H_3O^+_{(aq)} \qquad (6.5)$$

Carbonic acid, $H_2CO_{3(aq)}$, never exists as a pure compound; it only exists as a species in aqueous solution, where it dissociates in just the same way as ethanoic acid in Equation (6.1) to form a solvated proton and the $HCO^-_{3(aq)}$ ion. Note how we form a solvated proton $H_3O^+_{(aq)}$ by *splitting* a molecule of water, rather than merely donating a proton. Carbonic acid is, nevertheless, a Lowry–Brønsted acid.

The nitric acid in acid rain forms by a more complicated mechanism: $4NO_{(g)} + 2H_2O_{(l)} + O_{2(g)} \longrightarrow 4HNO_{3(aq)}$

The carbonic acid produced in Equation (6.5) is a proton donor, so the solution contains more solvated protons than hydroxide ions, resulting in rain that is (overall) an acid. To make the risk of pollution worse, 'acid rain' in fact contains a mixture of several water-borne acids, principally nitric acid, HNO_3 (from nitrous oxide in water), and sulphurous acid, H_2SO_3 (an aqueous solution of sulphur dioxide).

In summary, we see how the concentrations of H_3O^+ and OH^- are the same if water contains no dissolved solutes, but dissolving a solute such as NO increases the concentration of H_3O^+; in a similar way, the concentration of OH^- will increase if the water contains any species capable of consuming protons.

'Hydrolysis' means to split water, the word coming from the two Greek roots *hydro* meaning water, and *lysis* meaning 'to cleave or split'.

It is time to introduce a few new words. We say carbonic acid forms by *hydrolysis*, i.e. by splitting a molecule of water. We describe the extent of hydrolysis in Equation (6.5) by the following equilibrium constant:

$$K = \frac{[HCO_3^-][H_3O^+]}{[CO_2][H_2O]^2} \qquad (6.6)$$

We sometimes call Equation (6.6) the *hydrolysis constant* of carbon dioxide. In fact, the water term in the 'denominator' (the bottom line) is so large compared with all the other terms that it remains essentially constant. Therefore, we write Equation (6.6) in a different form:

Care: the values of K from these equations are only meaningful for concentrations at *equilibrium*.

$$K' = \frac{[HCO_3^-][H_3O^+]}{[CO_2]} \qquad (6.7)$$

Note how the two K terms, K in Equation (6.6) and K' in Equation (6.7), will have different values.

Why does cutting an onion make us cry?

Other aqueous acids in the environment

The reason why our eyes weep copiously when peeling an onion is because the onion contains minute pockets of sulphur trioxide, $SO_{3(g)}$. Cutting the onion releases this gas. A mammalian eye is covered with a thin film of water-based liquid ('tears') to minimize friction with the eyelid. The tears occur in response to SO_3 dissolving in this layer of water to form sulphuric acid:

$$SO_{3(g)} + H_2O(l) \longrightarrow H_2SO_{4(aq)} \tag{6.8}$$

The sulphuric acid produced dissociates in the water to form SO_4^{2-} and two protons. The eyes sting as a direct consequence of contact with this acid.

Why does splashing the hands with sodium hydroxide solution make them feel 'soapy'?

Proton abstraction

Sodium hydroxide in solution dissociates to yield solvated cations and anions, Na^+ and the hydroxide ion OH^- respectively:

$$NaOH_{(s)} \longrightarrow Na^+_{(aq)} + OH^-_{(aq)} \tag{6.9}$$

The solvated hydroxide ion in Equation (6.9) is formed in addition to the hydroxide ions produced during water autoprotolysis, so there are more hydroxide ions in solution than solvated protons, yielding excess hydroxide in solution. We say the solution is *alkaline*. As an alternative name, we say hydroxide is a *base* (see p. 241).

Oils in the skin react readily with the hydroxide ions via the same chemical process occurring when spray-on oven cleaner 'eats' into the grime in an oven, reacting principally by the $OH^-_{(aq)}$ ion consuming protons. Let us start, for example, with a molecule possessing a proton capable of being donated; call it HA, where 'A' is merely an anion of some sort. This proton must be *labile*. The hydroxide ion removes this labile proton to generate water, according to

$$HA + OH^-_{(aq)} \longrightarrow A^-_{(aq)} + H_2O \tag{6.10}$$

This proton-removing ability characterizes the reactions of hydroxide ions in aqueous solutions, and of *bases* in general. We

> Care: It is a common mistake to call the OH^- ion a 'hydroxyl'. It is not: a hydroxyl is correctly a *covalently bound* –OH group, for example in an alcohol.

> *Fullers' earth* is a type of clay named after a *fuller*, whose job was to clean cloth, e.g. stripping wool of its grease. Fullers' earth removes oils and grease from cloth because of its alkalinity, just like an oven cleaner solution.

> The Lowry–Brønsted theory says a *base* is a proton remover.

A *basic* chemical consumes protons.

go further by *defining* hydroxide as a base because it reacts with (i.e. consumes) labile protons. And any chemical capable of removing protons is said to be *basic*.

Aside

Saponification

The word 'saponify' comes from the Latin *sapo*, meaning soap.

Hydroxide ions react to split ('hydrolyse') natural esters in the skin to form glycerol (**II**) and palmitic or stearic acid – a reaction called *saponification*. Palmitic and stearic acids subsequently react with the base to form the respective long-chain carboxylate anions – which is soap.

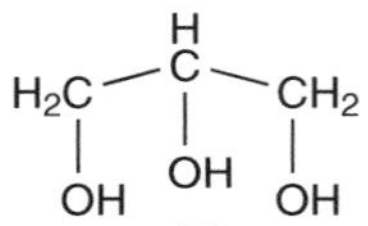

$$H_2C \overset{\overset{\displaystyle H}{|}}{\underset{\underset{\displaystyle OH}{|}}{C}} CH_2$$

$$\underset{OH}{|} \qquad \underset{OH}{|}$$

(**II**)

These cleansing properties of bases were appreciated in antiquity. For example, in a portion of the Bible probably written in about 1200 BC, a character called Job declares his desire to be clean, saying, 'If I washed myself with soap and snow, and my hands with washing soda . . .' (snow was thought to be especially pure and soda ($Na_2CO_3 \cdot 10H_2O$) is alkaline and has long been used as a soap). This quote may be found in full in the Bible, see Job 9:30.

The Jewish Prophet Jeremiah writing in about 700 BC says much the same thing: look at Jeremiah 2:21–23 in the Hebrew Bible.

Why is aqueous ammonia alkaline?

Lowry–Brønsted bases

All aqueous solutions naturally contain hydroxide ions in consequence of the auto-protolytic reaction in Equation (6.2). As we have seen, there will be equal numbers of solvated protons and solvated hydroxide ions unless we add an acid or base to it. A solution containing more solvated protons than hydroxide ions is said to be an 'acid' within the Lowry–Brønsted theory, and a solution comprising more hydroxide ions than solvated protons is said to be a *base*.

But a word of caution: species other than metal hydroxides can act as bases. Ammonia is such an example, since it can abstract protons in aqueous solution according to

$$NH_{3(aq)} + H_2O \longrightarrow NH^+_{4(aq)} + OH^-_{(aq)} \qquad (6.11)$$

To *abstract* a proton is to remove only the proton. The substantial extent of dissociation in Equation (6.11) helps explain why 'aqueous ammonia' is more properly called 'ammonium hydroxide', NH_4OH. We generate the solvated hydroxide ion $OH^-_{(aq)}$ by abstracting a proton from water. The $OH^-_{(aq)}$ ion in Equation (6.11) is chemically and physically identical to the solvated hydroxide ion generated by dissolving NaOH or KOH in water.

> We say a proton is *abstracted* when removed selectively. Similarly, we call a *selective* summary or précis of a piece of prose 'an abstract'.

Why is there no vinegar in crisps of salt and vinegar flavour?

Conjugate acids and bases

Potato crisps come in many flavours, perhaps the most popular being 'salt and vinegar'. Curiously, a quick glance at the packet's list of ingredients reveals how the crisps contain unhealthy amounts of salt, but no vinegar (ethanoic acid) at all. In fact, the manufacturer dusts the crisps with powdered sodium ethanoate ($NaCO_2CH_3$), because 'real' vinegar would soon make the crisps limp and soggy. Inside the mouth, acid from the saliva reacts with the ethanoate anion to form ethanoic acid:

$$CH_3CO^-_{2(aq)} + H_3O^+_{(aq)} \longrightarrow CH_3CO_2H_{(aq)} + H_2O \qquad (6.12)$$

This reaction proceeds inside the mouth, rapidly reaching its position of equilibrium, and allowing the ethanoic acid to impart its distinctive vinegary flavour.

The solvated proton on the left of Equation (6.12) acts as an acid, since it *donates* a proton at the same time as the ethanoate ion behaves as a base, because it *accepts* a proton. To complicate the situation, the reaction is one half of a dynamic equilibrium, i.e. it proceeds in both the forward and backward directions. In the backward direction, we notice how this time the ethanoic acid acts as an acid and the water acts as a base.

The reaction in Equation (6.12) illustrates the coexistence of two acids and two bases. We say the ethanoate ion and ethanoic acid represent a *conjugate pair*, and the solvated proton and the water form a second conjugate pair. Within the ethanoic–ethanoate pair, the ethanoic acid is the *conjugate acid* and the ethanoate anion is the *conjugate base*. Similarly, H_3O^+ is a conjugate acid to the

> The word 'conjugate' comes from the Latin *conjugare*, meaning 'to yoke together' (the prefix *con* means 'together' and *jugare* is 'to yoke'). Similarly, the English word 'conjugal' relates to marriage and concerns the joining of husband and wife.

conjugate base of H_2O. Other examples of conjugate acid–base pairs include nitric acid and nitrate ion, and ammonium ion and ammonia (the acid being cited first in each case).

We must treat with caution one further aspect of the Brønsted theory: multiple proton-donation reactions. Consider the example of the bicarbonate ion HCO_3^- in water. When titrating bicarbonate with a base such as hydroxide, the ion behaves as an acid to form the carbonate anion and water:

A substance like bicar-
bonate, which can react
as either an acid or as
a base, is said to be
amphoteric. The word
comes from the Greek
amphoteros, meaning
'both'.

$$HCO_3^- + OH^- \longrightarrow CO_3^{2-} + H_2O \qquad (6.13)$$

But, conversely, when titrating ions with an acid, the bicarbonate behaves as a base, losing its proton to form carbonic acid:

$$HCO_3^- + H_3O^+ \longrightarrow H_2CO_3 + H_2O \qquad (6.14)$$

We see how the same ion acts as an acid or as a base, depending on the other reagents in solution. We say the bicarbonate ion is *amphoteric*, since it reacts either as an acid or as a base.

SAQ 6.1 Consider the following pairs, and for each decide which is the conjugate acid and which the base: (a) carbonate and bicarbonate; (b) H_2EDTA^{2-} and H_3EDTA^-; (c) HNO_2 and NO_2^-.

Aside

Related models of acids and bases

The concept of acid and base can be generalized in several ways. In liquid ammonia, for example, the ammonium and amide ions (NH_4^+ and NH_2^- respectively) coexist. The roles of these ions are directly comparable with H_3O^+ and OH^- in water. In ammonia, the species NH_4Cl and $NaNH_2$ can be considered to be the respective acid and base conjugates, just as HCl and NaOH are an acid–base pair in water. This solvent-based classification of acids and bases derived from Franklin, in 1905. His ideas are worth careful thought, although we no longer use his terminology.

Brønsted's definition of acids and bases (see p. 234 and 240) emphasizes the complementary nature of acids and bases, but it is broader than Franklin's model because it does not require a solvent, and can even be applied to gas-phase reactions, e.g. $HCl_{(g)} + NH_{3(g)} \rightarrow NH_4Cl_{(s)}$.

How did soldiers avoid chlorine gas poisoning at the Second Battle of Ypres?

Neutralization reactions with acids and bases

The bloody Second Battle of Ypres was fought in France on 22 April 1915, and was the first time in modern warfare when poison gases were employed. At a crucial stage in the battle, the German forces filled the air above the enemy trenches with chlorine gas.

Elemental chlorine Cl_2 dissolves slightly in water, and hydrolyses some of the water to yield hypochlorous acid, HOCl, according to

$$Cl_{2(aq)} + H_2O_{(l)} \longrightarrow HCl_{(aq)} + HOCl_{(aq)} \qquad (6.15)$$

> Hypochlorous acid, HOCl, is one of the active components in household bleach.

The reaction in Equation (6.15) occurs readily in the lungs and eyes (the sensitive tissues of which are lined with water) to cause irreparable damage. Troops exposed to chlorine apparently experienced a particularly slow and nasty death.

The German troops did not advance, because they were not sure if the gas masks issued to their own troops could withstand the chlorine. They were also deterred by the incursion of a Canadian regiment. But one of the young Canadian soldiers knew a little chemistry: sniffing the gas, he guessed its identity correctly, and ordered the soldiers to cover their faces with handkerchiefs (or bandages) soaked in their own urine. The idea spread quickly, and the Canadians, together with two Yorkshire territorial battalions, were able to push back the German troops.

> After this battle, both sides showed reluctance to employ poisonous gases again, being afraid it would drift back and poison their own troops. Cl_2 gas also caused extensive corrosion of rifles and artillery breech blocks, making them unusable.

One of the major constituents of urine is the di-amine, urea (**III**). Each amine group in urea should remind us of ammonia in Equation (6.11). Solutions of urea in water are *basic* because the two amine moieties each abstract a proton from water, to generate an ammonium salt and a hydroxide ion:

> *Reminder*: to a chemist, the word *basic* does not mean 'elementary' or 'fundamental', but 'proton abstracting'.

$$H_2N-\overset{\overset{\textstyle O}{\|}}{C}-NH_2$$

(III)

$$H_2N-\overset{\overset{\textstyle O}{\|}}{C}-NH_2 + 2H_2O \longrightarrow \overset{\oplus}{H_3N}-\overset{\overset{\textstyle O}{\|}}{C}-\overset{\oplus}{NH_3} \quad 2OH^- \qquad (6.16)$$

The two OH^- ions formed during Equation (6.16) explain why aqueous solutions of urea are alkaline.

As we saw above, chlorine forms hypochlorous acid, HOCl. The hydroxide ions generated from urea react with the hypochlorous acid in a typical acid–base reaction, to form a salt and water:

$$HOCl_{(aq)} + OH^-_{(aq)} \longrightarrow ClO^-_{(aq)} + H_2O \qquad (6.17)$$

where ClO^- is the hypochlorite ion.

Equation (6.17) is an example of a *neutralization* reaction, a topic we discuss in more depth in Section 6.3.

How is sherbet made?

Effervescence and reactions of acids

> Malic acid (**IV**) occurs naturally, and is the cause of the sharp taste in over-ripe apples.

Sherbet and sweets yielding a fizzy sensation in the mouth generally contain two components, an acid and a simple carbonate or bicarbonate. A typical reaction of an acid with a carbonate is *effervescence*: the generation of gaseous carbon dioxide. In a well-known brand of British 'fizzy lolly', the base is sodium bicarbonate and the acid is malic acid (**IV**). Ascorbic acid (vitamin C) is another common acid included within sherbet.

HO $\underset{HOOC}{\overset{}{\diagdown}} \overset{H}{\underset{CH_2COOH}{\diagup}}$

(IV)

The acid and the bicarbonate dissolve in saliva as soon as the 'fizzy lolly' is placed in the mouth. If we abbreviate the malic acid to HM (M being the maliate anion), the 'fizzing' reaction in the mouth is described by

$$HM_{(aq)} + HCO^-_{3(aq)} \longrightarrow M^-_{(aq)} + CO_{2(g)} \uparrow + H_2O_{(l)} \qquad (6.18)$$

where in this case the subscript 'aq' means aqueous saliva. The subjective sensation of 'fizz' derives from the evolution of gaseous carbon dioxide.

> Soda is an old fashioned name for sodium carbonate, $Na_2CO_3 \cdot 10H_2O$.

This reaction between an acid and a carbonate is one of the oldest chemical reactions known to man. For example, it says in a portion of the Hebrew Bible written about 1000 BC, '... like vinegar poured on soda, is one who sings songs to a heavy heart', i.e. inappropriate and lighthearted singing can lead to a dramatic response! The quote may be found in full in the book of Proverbs 25:20.

Why do steps made of limestone sometimes feel slippery?

> *Ossified* means 'to make rigid and hard.' The word comes from the Latin *ossis*, meaning bone.

The typical reactions of an alkali

Limestone is an ossified form of calcium carbonate, $CaCO_3$. Limestone surfaces soon become slippery and can be quite dangerous if

Table 6.3 Typical properties of Lowry–Brønsted bases

Base property	Example from everyday life
Bases react with an acid to form a salt and water ('neutralization')	Rubbing a dock leaf (which contains an organic base) on the site of a nettle sting (which contains acid) will neutralize the acid and relieve the pain
Bases react with esters to form an alcohol and carboxylic acid ('saponification')	Aqueous solutions of base feel 'soapy' to the touch
Bases can be corrosive	Oven cleaner comprising caustic soda (NaOH) can cause severe burns to the skin

pools of stagnant rainwater collect. Although calcium carbonate is essentially insoluble in water, minute amounts do dissolve to form a dilute solution, which is alkaline.

The alkali in these water pools reacts with organic matter such as algae and moss growing on the stone. The most common of these reactions is *saponification* (see p. 240), which causes naturally occurring esters to split, to form the respective carboxylic acid and an alcohol. Once formed, this carboxylic acid reacts with more alkaline rainwater to form a metal carboxylate, according to

$$2RCOOH_{(aq)} + CaCO_{3(aq)} \longrightarrow Ca^{2+}(RCOO^-)_{2(aq)}$$
$$+ CO_{2(aq)} + H_2O \qquad (6.19)$$

The carbon dioxide generated by Equation (6.19) generally remains in solution as carbonic acid, although the rainwater can look a little cloudy because minute bubbles form.

Like most other metal carboxylates, the calcium carboxylate $(Ca(RCOO)_2)$ formed during Equation (6.19) readily forms a 'soap', the name arising since its aqueous solutions feel slippery and soapy to the touch. The other commonly encountered metal carboxylates are the major components of household soap, which is typically a mixture of potassium stearate and potassium palmitate (the salts of stearic and palmitic acids).

In summary, steps of limestone become slippery because the stagnant water on their surface is alkaline, thereby generating a solution of an organic soap. Other typical properties of bases and alkalis are listed in Table 6.3.

> Metal carboxylates are called *soaps* because they saponify oils in the skin and decrease the surface tension γ of water, which makes the surfaces more slippery.

> The serial TV programs known as 'soap operas' earned their name in the USA at a time when much of a program-maker's funding came from adverts for household soap.

Why is the acid in a car battery more corrosive than vinegar?

pH

Car batteries generally contain sulphuric acid at a concentration of about $10 \ \text{mol dm}^{-3}$. It is extremely corrosive, and can generate horrific chemical burns. By contrast, the

Care: the 'H' in pH derives from the symbol for hydrogen, and is always given a big letter. The 'p' is a mathematical operator, and is always small.

An acid's pH is defined as minus the logarithm (to the base ten) of the hydrogen ion concentration.

The 'p' in Equation (6.20) is the mathematical operator '−$\log_{10}$' of something. pH means we have applied the operator 'p' to [H$^+$]. The p is short for *potenz*, German for power.

The lower the pH, the more concentrated the acid.

concentration of the solvated protons in vinegar lies in the range 10^{-4}–10^{-5} mol dm^{-3}.

Between these two acids, there is up to a million-fold difference in the number of solvated protons per litre. We cannot cope with the unwieldy magnitude of this difference and tend to talk instead in terms of the *logarithm* of the concentration. To this end, we introduce a new concept: the *pH*. This is defined mathematically as 'minus the logarithm (to the base ten) of the hydrogen ion concentration':

$$pH = -\log_{10}[H^+/mol\ dm^{-3}] \qquad (6.20)$$

The concentrations of bench acids in an undergraduate laboratory are generally less than 1 mol dm^{-3}, so by corollary the minus sign to Equation (6.20) suggests we generally work with positive values of pH. Only if the solution has a concentration greater than 1 mol dm^{-3} will the pH be negative. Contrary to popular belief, a negative pH is not impossible. (Try inserting a concentration of 2.0 mol dm^{-3} into Equation (6.20) and see what happens!)

Notice how we generally infer the *solvated proton*, H_3O^+, each time we write a concentration as [H$^+$], which helps explain why the concept of pH is rarely useful when considering acids dissolved in non-aqueous solvents. When comparing the battery acid with the bench acid, we say that the battery acid has a lower pH than does the bench acid, because the number of solvated protons is greater and, therefore, it is more acidic. Figure 6.1 shows the relationship between the concentration of the solvated protons and pH. We now appreciate why the pH increases as the concentration decreases.

Apart from the convenience of the logarithmically compressed scale, the concept of pH remains popular because one of the most popular methods of measuring the acidity of an aqueous solution is the glass electrode (see p. 336), the measurement of which is directly proportional to pH, rather than to [H_3O^+].

We need to introduce a word of caution. Most modern calculators cite an answer with as many as ten significant figures, but we do not know the concentration to more than two or three significant figures. In a related way, we note how the pH of blood is routinely measured to within 0.001 of a pH unit, but most chemical applications

[H_3O^+]/mol dm^{-3}	10	1	10^{-1}	10^{-2}	10^{-3}	...	10^{-7}	...	10^{-10}	10^{-11}
pH		−1	0	1	2	3	...	7	...	10	11

Figure 6.1 The relationship between concentrations of strong acids and the solution pH

do not require us to cite the pH to more than 0.05 of a pH unit. In fact, we can rarely cite pH to a greater precision than 0.01 for most biological applications.

Worked Example 6.2   What is the pH of bench nitric acid having a concentration of 0.25 mol dm^{-3}?

Inserting values into Equation (6.20):

$$\text{pH} = -\log[0.25 \text{ mol dm}^{-3}/\text{mol dm}^{-3}] = 0.6$$

The acid has a pH of 0.6.

> The concentration of the solvated protons in Equation (6.18) needs to be expressed in the familiar (but non-SI) units of mol dm^{-3}; the SI unit of concentration is mol m^{-3}.

SAQ 6.2   What is the pH of hydrochloric acid having a concentration of 0.2 mol dm^{-3}?

SAQ 6.3   To highlight the point made above concerning numbers of significant figures, determine the (negligible) difference in $[H_3O^+]$ between two acid solutions, one having a pH of 6.31 and the other 6.32.

> We divide the concentration by its units to yield a dimensionless *number*.

Sometimes we know the pH and wish to know the concentration of the solvated protons. Hence, we need to rewrite Equation (6.20), making $[H^+]$ the subject, to obtain

$$[H^+/\text{mol dm}^{-3}] = 10^{-\text{pH}} \qquad (6.21)$$

> We are assuming the concentration of the proton H^+ is the same as the concentration of $H_3O^+{}_{(aq)}$.

Worked Example 6.3   What is the concentration of nitric acid having a pH of 3.5?

Inserting values into Equation (6.21):

$$[H^+] = 10^{-3.5}$$

so

$$[H^+] = 3.16 \times 10^{-4} \text{ mol dm}^{-3}$$

SAQ 6.4   What is the concentration of nitric acid of pH = 2.2?

Occasionally, we can merely *look* at a pH and say straightaway what is its concentration. If the pH is a whole number – call it x – the concentration will take the form $1 \times 10^{-x} \text{ mol dm}^{-3}$. As examples, if the pH is 6, the concentration is $10^{-6} \text{ mol dm}^{-3}$; if the concentration is $10^{-3} \text{ mol dm}^{-3}$ then the pH is 3, and so on.

> Sometimes, we refer to a whole number as an *integer*.

Worked Example 6.4   Without using a calculator, what is the concentration of HNO_3 solution if its pH is 4?

If the pH is x, then the concentration will be 10^{-x} mol dm^{-3}, so the acid concentration is 10^{-4} mol dm^{-3}.

SAQ 6.5 Without using a calculator, what is the pH of hydrochloric acid of concentration 10^{-5} mol dm^{-3}?

Aside

Units

We encounter problems when it becomes necessary to take the logarithm of a concentration (which has units), since it contravenes one of the laws of mathematics. To overcome this problem, we implicitly employ a 'dodge' by rewriting the equation as

$$pH = -\log_{10}\left(\frac{[H_3O^+]}{c^\ominus}\right)$$

where the term $c^\ominus$ is the standard state, which generally has the value of 1 mol dm^{-3}. The $c^\ominus$ term is introduced merely to allow the units within the bracketed term above to cancel. Throughout this chapter, concentrations will be employed relative to this standard, thereby obviating the problem inherent with concentrations having units.

Justification Box 6.1

Sørenson introduced this definition of pH in 1909.

The definition of pH is given in Equation (6.20) as

$$pH = -\log_{10}[H^+]$$

First we multiply both sides by '-1':

$$-pH = \log_{10}[H^+]$$

and then take the antilog, to expose the concentration term. Correctly, the function 'antilog' means a mathematical operator, which performs the opposite job to the original function. The opposite function to log is a type of exponential. As the log is written in base 10, so the exponent must also be in base 10.

If $y = \log_{10} x$, then $x = 10^y$. This way we obtain Equation (6.21).

Sometimes we need to know the pH of *basic* solutions.

Worked Example 6.5   What is the pH of a solution of sodium hydroxide of concentration $0.02 \, mol \, dm^{-3}$? Assume the temperature is 298 K.

At first sight, this problem appears to be identical to those in previous Worked Examples, but we soon appreciate how it is complicated because we need first to calculate the concentration of the free protons before we can convert to a pH. However, if we know the concentration of the alkali, we can calculate the pH thus:

$$pK_w = pH + pOH \qquad (6.22)$$

where pK_w and pH have their usual definitions, and we define pOH as

$$pOH = -\log_{10}[OH^-] \qquad (6.23)$$

Inserting the concentration $[NaOH] = 0.02 \, mol \, dm^{-3}$ into Equation (6.22) yields the value of pOH = 2. The value of pK_w at 298 K is 14 (see Table 6.2). Therefore:

$$pH = pK_w - pOH$$

$$pH = 14 - 2$$

$$pH = 12$$

> $$pK_w = -\log_{10} K_w$$
> $$pH = -\log_{10}[H_3O^+_{(aq)}]$$
> $$pOH = -\log_{10}[OH^-_{(aq)}]$$

> We must look up the value of K_w from Table 6.2 if the temperature differs from 298 K, and then calculate a different value using Equation (6.26).

SAQ 6.6   What is the pH of a solution of potassium hydroxide of concentration $6 \times 10^{-3} \, mol \, dm^{-3}$. Again, assume the temperature is 298 K.

Justification Box 6.2

Water dissociates to form ions according to Equation (6.2). The ionic product of the concentrations is the autoprotolysis constant K_w, according to Equation (6.4).

Taking logarithms of Equation (6.4) yields

$$\log_{10} K_w = \log_{10}[H_3O^+] + \log_{10}[OH^-] \qquad (6.24)$$

Next, we multiply each term by '−1' to yield

$$-\log_{10} K_w = -\log_{10}[H_3O^+] - \log_{10}[OH^-] \qquad (6.25)$$

The term '$-\log_{10}[H_3O^+]$' is the solution pH and the term '$-\log_{10} K_w$' is defined according to

$$pK_w = -\log_{10} K_w \qquad (6.26)$$

We give the name pOH to the third term '$-\log_{10}[OH^-]$'. We thereby obtain Equation (6.22).

Why do equimolar solutions of sulphuric acid and nitric acid have different pHs?

Mono-, di- and tri-basic acids

Nitric acid, HNO_3, readily dissolves in water, where it dissociates according to

$$HNO_{3(aq)} + H_2O \longrightarrow 1H_3O^+_{(aq)} + NO^-_{3(aq)} \qquad (6.27)$$

Equimolar means 'of equal molarity', so equimolar solutions have the same concentration.

The stoichiometry illustrates how each formula unit generates a single solvated proton. By contrast, sulphuric acid, H_2SO_4, dissociates in solution according to

$$H_2SO_{4(aq)} + 2H_2O \longrightarrow 2H_3O^+_{(aq)} + SO^{2-}_{4(aq)} \qquad (6.28)$$

so each formula unit of sulphuric acid generates *two* solvated protons. In other words, each mole of nitric acid generates only 1 mol of solvated protons but each mole of sulphuric acid generates 2 mol of solvated protons. We say nitric acid is a *mono-protic* acid and sulphuric acid is a *di-protic* acid. Tri-protic acids are rare. Fully protonated ethylene diamine tetra-acetic acid H_4EDTA (**V**) is a tetra-protic acid.

(V)

Equation (6.27) demonstrated how the concentration of the solvated protons equates to the concentration of a mono-protic acid from which it derived; but, from Equation (6.28), the concentration of the solvated protons will be *twice* the concentration if the parent acid is di-protic. These different stoichiometries affect the pH, as demonstrated now by Worked Examples 6.6 and 6.7.

Worked Example 6.6   Nitric acid of concentration 0.01 mol dm^{-3} is dissolved in water. What is its pH?

Since one solvated proton is formed per molecule of acid, the concentration $[H^+_{(aq)}]$ is also 0.01 mol dm^{-3}.

The pH of this acidic solution is obtained by inserting values into Equation (6.20):

$$pH = -\log_{10}[0.01]$$

$$pH = 2$$

Worked Example 6.7   What is the pH of sulphuric acid having the same concentration in water as the nitric acid in Worked Example 6.6?

This time, *two* solvated protons are formed per molecule of acid, so the concentration of $[H^+_{(aq)}]$ will be $0.02 \, mol \, dm^{-3}$.

The pH of this acidic solution is obtained by inserting values into Equation (6.20):

$$pH = -\log_{10}[2 \times 0.01]$$

$$pH = 1.68$$

A pH electrode immersed in turn into these two solutions would register a different pH despite the concentrations of the *parent* acids being the same.

We need to be careful with these calculations, because the extent of dissociation may also differ; see p. 255ff.

pH electrodes and pH meters are discussed in Chapter 7.

What is the pH of a 'neutral' solution?

pH and neutrality

A medicine or skin lotion is often described as 'pH neutral' as though it was obviously a good thing. A solution is defined as *neutral* if it contains neither an excess of solvated protons nor an excess of hydroxide ions. Equation (6.4) tells us the autoprotolysis constant K_w of super-pure water (water containing no additional solute) is $10^{-14} \, (mol \, dm^{-3})^2$. Furthermore, we saw in Worked Example 6.1 how the concentration of the solvated protons was $10^{-7} \, mol \, dm^{-3}$ at 298 K.

All neutral solutions have a pH of 7 at 298 K.

By considering both the definition of pH in Equation (6.20) and the concentration of the solvated protons from Worked Example 6.1, we see how a sample of super-pure water – which is *neutral* – has a pH of 7 at 298 K. We now go further and say *all neutral solutions have a pH of 7*. By corollary, we need to appreciate how an acidic solution always has a pH less than 7. If the pH is exactly 7, then the solution is neutral.

The maximum pH of an acid will be just less than 7, at 298 K.

The pH of a Lowry–Brønsted acid decreases as its concentration increases. Bench nitric acid of concentration $1 \, mol \, dm^{-3}$ has a $pH = 0$. An acid of higher concentration will, therefore, have a *negative* pH (the occasions when we need to employ such solutions are, thankfully, rare).

The pH of a Lowry–Brønsted acid DEcreases as its concentration INcreases.

What do we mean when we say blood plasma has a 'pH of 7.4'?

The pH of alkaline solutions

Table 6.4 lists the pH of many natural substances, and suggests human blood plasma, for example, should have a pH in the range 7.3–7.5. The pH of many natural

Table 6.4 The pHs of naturally occurring substances, listed in order of decreasing acidity

Human body		Common foodstuffs	
Gastric juices	1.0–3.0	Limes	1.8–2.0
Faeces	4.6–8.4	Rhubarb	3.1–3.2
Duodenal contents	4.8–8.2	Apricots	3.6–4.0
Urine	4.8–8.4	Tomatoes	4.0–4.4
Saliva	6.5–7.5	Spinach	5.1–5.7
Milk	6.6–7.6	Salmon	6.1–6.3
Bile	6.8–7.0	Maple syrup	6.5–7.0
Spinal fluid	7.3–7.5	Tap water	6.5–8.0
Blood plasma	7.3–7.5	Egg white (fresh)	7.6–8.0

Source: *Handbook of Chemistry and Physics* (66th Edition), R. C. Weast (ed.), CRC Press, Boca Raton, Florida, 1985, page D-146.

substances is higher than 7, so we cannot call them either 'acidic' or 'neutral'. The pHs range from 1.8 for limes (which explains why they taste so sour) to 7.8 or so for fresh egg white (albumen).

> The word 'product' is used here in its mathematical sense of 'multiplied by'.

Blood 'plasma' is that part of the blood remaining after removal of the haemoglobin cells that impart a characteristic 'blood-red' colour. According to Table 6.4, most people's plasma has a pH in the range 7.3–7.5. So, what is the concentration of solvated protons in such plasma? We met the autoprotolysis constant K_w in Equation (6.4). Although we discussed it in terms of super-pure water, curiously the relationship still applies to any aqueous system. The product of the concentrations of solvated protons and hydroxide ions is always 10^{-14} at 298 K.

If Equation (6.4) applies although the water contains dissolved solute, then we can calculate the concentration of solvated protons and the concentration of hydroxide ions, and hence ascertain what a pH of more than '7' actually means.

Worked Example 6.8 What is the concentration $[OH^-_{(aq)}]$ in blood plasma of pH 7.4?

Answer strategy. (1) We first calculate the concentration of solvated protons from the pH, via Equation (6.20). (2) Second, we compare the concentration $[H_3O^+_{(aq)}]$ with that of a neutral solution, via Equation (6.4).

(1) The concentration $[H_3O^+_{(aq)}]$ is obtained from Equation (6.20). Inserting values:

$$[H_3O^+_{(aq)}] = 10^{-7.4}$$

so

$$[H_3O^+_{(aq)}] = 4 \times 10^{-8} \text{ mol dm}^{-3}$$

(2) We see how the concentration obtained in part (1) is in fact *less* than the 1×10^{-7} mol dm^{-3} we saw for pure water, as calculated in Worked Example 6.1.

Equation (6.4) says $K_w = [H_3O^+_{(aq)}] \times [OH^-_{(aq)}]$, where the product of the two concentration has a value of 10^{-14} $(mol\,dm^{-3})^2$ at 298 K. Knowing the values of K_w and $[H_3O^+_{(aq)}]$, we can calculate the concentration of hydroxide ions in the blood plasma.

Rearranging to make $[OH^-_{(aq)}]$ the subject, we obtain

$$[OH^-_{(aq)}] = \frac{K_w}{[H_3O^+_{(aq)}]}$$

Inserting values,

$$[OH^-_{(aq)}] = \frac{10^{-14}\ (mol\ dm^{-3})^2}{4 \times 10^{-8}\ mol\ dm^{-3}}$$

so

$$[OH^-_{(aq)}] = 2.5 \times 10^{-7}\ mol\,dm^{-3}$$

We see how the pH of blood plasma is higher than 7, so the concentration of hydroxide ions exceeds their concentration in superpure water. We derive the generalization: aqueous solutions in which the concentration of hydroxide is greater than the concentration of solvated protons have a pH higher than 7. The pH is lower than 7 if the concentration of hydroxide is *less* than the concentration $[H_3O^+_{(aq)}]$.

> Aqueous solutions in which the concentration of hydroxide exceeds the concentration of solvated protons show a pH higher than 7.

SAQ 6.7 A solution of ammonia in water has a pH of 9. Without using a calculator, what is the concentration of solvated protons and hence what is the concentration of hydroxide ions?

SAQ 6.8 What is the pH of sodium hydroxide solution of concentration 10^{-2} $mol\,dm^{-3}$?

6.2 'Strong' and 'weak' acids and bases

Why is a nettle sting more painful than a burn from ethanoic acid?

Introducing 'strong' and 'weak' acids

Brushing against a common nettle *Urtica dioica* can cause a painful sting. The active component in a nettle sting is methanoic acid (**VI**), also called 'formic acid'. The sting of a nettle also contains natural additives to ensure that the methanoic acid stays on the skin, thereby maximizing the damage to its sensitive underlying tissue known as the *epidermis*.

> The sting of a red ant also contains methanoic acid.

(VI)

The chemical structures of **I** and **VI** reveal the strong similarities between ethanoic and methanoic acids, yet the smaller molecule is considerably nastier to the skin. Why? Methanoic acid dissociates in water to form the solvated methanoate anion $HCOO^-_{(aq)}$ and a solvated proton in a directly analogous fashion to ethanoic acid dissolving in water; Equation (6.1). In methanoic acid of concentration $0.01 \, mol \, dm^{-3}$, about 0.14 per cent of the molecules have dissociated to yield a solvated proton. By contrast, in ethanoic acid of the same concentration, only 0.04 per cent of the molecules have dissociated. We say the methanoic acid is a *stronger* acid than ethanoic since it yields more protons per mole. Conversely, ethanoic acid is *weaker*.

We might rephrase this statement, and say an acid is *strong* if its extent of ionization is high, and *weak* if the extent of ionization is small. Within this latter definition, both **I** and **VI** are weak acids.

In summary, the word 'acid' is better applied to methanoic acid than to ethanoic acid, since it is more acidic, and so methanoic acid in a nettle sting is more able to damage the skin than the ethanoic acid in vinegar.

But we need to be careful. In everyday usage, we say often something is 'strong' when we mean its concentration is large; similarly, we say something is 'weak' if its concentration is small. As a good example, when a strong cup of tea has a dark brown colour (because the compounds imparting a colour are concentrated) we say the tea is 'strong'. To a chemist, the words 'strong' and 'weak' relate only to the extent of *ionic* dissociation.

The word 'epidermis' derives from two Greek words, *derma* meaning skin, and *epi* meaning 'at', 'at the base of', or 'in additional to'. The same root *epi* occurs in 'epidural', a form of pain relief in which an injection is made *at the base of* the dura, located in the spine.

We say an acid is *strong* if the extent of its ionization is high, and *weak* if the extent of its ionization is small.

Care: do not confuse the words *strong* and *weak* acids with everyday usage, where we usually say something is 'strong' if its concentration is large, and 'weak' if its concentration is small.

To a chemist, the words 'strong' and 'weak' relate *only* to the extent of *ionic* dissociation.

Why is 'carbolic acid' not in fact an acid?

Acidity constants

'Carbolic acid' is the old-fashioned name for hydroxybenzene (**VII**), otherwise known as *phenol*. It was first used as an antiseptic to prevent the infection of post-operative wounds. The British surgeon Joseph (later 'Lord') Lister (1827–1912) discovered these antiseptic qualities in 1867 while working as Professor of Medicine at the Glasgow Royal Infirmary. He squirted a dilute aqueous solution of **VII** directly onto a post-operative wound and found that the phenol killed all the bacteria, thereby yielding the first reliable antiseptic in an era when medical science was in its infancy.

(VII)

The antibacterial properties of **VII** are no longer utilized in modern hospitals because more potent antiseptics have now been formulated. But its memory persists in the continued use of 'carbolic soap', which contains small amounts of phenol.

Phenol in water is relatively reactive, thereby explaining its potency against bacteria. But phenol dissolved in water contains relatively few solvated protons, so it is not particularly acidic. But its old name is carbolic *acid*!

Phenol (**VII**) can dissociate according to

$$PhOH_{(aq)} + H_2O_{(l)} \longrightarrow PhO^-_{(aq)} + H_3O^+_{(aq)} \qquad (6.29)$$

where Ph is a phenyl ring and $PhO^-_{(aq)}$ is the phenolate anion. An equilibrium constant may be written to describe this reaction:

$$K = \frac{[PhO^-_{(aq)}][H_3O^+_{(aq)}]}{[PhOH_{(aq)}][H_2O]} \qquad (6.30)$$

In fact, the water term in the denominator remains essentially constant, since it is always huge compared with all the other terms. Accordingly, we usually write a slightly altered version of K, cross-multiplying both sides of the equation with the concentration of water to yield

$$K_a = \frac{[PhO^-_{(aq)}][H_3O^+_{(aq)}]}{[PhOH_{(aq)}]} \qquad (6.31)$$

The resultant (modified) equilibrium constant is called the *acidity constant* of phenol, and has the new symbol K_a, which has a value is 10^{-10} for phenol. K_a is also called the *acid constant*, the *acid dissociation constant* or just the *dissociation constant*. The value of K_a for phenol is clearly tiny, and quantifies just how small the extent is to which it dissociates to form a solvated proton.

Phenol is a rare example of a stable *enol* (pronounced 'ene-ol'), with a hydroxyl bonded to a C=C bond. Most enols *tautomerize* to form a ketone.

The word 'antiseptic' comes from the Latin prefix *anti* meaning 'before' or 'against', and 'septic' comes from the Latin *septis*, meaning a bacterial infection. An 'antiseptic', therefore, prevents the processes or substances causing an infection.

Historically, carbolic *acid* was so called because solid phenol causes nasty chemical burns to the skin. The root *carbo* comes from the French for 'coal'.

Care: Do not confuse 'Ph' (a common abbreviation for a phenyl ring) with 'pH' (which is a mathematical operator meaning $-\log_{10}[H^+_{(aq)}]$).

K_a for phenol is 10^{-10} when expressing the concentrations with the units of mol dm^{-3}.

The value of K_a for ethanoic acid is a hundred thousand times larger at 1.8×10^{-5}, and K_a for methanoic acid is ten times larger still, at 1.8×10^{-4}; so methanoic acid generates more solvated protons per mole of acid than either phenol or ethanoic acid.

We discover the relative differences in K_a when walking in the country, for a nettle can give a nasty sting (i.e. a chemical burn) but vinegar does not burn the skin. We say methanoic acid is a *stronger* acid than ethanoic acid because its value of K_a is larger. A mole of phenol yields few protons, so we say it is a *weak* acid, because its value of K_a is tiny.

A *strong* acid has a large value of K_a, and a *weak* acid has a low value of K_a.

A crude generalization suggests that inorganic acids are strong and organic acids are weak.

The values of K_a generally increase with increasing temperature, causing the acid to be stronger at high T.

These descriptions of 'strong' and 'weak' acid are no longer subjective, but depend on the magnitude of K_a: a *strong* acid has a large value of K_a and a *weak* acid has a low value of K_a. Stated another way, the position of the acid-dissociation equilibrium lies close to the reactants for a weak acid but close to the products for a strong acid, as shown schematically in Figure 6.2.

Carboxylic acids such as ethanoic acid are generally weak because their values of K_a are small (although see p. 261). By contrast, so called mineral acids such as sulphuric or nitric are classed as strong because their respective values of K_a are large. Although there is little consensus, a simplistic rule suggests we class an acid as weak if its value of K_a drops below about 10^{-3}. The acid is strong if $K_a > 10^{-3}$.

Table 6.5 contains a selection of K_a values. Acids characterized by large values of K_a are stronger than those with smaller values of K_a. Each K_a value in Table 6.5 was obtained at 298 K. Being an equilibrium constant, we anticipate temperature-dependent values of K_a, with K_a generally increasing slightly as T increases.

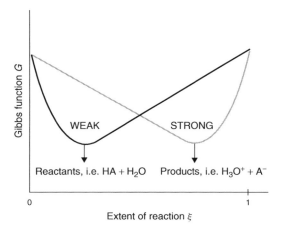

Figure 6.2 Graph of Gibbs function G (as 'y') against the extent of reaction ξ (as 'x'). The minimum of the graph corresponds to the position of equilibrium: the position of equilibrium for a weak acid, such as ethanoic acid, lies near the un-ionized reactants; the position of equilibrium for a strong acid, like sulphuric acid, lies near the ionized products

Table 6.5 Acidity ('dissociation') constants K_a for inorganic Lowry–Brønsted acids in water at 298 K. Values of K_a are dimensionless: all values presuppose equilibrium constants such as Equation (6.35), and were calculated with concentrations expressed in mol dm^{-3}

Acid	$K_{a(1)}$	$K_{a(2)}$	$K_{a(3)}$
Hypochlorous, HOCl	4.0×10^{-8}		
Hydrochloric, HCl	1.0×10^{7}		
Nitrous, HNO$_2$	4.6×10^{-4}		
Sulphuric, H$_2$SO$_4$	1.0×10^{2}	1.2×10^{-2}	
Sulphurous, H$_2$SO$_3$	1.4×10^{-2}	1.02×10^{-7}	
Carbonic, H$_2$CO$_3$	4.3×10^{-7}	5.61×10^{-11}	
Phosphoric, H$_3$PO$_4$	7.53×10^{-3}	6.23×10^{-8}	2.2×10^{-11}

Table 6.6 As for Table 6.5, but for inorganic acids and showing the effects of various structural changes

Acid	$10^5\,K_a$
Effect of extent of halogenation	
CH$_3$COOH	1.75
ClCH$_2$COOH	136
Cl$_2$CHCOOH	5530
Cl$_3$COOH	23 200
Effect of halide	
FCH$_2$COOH	260
ClCH$_2$COOH	136
BrCH$_2$COOH	125
ICH$_2$COOH	67
Effect of chain length	
HCOOH	17.7
CH$_3$COOH	1.75
CH$_3$CH$_2$COOH	1.35
CH$_3$CH$_2$CH$_2$COOH	1.51
Effect of substituent in benzoic acids	
C$_6$H$_5$COOH	6.3
p-NO$_2$–C$_6$H$_4$COOH	36.0
p-CH$_3$O–C$_6$H$_4$COOH	3.3
p-NH$_2$–C$_6$H$_4$COOH	1.4

In summary, carbolic acid (phenol, **VII**) is an extremely weak acid because its value of K_a is 10^{-10}, quantifying how small is the concentration of solvated protons in its solutions.

Basicity constants

Having categorized acids into 'strong' and 'weak' via the concept of acidity constants K_a, we now look at the strengths of various

> The cause of phenol's corrosive properties does not relate to its ability to form solvated protons (as indicated by the value of K_a) but its ability to penetrate the skin and disrupt the chemical processes occurring within the *epidermis*, to painful effect.

bases. It is possible to write an equilibrium constant K to describe the hydrolysis of bases such as ammonia (see Equation (6.12)). We write the appropriate equilibrium constant in just the same way as we wrote an expression for K to describe the acidic behaviour of phenol:

$$K = \frac{[NH_4^+][OH^-]}{[NH_3][H_2O]} \tag{6.32}$$

> We sometimes call the equilibrium constant in Equation (6.33) a *basicity constant*, and symbolize it as K_b.

As with the expression in Equation (6.6), this equilibrium constant can be simplified by incorporating the water term into K, thereby yielding a new constant which we will call K_b, the *basicity constant*:

$$K_b = \frac{[NH_4^+][OH^-]}{[NH_3]} \tag{6.33}$$

where K_b in Equation (6.33) is quite different from the K in Equation (6.32). The value of K_b for ammonia is 1.74×10^{-5}, which is quite small, causing us to say ammonia is a *weak base*. The value of K_b for sodium hydroxide is much larger at 0.6, so we say NaOH is a *strong base*.

But, curiously, this new equilibrium constant K_b is redundant because we could have *calculated* its value from known values of K_a according to

$$K_a \times K_b = K_w \tag{6.34}$$

where K_w is the autoprotolysis constant of water from p. 236. Older textbooks sometimes cite values of K_b, but we really do not need to employ two separate K constants.

SAQ 6.9 What is the value of K_a for the ammonium ion, NH_4^+? Take K_b from the paragraphs immediately above, and $K_w = 10^{-14}$.

Justification Box 6.3

Consider a weak acid, HA, dissociating: $HA \rightarrow H_3O^+ + A^-$. Its acidity constant K_a is given by

$$K_a = \frac{[A^-][H_3O^+]}{[HA]} \tag{6.35}$$

and then consider a weak base (the conjugate of the weak acid) forming a hydroxide ion in solution, $H_2O + A^- \rightarrow OH^- + HA$. Its basicity constant is given by

$$K_b = \frac{[HA][OH^-]}{[A^-]} \tag{6.36}$$

Multiplying the expressions for K_a and K_b yields

$$K_a \times K_b = \frac{[\text{A}^-][\text{H}_3\text{O}^+]}{[\text{HA}]} \times \frac{[\text{HA}][\text{OH}^-]}{[\text{A}^-]}$$

The HA and A$^-$ terms clearly cancel to yield $[\text{H}_3\text{O}^+][\text{OH}^-]$, which is K_w.

Why does carbonic acid behave as a mono-protic acid?

Variations in the value of K_a

Carbonic acid, H_2CO_3, is naturally occurring, and forms when carbon dioxide from the air dissolves in water. From its formula, we expect it to be a di-protic acid, but it is generally classed as mono-protic. Why?

In water at 298 K, the ionization reaction follows the equation

$$\text{H}_2\text{CO}_{3(aq)} + \text{H}_2\text{O} \longrightarrow \text{HCO}_{3(aq)}^- + \text{H}_3\text{O}^+_{(aq)} \tag{6.37}$$

The value of K_a for the reaction in Equation (6.37) is 4.3×10^{-7}, so carbonic acid is certainly a very weak acid. The hydrogen carbonate anion HCO_3^- could dissociate further, according to

$$\text{HCO}^-_{3(aq)} + \text{H}_2\text{O} \longrightarrow \text{CO}^{2-}_{3(aq)} + \text{H}_3\text{O}^+_{(aq)} \tag{6.38}$$

but its value of K_a is low at 5.6×10^{-11}, so we conclude that the HCO_3^- ion is too weak an acid to shed its proton under normal conditions. Thus, carbonic acid has two protons: the loss of the first one is relatively easy, but the proportion of molecules losing *both* protons is truly minute. Only one of the protons is labile.

This situation is relatively common. If we look, for example, at the values of K_a in Table 6.5, we see that phosphoric acid is a strong acid insofar as the loss of the first proton occurs with $K_a = 7.5 \times 10^{-3}$, but the loss of the second proton, to form HPO_4^{2-}, is difficult, as characterized by $K_a = 6.2 \times 10^{-8}$. In other words, the dihydrogen phosphate anion $H_2PO_4^-$ is a very weak acid. And the hydrogen phosphate di-anion HPO_4^{2-} has a low a value of $K_a = 2.2 \times 10^{-11}$, causing us to say that the PO_4^{3-} anion does not normally exist. Even the loss of the second proton of sulphuric acid is characterized by a modest value of $K_a = 10^{-2}$.

This formal definition of K_a can be extended to multi-protic acids. We consider the dissociation to occur in a step-wise manner, the acid losing one proton at a time. Consider, for example, the two-proton donation reactions of sulphuric acid:

Multi-protic acids have a different value of K_a for each proton donation step, with the values of K_a decreasing with each proton donation step.

$$(1) \quad \text{H}_2\text{SO}_4 \longrightarrow \text{H}^+ + \text{HSO}_4^-$$

$$(2) \quad \text{HSO}_4^- \longrightarrow \text{H}^+ + \text{SO}_4^{2-}$$

The equilibrium constant for the dissociation of H_2SO_4:

> The subscript (1) tells us we are considering the *first* proton to be lost.

$$K_{a(1)} = \frac{[HSO_4^-][H^+]}{[H_2SO_4]}$$

Similarly,

> $K_{a(1)}$ is always bigger than $K_{a(2)}$.

$$K_{a(2)} = \frac{[SO_4^{2-}][H^+]}{[HSO_4^-]}$$

In fact, we can also extend this treatment to bases, looking at the step-base addition of protons.

SAQ 6.10 The tetra-protic acid H_4EDTA (**V**) has four possible proton equilibrium constants. Write an expression for each, for $K_{a(1)}$ to $K_{a(4)}$.

Why is an organic acid such as trichloroethanoic acid so strong?

Effect of structure on the K_a of a weak acid

The value of K_a for trichloroacetic acid CCl_3COOH (**VIII**) is very large at 0.23. Indeed, it is stronger as an acid than the HSO_4^- ion – quite remarkable for an organic acid!

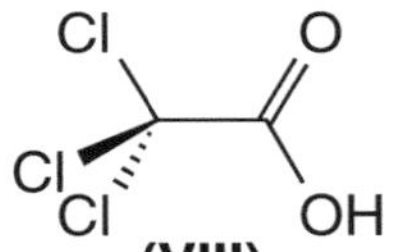

(VIII)

Let us return to the example of ethanoic acid (**I**). The principal structural difference between **I** and **VIII** is the way we replace each of the three methyl protons in ethanoic acid with chlorine atoms.

The three methyl protons in **I** are slightly *electropositive*, implying that the central carbon of the $-CH_3$ group bears a slight negative charge. This excess charge is not large, but it is sufficient to disrupt the position of the acid-dissociation equilibrium, as follows. Although the undissociated acid has no formal charge, the ethanoate anion has a full negative charge, which is located principally on the carboxyl end of the anion. It might be easier to think of this negative charge residing on just one of the oxygen atoms within the anion, but in fact both oxygen atoms and the central carbon each bear some of the charge. We say the charge is *delocalized*, according to structure **IX**, which is a more accurate representation of the carboxylate *anion* than merely $-COO^-$. The

> Delocalization is a means of *stabilizing* an ion.

right-hand structure of **IX** is effectively a *mixture* of the two structures to its left. Note how the name *resonance* implies charge *delocalization*.

$$R-\overset{\displaystyle O}{\underset{\displaystyle O^{\ominus}}{C}} \longleftrightarrow R-\overset{\displaystyle O^{\ominus}}{\underset{\displaystyle O}{C}} \longleftrightarrow R-\overset{\displaystyle O}{\underset{\displaystyle O}{C{\ominus}}}$$

(IX)

> A double-headed arrow '↔' indicates *resonance*.

To reiterate, the hydrogen atoms in the methyl group are slightly electropositive, with each seeking to relocate their own small amounts of charge onto the central carboxyl carbon. In consequence, the ethanoic anion (cf. structure **IX** with R = CH_3) has a central carbon bearing a larger negative charge – both from the ionization reaction but also from the hydrogen atoms of the $-CH_3$ group. In consequence, the central carbon of the ethanoate anion is slightly destabilized; and any chemical species is less likely to form if it is unstable.

> Ions are more likely to form if they are stable, and less likely to form if unstable.

Next we look at the structure of trichloroethanoic acid (**VIII**). In contrast to the hydrogen atoms of ethanoic acid, the three chlorine atoms are powerfully electron withdrawing. The chlorine atoms cause extensive delocalization of the negative charge on the Cl_3COO^- anion, with most of the negative charge absorbed by the three chlorine atoms and less on the oxygen atoms of the carboxyl. Such a relocation of charge stabilizes the anion; and any chemical species is more likely to form if it is stable.

Statistically, we find fewer ethanoate anions than trichloroethanoate anions in the respective solutions of the two acids. And if there are fewer ethanoate anions in solution per mole of ethanoic acid, then there will be fewer solvated protons. In other words, the extent to which ethanoic acid dissociates is less than the corresponding extent for trichloroethanoic acid. **I** is a weak acid and **VIII** is strong; dipping a simple pH electrode into a solution of each of the two acids rapidly demonstrates this truth.

This sort of delocalization stabilizes the ion; in fact, the Cl_3COO^- anion is more stable than the parent molecule, Cl_3COOH. For this reason, the solvated anion resides in solution in preference to the acid. K_a is therefore large, making trichloroacetic acid one of the strongest of the common organic acids.

Trifluoroethanoic acid (probably better known as trifluoro*acetic* acid, TFA) is stronger still, with a value of $K_a = 1.70$.

5.3 Titration analyses

Why does a dock leaf bring relief after a nettle sting?

Introducing titrations

We first met nettle stings on p. 253, where methanoic ('formic') acid was identified as the active toxin causing the pain. Like its

> The common dock leaf, *Rumex obtusifolia*, and the yellow dock leaf, *Rumex crispus*, are in fact equally common.

The naturally occurring substance *histamine* causes blood capillaries to dilate and smooth muscle to contract. Most cells release it in response to wounding, allergies, and most inflammatory conditions. *Antihistamines* block the production of this substance, thereby combating a painful swelling.

structurally similar sister, ethanoic acid (**I**), methanoic acid dissociates in water to yield a solvated proton, $H_3O^+{}_{(aq)}$.

Rubbing the site of the sting with a crushed dock leaf is a simple yet rapid way of decreasing the extent of the pain. In common with many other weeds, the sap of a dock leaf contains a mixture of natural amines (e.g. urea (**III**) above), as well as natural antihistamines to help decrease any inflammation. The amines are solvated and, because the sap is water based, are alkaline. Being alkaline, these amines react with methanoic acid to yield a neutral salt, according to

$$\text{(6.39)}$$

where R is the remainder of the amine molecule. We see how the process of pain removal involves a neutralization process.

Notice how the lone pair on nitrogen of the amine attracts a proton from the carboxylic acid.

How do indigestion tablets work?

Calculations concerning neutralization

Excess acid in the stomach is one of the major causes of indigestion, arising from a difficulty in digesting food. The usual cause of such indigestion is the stomach simply containing too much hydrochloric acid, or the stomach acid having too high a concentration (its pH should be about 3). These failures cause acid to remain even when all the food has been digested fully. The excess acid is not passive, but tends to digest the lining of the stomach to cause an ulcer, or reacts by alternative reaction routes, generally resulting in 'wind', the gases of which principally comprises methane.

Some indigestion tablets contain chalk ($CaCO_3$) but the large volume of CO_2 produced (cf. Equation (6.19)) can itself cause dyspepsia.

Most indigestion tablets are made of aluminium or magnesium hydroxides. The hydroxide in the tablet removes the excess stomach acid via a simple acid–base neutralization reaction:

$$3HCl_{(stomach)} + Al(OH)_{3(tablet)} \longrightarrow 3H_2O + AlCl_{3(aq)} \quad \text{(6.40)}$$

The cause of the indigestion is removed because the acid is consumed. Solid (unreacted) aluminium hydroxide is relatively insoluble in the gut, and does not dissolve to generate an alkaline solution. Rather, the outer layer of the tablet dissolves slowly, with just sufficient entering solution to neutralize the acid. Tablet dissolution stops when the neutralization reaction is complete.

A similar process occurs when we spread a thick paste of zinc and castor oil on a baby's bottom each time we change its nappy. The 'zinc' is in fact zinc oxide, ZnO, which, being amphoteric, reacts with the uric acid in the baby's urine, thereby neutralizing it.

Worked Example 6.9   But how much stomach acid is neutralized by a single indigestion tablet? The tablet contains 0.01 mol of MOH, where 'M' is a monovalent metal and M^+ its cation.

We first consider the reaction in the stomach, saying it proceeds with 1 mol of hydrochloric acid reacting with 1 mol of alkali:

$$MOH_{(s)} + H_3O^+{}_{(aq)} \longrightarrow M^+{}_{(aq)} + 2H_2O \qquad (6.41)$$

M^+ is merely a cation. We say Equation (6.41) is a 1:1 reaction, occurring with a 1:1 stoichiometry. Such a stoichiometry simplifies the calculation; the 3:1 stoichiometry in Equation (6.40) will be considered later.

From the stoichiometry of Equation (6.41), we say the neutralization is complete after equal amounts of acid and alkali react. In other words, we neutralize an amount n of hydrochloric acid with exactly the same amount of metal hydroxide, i.e. with 1×10^{-2} mol.

The tablet can neutralize 0.01 mol of stomach acid.

This simple calculation illustrates the fundamental truth underlying neutralization reactions: complete reaction requires equal amounts of acid and alkali. In fact, the primary purpose of a *titration* is to measure an unknown amount of a substance in a sample, as determined via a chemical reaction with a known amount of a suitable reagent. We perform the titration to ascertain when an equivalent amount of the reagent has been added to the sample. When the amount of acid and alkali are just equal, we have the *equivalence point*, from which we can determine the unknown amount.

In a typical titration experiment, we start with a known volume of sample, call it $V_{(sample)}$. If we know its concentration $c_{(sample)}$, we also know the amount of it, as $V_{(sample)} \times c_{(sample)}$. During the course of the titration, the unknown reagent is added to the solution, usually drop wise, until the equivalence point is reached (e.g. determining the endpoint by adding an indicator; see p. 273ff). At equivalence, the amounts of known and unknown reagents are the same, so $n_{(sample)} = n_{(unknown)}$. Knowing the amount of sample and the volume of solution of the unknown, we can calculate the concentration of the unknown.

Some campaigners believe the $AlCl_3$ produced by Equation (6.40) hastens the onset of Alzheimer's disease. Certainly, the brains of people with this nasty condition contain too much aluminium.

The experimental technique of measuring out the amount of acid and alkali needed for neutralization is termed a *titration*.

The amounts of acid and alkali are *equal* at the *equivalence* point. The linguistic similarity between these two words is no coincidence!

We need equal numbers of moles of acid and alkali to effect *neutralization*.

Worked Example 6.10   The methanoic acid from a nettle sting is extracted into 50 cm^3 of water and neutralized in the laboratory by titrating with sodium hydroxide solution. The concentration of NaOH is 0.010 mol dm^{-3}. The volume of NaOH solution needed to neutralize the acid is 34.2 cm^3. What is the concentration c of the acid?

Unlike the Worked Example 6.9, we do not know the number of moles n of either reactant, we only know the volumes of each. But we do know one of the concentrations.

Answer strategy. (1) First, we calculate the amount n of hydroxide required to neutralize the acid. (2) We equate this amount n with the amount of acid neutralized by the alkali. (3) Knowing the amount of acid, we finally calculate its concentration.

(1) To determine the amount of alkali, we first remember the definition of concentration c as

$$\text{concentration, } c = \frac{\text{amount, } n}{\text{volume, } V} \qquad (6.42)$$

and rearrange Equation (6.41) to make amount n the subject, i.e.

$$n = c \times V$$

The volume V of alkali is 34.2 cm^3. As there are 1000 cm^3 in a litre, this volume equates to (34.2 ÷ 1000) dm^3 = 0.0342 dm^3. Accordingly

$$n = 0.0342 \text{ dm}^3 \times 0.010 \text{ mol dm}^{-3} = 3.42 \times 10^{-4} \text{ mol}$$

(2) The reaction between the acid and alkali is a simple 1:1 reaction, so 3.42 × 10^{-4} mol of alkali reacts with exactly 3.42 × 10^{-4} mol of acid.

(3) The concentration of the acid is given by Equation (6.42) again. Inserting values:

$$\text{concentration } c = \frac{3.42 \times 10^{-4} \text{ mol}}{0.05 \text{ dm}^3}$$

$$c = 6.8 \times 10^{-3} \text{ mol dm}^{-3}$$

After extraction, the concentration of the methanoic acid is 0.068 mol dm^{-3}.

> We could have achieved this conversion with *quantity calculus*: knowing there are 1000 cm^3 per dm^3 (so 10^{-3} dm^3 cm^{-3}). In SI units, we write the volume as 34.2 cm^3 × 10^{-3} dm^3 cm^{-3}. The units of cm^3 and cm^{-3} cancel to yield V = 0.0342 dm^3.

> Notice now the units of dm^3 and dm^{-3} cancel out here.

> Notice how we converted the volume of acid solution (50 cm^3) to 0.05 dm^3.

An altogether simpler and quicker way of calculating the concentration of an acid during a titration is to employ the equation

$$c_{(\text{acid})} \times V_{(\text{acid})} = c_{(\text{alkali})} \times V_{(\text{alkali})} \qquad (6.43)$$

where the V terms are volumes of solution and the c terms are concentrations.

Worked Example 6.11   A titration is performed with 25 cm^3 of NaOH neutralizing 29.4 cm^3 of nitric acid. The concentration of NaOH is 0.02 mol dm^{-3}. Calculate the concentration of the acid.

We rearrange Equation (6.43), to make $c_{(acid)}$ the subject:

$$c_{(acid)} = \frac{c_{(alkali)} V_{(alkali)}}{V_{(acid)}} \qquad (6.44)$$

We then insert values into Equation (6.44):

$$c_{(acid)} = \frac{0.02 \text{ mol dm}^3 \times 25 \text{ cm}^3}{29.4 \text{ cm}^3}$$

$$c_{(acid)} = 0.017 \text{ mol dm}^{-3}$$

> We obtain here a *ratio* of volumes ($V_{(alkali)} \div V_{(acid)}$), enabling us to cancel the units of the two volumes. Units are irrelevant if both volumes have the same units.

Justification Box 6.4

A definition of the point of 'neutralization' in words says, 'at the neutralization point, the number of moles of acid equals the number of moles of hydroxide'. We re-express the definition as

$$n_{(acid)} = n_{(alkali)} \qquad (6.45)$$

Next, from Equation (6.42), we recall how the concentration of a solution c when multiplied by its respective volume V equals the number of moles of solute: $n = c \times V$. Clearly, $n_{(acid)} = c_{(acid)} \times V_{(acid)}$, and $n_{(alkali)} = c_{(alkali)} \times V_{(alkali)}$.

Accordingly, substituting for $n_{(acid)}$ and $n_{(alkali)}$ into Equation (6.45) yields Equation (6.43).

SAQ 6.11   What volume of NaOH (of concentration 0.07 mol dm^{-3}) is required to neutralize 12 cm^3 of nitric acid of concentration 0.05 mol dm^{-3}?

Aside

Equation (6.43) is a simplified version of a more general equation:

$$c_{(acid)} = s \frac{c_{(alkali)} V_{(alkali)}}{V_{(acid)}} \qquad (6.46)$$

where s is the so-called *stoichiometric ratio*.

For the calculation of a mono-protic acid with a mono-basic base, the stoichiometry is simply 1:1 because 1 mol of acid reacts with 1 mol of base. We say the stoichiometric ratio $s = 1$. The value of s will be two if sulphuric acid reacts with NaOH since 2 mol of base are required to react fully with 1 mol of acid. For the reaction of NaOH with citric acid, $s = 3$; and $s = 4$ if the acid is H_4EDTA.

The value of s when $Ca(OH)_2$ reacts with HNO_3 will be $\frac{1}{2}$, and the value when citric acid reacts with $Ca(OH)_2$ will be $\frac{3}{2}$.

SAQ 6.12 What volume of $Ca(OH)_2$ (of concentration $0.20\ mol\,dm^{-3}$) is required to neutralize $50\ cm^3$ of nitric acid of concentration $0.10\ mol\ dm^{-3}$?

Sigmoidal literally means 'shaped like a Greek sigma ς'. The name derives from the Greek word *sigmoides*, meaning 'sigma-like'. (There are two Greek letters called sigma, used differently in word construction. The other has the shape σ.)

An alternative way of determining the endpoint of a titration is to monitor the pH during a titration, and plot a graph of pH (as 'y') against volume V of alkali added (as 'x'). Typically, the concentration of the acid is unknown, but we know accurately the concentration of alkali. Figure 6.3 shows such as graph – we call it a *pH curve* – in schematic form. The shape is *sigmoidal*, with the pH changing very rapidly at the *end point*.

In practice, we obtain the end point by extrapolating the two linear regions of the pH curve (the extrapolants should be parallel). A third parallel line is drawn, positioned exactly midway between the two extrapolants. The volume at which this third line crosses the pH curve indicates the end point. Knowing the volume $V_{(end\ point)}$, we can calculate the concentration of the acid via a calculation similar to Worked Example 6.11.

Incidentally, the end point also represents the volume at which the pH changes most dramatically, i.e. the steepest portion of the graph. For this reason, we occasionally plot a different graph of gradient (as 'y') against volume V (as 'x'); see Figure 6.4. We obtain the gradient as '$\Delta pH \div \Delta V$'. The end point in Figure 6.4 relates to the graph maximum.

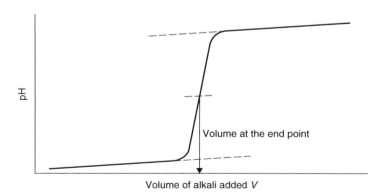

Volume at the end point

Volume of alkali added V

Figure 6.3 A schematic pH curve for the titration of a strong acid with a strong base. At the equivalence point, the amount of alkali added is the same as the amount of acid in solution initially, allowing for an accurate calculation of the acid's concentration. Note how the end point is determined by extrapolating the linear regions, and drawing a third parallel line between them

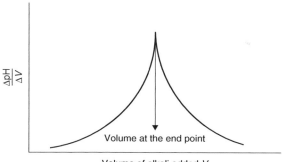

Figure 6.4 A schematic of the *first derivative* of the pH curve in Figure 6.3. The end-point volume is determined as the volume at the peak. A first derivative plot such as this can yield a more accurate end point than drawing parallel lines on Figure 6.3

6.4 pH buffers

Why does the pH of blood not alter after eating pickle?

Introduction to buffers

A 'pickle' is a food preserved in vinegar (ethanoic acid). Pickles generally have a sharp, acidic flavour in consequence of the acid preservative. Many systems – especially living cells – require their pH to be maintained over a very restricted range in order to prevent catastrophic damage to the cell. Enzymes and proteins denature, for example, if the pH deviates by more than a fraction. Traces of the food we eat are readily detected in the blood quite soon after eating, so why does the concentration in the blood remain constant, rather than dropping substantially with the additional acid in our diet?

Before we attempt an answer, look again at Figure 6.3, which clearly shows an almost invariant pH after adding a small volume of alkali. Similarly, at the right-hand side of the graph the pH does not vary much. We see an insensitivity of the solution pH to adding acid or alkali; only around the end point does the pH alter appreciably. The parts of the titration graph having an invariant pH are termed the *buffer regions*, and we call the attendant pH stabilization a *buffer* action.

In a similar way, blood does not change its pH because it contains suitable concentrations of carbonic acid and bicarbonate ion, which act as a buffer, as below.

Why are some lakes more acidic than others?

Buffer action

Acid rain is the major cause of acidity in open-air lakes and ponds (see p. 237). Various natural oxides such as CO_2 dissolve in water

Pollutant gases include SO_2, SO_3, NO and NO_2. It is now common to write SO_x and NO_x to indicate this variable valency within the mixture.

to generate an acid, so the typical pH of normal rainwater is about 5.6; but rainwater becomes more acidic if pollutants, particularly SO_x and NO_x, in addition to natural CO_2, dissolve in the water. As an example, the average pH of rain in the eastern United States of America (which produces about one-quarter of the world's pollution) lies in the range 3.9–4.5. Over a continental landmass, the partial pressure of SO_2 can be as high as $5 \times 10^{-9} \times p^{\ominus}$, representing a truly massive amount of pollution.

> *Remember:* 'weak' in this sense indicates the extent to which a weak acid dissociates, and does not relate to its concentration.

After rainfall, the pH of the water in some lakes does not change, whereas others rapidly become too acidic to sustain aquatic life. Why? The difference arises from the *buffering* action of the water. Some lakes resist gross change in pH because they contain other chemicals that are able to take up or release protons into solution following the addition of acid (in the rain). These chemicals in the lake help stabilize the water pH, to form a buffer. Look at Figure 6.5, which shows a pH curve for a weak acid titrated with an alkali. Figure 6.5 is clearly similar to Figure 6.3 after the end-point volume, but it has a much shallower curve at lower volumes. In fact, we occasionally have difficulty ascertaining a clear end point because the curvature is so pronounced.

A *buffer* comprises (1) a weak acid and a salt of that acid, (2) a weak base and a salt of that base, or (3) it may contain an acid salt. We define an acid–base buffer as

> A buffer is a solution of constant pH, which resists changes in pH following the addition of small amounts of acid or alkali.

'a solution whose pH does not change after adding (small amounts of) a strong acid or base'. Sodium ascorbate is a favourite buffer in the food industry.

We can think of water entering the lake in terms of a titration. A solution of alkali enters a fixed volume of acid: the alkaline solution is water entering from the lake's tributary rivers, and the acid is the lake, which contains the weak acid H_2CO_3 (carbonic acid) deriving from atmospheric carbon dioxide. The alkali in the tributary rivers is calcium hydroxide $Ca(OH)_2$, which enters the

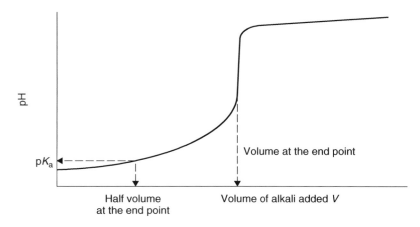

Figure 6.5 A typical pH curve for the titration of carbonic acid (a weak acid) with a strong base. The concentration of H_2CO_3 and HCO_3^- are the same after adding half the neutralization volume of alkali. At this point, $pH = pK_a$

water as the river passes over the limestone floor of river basins. Calcium hydroxide is a fairly strong base.

Figure 6.5 shows a buffering action since the pH does not change particularly while adding alkali to the solution. In fact, as soon as the alkali mixes with the acid in the lake, its hydroxide ions are neutralized by reaction with solvated protons in the lake, thereby resisting changes in the pH. Figure 6.5 shows how little the lake pH changes; we term the relatively invariant range of constant pH the *buffer region* of the lake water. The mid pH of the buffer region corresponds quite closely to the pK_a of the weak acid (here H_2CO_3), where the pK_a is a mathematical function of K_a, as defined by

$$pK_a = -\log_{10} K_a \qquad (6.47)$$

As a good generalization, the buffer region extends over the range of $pK_a \pm 1$.

Only when all the acid in the lake has been consumed will the pH rise significantly. In fact, the *end point* of such a titration is gauged when the pH rises above pH 7, i.e. the pH of acid–base neutrality.

The pH of the lake water fluctuates when not replenished by the alkaline river water in the tributary rivers. In fact, the pH of the lake water drops significantly each time it rains (i.e. when more H_2CO_3 enters the lake). If the amounts of acid and alkali in the water remain relatively low, then the slight fluctuations in water pH will not be great enough to kill life forms in the lake.

The system above describes the addition of alkali to a lake containing a weak acid. The reverse process also occurs, with acid being added to a base, e.g. when the tributary rivers deliver acid rain to a lake and the lake basin is made of limestone or chalk. In such a case, the lake pH drops as the acid rain from the rivers depletes the amounts of natural $Ca(OH)_2$ dissolved in the lake.

As a further permutation, adding a strong acid to a weak base also yields a buffer solution, this time with a buffer region centred on the pK_a of the base. The pH at the end point will be lower than 7.

Limestone or chalk dissolve in water to a limited extent. The $CaCO_3$ decomposes naturally to form $Ca(OH)_2$, thereby generating alkaline water.

The natural buffers in the lake 'mop up' any additional alkali entering the lake from the tributary rivers, thereby restricting any changes to the pH.

A buffer is only really effective at restricting changes if the pH remains in the range $pK_a \pm 1$.

Buffers

Each species within a buffer solution participates in an equilibrium reaction, as characterized by an equilibrium constant K. Adding an acid (or base) to a buffer solution causes the equilibrium to shift, thereby preventing the number of protons from changing, itself preventing changes in the pH. The change in the reaction's position of equilibrium is another manifestation of Le Chatelier's principle (see p. 166).

One of the most common buffers in the laboratory is the so-called 'phosphate buffer', which has a pH of 7.0. It comprises salts

A *buffer* is a solution of a weak acid mixed with its conjugate base, which restricts changes to the pH.

of hydrogen phosphate and dihydrogen phosphate, in the following equilibrium:

$$\boxed{H_2PO_4{}^{2-}} \rightarrow \boxed{HPO_4{}^-} + H^+ \tag{6.48}$$

conjugate acid   **conjugate base**

If this example were to proceed in the reverse direction, then the hydrogen phosphate (on the right) would be the base, since it receives a proton, and the dihydrogen phosphate (on the left) would be the conjugate the acid.

The equilibrium constant of the reaction in Equation (6.48) is given by

$$K_a = \frac{[H^+][HPO_4^{2-}]}{[H_2PO_4^-]} \tag{6.49}$$

Notice how the equilibrium constant K in Equation (6.49) is also an acidity constant, hence the subscripted 'a'. The value of K remains constant provided the temperature is not altered.

Now imagine adding some acid to the solution – either by mistake or deliberately. Clearly, the concentration of H^+ will increase. To prevent the value of K_a changing, some of the hydrogen phosphate ions combine with the additional protons to form *di*hydrogen phosphate (i.e. Equation (6.48) in reverse). The position of the equilibrium adjusts quickly and efficiently to 'mop up' the extra protons in the buffer solution. In summary, the pH is prevented from changing because protons are consumed.

How do we make a 'constant-pH solution'?

The Henderson–Hasselbach equation

We often need to prepare a solution having a constant pH. Such solutions are vital in the cosmetics industry, as well as when making foodstuffs and in the more traditional experiments performed by the biologist and physical chemist.

To make such a solution, we could calculate exactly how many moles of acid to add to water, but this method is generally difficult, since even small errors in weighing the acid can cause wide fluctuations in the pH. Furthermore, we cannot easily weigh out one of acid oxides such as NO. Anyway, the pH of a weak acid does not clearly follow the acid's concentration (see p. 254).

In some texts, Equation (6.50) is called the Henderson–HasselbaLch equation.

The Henderson–Hasselbach equation, Equation (6.50), relates the pH of a buffer solution to the amounts of conjugate acid and conjugate base it contains:

$$pH = pK_a + \log_{10}\left(\frac{[A^-]}{[HA]}\right) \tag{6.50}$$

We follow the usual pattern here by making a buffer with a weak acid HA and a solution of its conjugate base, such as the sodium salt of the respective anion, A^-.

We can prepare a buffer of almost any pH provided we know the pK_a of the acid; and such values are easily calculated from the K_a values in Table 6.5 and in most books of physical chemistry and Equation (6.50). We first choose a weak acid whose pK_a is relatively close to the buffer pH we want. We then need to measure out accurately the volume of acid and base solutions, as dictated by Equation (6.50).

Worked Example 6.12   We need to prepare a buffer of pH 9.8 by mixing solutions of ammonia and ammonium chloride solution. What volumes of each are required? Take the K_a of the ammonium ion as 6×10^{-10}. Assume the two solutions have the same concentration before mixing.

Strategy: (1) We calculate the pK_a of the acid. (2) We identify which component is the acid and which the base. (3) And we calculate the proportions of each according to Equation (6.50).
   (1) From Equation (6.50), we define the pK_a as $-\log_{10} K_a$. Inserting values, we obtain a pK_a of 9.22.
   (2) The action of the buffer represents the balanced reaction, $NH_4Cl \rightarrow NH_3 + HCl$, so NH_4Cl is the acid and NH_3 is the base.
   (3) To calculate the ratio of acid to base, we insert values into Equation (6.50):

$$9.8 = 9.2 + \log_{10} \left(\frac{[NH_3]}{[NH_4^+]} \right)$$

$$0.6 = \log_{10} \left(\frac{[NH_3]}{[NH_4^+]} \right)$$

Taking antilogs of both sides to remove the logarithm, we obtain

$$10^{0.6} = \frac{[NH_3]}{[NH_4^+]}$$

$$\frac{[NH_3]}{[NH_4^+]} = 4$$

> We are permitted to calculate a *ratio* like this if the concentrations of acid and conjugate base are the same.

So, we calculate the buffer requires four volumes of ammonia solution to one of ammonium (as the chloride salt, here).

SAQ 6.13 What is the pH of ammonia–ammonium buffer if three volumes of NH_4Cl are added to two volumes of NH_3?

   We want a buffer solution because its pH stays constant after adding small amounts of acid or base. Consider the example of adding hydrochloric acid to a buffer, as described in the following Worked Example.

Worked Example 6.13   Consider the so-called 'acetate buffer', made with equal volumes of sodium ethanoate and ethanoic acid solutions. The concentration of each solution is $0.1 \, mol \, dm^{-3}$. A small volume ($10 \, cm^3$) of strong acid (HCl of concentration $1 \, mol \, dm^{-3}$) is added to a litre of this buffer. The pH before adding HCl is 4.70. What is its new pH?

The acetate buffer is an extremely popular choice in the food industry. The buffer might be described on a food packet as an *acidity regulator*.

Strategy: we first calculate the number of moles of hydrochloric acid added. Second, we calculate the new concentrations of ethanoic acid and ethanoate. And third, we employ the Henderson–Hasselbach equation once more.

(1) 10 cm^3 represents one-hundredth of a litre. From Equation (6.42), the number of moles is 0.01 mol.

(2) Before adding the hydrochloric acid, the concentrations of ethanoate and ethanoic acid are constant at 0.1 mol dm^{-3}. The hydrochloric acid added reacts with the conjugate base in the buffer (the ethanoate anion) to form ethanoic acid. Accordingly, the concentration $[CH_3COO^-]$ decreases and the concentration $[CH_3COOH]$ increases. (We assume the reaction is quantitative.) Therefore, the concentration of ethanoate is $(0.1 - 0.01) \text{ mol dm}^{-3} = 0.09 \text{ mol dm}^{-3}$. The concentration of ethanoic acid is $(0.1 + 0.01) \text{ mol dm}^{-3} = 0.11 \text{ mol dm}^{-3}$.

(3) Inserting values into Equation (6.50):

$$pH = 4.70 + \log_{10}\left(\frac{0.09}{0.11}\right)$$

$$pH = 4.70 + \log\ (0.818)$$

$$pH = 4.70 + (-0.09)$$

so

$$pH = 4.61$$

So, we see how the pH shifts by less than one tenth of a pH unit after adding quite a lot of acid. Adding this same amount of HCl to distilled water would change the pH from 7 to 2, a shift of five pH units.

SAQ 6.14 Consider the ammonia–ammonium buffer in Worked Example 6.12. Starting with 1 dm^3 of buffer solution containing 0.05 mol dm^{-3} each of NH_3 and NH_4Cl, calculate the pH after adding 8 cm^3 of NaOH solution of concentration 0.1 mol dm^{-3}.

Justification Box 6.5

We start by writing the equilibrium constant for a weak acid HA dissociating in water, $HA + H_2O \rightarrow H_3O^+ + A^-$, where each ion is solvated. The dissociation constant for the acid K_a is given by Equation (6.35):

$$K_a = \frac{[H_3O^+][A^-]}{[HA]}$$

where, as usual, we ignore the water term. Taking logarithms of Equation (6.35) yields

$$\log_{10}\ K_a = \log_{10} \frac{[H_3O^+][A^-]}{[HA]} \tag{6.51}$$

We can split the fraction term in Equation (6.51) by employing the laws of logarithms, to yield

$$\log_{10} K_a = \log_{10}[H_3O^+] + \log_{10} \frac{[A^-]}{[HA]} \qquad (6.52)$$

The term '$\log_{10} K_a$' should remind us of pK_a (Equation 6.52), and the term $\log_{10}[H_3O^+]$ will remind us of pH in Equation (6.20), so we rewrite Equation (6.52) as

$$-pK_a = -pH + \log_{10} \frac{[A^-]}{[HA]} \qquad (6.53)$$

which, after a little rearranging, yields the *Henderson–Hasselbach equation*, Equation (6.50).

6.5   Acid–base indicators

What is 'the litmus test'?

pH indicators

Litmus is a naturally occurring substance obtained from lichen. It imparts an intense colour to aqueous solutions. In this sense, the indicator is a dye whose colour is sensitive to the solution pH. If the solution is rich in solvated protons (causing the pH to be less than 7) then litmus has an intense red colour. Conversely, a solution rich in hydroxide ions (with a pH greater than 7) causes the litmus to have a blue colour.

To the practical chemist, the utility of litmus arises from the way its colour changes as a function of pH. Placing a single drop of litmus solution into a beaker of solution allows us an instant test of the acidity (or lack of it). It indicates whether the pH is less than 7 (the litmus is red, so the solution is acidic), or the pH is greater than 7 (the litmus is blue, so the solution is alkaline). Accordingly, we call litmus a pH *indicator*.

In practical terms, we generally employ litmus during a titration. The flask will contain a known volume of acid of unknown concentration, and we add alkali from a burette. We know we have reached neutralization when the Litmus changes from red (acid in excess) and *just* starts changing to blue. We know the pH of the solution is exactly 7 when neutralization is complete, and then note the volume of the alkali, and perform a calculation similar to Worked Example 6.11.

The great English scientist Robert Boyle (1627–1691) was the first to document the use of natural vegetable dyes as acid–base indicators.

The name 'litmus' comes from the Old Norse *litmosi*, which derives from *litr* and *mosi*, meaning dye and moss respectively.

Much of the litmus in a laboratory is pre-impregnated on dry paper.

Litmus is an *indicator*. To avoid ambiguity, we shall call it an 'acid–base indicator' or a 'pH indicator'.

Litmus often looks purple–grey at the neutralization point. This colour tells us we have a mixture of both the red and blue forms of litmus.

Why do some hydrangea bushes look red and others blue?

The chemical basis of acid–base indicators

> The name 'hydrangea' derives from classical Greek mythology, in which the 'hydra' was a beast with many heads.

Hydrangeas (genus *Hydrangea*) are beautiful bushy plants having multiple flower heads. In soils comprising much compost the flowers have a blue colour, but in soils with much lime or bone meal the heads are pink or even crimson–pink in colour. Very occasionally, the flowers are mauve. 'Lime' is the old-fashioned name for calcium oxide, and is alkaline; bone meal contains a lot of phosphate, which is also likely to raise the soil pH. The colour of the hydrangea is therefore an indication of the acid content of the soil: the flower of a hydrangea is blue in acidic soil because the plant sap is slightly acidic; red hydrangeas exist in alkaline soil because the sap transports alkali from the soil to the petals. The rare mauve hydrangea indicates a soil of neutral pH. We see how the chromophore in the flower is an acid–base *indicator*.

> The word 'chromophore' comes from two Greek words, *'khromos'* meaning colour and *'phoro'*, which means 'to give' or 'to impart'. A chromophore is therefore a species imparting colour.

The chromophore in hydrangeas is *delphinidin* (**X**), which is a member of the anthrocyanidin class of compounds. Compound **X** reminds us of phenol (**VII**), indicating that delphinidin is also a weak acid. In fact, all pH indicators are weak acids or weak bases, and the ability to change colour is a visible manifestation of the indicator's ability to undergo *reversible* changes in structure. In the laboratory, only a tiny amount of the pH indicator is added to the titration solution, so it is really just a *probe* of the solution pH. It does not participate in the acid–base reaction, except insofar as its own structure changes with the solution pH.

> All pH indicators are weak acids or weak bases.

As an example, whereas the anthracene-based core of molecular **X** is relatively inert, the side-chain 'X' is remarkably sensitive to the pH of its surroundings (principally, to the pH of the solution in which it dissolves).

(X)

Figure 6.6 shows the structure of the side substituent as a function of pH.

The hydroxyl group placed para to the anthracene core is protonated in acidic solutions (i.e. when the hydrangea sap is slightly acidic). The proton is abstracted in alkaline sap, causing molecular rearrangement to form the quinone moiety.

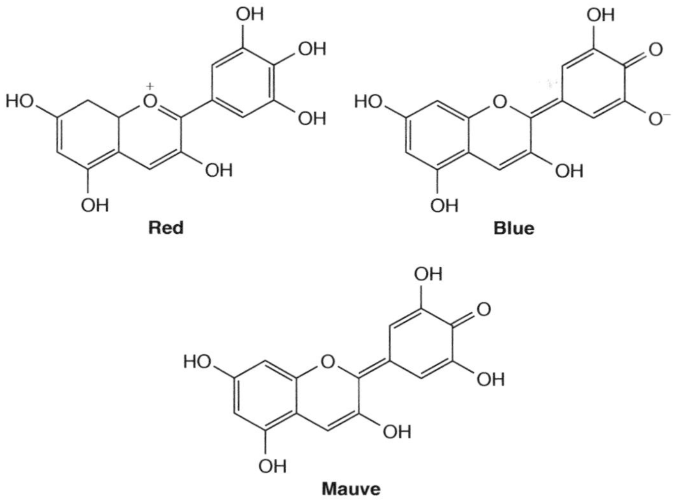

Figure 6.6 Anthrocyanidins impart colour to many natural substances, such as strawberries and cherries. The choice of side chains can cause a huge change in the anthrocyanidin's colour. If the side chain is pH sensitive then the anthrocyanidin acts as an acid–base indicator: structures of an anthrocyanidin at three pHs (red in high acidity and low pH, blue in low acidity and high pH and mauve in inter-midiate pHs)

It is also astonishing how the rich blue of a cornflower (*Centaurea cyanus*) and the majestic red flame of the corn poppy (*Papaver rheas*) each derive from the same chromophore – again based on an *anthrocyanidin*. The pH of cornflower and poppy sap does not vary with soil composition, which explains why we see neither red cornflowers nor blue corn poppies.

Aside

It is fascinating to appreciate the economy with which nature produces colours (elementary colour theory is outlined in Chapter 9). The trihydroxyphenyl group of the anthrocyanidin (**X**) imparts a colour to both hydrangeas and delphiniums. The dihydroxyphenyl group (**XI**) is remarkably similar, and imparts a red or blue colour to roses, cherries and blackberries. The singly hydroxylated phenyl ring in **XII** is the chromophore giving a red colour to raspberries, strawberries and geraniums, but it is not pH sensitive.

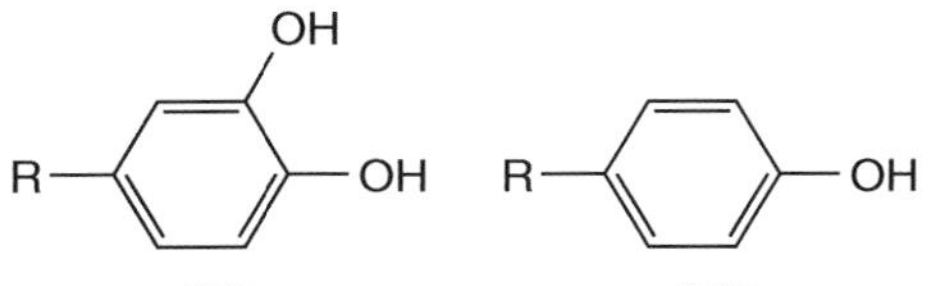

Why does phenolphthalein indicator not turn red until pH 8.2?

Which acid–base indicator to use?

Litmus was probably the most popular choice of acid–base indicator, but it is not a good choice for colour-blind chemists. The use of phenolphthalein as an acid–base indicator comes a close second. Phenolphthalein (**XIII**) is another weak organic acid. It is not particularly water soluble, so we generally dissolve it in aqueous ethanol. The ethanol explains the pleasant, sweet smell of phenolphthalein solutions.

(XIII)

Phenolphthalein is colourless and clear in acidic solutions, but imparts an intense puce pink colour in alkaline solutions of higher pH, with $\lambda_{(max)} = 552$ nm. The coloured form of phenolphthalein contains a quinone moiety; in fact, any chromophore based on a quinone has a red colour. But if a solution is prepared at pH 7 (e.g. as determined with a pH meter), we find the phenolphthalein indicator is still colourless, and the pink colour only appears when the pH reaches 8.2. Therefore, we have a problem: *the indicator has not detected neutrality*, since it changes colour at too

Table 6.7   Some common pH indicators, their useful pH ranges and the changes in colour occurring as the pH increases. An increasing pH accompanies a decreasing concentration of the solvated proton

Indicator	pH range	Colour change
Methyl violet	0.0–1.6	Yellow → blue
Crystal violet	0.0–1.8	Yellow → blue
Litmus	6.5–7.5	Red → blue
Methyl orange	3.2–4.4	Red → yellow
Ethyl red	4.0–5.8	Colourless → red
Alizarin red S	4.6–8.0	Yellow → red
3-Nitrophenol	6.8–8.6	Colourless → yellow
Phenolphthalein	8.2–10.0	Colourless → pink

high a value of pH. However, the graph in Figure 6.3 shows how the pH of the titration solution changes dramatically near the end point: in fact, only a tiny incremental addition of alkali solution is needed to substantially increase the solution pH by several pH units. In other words, a fraction of a drop of alkali solution is the only difference between pH 7 (at the true volume at neutralization) and pH 8 when the phenolphthalein changes from colourless to puce pink.

Table 6.7 lists the pH changes for a series of common pH indicators. The colour changes occur over a wide range of pHs, the exact value depending on the indicator chosen. Methyl violet changes from yellow to blue as the pH increases between 0 and 1.6. At the opposite extreme, phenolphthalein responds to pH changes in the range 8.2 to 10.

7

Electrochemistry

Introduction

This chapter commences by describing cells and redox chemistry. Faraday's laws of electrolysis describe the way that charge and current passage necessarily consume and produce redox materials. The properties of each component within a cell are described in terms of potential, current and composition.

Next, the nature of half-cells is explained, together with the necessary thermodynamic backgrounds of the theory of activity and the Nernst equation.

In the final sections, we introduce several key electrochemical applications, such as the pH electrode (a type of concentration cell), nerve cells (which rely on junction potentials) and batteries.

7.1 Introduction to cells: terminology and background

Why does putting aluminium foil in the mouth cause pain?

Introduction to electrochemistry

Most people have at some time experienced a severe pain in their teeth after accidentally eating a piece of sweet wrapper. Those teeth that hurt are usually nowhere near the scrap of wrapper. The only people who escape this nasty sensation are those without metal fillings in their teeth.

The type of sweet wrapper referred to here is generally made of aluminium metal, even if we call it 'silver paper'. Such aluminium dissolves readily in acidic, conductive electrolytes; and the pH of saliva is about 6.5–7.2.

The dissolution of aluminium is an oxidative process, so it generates several electrons. The resultant aluminium ions stay in solution next to the metal from which they came. We generate a *redox couple*, which we define as 'two redox states of the same material'.

> Two redox states of the same material are called a *redox couple*.

The word 'amalgam' probably comes from the Greek *malagma* meaning 'to make soft', because a metal becomes pliable when dissolved in mercury. Another English word from the same root is 'malleable'.

A *cell* comprises two or more half-cells in contact with a common electrolyte. The cell is the cause of the pain.

Oxidation reactions occur at the *anode*.

Reduction reactions occur at the *cathode*.

While it feels as though all the mouth fills with this pain, in fact the pain only manifests itself through those teeth filled with metal, the metal being silver dissolved in mercury to form a solid – we call it a *silver amalgam*. Corrosion of the filling's surface causes it to bear a layer of oxidized silver, so the tooth filling also represents a redox couple, with silver and silver oxide coexisting.

An electrochemical *cell* is defined as 'two or more half-cells in contact with a common electrolyte'. We see from this definition how a cell forms within the mouth, with aluminium as the more positive pole (the *anode*) and the fillings acting as the more negative pole (the *cathode*). Saliva completes this cell as an *electrolyte*. All the electrochemical processes occurring are contained within the boundaries of the cell.

Oxidation proceeds at the *anode* of the cell according to

$$Al_{(s)} \longrightarrow Al^{3+}_{(aq)} + 3e^- \qquad (7.1)$$

and occurs concurrently with a reduction reaction at the *cathode*:

$$Ag_2O_{(s)} + 2e^- \longrightarrow 2Ag^0_{(s)} + O^{2-} \qquad (7.2)$$

The origins of the words 'anode' and 'cathode' tell us much. 'Anode' comes from the Greek words *ana*, meaning 'up', and *hodos* means 'way' or 'route', so the anode is the electrode to which electrons travel from oxidation, travelling to higher energies (i.e. energetically 'uphill'). The word 'cathode' comes from the Greek *hodos* (as above), and *cat* meaning 'descent'. The English word 'cascade' comes from this same source, so a cathode is the electrode to which the electrons travel (energetically downhill) during reduction.

The oxidation and reduction reactions must occur concurrently because the electrons released by the dissolution of the aluminium are required for the reduction of the silver oxide layer on the surface of the filling. For this reason, we need to balance the two electrode reactions in Equations (7.1) and (7.2) to ensure the same number of electrons appear in each. The pain felt at the tooth's nerve is a response to this flow of electrons. The paths of electron flow are depicted schematically in Figure 7.1.

Each electron has a 'charge' Q. When we quantify the number of electrons produced or consumed, we measure the overall charge flowing. Alternatively, we might measure the *rate* at which the electrons flow (how many flow per unit time, t): this rate is termed the current I. Equation (7.3) shows the relationship between current I and charge Q:

$$I = \frac{dQ}{dt} \qquad (7.3)$$

So ultimately the pain we feel in our teeth comes from a flow of *current*.

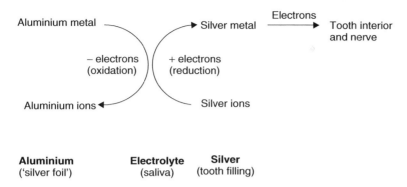

Figure 7.1 Schematic illustration of the electron cycles that ultimately cause a sensation of pain in the teeth in people who have metallic fillings and who have inadvertently eaten a piece of aluminium ('silver') foil, e.g. while eating sweets

Why does an electric cattle prod cause pain?

The magnitude of a current

An electric cattle prod looks a little like a walking stick with an attached battery. A potential from the tip of the stick is applied to a cow's flank, and the induced current hurts the animal. The cow moves where prompted to avoid a reapplication of the pain, thereby simplifying the job of cowherding.

Ohm's law says that applying a potential V across an electrical resistor R induces a proportional current I. We can state this relationship mathematically as

$$V = IR \qquad (7.4)$$

It is reported that the great Victorian scientist Michael Faraday discovered a variation of Ohm's law some 20 years before Ohm himself published the law that today bears his name. Faraday, with his typical phlegm and ingenuity, grasped a resistor between his two hands, immersed them both in a bowl of tepid water, and applied a voltage between them. It hurt. He found the empirical relationship 'pain $\propto I \times R$'.

The pain Faraday induced was in direct proportion to the magnitude of the current passed. He discovered the principle underlying the action of an electric *cattle prod*. It is sobering to realize how Faraday's result is today employed through most of the world as the basis of torture. Despite the explicit banning of torture by the UN Charter on Human Rights (of 1948), it is common knowledge that giving an electric shock is one of the most effective known means of causing pain.

In summary, the cattle prod causes pain because of the current formed in response to applying a voltage.

> The word 'empirical' implies a result derived from experiment rather than theory. Similarly, a chemist calculates a compound's 'empirical formula' while fully aware its value is based on experiment rather than theory.

What is the simplest way to clean a tarnished silver spoon?

Electrochemistry: the chemistry of electron transfer

> *Oxidation* is loss of electrons. *Reduction* is gain of electrons.

Cutlery or ornaments made of silver tarnish and become black; this is a shame, because clean, shiny silver is very attractive. The 'tarnish' comprises a thin layer of silver that has oxidized following contact with the air to form black silver(I) oxide:

$$4Ag_{(s)} + O_{2(g)} \longrightarrow 2Ag_2O_{(s)} \qquad (7.5)$$

> In fact, the tarnish on silver usually comprises both silver(I) oxide and a little silver(I) sulphide.

Such silver can be rather difficult to clean without abrasives (which wear away the metal). The following is a simple *electrochemical* means of cleaning the silver: immerse the tarnished silver in a saucer of electrolyte, such as salt solution or vinegar, and wrap it in a piece of aluminium foil. Within a few minutes the silver is cleaner and bright, whereas the aluminium has lost some of its shininess.

The shine from the aluminium is lost as atoms on the surface of the foil are oxidized to form Al^{3+} ions (Equation (7.1)), which diffuse into solution. Because the aluminium touches the silver, the electrons generated by Equation (7.1) enter the silver and cause *electro-reduction* of the surface layer of Ag_2O (Equation (7.2)).

In summary, we construct a simple electrochemical cell in which the silver to be cleaned is the cathode (Equation (7.2)) and aluminium foil as the anode supplies the electrons via Equation (7.1).

The salt or vinegar acts as an electrolyte, and is needed since the product Al^{3+} requires counter ions to ensure electro-neutrality (so aluminium ethanoate forms). The oxide ions combine with protons from the vinegar to form water. Figure 7.2 illustrates these processes occurring in schematic form.

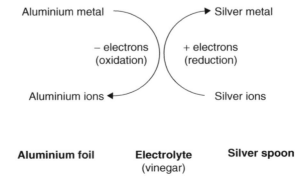

Figure 7.2   Illustration of the electron cycles that allow for the trouble-free cleaning of silver: we immerse the tarnished silver in an electrolyte, such as vinegar, and touch the silver with aluminium foil

And finally a word of caution: *aluminium ethanoate is toxic*, so wash the silver spoon thoroughly after removing its tarnish.

How does 'electrolysis' stop hair growth?

Electrochemical reactions and electrolysis

To many people, particularly the image conscious, *electrolysis* means removing hairs from the arms and legs – a practice sometimes called 'electrology'. The purpose of such electrolysis is to remove the hair follicles temporarily, thereby avoiding the need to shave. We 'treat' each individual hair by inserting a tiny surgical 'probe' (in reality, an electrode) into the hair follicle. Applying a voltage to the hair root for a fraction of a second kills the root. The electrolysed hair is then removed, and will not regrow for some time. This procedure is performed repeatedly until the desired area is cleared. But how does it work?

Electrochemical means the chemistry of the electron.

A voltage is applied to the electrode: we say it is *polarized*. A current flows in response to the voltage, and electrons are consumed by electrochemical reactions around the electrodes. *Electrolysis* occurs. Each electron that flows through the electrode must be involved in a redox reaction, either oxidation or reduction. The electrons *entering* or *leaving* the electrode move as a result of reactions occurring in the immediate vicinity of the electrode. Conversely, electrons can travel in the opposite direction (*leaving* the electrode) to facilitate reduction reactions.

It is wrong to claim a *permanent* method of hair removal: after electrolysis, the hair *will* grow back, albeit thinner and finer.

In summary, this form of electrolysis is effective because the charge passing through the electrode generates chemicals inside the hair follicle. The resultant trauma kills the hair root. A leg or arm treated in this way remains hairless until a new, healthy root regrows later in the previously damaged follicle.

The word 'electrolysis' derives from the Greek words *lysis*, meaning 'splitting' or 'cleavage', and the root *electro*, meaning 'charge' or 'electricity'. Strictly, then, electrolysis involves electrochemical bond cleavage.

Why power a car with a heavy-duty battery yet use a small battery in a watch?

Faraday's laws of electrolysis

A battery is a device for converting chemical energy into electrical energy. The amount of energy required by the user varies according to the application in mind. For example, a watch battery only powers the tiny display on the face of a digital watch. For this purpose,

Batteries are described in more detail in Section 7.7.

Table 7.1 Faraday's laws of electrolysis

Faraday's first law

The number of moles of a species formed at an electrode during electrolysis is proportional to the electrochemical charge passed: $Q = I \times t$

Faraday's second law

A given charge liberates different species in the ratio of their relative formula masses, divided by the number of electrons in the electrode reaction

it only needs to deliver a tiny current of a micro-amp or so. Conversely, a car battery (usually a 'lead–acid cell', as described on p. 347) is bulky and heavy because it must deliver a massive amount of electrical energy, particularly when starting the car. Other batteries generate currents of intermediate magnitude, such as those needed in torches, mobile phones and portable cassette and CD players.

The amount of charge generated or consumed by a battery is in direct proportion to the number of electrons involved, according to *Faraday's laws*, which are given in Table 7.1. Both electrons and ions possess charge. When a current is drawn through a cell, the charged electrons move through the conductive electrodes (as defined on p. 300) concurrently with charged ions moving through the electrolyte. The ions are *anions* (which bear a negative charge) and *cations* (which are positive).

Underlying both of Faraday's laws lies the fundamental truth that each electron possesses the same charge.

Worked Example 7.1   What is the charge on 1 mol of electrons?

> The coulomb, C, is the SI unit of charge.

The charge e on a single electron is 1.6×10^{-19} C and there are 6.022×10^{23} electrons per mole (the Avogadro number L), so the charge on a mole of electrons is given by the simple expression

$$\text{charge on one electron} = L \times e \qquad (7.6)$$

Inserting numbers into Equation (7.6), we obtain

> The charge on 1 mol of electrons is termed 'a faraday' F.

$$\text{charge on 1 mole of electrons}$$
$$= 1.6 \times 10^{-19} \text{ C} \times 6.026 \times 10^{23} \text{ mol}^{-1}$$

We see that 1 mol has a charge of $96\,487$ C mol^{-1}. This quantity of charge is known as the 'Faraday' F.

SAQ 7.1 An electrolysis needle (i.e. an electrode) delivers 1 nmol of electrons to a hair root. How many faraday's of charge are consumed, and how many coulombs does it represent?

Worked Example 7.2   How much silver is generated by reductively passing $1F$ of charge through a silver-based watch battery?

We will answer this question in terms *Faraday's first law*, which was first formulated in 1834 (see Table 7.1).

Each electron is required to effect the reduction reaction $Ag^+_{(aq)} + e^- \rightarrow Ag_{(s)}$, so one electron generates one atom of Ag, and $1F$ of charge (i.e. 1 mol of electrons) generates 1 mol of Ag atoms. The metal forms at the expense of 1 mol of Ag^+ ions. Similarly, 1 mol of electrons, if passed oxidatively, would generate 1 mol of Ag^+ ion from Ag metal.

We see a direct proportionality between the charge passed and the amount of material formed during electrolysis, as predicted by Faraday's first law.

> *Silver* has the symbol 'Ag' because its Latin name is *argentium*, itself derived from the Greek for money, *argurion*. The Spanish colonized parts of South America in the 16th century. They named it Argentina (dog-Latin for 'silver land') when they discovered its vast reserves of silver.

SAQ 7.2 $10^{-10}F$ of charge are inserted into a hair pore through a fine needle electrode. Each electron generates one molecule of a chemical to poison the hair root. How much of the chemical is formed?

Worked Example 7.3   A Daniell cell (see p. 345) is constructed, and $1F$ of charge is passed through a solution containing copper(II) ions. What mass of copper is formed? Assume that the charge is only consumed during the reduction reaction, and is performed with 100 percent efficiency.

We will answer this question by introducing *Faraday's second law* (see Table 7.1).

The reduction of *one* copper(II) ion requires two electrons according to the reaction

$$Cu^{2+}_{(aq)} + 2e^- \longrightarrow Cu_{(s)} \tag{7.7}$$

We need $2F$ of charge to generate 1 mol of $Cu_{(s)}$, and $1F$ of charge will form 0.5 mol of copper.

1 mol of copper has a mass of 64 g, so $1F$ generates $(64 \div 2)$ g of copper. The mass of copper generated by passing $1F$ is 32 g.

SAQ 7.3   How much aluminium metal is formed by passing $2F$ of charge through a solution of Al^{3+} ions? [Hint: assume the reaction at the electrode is $Al^{3+}_{(aq)} + 3e^- \rightarrow Al_{(s)}$.]

How is coloured ('anodized') aluminium produced?

Currents generate chemicals: dynamic electrochemistry

Saucepans and other household implements made of aluminium often have a brightly coloured, shiny coating. This outer layer comprises aluminium oxide incorporating a small amount of dye.

Al$_2$O$_3$ is also called *alumina*.

The layer is deposited with the saucepan immersed in a vat of dye solution (usually acidified to pH 1 or 2), and made the positive terminal of a cell. As the electrolysis proceeds, so the aluminium on the surface of the saucepan is oxidized:

$$2Al_{(s)} + 3H_2O \longrightarrow Al_2O_{3(s)} + 6H^+_{(aq)} \tag{7.8}$$

The aluminium is white and shiny before applying the potential. A critical potential exists below which no electro-oxidation will commence. At more extreme potentials, the surface atoms of the aluminium oxidize to form Al^{3+} ions, which combine with oxide ions from the water to form Al_2O_3. This *electro-precipitation* of solid aluminium oxide is so rapid that molecules of dye get trapped within it, and hence its coloured aspect.

We say the dye *occludes* within a *matrix* of solid aluminium oxide.

The dye resides *inside* the layer of alumina. Its colour persists because it is protected from harmful UV light, as well as mechanical abrasion and chemical attack.

A cell must comprise a *minimum of two electrodes*.

But the chemical reaction forming this coloured layer of oxide represents only one part of the cell. A cell contains a minimum of two electrodes, so a cell comprises two reactions – we call them *half-reactions*: one describes the chemical changes at the positive electrode (the *anode*) and the other describes the changes that occur at the negative *cathode*.

The same number of electrons conducts through (i.e. are conducted by) each of the two electrodes. If we think in terms of charge flowing per unit time, we would say the same 'current' I flows through each electrode. The electrons travel in opposite directions, insofar as they leave or enter an electrode, which explains why the current through the anode is oxidative and the current through the cathode is reductive. We say

$$I_{(anode)} = -I_{(cathode)} \tag{7.9}$$

where the minus sign reminds us that the electrons either move in or out of the electrode.

Because these two currents are equal (and opposite), the same amount of reaction will occur at either electrode. We see how an electrode reaction must also occur at the cathode as well as the desired oxidative formation of alumina at the anode. (The exact nature of the reaction at the anode will depend on factors such as the choice of electrode material.)

How do we prevent the corrosion of an oil rig?

Introduction to electrochemical equilibrium

Oil rigs are often built to survive in some of the most inhospitable climates in the world. For example, the oil rigs in the North Sea between the UK and Scandinavia frequently withstand force-10 gales. Having built the rig, we appreciate how important is the need of maximizing its lifetime. And one of the major limits to its life span is corrosion.

Oil rigs are made of steel. The sea in which they stand contains vast quantities of dissolved salts such as sodium chloride, which is particularly 'aggressive' to ferrous metals. The corrosion reaction generally involves oxidative dissolution of the iron, to yield ferric salts, which dissolve in the sea:

$$Fe_{(s)} \longrightarrow Fe^{3+}{}_{(aq)} + 3e^- \qquad (7.10)$$

If left unchecked, dissolution would cause thinning and hence weakening of the legs on which the rig stands.

One of the most ingenious ways in which corrosion is inhibited is to strap a power pack to each leg (just above the level of the sea) and apply a continuous reductive current. An electrode *couple* would form when a small portion of the iron oxidizes. The couple would itself set up a small voltage, itself promoting further dissolution. The reductive current coming from the power pack reduces any ferric ions back to iron metal, which significantly decreases the rate at which the rig leg corrodes.

> In reality, several of the iron compounds are solid, such as rust. This clever method of averting corrosion can also arrest the corrosion of rails and the undersides of boats.

Clearly, we want the *net* current at the iron to be zero (hence no overall reaction). The rate of corrosion would be enhanced if the power pack supplied an oxidative current, and wasteful side reactions involving the seawater itself would occur if the power pack produced a *large* reductive current. The net current through the iron can be positive, negative or zero, depending on the potential applied to the rig's leg. The conserver of the rig wants *equilibrium*, implying no change.

> The simplest definition of *equilibrium* in an electrochemistry cell is that *no concentrations change*.

All the discussions of electrochemistry so far in this chapter concern current – the flow of charged electrons. We call this branch of electrochemistry *dynamic*, implying that compositions change in response to the flow of electrons. Much of the time, however, we wish to perform electrochemical experiments at *equilibrium*.

One of our best definitions of 'equilibrium electrochemistry' says the *net current is zero*; and from Faraday's laws (Table 7.1), a zero current means that no material is consumed and no products are formed at the electrode.

> Electrochemical measurements at *equilibrium* are made at zero current.

But this equilibrium at the oil rig is *dynamic*: the phrase 'dynamic equilibrium' implies that currents do pass, but the current of the forward reaction is equal and opposite to the current of the back reaction, according to

$$I_{(forward,\ eq)} = -I_{(backward,\ eq)} \qquad (7.11)$$

and the overall (net) current is the sum of these two:

$$I_{(net)} = I_{(forward)} + I_{(backward)} \qquad (7.12)$$

Equation (7.11) is important, since it emphasizes how currents flow even at equilibrium.

But the value of $I_{(net)}$ is only ever zero at equilibrium because $I_{(forward)} = -I_{(backward)}$, which can only happen at one particular energy, neither too reductive nor too oxidative. The voltage around the legs of the oil rig needs to be chosen carefully.

What is a battery?

The emf of cells

> A *battery* is a device for converting chemical energy into electrical energy.

A battery is an electrochemical cell, and is defined as 'a device comprising two or more redox couples' (where each *couple* comprises two redox states of the same material). An oxidation reaction occurs at the negative pole of the battery in tandem with a reduction reaction at the positive pole. Both reactions proceed with the passage of current. The two redox couples are separated physically by an electrolyte.

> The word 'cell' comes from the Latin for 'small room', which explains why a prisoner is kept in a 'cell'.

The battery requires two redox couples because it is a cell. Each couple could be thought of as representing half of a complete cell. This sort of reasoning explains why the two redox couples are called *half-cells*. We could, therefore, redefine a cell as a device comprising two half-cells separated with an electrolyte.

In practice, the voltage of a battery is measured when its two ends are connected to the two terminals of a voltmeter, one contact secured to the positive terminal of the battery and the other at the negative. But a voltmeter is a device to measure *differences* in potential, so we start to see how the 'voltage' cited on a battery label is simply the difference in potential between the two poles of the battery.

> The cell's *emf* is a primary physicochemical property, and is measured with a voltmeter or potentiometer.

While the voltage of the cell represents the potential difference between the two 'terminals' of the battery, in reality it relates to the separation in energy between the two half-cells. We call this separation the *emf*, where the initials derive from the archaic phrase *electromotive force*. An *emf* is defined as always being positive.

> An *emf* is *always* defined as being positive.

We have already seen from Faraday's laws how a zero current implies that no redox chemistry occurs. Accordingly, we stipulate that the meter must draw absolutely no current if we want to measure the battery's *emf* at *equilibrium*. Henceforth, we will assume that all values of *emf* were determined at zero current.

Aside

The term '*emf*' follows from the archaic term 'electromotive force'. Physicists prefer to call the *emf* a 'potential difference' or symbolize it as a 'p.d.'.

Confusingly, potential also has the symbols of U, V and E, depending on the context.

Why do hydrogen fuel cells sometimes 'dry up'?

Cells and half-cells

Hydrogen fuel cells promise to fuel prototype cars in the near future. We define such a fuel cell as a machine for utilizing the energies of hydrogen and oxygen gases, hitherto separated, to yield a usable electric current without combustion or explosion. Unlike the simple batteries above, the oxygen and hydrogen gases fuelling these cells are transported from large, high-pressure tanks outside the cell. The gases then feed through separate pipes onto the opposing sides of a semi-permeable membrane (see Figure 7.3), the two sides of which are coated with a thin layer of platinum metal, and represent the anode and cathode of the fuel cell. This membrane helps explain why such cells are often called PEM fuel cells, where the acronym stands for 'proton exchange membrane'.

When it reaches the polymer membrane, hydrogen gas is oxidized at the *negative* side of the cell (drawn on the left of Figure 7.3), forming protons according to

$$H_{2(g)} \longrightarrow 2H^+ + 2e^-{}_{(Pt)} \qquad (7.13)$$

The subscripted 'Pt' helps emphasize how the two electrons conduct away from the membrane through the thin layer of platinum

> The energy necessary to cleave the H–H bond is provided by the energy *liberated* when forming the two H–Pt bonds after molecular dissociation.

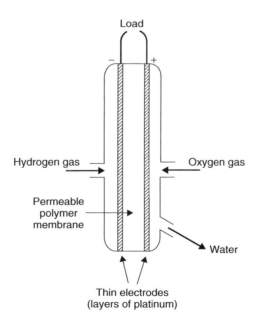

Figure 7.3 A hydrogen–oxygen fuel cell. The water formed at the cathode on the right-hand side of the cell condenses and collects at the bottom of the cell, and drains through a channel at the bottom right-hand side

metal surrounding the electrolyte, and enter into the external circuit where they perform *work*. The platinum also catalyses the dissociation of diatomic H_2 gas to form reactive $H^\bullet$ atoms.

> The membrane, made from a perfluorinated polymer bearing sulphonic acid groups, is known in the trade as Nafion^{TM}.

Once formed, the protons diffuse through the platinum layer and enter deep into the layer of semi-permeable membrane. They travel from the left-hand side of the membrane to its right extremity in response to a gradient in concentration. (Movement caused by a concentration gradient will remind us of dye diffusing through a saucer of water, as described on p. 129.)

At the *positive* side of the cell (drawn here on the right), oxygen gas is reduced to oxide ions, according to

> *Note*: the electron count in Equations (7.13) and (7.14) should balance in reality

$$O_{2(g)} + 4e^-_{(Pt)} \longrightarrow 2O^{2-}_{(aq)} \qquad (7.14)$$

The electrons necessary to effect the reduction of gaseous oxygen come from the external circuit, and enter the oxygen half-cell through the layer of platinum coating the cathode, thereby explaining why the electrons in Equation (7.14) are subscripted with 'Pt'.

The O^{2-} ions combine chemically with protons that have traversed the Nafion membrane and form water, which collects at the foot of the cell.

Because the overall cell reaction is exothermic, the value of the cell *emf* decreases with increasing temperature, so the temperature is generally kept relatively low at about 200 °C. The cell *emf* is 1.23 V at this temperature.

> The rate of oxygen reduction can be accelerated by finely dividing the platinum catalyst, thereby increasing its effective area.

One of the main advantages of this hydrogen fuel cell is the rapid rate at which hydrogen is oxidized at the platinum surface. Most of the cell's operational difficulties relate to the oxygen side of the device. Firstly, the reduction of gaseous oxygen in Equation (7.14) is relatively slow, so the rate at which the cell operates is somewhat limited. But more serious is the way the cell requires a *continual* flow of gas, as below. The hydrogen half-cell comprises an electrode couple because two redox states of the same material coexist there (H_2 and H^+). In a similar way, the oxygen half-cell also comprises an electrode couple, but this time of O_2 and O^{2-}.

If the flow of oxygen falters, e.g. when the surface of the cathode is covered with water, then no gaseous O_2 can reach the platinum outer layer. In response, firstly no electrons are consumed to yield oxide ions, and secondly the right-hand side of the cell 'floods' with the excess protons that have traversed the polymer membrane and not yet reacted with O^{2-}. Furthermore, without the reduction of oxygen, there is no redox couple at the cathode. The fuel cell ceases to operate, and can produce no more electrical energy.

This simple example helps explain why a cell requires no fewer than two half-cells. A half-cell on its own cannot *exchange* electrons, and cannot truly be termed a cell.

Why bother to draw cells?

Cells and the 'cell schematic'

Having defined a cell, we now want to know the best way of representing it. Undoubtedly, the simplest is to draw it diagrammatically – no doubt each picture of a cell being a work of art in miniature.

But drawing is laborious, so we generally employ a more sensible alternative: we write a *cell schematic*, which is a convenient *abbreviation* of a cell. It can be 'read' as though it was a cross-section, showing each interface and phase. It is, therefore, simply a shorthand way of saying which components are incorporated in the cell as cathode, anode, electrolyte, etc., and where they reside.

Most people find that a correct understanding of how to write a cell schematic also helps them understand the way a cell works. Accordingly, Table 7.2 contains a series of simple rules for constructing the schematic.

Worked Example 7.4   We construct a cell with the copper(II) | copper and zinc(II) | zinc redox couples, the copper couple being more positive than the zinc couple. What is the cell schematic?

Answer strategy: we will work sequentially through the rules in Table 7.2.

First, we note how the copper couple is the most positive, so we write it on the right. The zinc is, therefore, the more negative and we write it on the left. We commence the schematic by writing, $\ominus Zn_{(s)} \ldots Cu_{(s)} \oplus$.

> For convenience, we often omit the subscript descriptors and the '$\ominus$' and '$\oplus$' signs.

Table 7.2   Rules for constructing a cell schematic

1. We always write the redox couple associated with the positive electrode on the right-hand side
2. We write the redox couple associated with the negative electrode on the left-hand side
3. We write a salt bridge as a double vertical line: ‖
4. If one redox form is conductive and can function as an electrode, then we write it on one extremity of the schematic.
5. We represent the phase boundary separating this electrode and the solution containing the other redox species by a single vertical line:[a]|
6. If both redox states of a couple reside in the same solution (e.g. Pb^{2+} and Pb^{4+}), then they share the same phase. Such a couple is written conventionally with the two redox states separated by only a comma: Pb^{4+}, Pb^{2+}
7. Following from 6: we see that no electrode is in solution to measure the energy at equilibration of the two redox species. Therefore, we place an inert electrode in solution; almost universally, platinum is the choice

[a]We write a single line | or, better, a dotted vertical line, if the salt bridge is replaced by a simple porous membrane.

There is a *phase boundary* between the Zn and Zn^{2+} because the Zn is solid but the Zn^{2+} is dissolved within a liquid electrolyte. A similar boundary exists in the copper half-cell.

To be a redox couple, the zinc ions will be in contact with the zinc electrode, which we write as $Zn^{2+}_{(aq)}|Zn_{(s)}$, the vertical line emphasizing that there is a *phase boundary* between them. We can write the other couple as $Cu^{2+}_{(aq)}|Cu_{(s)}$, with similar reasoning. Note that if the two electrodes are written at the extreme ends of the cell schematic, then the redox ionic states must be located somewhere between them. The schematic now looks like $\ominus Zn_{(s)}|Zn^{2+}_{(aq)} \cdots Cu^{2+}_{(aq)}|Cu_{(s)}\oplus$.

Finally, we note that the two half-cells must 'communicate' somehow – they must be connected. It is common practice to assume that a salt bridge has been incorporated, unless stated otherwise, so we join the notations for the two half-cell with a double vertical line, as $\ominus Zn_{(s)}|Zn^{2+}_{(aq)}||Cu^{2+}_{(aq)}|Cu_{(s)}\oplus$.

An alternative way of looking at the schematic is to consider it as 'the path taken by a charged particle during a walk from one electrode to the other'.

SAQ 7.4 Write the cell schematic for a cell comprising the Fe^{3+},Fe (positive) and Co^{2+},Co (negative) couples.

Worked Example 7.5 Write a cell schematic for a cell comprising the couples Br_2, Br^- and H^+, H_2. The bromine | bromide couple is the more positive. Assume that all solutions are aqueous.

This is a more complicated cell, because we have to consider the involvement of more phases than in the previous example.

Right-hand side: the bromine couple is the more positive couple, so we write it on the right of the schematic. Neither Br_2 nor Br^- is metallic, so we need an inert electrode. By convention, we employ platinum if no other choice is stipulated. The electrode at the far right of the schematic is therefore Pt, as $\ldots Pt_{(s)}\oplus$.

We write the oxidized form first if both redox states reside in the same phase, separating them with a comma.

Br_2 and Br^- are both soluble in water – indeed, they are mutually soluble, forming a single-phase solution. Being in the same phase, we cannot write a phase boundary (as either '$Br_2|Br^-$' or as '$Br^-|Br_2$'), so we write it as '$Br_2, Br^-_{(aq)}$'. Note how we write the oxidized form first and separate the two redox states with a comma. The right-hand side of the schematic is therefore '$Br_2, Br^-_{(aq)}|Pt_{(s)}\oplus$'.

Left-hand side: neither gaseous hydrogen nor aqueous protonic solutions will conduct electrons, so again an inert electrode is required on the extremity of the schematic. We again choose platinum. The left-hand side of the cell is: $\ominus Pt_{(s)}$.

Hydrogen gas is in immediate contact with the platinum inert electrode. (We bubble it through an acidic solution.) Gas and solution are different phases, so we write the hydrogen couple as $H_{2(g)}|H^+_{(aq)}$, and the left-hand side of the schematic becomes '$\ominus Pt_{(s)}|H_{2(g)}|H^+_{(aq)}$'.

Finally, we join the two half-cells via a salt bridge, as

$$\ominus \text{Pt}_{(s)} | \text{H}_{2(g)} | \text{H}^+{}_{(aq)} \| \text{Br}_2, \text{Br}^-{}_{(aq)} | \text{Pt}_{(s)} \oplus .$$

From now on, we will omit both the symbols $\ominus$ and $\oplus$, and merely assume that the right-hand side is the positive pole and the left-hand side the negative.

> As an extra check, we note how the salt bridge dips into the proton solution, so the term for H^+ needs to be written adjacent to the symbol '$\|$'.

Why do digital watches lose time in the winter?

The temperature dependence of emf

A digital watch keeps time by applying a tiny potential (voltage) across a crystal of quartz, causing it to vibrate at a precise frequency of ν cycles per second. The watch keeps time by counting off 1 s each time the quartz vibrates ν times, explaining why the majority of the components within the watch comprise a counting mechanism.

> Frequency ν has the SI unit of hertz (Hz). 1 Hz represents 1 cycle or vibration per second, so a frequency of ν Hz means ν cycles per second.

Unfortunately, the number of vibrations of the crystal per second is dictated by the potential applied to the quartz, so a larger voltage makes the frequency ν increase, and a smaller voltage causes ν to slow. For this reason, the potential of the watch battery must be constant.

In Section 4.6, we saw how the value of ΔG is never independent of temperature, except in those rare cases when $\Delta S_{(cell)} = 0$. Accordingly, the value of $\Delta G_{(cell)}$ for a battery depends on whether someone is wearing the watch while playing outside in the cold snow or is sunbathing in the blistering heat of a tropical summer.

> The voltage from the battery induces a minute mechanical strain in the crystal, causing it to vibrate – a property known as the *piezo-electric effect*.

And the *emf* of the watch battery is itself a function of the change in Gibbs function, $\Delta G_{(cell)}$ according to

$$\Delta G_{(cell)} = -nF \times emf \qquad (7.15)$$

where F is the Faraday constant and n is the number of moles of electrons transferred per mole of reaction. The value of $\Delta G_{(cell)}$ is negative if the reaction proceeds reversibly (see Section 4.3), so the *emf* is *defined* as positive to ensure that $\Delta G_{(cell)}$ is always negative. In other words, the value of $\Delta G_{(cell)}$ relates to the *spontaneous* cell reaction.

> Care: The output voltage of a battery is only an *emf* when measured at zero current, i.e. when not operating the watch.

So we understand that as the *emf* changes with temperature, so the quartz crystal vibrates at a different frequency – all because

$\Delta G_{(cell)}$ is a function of temperature. Ultimately, then, a digital watch loses time in the winter as a simple result of the cold.

Worked Example 7.6   The *emf* of a typical 'lithium' watch battery (which is a cell) is 3.0 V. What is $\Delta G_{(cell)}$?

Look at the units and note how $1\,J = 1\,C \times 1\,V$.	The number of electrons transferred n in a lithium battery is one, since the redox couple is Li^+, Li. Inserting values into Equation (7.15) yields:

$$\Delta G_{(cell)} = -1 \times 96\,485\ C\,mol^{-1} \times 3.0\ V$$

$$\Delta G_{(cell)} = -289\,500\ J\,mol^{-1} = -290\ kJ\,mol^{-1}$$

Notice how the molar energy released by a simply battery is enormous.

SAQ 7.5   A manganese dioxide battery has an *emf* of 1.5 V and $n = 2$. Calculate $\Delta G_{(cell)}$.

In the absence of any pressure–volume work, the value of $\Delta G_{(cell)}$ is equal to the *work* needed to transfer charge from the negative end of the cell to the positive. In practice, $\Delta G_{(cell)}$ equates to the amount of charge passed, i.e. the number of charged particles multiplied by the magnitude of that charge.

Why is a battery's potential not constant?

Non-equilibrium measurements of emf

We *define* the *emf* as having a positive value and, strictly, it is always determined at equilibrium.	A healthy battery for powering a Walkman or radio has a voltage of about 1.5 V. In the terminology of batteries, this value is called its *open-circuit potential*, but an electrochemist talking in terms of cells will call it the *emf*. This voltage is read on a voltmeter when we remove the battery from the device before measurement. But the voltage would be different if we had measured it while the battery was, for example, powering a torch.

	We perform *work* whenever we connect the two poles of the battery across a *load*. The 'load' in this respect might be a torch, calculator, car, phone or watch – anything which causes a current to pass. And this flow of current causes the voltage across the battery or cell to decrease; see Figure 7.4. We call this voltage the 'voltage under load'.
The two currents (for anode and cathode) are generally different. Neither of them is related to potential in a linear way.	A similar graph to Figure 7.4 could have been drawn but with the x axis being the resistance between the two electrodes: if the

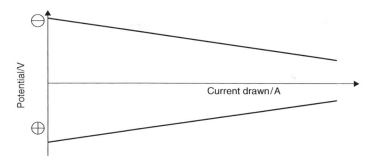

Figure 7.4 Schematic diagram showing how a cell's potential decreases with current. We call the cell potential the *emf* only when the current is zero

resistance between the two electrodes is zero, which is clearly the case if they should touch, then the cell potential is zero – we say the cell has 'shorted'.

Ohm's law, Equation (7.16), describes the difference between the *emf* and a voltage under load:

$$V = IR \qquad (7.16)$$

where I is the current flowing, R is the resistance of the load and V is the decrease in the voltage of the cell. When a current is drawn, the potential of the cell decreases by the amount V in Equation (7.16). We will call this new (smaller) voltage $E_{(\text{load})}$, and its magnitude is given by

$$E_{(\text{load})} = emf - IR \qquad (7.17)$$

In summary, we say the voltage of a cell is the same as a cell's *emf* if determined at zero current. From Faraday's laws of electrolysis, this criterion implies that none of the compositions within the cell can change. In other words, a cell *emf* is an *equilibrium* quantity.

> The *emf* can only ever be determined at zero current.

For this reason, it is not wise to speak of terms such as 'anode' of 'cathode' for a cell at equilibrium, because these terms relate to electrodes that give or receive charge during current flow; and our definition of equilibrium implies that no current does flows. We therefore adopt the convention: *the terms 'anode' or 'cathode' will no longer be employed in our treatment of equilibrium electrochemistry.*

> Why a battery's *emf* decreases permanently *after* a current has flowed is explained on p. 328.

What is a 'standard cell'?

The thermodynamics of cells

A *standard cell* produces a precise voltage and, before the advent of reliable voltmeters, was needed to calibrate medical and laboratory equipment. It is generally agreed that the first standard cell was the Clark cell (see p. 299), but the most popular was the Weston saturated cadmium cell, patented in 1893.

Edward Weston (1850–1936) was a giant in the history of electrical measuring instruments. In the field of measurement, he developed three important components: the standard cell, the manganin resistor and the electrical indicating instrument.

The main advantage of Weston's cell was its insensitivity to temperature, and the *emf* of almost 1 V: to be precise, 1.0183 V at 20 °C. It is usually constructed in an H-shaped glass vessel. One arm contains a cadmium amalgam electrode beneath a paste of hydrated cadmium sulphate ($3CdSO_4 \cdot 5H_2O$) cadmium sulphate and mercury(I) sulphate. The other arm contains elemental mercury. Its schematic is $Cd(Hg)|CdSO_{4(aq)}$, $Hg_2SO_4|Hg$.

> The Clark cell was patented by Latimer Clark in the 1880s, and was the first standard cell.

> *Remember*: the small subscripted 'p' indicates that the quantity is measured at *constant pressure*. It does not mean 'multiplied by p'.

The Weston saturated cadmium cell became the international standard for *emf* in 1911. Weston waived his patent rights shortly afterward to ensure that anyone was allowed to manufacture it.

Weston's cell was much less temperature sensitive than the previous standard, the Clark cell. We recall how the value of ΔG changes with temperature according to Equation (4.38). In a similar way, the value of $\Delta G_{(cell)}$ for a cell relates to the *entropy change* $\Delta S_{(cell)}$ such that the change of *emf* with temperature follows

$$\Delta S_{(cell)} = nF \left(\frac{d(emf)}{dT} \right)_p \tag{7.18}$$

the value of $(d(emf)/dT)_p$ is virtually zero for the Weston cell.

If we assume the differential $(d(emf)/dT)$ is a constant, then Equation (7.18) has the form of a straight line, $y = mx$, and a graph of *emf* (as '*y*') against T (as '*x*') should be linear. Figure 7.5 shows such a graph for the Clark cell, $Hg|HgSO_4$, $ZnSO_4$(sat'd)| Zn. Its gradient represents the extent to which the cell *emf* varies with temperature, and is called the *temperature voltage coefficient*. The gradient may be either positive or negative depending on the cell, and typically has a magnitude in the range 10^{-5} to 10^{-4} $V\,K^{-1}$. We want a smaller value of $(d(emf)/dT)$ if the *emf* is to be insensitive to temperature.

> The *temperature voltage coefficient* has several names: 'temperature coefficient', 'voltage coefficient' or 'temperature coefficient of voltage'. Table 7.3 contains a few values of $(d(emf)/dT)$.

Having determined the temperature dependence of *emf* as the gradient of a graph of *emf* against temperature, we obtain the value of $\Delta S_{(cell)}$ as 'gradient $\times n \times F$'.

> That this value of d(emf)/dT is negative tells us that the *emf* DEcreases when the temperature INcreases.

Worked Example 7.7   The temperature voltage coefficient for a simple alkaline torch battery is -6.0×10^{-4} $V\,K^{-1}$. What is the entropy change associated with battery discharge? The number of electrons transferred in the cell reaction $n = 2$.

Inserting values into Equation (7.18):

$$\Delta S_{(cell)} = 2 \times 96\,485 \text{ C mol}^{-1} \times -6.0 \times 10^{-4} \text{ V K}^{-1}$$

$$\Delta S_{(cell)} = -116 \text{ J K}^{-1} \text{ mol}^{-1}$$

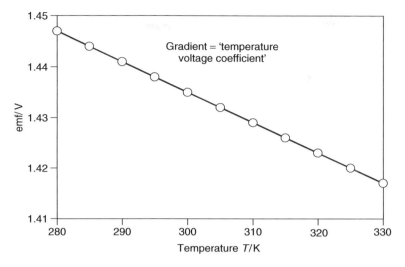

Figure 7.5 Graph of cell *emf* against temperature for the Clark cell $Hg|HgSO_4$, $ZnSO_4$(sat'd)|Zn. We call the gradient of this graph the 'temperature voltage coefficient'

Table 7.3 Temperature voltage coefficients for various cells and half cells

Cell[a]	$(d(emf)/dT)_p/V\ K^{-1}$
Standard hydrogen electrode	0 (by definition)[b]
Clark standard cell	6.0×10^{-4}
Saturated calomel electrode	$+7 \times 10^{-4}$
Silver–silver chloride	-8.655×10^{-5}
Silver–silver bromide	-4.99×10^{-4}
Weston standard cell	-5×10^{-5}

[a] Individual electrodes are cited with the SHE as the second electrode of the cell.
[b] The potential of the SHE is *defined* as zero at all temperatures.

SAQ 7.6 The *emf* of a lithium watch battery is exactly 3.000 V at 298 K, but the value decreases to 2.985 V at 270 K. Calculate the temperature voltage coefficient and hence the change in entropy $\Delta S_{(cell)}$ during cell discharge. (Take $n = 1$.)

Occasionally, the temperature voltage coefficient is not expressed as a simple number, but as a power series in T (we generally call it a *virial* series, or expansion). For example, Equation (7.19) cites such a series for the cell $Pt_{(s)}|H_{2(g)}|HBr_{(aq)}|AgBr_{(s)}|Ag_{(s)}$:

$$emf/V = 0.071\,31 - 4.99 \times 10^{-4}(T/K - 298) - 3.45 \times 10^{-6}(T/K - 298)^2 \quad (7.19)$$

We insert values of temperature T into the expression to obtain a value for *emf*. Values of $\Delta S_{(cell)}$ are obtained by performing two calculations, inserting first one temperature

T_1 to obtain the *emf* at T_1, and then a second T_2 to obtain another value of *emf*. We calculate a value of $(d(emf)/dT)$ using

$$\frac{d(emf)}{dT} = \frac{emf \text{ at } T_2 - emf \text{ at } T_1}{T_2 - T_1} \tag{7.20}$$

The value of $\Delta S_{(cell)}$ is then determined in the usual way via Equation (7.18).

SAQ 7.7 Insert values of $T = 310$ K into Equation (7.19) to calculate the potential of the cell $Pt_{(s)}|H_{2(g)}|HBr_{(aq)}|AgBr_{(s)}|Ag_{(s)}$.

SAQ 7.8 Repeat the calculation in SAQ 7.7, this time with $T = 360$ K, and hence determine $\Delta S_{(cell)}$.

Justification Box 7.1

The relationship between changes in Gibbs function and temperature (at constant pressure p) is defined using Equation (4.38):

$$-\Delta S = \left(\frac{\partial \Delta G}{\partial T}\right)_p$$

We know from Equation (7.15) that the change in $\Delta G_{(cell)}$ with temperature is '$-nF \times emf$'. The entropy change of the cell $\Delta S_{(cell)}$ is then obtained by substituting for $\Delta G_{(cell)}$ in Equation (7.18):

$$-\Delta S_{(cell)} = \left(\frac{\partial(-nF \times emf)}{\partial T}\right)_p \tag{7.18}$$

Firstly, the two minus signs cancel; and, secondly, n and F are both constants. Taking them out of the differential yields Equation (7.18) in the form above.

These values of ΔG, ΔH and ΔS relate to a complete cell, because thermodynamic data cannot be measured experimentally for *half-cells* alone.

To obtain the change in *enthalpy* during the cell reaction, we recall from the second law of thermodynamics how $\Delta H = \Delta G + T\Delta S$ (Equation (4.21)). In this context, each term relates to the cell. We substitute for $\Delta G_{(cell)}$ and $\Delta S_{(cell)}$ via Equations (7.15) and (7.18) respectively, to yield

$$\Delta H_{(cell)} = -nF \times emf + TnF\left(\frac{d(emf)}{dT}\right)_p \tag{7.21}$$

so, knowing the *emf* as a function of temperature, we can readily obtain a value of $\Delta H_{(cell)}$.

Worked Example 7.8   The Clark cell $Zn|Zn^{2+}, Hg_2SO_4|Hg$ is often employed as a *standard cell* since its *emf* is known exactly as a function of temperature. The cell *emf* is 1.423 V at 298 K and its temperature coefficient of voltage is -1.2×10^{-4} V K^{-1}. What are $\Delta G_{(cell)}$, $\Delta S_{(cell)}$ and thence $\Delta H_{(cell)}$ at 298 K?

Before we commence, we note that the spontaneous cell reaction is

$$Zn + Hg_2SO_4 + 7H_2O \longrightarrow ZnSO_4 \cdot 7H_2O + 2Hg^0$$

so the cell reaction is a two-electron process.

Next, we recall from Equation (7.15) that $\Delta G_{(cell)} = -nF \times emf$. Inserting values for the cell at 298 K gives

$$\Delta G_{(cell)} = -2 \times 96\,485 \text{ C mol}^{-1} \times 1.423 \text{ V}$$

$$\Delta G_{(cell)} = -275 \text{ kJ mol}^{-1}$$

Then, from Equation (7.18), we recall that

$$\Delta S_{(cell)} = nF \left(\frac{\mathrm{d}(emf)}{\mathrm{d}T} \right)_p$$

Inserting values:

$$\Delta S_{(cell)} = 2 \times 96\,485 \text{ C mol}^{-1} \times (-1.2 \times 10^{-4} \text{ V K}^{-1})$$

$$\Delta S_{(cell)} = -23.2 \text{ J K}^{-1} \text{ mol}^{-1}$$

Finally, from Equation (4.21), we say that $\Delta H_{(cell)} = \Delta G_{(cell)} + T \Delta S_{(cell)}$. We again insert values:

$$\Delta H_{(cell)} = (-275 \text{ kJ mol}^{-1}) + (298 \text{ K} \times -23.2 \text{ J K}^{-1} \text{ mol}^{-1})$$

so

$$\Delta H_{(cell)} = -282 \text{ kJ mol}^{-1}$$

SAQ 7.9   A different cell has an *emf* of 1.100 V at 298 K. The temperature voltage coefficient is $+0.35$ mV K^{-1}. Calculate $\Delta G_{(cell)}$, $\Delta S_{(cell)}$ and hence $\Delta H_{(cell)}$ for the cell at 298 K. Take $n = 2$.

Remember: 5 mV (5 *milli*-volts) is the same as 5×10^{-3} V.

Having performed a few calculations, we note how values of $\Delta G_{(cell)}$ and for $\Delta H_{(cell)}$ tend to be rather large. Selections of $\Delta G_{(cell)}$ values are given in Table 7.4.

Table 7.4   Table of values of $\Delta G_{(cell)}$ as a function of *emf* and *n*

emf/V	$\Delta G_{(cell)}$/kJ mol^{-1} (when $n = 1$)	$\Delta G_{(cell)}$/kJ mol^{-1} (when $n = 2$)
1.1	106	212
1.2	116	232
1.3	125	251
1.4	135	270
1.5	145	289
1.6	154	309
1.7	164	328
1.8	174	347
1.9	183	367
2.0	193	386

Aside

None of these thermodynamic equations is reliable if we fail to operate the cell reversibly, since the *emf* is no longer an exact thermodynamic quantity.

While electrochemical methods are experimentally easy, the practical difficulties of obtaining *accurate* thermodynamic data are so severe that the experimentally determined values of $\Delta G_{(cell)}$, $\Delta H_{(cell)}$ and $\Delta S_{(cell)}$ can only be regarded as 'approximate' unless we perform a daunting series of precautions. The two most common errors are: (1) allowing current passage to occur, causing the value of the cell *emf* to be too small; and (2) not performing the measurement *reversibly*.

The most common fault under (2) is changing the temperature of the cell too fast, so the temperature inside the cell is not the same as the temperature of, for example, the water bath in which a thermometer is placed.

Why aren't electrodes made from wood?

Electrodes: redox, passive and amalgam

The impetus for electronic motion is the chance to lose energy (see p. 60), so electrons move from high energy to low.

Electrochemical cells comprise a minimum of two *half-cells*, the energetic separation between them being proportional to the cell *emf*. Since this energy is usually expressed as a voltage, we see that the energy needs to be measured electrically as a *voltage*.

We have seen already how a cell's composition changes if a charge flows through it – we argued this phenomenon in terms of Faraday's laws. We cause electrochemical reactions to occur whenever a cell converts chemical energy into electrical energy.

For this reason, we say a battery or cell *discharges* during operation, with each electron from the cell flowing from high energy to low.

All electrochemical cells (including batteries) have two poles: one relates to the half-cell that is positively charged and the other relates to the negatively charged pole. Negatively charged electrons are produced at the *anode* as one of the products of the electrochemical reaction occurring at there. But if the electrons are to move then we need something through which they can *conduct* to and from the terminals of the cell: we need an *electrode*.

> Currents conduct through an electrode by means of *electrons*.

The phenomenon we call electricity comprises a *flow* of charged electrons. Wood is a poor conductor of electricity because electrons are inhibited from moving freely through it: we say the wood has a high electrical 'resistance' R. By contrast, most metals are good conductors of charge. We see how an electrode needs to be electrically conductive if the electrons are to move.

Most electrodes are metallic. Sometimes the metal of an electrode can also be one component part of a redox couple. Good examples include metallic iron, copper, zinc, lead or tin. A tin electrode forms a couple when in contact with tin(IV) ions, etc. Such electrodes are called *redox* electrodes (or *non-passive*). In effect, a redox electrode has two roles: first, it acts as a reagent; and, secondly, it measures the energy of the redox couple of which it forms one part when connected to a voltmeter.

> A *redox electrode* acts as a reagent as well as an electron conductor, as the metal of an electrode can also be one component part of a redox couple.

Some metals, such as aluminium or magnesium, cannot function as redox electrodes because of a coating of passivating oxide. Others, such as calcium or lithium, are simply too reactive, and would dissolve if immersed in solution.

> Metallic *mercury* is a poor choice of inert electrode at *positive* potentials because it oxidizes to form Hg(II) ions.

But it is also extremely common for both redox states of a redox couple to be non-conductive. Simple examples might include dissolving bromine in an aqueous solution of bromide ions, or the oxidation of hydrogen gas to form protons, at the heart of a hydrogen fuel cell; see Equation (7.13). In such cases, the energy of the couple must be determined through a different sort of electrode, which we call an *inert electrode*. Typical examples of inert electrodes include platinum, gold, glassy carbon or (at negative potentials) mercury. The metal of an inert electrode itself does not react in any chemical sense: such electrodes function merely as a *probe* of the electrode potential for measurements at zero current, and as a source, or sink, of electrons for electrolysis processes if current is to flow.

> We require an *inert electrode* when both parts of a redox couple reside in solution, or do not conduct: the electrode measures the *energy* of the couple.

The final class of electrodes we encounter are *amalgam electrodes*, formed by 'dissolving' a metal in elemental (liquid) mercury, generally to yield a solid. We denote an amalgam with brackets, so the amalgam of sodium in mercury is written as Na(Hg). The properties of such amalgams can be surprisingly different from their

> We denote an *amalgam* by writing the 'Hg' in brackets after the symbol of the dissolved element; cobalt amalgam is symbolized as Co(Hg).

constituent parts, so Na(Hg) is solid and, when prepared with certain concentrations of sodium, does not even react with water.

Why is electricity more dangerous in wet weather?

Electrolytes for cells, and introducing ions

Most electrical apparatus is safe when operated in a dry environment, but everyone should know that water and electricity represent a lethal combination. Only a minimal amount of charge conducts through air, so cutting dry grass with an electric mower is safe. Cutting the same grass during a heavy downpour risks electrocution, because water is a good *conductor* of electricity.

> Currents conduct through an electrolyte by means of *ions*.

But water does not conduct electrons, so the charges must move through water by a wholly different mechanism than through a metallic electrode. In fact, the charge carriers through solutions – aqueous or otherwise – are solvated *ions*. The 'mobility' μ of an ion in water is sufficiently high that charge conducts rapidly from a wet electrical appliance toward the person holding it: it behaves as an electrolyte.

> Ultimately, the word 'ion' derives from the Greek *eimi* 'to go', implying the arrival of someone or something. We get the English word 'aim' from this root.

All cells comprise half-cells, electrodes and a conductive *electrolyte*; the latter component separates the electrodes and conducts ions. It is usually, although not always, a liquid and normally has an ionic substance dissolved within it, the solid dissociating in solution to form ions. Aqueous electrolytes are a favourite choice because the high 'dielectric constant' ϵ of water imparts a high 'ionic conductivity' κ to the solution.

> *Ionic conductivity* is often given the Greek symbol kappa (κ) whereas *electrical conductivity* is given the different Greek symbol sigma (σ).

Sometimes electrochemists are forced to construct electrochemical cells without water, e.g. if the analyte is water sensitive or merely insoluble. In these cases, we construct the cell with an *organic* solvent, the usual choice being the liquids acetonitrile, propylene carbonate (**I**), *N,N*-dimethylformamide (DMF) or dimethylsulphoxide (DMSO), each of which is quite polar because of its high dielectric constant ϵ.

(I)

In some experiments, we need to enhance the ionic conductivity of a solution, so we add an additional ionic compound to it. Rather confusingly, we call both the compound and the resultant solution 'an electrolyte'.

The preferred electrolytes if the solvent is water are KCl and $NaNO_3$. If the solvent is a non-aqueous organic liquid, then we prefer salts of tetra-alkyl ammonium, such as tetra-n-butylammonium tetrafluoroborate, $^nBu_4N^+BF_{4-}$.

.2 Introducing half-cells and electrode potentials

Why are the voltages of watch and car batteries different?

Relationships between emf and electrode potentials

Being a cell, a battery contains two half-cells separated by an electrolyte. The electrodes are needed to connect the half-cells to an external circuit. Each electrode may act as part of a redox couple, but neither has to be.

The market for batteries is huge, with new types and applications being developed all the time. For example, a watch battery is a type of 'silver oxide' cell: silver in contact with silver oxide forms one half-cell while the other is zinc metal and dications. Conversely, a car battery is constructed with the two couples lead(IV)|lead and lead(IV)|lead(II). The electrolyte is sulphuric acid, hence this battery's popular name of *'lead–acid' cell* (see further discussion on p. 347).

> An 'electrode potential' E is the energy (expressed as a voltage) when a redox couple is at equilibrium. The value of E cannot be measured directly and must be *calculated* from an experimental *emf*.

The first difference between these two batteries is the voltage they produce: a watch battery produces about 3 V and a lead–acid cell about 2 V. The obvious cause of the difference in *emf* are the different half-cells. The 'electrode potential' E is the energy, expressed as a voltage, when a redox couple is at equilibrium. As a cell comprises two half-cells, we can now define the *emf* according to

> Two redox states of the same material form a *redox couple*.

$$emf = E_{(\text{positive half-cell})} - E_{(\text{negative half-cell})} \quad (7.22)$$

This definition is absolutely crucial. It does not matter if the values of E for both half-cells are negative or both are positive: $E_{(\text{positive})}$ is defined as being *the more positive* of the two half-cells, and $E_{(\text{negative})}$ is *the more negative*.

We now consider the *emf* in more detail, and start by saying that it represents the *separation* in potential between the two half-cell potentials; See Equation (7.22). In order for $\Delta G_{(\text{cell})}$ to remain positive for all thermodynamically spontaneous cell discharges, the *emf* is defined as always being positive.

> It is impossible to determine the potential of a single electrode: only its potential *relative to another electrode* can be measured.

Another convention dictates that we write the more positive electrode on the right-hand side of a cell, so we often see Equation (7.22) written in a slightly different form:

$$emf = E_{(RHS)} - E_{(LHS)} \qquad (7.23)$$

Being a potential, the electrode potential has the symbol E. We must exercise care in the way we cite it. E is the energy of a redox *couple*, since it relates to two redox species, both an oxidized and reduced form, 'O' and 'R' respectively. We supplement the symbol E with appropriate subscripts, as $E_{O,R}$.

Worked Example 7.9 Consider the electrode potentials for metallic lead within a lead–acid battery. The lead has three common redox states, Pb^{4+}, Pb^{2+} and Pb^{0}, so there are three possible equilibria to consider:

$$Pb^{4+}{}_{(aq)} + 2e^- = Pb^{2+}{}_{(aq)} \quad \text{for which } E_{(equilibrium)} = E_{Pb^{4+},Pb^{2+}}$$

$$Pb^{4+}{}_{(aq)} + 4e^- = Pb^{0}{}_{(s)} \quad \text{for which } E_{(equilibrium)} = E_{Pb^{4+},Pb^{0}}$$

$$Pb^{2+}{}_{(aq)} + 2e^- = Pb^{0}{}_{(s)} \quad \text{for which } E_{(equilibrium)} = E_{Pb^{2+},Pb^{0}}$$

We now see why it is so misleading to say merely E_{Pb} or, worse, 'the electrode potential of lead'.

> We cite the oxidized form first, as $E_{O,R}$.

We conventionally cite the oxidized form first within each symbol, which is why the general form is $E_{O,R}$, so $E_{Pb^{4+},Pb^{2+}}$ is correct, but $E_{Pb^{2+},Pb^{4+}}$ is not. Some people experience difficulty in deciding which redox state is oxidized and which is the reduced. A simple way to differentiate between them is to write the balanced redox reaction as a reduction. For example, consider the oxidation reaction in Equation (7.1). On rewriting this as a reduction, i.e. $Al^{3+}{}_{(aq)} + 3e^- = Al_{(s)}$, the oxidized redox form will automatically precede the reduced form as we read the equation from left to right, i.e. are written in the correct order. For example, $E_{O,R}$ for the couple in Equation (7.1) is $E_{Al^{3+},Al}$.

> We usually omit the superscripted 'zero' on uncharged redox states.

We usually cite an uncharged participant without a superscript. Considering the reaction $Pb^{2+} + 2e^- = Pb$, the expression $E_{Pb^{2+},Pb}$ is correct but the '0' in $E_{Pb^{2+},Pb^{0}}$ is superfluous.

> When choosing between two ionic valences, the name of the higher (more oxidized) state ends with *-ic* and the lower (less oxidized) form ends with *-ous*.

SAQ 7.10 Consider the cobaltous ion | cobalt redox couple. Write an expression for its electrode potential.

With more complicated redox reactions, such as $2H^+{}_{(aq)} + 2e^- = H_{2(g)}$, we would not normally write the stoichiometric number, so we prefer E_{H^+,H_2} to E_{2H^+,H_2}; the additional '2' before H^+ is superfluous here.

SAQ 7.11 Write down an expression for the electrode potential of the bromine | bromide couple. [Hint: it might help to write the balanced redox reaction first.]

How do 'electrochromic' car mirrors work?

Introducing an orbital approach to dynamic electrochemistry

It's quite common when driving at night to be dazzled by the lights of the vehicle behind as they reflect from the driver's new-view or door mirror. We can prevent the dazzle by forming a layer of coloured material over the reflecting surface within an *electrochromic mirror*. Such mirrors are sometimes called 'smart mirrors' or electronic 'anti-dazzle mirrors'.

> Electrochromic mirrors are now a common feature in expensive cars.

These mirrors are electrochromic if they contain a substance that changes colour according to its redox state. For example, *methylene blue*, MB^+ (**II**), is a chromophore because it has an intense blue colour. **II** is a popular choice of electrochromic material for such mirrors: it is blue when fully oxidized, but it becomes colourless when reduced according to

$$MB^0 \longrightarrow MB^+ + e^-$$

$$\text{colourless} \qquad \text{blue}$$

(7.24)

(II)

We can now explain how an electrochromic car mirror operates. The mirror is constructed with **II** in its colourless form, so the mirror functions in a normal way. The driver 'activates' the mirror when the 'anti-dazzle' state of the mirror is required, and the coloured form of methylene blue (MB^+) is generated oxidatively according to Equation (7.24). Coloured MB^+ blocks out the dazzling reflection at the mirror by absorbing about 70 per cent of the light. After our vehicle has been overtaken and we require the mirror to function normally again, we reduce MB^+ back to colourless MB^0 via the reverse of Equation (7.24), and return the mirror to its colourless state. These two situations are depicted in Figure 7.6.

We discuss 'colour' in Chapter 9, so we restrict ourselves here to saying the colour of a substance depends on the way its electrons interact with light; crucially, absorption of a photon causes an electron to promote between the two *frontier orbitals*. The separation in energy between these two orbitals is E, the magnitude of which relates to the wavelength of the light absorbed λ according to the *Planck–Einstein* equation, $E = hc/\lambda$, where h is the Planck constant and c is the

> An electron is donated *to* an orbital during reduction. The electron removed during oxidation is taken *from* an orbital.

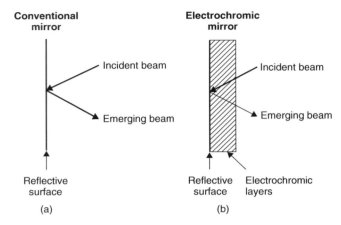

Figure 7.6 Mirrors: (a) an ordinary car driver's mirror reflects the lights of a following car, which can dazzle the driver; (b) in an electrochromic mirror, a layer of optically absorbing chemical is electro-generated in front of the reflector layer, thereby decreasing the scope for dazzle. The width of the arrows indicates the relative light intensity

speed of light in a vacuum. The value of E for MB^0 corresponds to an absorption in the UV region, so MB^0 appears colourless. Oxidation of MB^0 to MB^+ causes a previously occupied orbital to become empty, itself changing the energy separation E between the two frontier orbitals. And if E changes, then the Planck–Einstein equation tells us the wavelength λ of the light absorbed must also change. E for MB^+ corresponds to λ of about 600 nm, so the ion is blue.

The reasoning above helps explain why MB^0 and MB^+ have different colours. To summarize, we say that the colours in an electrochromic mirror change following oxidation or reduction because different orbitals are occupied before and after the electrode reaction.

Why does a potential form at an electrode?

Formation of charged electrodes

For convenience, we will discuss here the formation of charges with the example of copper metal immersed in a solution of copper sulphate (comprising Cu^{2+} ions). We consider first the situation when the positive pole of a cell is, say, bromine in contact with bromide ions, causing the copper to be the negative electrode.

Let's look at the little strip cartoon in Figure 7.7, which shows the surface of a copper electrode. For clarity, we have drawn only one of the trillion or so atoms on its surface. When the cell of which it is a part is permitted to discharge spontaneously, the copper electrode acquires a negative charge in consequence of an *oxidative* electron-transfer reaction (the reverse of Equation (7.7)). During the oxidation, the surface-bound atom loses the two electrons needed to bond the atom to the electrode surface, becomes a cation and diffuses into the bulk of the solution.

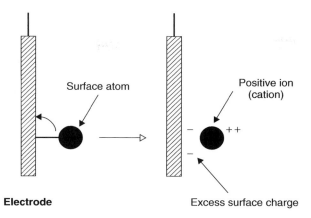

Figure 7.7 Schematic drawing to illustrate how an electrode acquires its negative charge

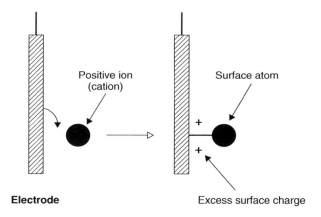

Figure 7.8 Schematic drawing to illustrate how an electrode acquires its positive charge

The two electrons previously 'locked' into the bond remain on the electrode surface, imparting a negative potential.

We now consider a slightly different cell in which the copper half-cell is the positive pole. Perhaps the negative electrode is zinc metal in contact with Zn^{2+} ions. If the cell discharges spontaneously, then the electron-transfer reaction is the reduction reaction in Equation (7.7) as depicted in the strip cartoon in Figure 7.8. A bond forms between the surface of the copper electrode and a Cu^{2+} cation in the solution The electrons needed to reduce the cation come from the electrode, imparting a net positive charge to its surface.

Finally, we should note that the extent of oxidation or reduction needed to cause a surface charge of this type need not be large; and the acquisition of charge, whether positive or negative, is fast and requires no more than a millisecond after immersing the electrodes in their respective half-cells.

7.3 Activity

Why does the smell of brandy decrease after dissolving table salt in it?

Real and 'perceived' concentrations

At the risk of spoiling a good glass of brandy, try adding a little table salt to it and notice how the intensity of the smell is not so strong after the salt dissolves.

> We mention the volatile alcohol here because it is responsible for the smell.

We recall from Chapter 5 how the intensity of a smell we detect with our nose is proportional to the vapour pressure of the substance causing it. The vapour pressure of ethanol is $p_{\text{(ethanol)}}$, its magnitude being proportional to the mole fraction of ethanol in the brandy; brandy typically contains about 40 per cent (by volume) of alcohol.

Although adding table salt does not decrease the proportion of the alcohol in the brandy, it does decrease the *apparent* amount. And because the perceived proportion is lowered, so the vapour pressure drops, and we discern the intensity of the smell has decreased. We are entering the world of 'perceived' concentrations.

> The 'activity' *a* is the *thermodynamically perceived* concentration.

Although the *actual* concentrations of the volatile components in solution remain unchanged after adding the salt, the system *perceives* a decrease in the concentration of the volatile components. This phenomenon – that the perceived concentration differs from the real concentration – is quite common in the thermodynamics of solution-phase electrochemistry. We say that the concentration persists, but the 'activity' *a* has decreased by adding the salt.

As a working definition, the activity may be said to be 'the *perceived* concentration' and is therefore somewhat of a 'fudge factor'. More formally, the activity *a* is defined by

$$a = \frac{c}{c^{\ominus}}\gamma \tag{7.25}$$

where *c* is the *real* concentration. The concluding term γ, termed the *activity coefficient*, is best visualized as the ratio of a solute's 'perceived' and 'real' concentrations.

> We only add the term $c^{\ominus}$ in order to render the activity dimensionless.

The activity *a* and the activity coefficient γ are both dimensionless quantities, which explains why we must include the additional '$c^{\ominus}$' term, thereby ensuring that *a* also has no units. We say the value of $c^{\ominus}$ is 1 mol dm^{-3} when *c* is expressed in the usual units of mol dm^{-3}, and 1 mol m^{-3} if *c* is expressed in the SI units of mol m^{-3}, and so on.

Why does the smell of gravy become less intense after adding salt to it?

The effect of composition on activity

Gravy is a complicated mixture of organic chemicals derived from soluble meat extracts. Its sheer complexity forces us to simplify our arguments, so we will approximate and say it contains just one component in a water-based solution. Any incursions into reality, achieved by extending our thoughts to encompass a *multicomponent* system, will not change the *nature* of these arguments at all.

Adding table salt to gravy causes its lovely smell to become less intense. This is a general result in cooking: adding a solute (particularly if the solute is *ionic*) decreases the smell, in just the same way as adding table salt decreased the smell of brandy in the example directly above.

The ability to smell a solute relies on it having a vapour pressure above the solution. Analysing the vapour above a gravy dish shows that it contains molecules of both solvent (water) and solute (gravy), hence its damp aroma. The vapour pressures above the gravy dish do not alter, provided that we keep the temperature constant and maintain the equilibrium between solution and vapour. The proportion of the solute in the vapour is always small because most of it remains in solution, within the heavier liquid phase.

> The pressure above a solution relates to the composition of solution, according to Henry's law; see Section 5.6.

As a good approximation, the vapour pressure of each solute in the vapour above the dish is dictated by the respective *mole fractions* in the gravy beneath. As an example, adding water to the gravy solution dilutes it and, therefore, decreases the gravy smell, because the mole fraction of the gravy has decreased.

Putting ionic NaCl in the gravy increases the number of ions in solution, each of which can then *interact* with the water and the solute, which decreases the 'perceived concentration' of solute. In fact, we can now go further and say the thermodynamic activity a represents the concentration of a solute in the presence of interactions.

> An electrochemist assesses the number of ions and their relative influence by means of the 'ionic strength' I (as defined below).

Why add alcohol to eau de Cologne?

Changing the perceived concentration

Fragrant *eau de Cologne* is a dilute perfume introduced in Cologne (Germany) in 1709 by Jean Marie Farina. It was probably a modification of a popular formula made before 1700 by Paul Feminis, an Italian in Cologne, and was based on bergamot and other citrus oils. The water of Cologne was believed to have the power to ward off bubonic plague.

> The word 'perfume' comes from the Latin *per fumem*, meaning 'through smoke'.

Eau de Cologne perfume is made from about 80–85 per cent water and 12–15 per cent ethanol. Volatile esters make up the remainder, and provide both the smell and colour.

The vapour pressure of alcohol is higher than that of water, so adding alcohol to an aqueous perfume increases the pressure of the gases above the liquid. In this way, the activity a of the organic components imparting the smell will increase and thereby increase the perceived concentration of the esters. And increasing $a_{(ester)}$

> These esters are stable in the dark, but degrade in strong sunlight, which explains why so many perfumes are sold in bottles of darkened or frosted glass.

has the effect of making the *eau de Cologne* more pungent. Stated another way, the product requires less ester because the alcohol increases its perceived concentration. Incidentally, the manufacturer also saves money this way.

Thermodynamic activity *a*

Every day, electrochemists perform measurements that require a knowledge of the activity *a*. Measurements can be made in terms of straightforward concentrations if solutions are very dilute, but 'very dilute' in this context implies $c \approx 10^{-4}$ mol dm^{-3}, or less. Since most solutions are far more concentrated than millimoles per litre, from now on we will write all equations in terms of activities *a* instead of concentration *c*.

> The concept of activity was introduced in the early 20th century by one of the giants of American chemistry, G. N. Lewis.

The values of activity *a* and concentration *c* are the same for very dilute solutions, so the ratio of *a* and *c* is one because the real and perceived concentrations are the same. If $a = c$, then Equation (7.25) shows how the activity coefficient γ has a value of unity at low concentration.

By contrast, the perceived concentration is usually less than the real concentration whenever the solution is more concentrated, so $\gamma < 1$. To illustrate this point, Figure 7.9 shows the relationship between the activity coefficient γ (as '*y*') and concentration (as '*x*') for a few simple solutes in water. The graph shows clearly how the value of γ can drop quite dramatically as the concentration increases.

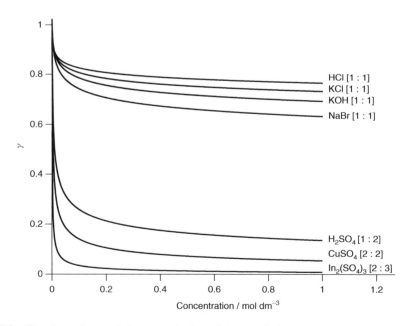

Figure 7.9   The dependence of the mean ionic activity coefficient $\gamma_{\pm}$ on concentration for a few simple solutes

The activity of a solid   The activity of a *pure* solid in its standard state is unity, so the activity of pure copper or of zinc metal electrodes is one. We write this as $a_{(Cu)}$ or $a_{(Zn)} = 1$.

> *Unit* and *unity* here both mean 'one', so unit activity means $a = 1$.

The activity of an *impure* solid is more complicated. Such an 'impure' system might be represented by a solid metal with a dirty surface, or it might represent a mixture of two metals, either as an alloy or an amalgam with a metal 'dissolved' in mercury.

For example, consider the bi-metallic *alloy* known as bronze, which contains tin (30 mol%) and copper (70 mol%). There are two activities in this alloy system, one each for tin and copper. The activity of each metal is obtained as its respective mole fraction x, so $x_{(Sn)} = a_{(Sn)} = 0.3$, and $a_{(Cu)} = 0.7$.

> An 'alloy' is a *mixture* of metals, and is not a compound.

Worked Example 7.10   A tooth filling is made of a silver amalgam that comprises 37 mol% silver. What is the activity of the mercury, $a_{(Hg)}$?

The activity of the mercury $a_{(Hg)}$ is the same as its mole fraction, $x_{(Hg)}$. By definition

$$x_{(Hg)} + x_{(Ag)} = 1$$

> The sum of the mole fractions x must always add up to one because 'the sum of the constituents adds up to the whole'.

so

$$x_{(Hg)} = 1 - x_{(Ag)} = 0.63$$

The activity is therefore

$$x_{(Hg)} = a_{(Hg)} = 0.63$$

The activity of a gas   The activity of a pure gas is its pressure (in multiples of $p^{\ominus}$), so $a_{(H_2)} = p_{(H_2)} \div p^{\ominus}$. The activity of pure hydrogen gas $a_{(H_2)}$ at $p^{\ominus}$ is therefore unity.

> *Reminder*: the value of $p^{\ominus}$ is 10^5 Pa.

In fact, for safety reasons it is not particularly common to employ *pure* gases during electrochemical procedures, so mixtures are preferred. As an example, the hydrogen gas at the heart of the standard hydrogen electrode (SHE) is generally mixed with elemental nitrogen, with no more than 10 per cent of H_2 by pressure. We call the

> In a mixture of gases, we call the inert gas a *base* or *bath* gas.

other gas a *base* or *bath* gas. Conversely, we might also say that hydrogen dilutes the nitrogen, and so is a *diluent*.

In such cases, we can again approximate the activity to the mole fraction x.

Worked Example 7.11   Hydrogen gas is mixed with a nitrogen 'bath gas'. The overall pressure is $p^{\ominus}$. If the mole fraction of the hydrogen is expressed as 10 per cent, what is its activity?

By definition, $x_{(X)} = $ partial pressure, $p_{(X)}$, so

$$a_{(H_2)} = p_{(H_2)} \div p_{(total)} = 0.1$$

The activity of a solution   It is unwise to speak in broad terms of 'the activity of a solution' because so many different situations may be considered. For example, consider the following two examples.

(1) *The activity of a mixture of liquids.* It is rarely a good idea to suggest that the activity of a liquid in a mixture is equal to its mole fraction x because of complications borne of intermolecular interactions (e.g. see Chapter 2 and Section 5.6 concerning Raoult's law). Thankfully, it is generally rare that an electrochemist wants to study liquid mixtures of this sort (except amalgams diluted to a maximum mole fraction of about 1 per cent metal in Hg), so we will not consider such a situation any further.

> Amalgams are liquid when very dilute, but are solid if the mole fraction of mercury drops below about 70 per cent.

(2) *The activity of a solute in a liquid solvent.* The activity a and concentration c may be considered to be wholly identical if the concentration is tiny (to a maximum of about 10^{-3} $mol\,dm^{-3}$), provided the solution contains no other solutes. Such a concentration is so tiny, however, as to imply slightly polluted distilled water, and is not particularly useful.

For all other situations, we employ the Debye–Hückel laws (as below) to calculate the activity coefficient γ. And, knowing the value of γ, we then say that $a = (c \div c^{\ominus}) \times \gamma$ (Equation (7.25)), remembering to remove the concentration units because a is dimensionless.

Why does the cell emf alter after adding LiCl?

Ionic 'screening'

Consider the Daniell cell $Zn|Zn^{2+}||Cu^{2+}|Cu$. The cell *emf* is about 1.1 V when prepared with clean, pure electrodes and both solutions at unit activity. The *emf* decreases to about 1.05 V after adding lithium chloride to the copper half-cell. Adding more LiCl, but this time to the zinc solution, increases the *emf* slightly, to about 1.08 V.

> In fact, a similar result is obtained when adding most ionic electrolytes.

No redox chemistry occurs, so no copper ions are reduced to copper metal nor is zinc metal oxidized to form Zn^{2+}. No complexes form in solution, so the changes in *emf* may be attributed entirely to changing the composition of the solutions.

Lithium and chloride ions are not wholly passive, but interact with the ions originally in solution. Let us look at the copper ions, each of which can associate electrostatically with chloride ions, causing it to resemble a dandelion 'clock' with the central copper ion looking as though it radiates chloride ions. All the ions are solvated with water. These interactions are *coulombic* in nature, so negatively charged chloride anions interact attractively with the positive charges of the copper cations. Copper and lithium cations repel. Conversely, the additional Li^+ ions attract the negatively charged sulphates from the original solution; again, Cl^- and SO_4^{2-} anions repel.

> We need a slightly different form of γ when working with electrolyte solutions: we call it the *mean ionic* activity coefficient $\gamma_\pm$, as below.

The ionic atmosphere moves continually, so we consider its composition *statistically*. Crystallization of solutions would occur if the ionic charges were static, but association and subsequent dissociation occur all the time in a *dynamic* process, so even the ions in a dilute solution form a three-dimensional structure similar to that in a solid's repeat lattice. Thermal vibrations free the ions by shaking apart the momentary interactions.

> 'Associated ions' in this context means an association species held together (albeit transiently) via electrostatic interactions.

The ions surrounding each copper cation are termed the *ionic atmosphere*. In the neighbourhood of any positively charged ion (such as a copper cation), there are likely to be more negative charges than positive (and vice versa). We say the cations are surrounded with a *shell* of anions, and each anion is surrounded by a shell of cations. The ionic atmosphere can, therefore, be thought to look much like an onion, or a Russian doll, with successive layers of alternate charges, with the result that charges effectively 'cancel' each other out when viewed from afar.

Having associated with other ions, we say the copper ion is *screened* from anything else having a charge (including the electrode), so the full extent of its charge cannot be 'experienced'. In consequence, the magnitude of the electrostatic interactions between *widely separated ions* will decrease.

The electrode potential measured at an electrode relates to the 'Coulomb potential energy' V 'seen' by the electrode due to the ions in solution. V relates to two charges z_1 and z_2 (one being the electrode here) separated by a distance r, according to

> The 'Coulomb potential energy' V is equal to the *work* that must be done to bring a charge z^+ from infinity to a distance of r from the charge z^-.

$$V = \frac{z^+ z^-}{4\pi \epsilon_0 \epsilon_r r} \qquad (7.26)$$

where ϵ_0 is the permittivity of free space and ϵ_r is the relative permittivity of the solvent. In water at $25\,^\circ\text{C}$, ϵ_r has a value of 78.54.

The magnitude of V relates to interactions between the electrode and nearby ions nestling within the interface separating the electrode and the ionic solution. Since the 'effective' (visible) charge on the ions decreases, so the electrode perceives there to be fewer of them. In other words, it *perceives* the concentration to have dipped below the *actual* concentration. This perceived decrease in the number of charges then causes the voltmeter to read a different, smaller value of $E_{Cu^{2+},Cu}$.

The zinc ions in the other half of the Daniell cell can similarly interact with ions added to solution, causing the zinc electrode to 'see' fewer Zn^{2+} species, and the voltmeter again reads a different, smaller value of $E_{Zn^{2+},Zn}$. Since the *emf* represents the *separation* between the electrode potentials of the two half-cells, any changes in the *emf* illustrate the changes in the constituent electrode potentials.

Background to the Debye–Hückel theory

The interactions between the ions originally in solution and any added LiCl are best treated within the context of the *Debye–Hückel* theory, which derives from a knowledge of electrostatic considerations.

Firstly, we assume the ions have an energy distribution as defined by the Boltzmann distribution law (see p. 35). Secondly, we say that electrostatic forces affect the behaviour and the mean positions of all ions in solution. It should be intuitively clear that ions having a larger charge are more likely to associate strongly than ions having a smaller charge. This explains why copper ions are more likely to associate than are sodium ions. The magnitude of the force exerted by an ion with a charge z_1 on another charge z_2 separated by an inter-ion distance of r in a medium of relative permittivity ϵ_r is the 'electrostatic interaction' ϕ, as defined by

$$\phi = \frac{z^+ z^-}{4\pi \epsilon_0 \epsilon_r r^2} \tag{7.27}$$

Note how this equation states that the force is inversely proportional to the *square* of the distance between the two charges r, so the value of ϕ decreases rapidly as r increases.

Since cations and anions have opposite charges, ϕ is negative. The force between two anions will yield a positive value of ϕ. We see how a positive value of ϕ implies an inter-ionic repulsion and a negative value implies an inter-ionic attraction.

> Positive values of ϕ imply repulsion, and a negative value attraction.

The Debye–Hückel theory suggests that the probability of finding ions of the opposite charge within the ionic atmosphere increases with increasing attractive force.

Why does adding NaCl to a cell alter the emf, but adding tonic water doesn't?

The effects of ion association and concentration on γ

Sodium chloride – table salt – is a 'strong' ionic electrolyte because it dissociates fully when dissolved in water (see the discussion of weak and strong acids in Section 6.2). The only electrolytes in tonic water are sugar (which is not ionic) and sodium carbonate, which is a weak electrolyte, so very few ions are formed by adding the tonic water to a cell.

The ratio of perceived to real concentrations is called the activity coefficient γ (because, from Equation (7.25), $\gamma = a \div c$). Furthermore, from the definition of activity in Equation (7.20), γ will have a value in the range zero to one. The diagram in Figure 7.9 shows the relationship between γ and concentration c for a few ionic electrolytes.

Adding NaCl to solution causes γ to decrease greatly because the number of ions in solution increases. Adding tonic water does *not* decrease the activity coefficient much because the concentration of the ions remains largely unchanged. The change in γ varies more with ionic electrolytes because the interactions are far stronger. And if the value of γ does not change, then the real and perceived concentrations will remain essentially the same.

The extent of ionic screening depends on the extent of association. The only time that association is absent, and we can treat ions as though free and visible ('unscreened'), is at *infinite dilution*.

> *Infinite dilution* (extrapolation to zero concentration) means so small a concentration that the possibility of two ions meeting, and thence associating, is tiny to non-existent.

Why does MgCl$_2$ cause a greater decrease in perceived concentration than KCl?

The mean ionic activity coefficient $\gamma_\pm$

> The value of γ depends on the solute employed.

The extent of ionic association depends on the ions we add to the solution. And the extent of association will effect the extent of screening, itself dictating how extreme the difference is between perceived and real concentration. For these reasons, the value of $\gamma (= a \div c)$ depends on the choice of solute as well as its concentration, so we ought to cite the solute whenever we cite an activity coefficient.

The value of γ is even more difficult to predict because solutes contain both anions and cations. In fact, it is *impossible* to differentiate between the effects of each, so we measure a *weighted average*. Consider a simple electrolyte such as KCl, which has one anion per cation. (We call it a '1:1 electrolyte'.) In KCl, the activity coefficient of the anions is called $\gamma_{(Cl^-)}$ and the activity coefficient of the cations is $\gamma_{(K^+)}$. We cannot know either γ_+ or γ_-; we can only know the value of $\gamma_\pm$. Accordingly, we modify Equation (7.25) slightly by writing

> We cannot know either γ_+ or γ_-; we can only know the value of their geometric mean $\gamma_\pm$.

> We call KCl a *1:1 electrolyte*, since the ratio of anions to cations is 1:1.

$$a = \frac{c}{c^{\ominus}}\gamma_\pm \qquad (7.28)$$

where the only change is the incorporation of the *mean ionic* activity coefficient $\gamma_\pm$.

The mean ionic activity coefficient is obtained as a *geometric mean* via

$$\gamma_\pm = \sqrt{\gamma_{(K^+)} \times \gamma_{(Cl^-)}} \qquad (7.29)$$

By analogy, the expression for the mean ionic activity coefficient $\gamma_\pm$ for a 2:1 electrolyte such as K$_2$SO$_4$ is given by

$$\gamma_\pm = \sqrt[3]{\gamma_+^2 \times \gamma_-} \qquad (7.30)$$

where the cube root results from the stoichiometry, since K$_2$SO$_4$ contains three ions (we could have written the root term alternatively as $\sqrt[3]{\gamma_+ \times \gamma_+ \times \gamma_-}$, with one γ term per ion). Again, a 1:3 electrolyte such as FeCl$_3$ dissolves to form four ions, so an expression for its mean ionic activity coefficient $\gamma_\pm$ will include a fourth root, etc.

SAQ 7.12 Write an expression similar to Equation (7.29) for the 2:3 electrolyte $Fe_2(SO_4)_3$.

Why is calcium better than table salt at stopping soap lathering?

Ionic strength I and the Debye–Hückel laws

People whose houses are built on chalky ground find that their kettles and boilers become lined with a hard 'scale'. We say that the water in the area is 'hard', meaning that minute amounts of chalk are dissolved in it. The hard layer of 'scale' is chalk that precipitated onto the inside surface of the kettle or boiler during heating.

We look at the actions of soaps in Chapter 10.

It is difficult to get a good soapy froth when washing the hands in 'hard water' because the ions from chalk in the water associate with the long-chain fatty acids in soap, preventing it from ionizing properly. Conversely, if the water contains table salt – for example, when washing the dishes after cooking salted meat – there is less of a problem with forming a good froth. Although the concentrations of sodium and calcium ions may be similar, the larger charges on the calcium and carbonate ions impart a disproportionate effect, and strongly inhibit the formation of frothy soap bubbles.

In 'dynamic' electrochemistry (when currents flow) we need to be careful not to mistake ionic strength and current, since both have the symbol I.

Having discussed ionic screening and its effects on the value of $\gamma_\pm$, we now consider the ionic charge z. When assessing the influence of z, we first define the extent to which a solute promotes association, and thus screening. The preferred parameter is the 'ionic strength' I, as defined by

$$I = \frac{1}{2} \sum_{i=1}^{i=i} c_i z_i^2 \qquad (7.31)$$

where z_i is the charge on the ion i in units of electronic charge, and c_i is its concentration. We will consider three simple examples to demonstrate how ionic strengths I are calculated.

Worked Example 7.12 Calculate the ionic strength of a simple 1:1 electrolyte, such as NaCl, that has a concentration of $c = 0.01 \text{ mol dm}^{-3}$.

Inserting values into Equation (7.31) we obtain

$$I = \frac{1}{2} \sum \left\{ \boxed{[Na^+] \times (+1)^2} + \boxed{[Cl^-] \times (-1)^2} \right\}$$

terms for the sodium ions   **terms for the chloride ions**

We next insert concentration terms, noting that one sodium ion and one chloride are formed per formula unit of sodium chloride (which is why we call it a 1:1 electrolyte). Accordingly, the concentrations of the two ions, $[Na^+]$ and $[Cl^-]$, are the same as $[NaCl]$, so

$$I = \frac{1}{2}\{([NaCl] \times 1) + ([NaCl] \times 1)\}$$

> NaCl is called a '1:1 electrolyte' because the formula unit contains one anion and one cation.

so we obtain the result for a 1:1 electrolyte that $I_{(NaCl)} = [NaCl]$.

Note that I has the same units as concentration: inserting the NaCl concentration $[NaCl] = 0.01$ mol dm^{-3}, we obtain $I = 0.01$ mol dm^{-3}.

> We obtain the result $I = c$ only for 1:1 (univalent) electrolytes.

Worked Example 7.13   Calculate the ionic strength of the 2:2 electrolyte FeSO$_4$ at a concentration $c = 0.01$ mol dm^{-3}.

Inserting charges in Equation (7.31):

$$I = \frac{1}{2}\{[Fe^{2+}] \times (+2)^2 + [SO_4^{2-}] \times (-2)^2\}$$

We next insert concentrations, again noting that one ferrous ion and one sulphate ion are formed per formula unit:

$$I = \frac{1}{2}\{([FeSO_4] \times 4) + ([FeSO_4] \times 4)\}$$

so we obtain the result $I = 4 \times c$ for this, a 2:2 electrolyte.

Inserting the concentration c of $[FeSO_4] = 0.01$ mol dm^{-3}, we obtain $I = 0.04$ mol dm^{-3}, which explains why hard water containing FeSO$_4$ has a greater influence than table salt of the same concentration.

Worked Example 7.14   Calculate the ionic strength of the 1:2 electrolyte CuCl$_2$, again of concentration 0.01 mol dm^{-3}.

Inserting charges into Equation (7.31):

$$I = \frac{1}{2}\{[Cu^{2+}] \times (+2)^2 + [Cl^-] \times (-1)^2\}$$

We next insert concentrations. In this case, there are two chloride ions formed per formula unit of salt, so $[Cl^-] = 2 \times [CuCl_2]$, but only one copper, so $[Cu^{2+}] = [CuCl_2]$.

> Note how the calculation requires the charge *per* anion, rather than the *total* anionic charge.

$$I = \frac{1}{2}\{([CuCl_2] \times 4) + (2[CuCl_2] \times 1)\}$$

so we obtain the result $I = 3 \times c$ for this, a 1:2 electrolyte. And, $I = 0.03$ mol dm^{-3} because $[CuCl_2] = 0.01$ mol dm^{-3}.

SAQ 7.13 Calculate the relationship between concentration and ionic strength for the 1:3 electrolyte $CoCl_3$.

Ionic strength I is an integral multiple of concentration c, where integer means *whole* number. A calculation of I not yielding a whole number is *wrong*.

Table 7.5 summarizes all the relationships between concentration and ionic strength I for salts of the type $M^{x+}X^{y-}$, listed as a function of electrolyte concentration. Notice that the figures in the table are all integers. A calculation of I not yielding a whole number is wrong.

Ions with large charges generally yield weak electrolytes, so the numbers of ions in solution are often smaller than predicted. For this reason, values of I calculated for salts represented by the bottom right-hand corner of Table 7.5 might be too high.

Why does the solubility of AgCl change after adding $MgSO_4$?

Calculating values of $\gamma_\pm$

We obtain the concentration $[Ag^+] = 1.3 \times 10^{-5}$ as the square root of 1.74×10^{-10} mol^2 dm^{-6}.

Silver chloride is fairly insoluble (see p. 332), with a solubility product K_{sp} of 1.74×10^{-10} mol^2 dm^{-6}. Its concentration in pure distilled water will, therefore, be 1.3×10^{-5} mol dm^{-3}, but adding magnesium sulphate to the solution increases it solubility appreciably; see Figure 7.10.

This increase in solubility is not an example of the common ion effect, because there *are* no ions in common. Also impossible is the idea that the equilibrium constant has changed, because it is a *constant*.

Strictly, we should formulate all equilibrium constants in terms of activities rather than concentrations, so Equation (7.32) describes K_{sp} for dissolving partially soluble AgCl in water:

$$K_{sp} = a_{(Ag^+)}a_{(Cl^-)} = [Ag^+][Cl^-] \times \gamma_{(Ag^+)}\gamma_{(Cl^-)} \qquad (7.32)$$

Table 7.5 Summary of the relationship between ionic strength I and concentration c. As an example, sodium sulfate (a 1:2 electrolyte) has an ionic strength that is three times larger than c

	X^-	X^{2-}	X^{3-}	X^{4-}
M^+	1	3	6	10
M^{2+}	3	4	15	12
M^{3+}	6	15	9	42
M^{4+}	10	12	42	16

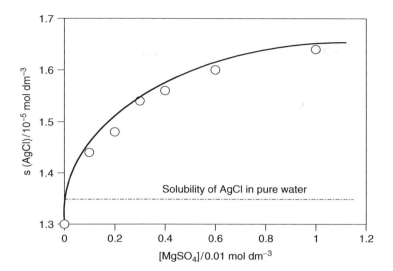

Figure 7.10 The solubility s of AgCl (as 'y') in aqueous solutions of MgSO$_4$ against its concentration, [MgSO$_4$] (as 'x'). $T = 298.15$ K

The exact structure of the equilibrium constant on the right-hand side of Equation (7.32) follows from the definition of activity a in Equation (7.25). The product of the two γ terms is $\gamma_{\pm}^{2}$.

The values of the activity coefficients decrease with increasing ionic strength I (as below). The only way for K_{sp} to remain constant at the same time as the activity coefficient $\gamma_{\pm}$ decreasing is for the concentrations c to *in*crease. And this is exactly what happens: the concentration of AgCl has increased by about 50 per cent when the concentration of MgSO$_4$ is 1.2 mol dm^{-3}.

Changes in solubility product are one means of experimentally determining a value of activity coefficient, because we can independently determine the concentrations (e.g. via a titration) and the values of all $\gamma_{\pm}$ will be 'one' at zero ionic strength.

Alternatively, we can calculate a value of $\gamma_{\pm}$ with the Debye–Hückel laws. There are two such laws: the *limiting* and the *simplified* laws. Calculations with the *limiting* law are only valid at very low ionic strengths (i.e. $0 < I \leq 10^{-3}$ mol dm^{-3}), which is very dilute. The limiting law is given by

> An ionic strength of 10^{-3} mol dm^{-3} could imply a *concentration* as low as 10^{-4} mol dm^{-3}, because $I > c$.

$$\log_{10} \gamma_{\pm} = -A|z^{+}z^{-}|\sqrt{I} \qquad (7.33)$$

where A is the so-called Debye–Hückel 'A' constant (or factor), which has a value of 0.509 mol$^{-1/2}$ dm$^{3/2}$ at 25 °C. z^{+} and z^{-} are the charges per cation and per anion respectively. The vertical modulus lines '|' signify that the charges on the ions have magnitude, but we need to ignore their signs (in practice, we call them both positive).

> The quantities between the two vertical *modulus* lines '|' have magnitude alone, so we ignore the signs on the charges z^{+} and z^{-}.

From Equation (7.28), we expect a plot of $\log_{10} \gamma_{\pm}$ (as 'y') against $\sqrt{I}$ (as 'x') to be linear. It generally *is* linear, although it deviates appreciably at higher ionic strengths.

Worked Example 7.15   What is the activity coefficient of copper in a solution of copper sulphate of concentration 10^{-4} mol dm^{-3}?

> Note how we ignore the *sign* of the negative charge here.

Copper sulphate is a 2:2 electrolyte so, from Table 7.5, the ionic strength I is four times its concentration. We say $I = 4 \times 10^{-4}$ mol dm^{-3}.

Inserting values into Equation (7.33):

$$\log_{10} \gamma_{\pm} = -0.509 \mid +2 \times -2 \mid (4 \times 10^{-4})^{1/2}$$

$$\log_{10} \gamma_{\pm} = -2.04 \times (2 \times 10^{-2})$$

$$\log_{10} \gamma_{\pm} = -4.07 \times 10^{-2}$$

> When calculating with Equation (7.33), be sure to use 'log' (in base 10) rather than 'ln' (log in base e).

Taking the anti-log:

$$\gamma_{\pm} = 10^{-0.0407}$$

$$\gamma_{\pm} = 0.911$$

We calculate that the perceived concentration is 91 percent of the real concentration.

> At extremely low ionic strengths, the simplified law becomes the limiting law. This follows since the denominator '$1 + b\sqrt{I}$' tends to one as ionic strength I tends to zero, causing the numerator to become one.

For solutions that are more concentrated (i.e. for ionic strengths in the range $10^{-3} < I < 10^{-1}$), we need the Debye–Hückel *simplified* law:

$$\log_{10} \gamma_{\pm} = -\frac{A|z^+ z^-|\sqrt{I}}{1 + b\sqrt{I}} \qquad (7.34)$$

where all other terms have the same meaning as above, and b is a constant having an approximate value of one. We include b because its units are mol$^{-1/2}$dm$^{3/2}$. It is usual practice to say $b = 1$ mol$^{-1/2}$ dm$^{3/2}$, thereby making the denominator dimensionless.

SAQ 7.14   Prove that the simplified law becomes the limiting law at very low I.

Worked Example 7.16   What is the activity coefficient of a solution of CuSO$_4$ of concentration 10^{-2} mol dm^{-3}?

Again, we start by saying that $I = 4 \times c$, so $I = 4 \times 10^{-2}$ mol dm^{-3}. Inserting values into Equation (7.34):

$$\log_{10} \gamma_{\pm} = -\frac{0.509 \mid 2 \times -2 \mid \sqrt{4 \times 10^{-2}}}{1 + \sqrt{4 \times 10^{-2}}}$$

Table 7.6 Typical activity coefficients $\gamma_{\pm}$ for ionic electrolytes as a function of concentration c in water

Electrolyte/ mol dm^{-3}	$\gamma_{\pm}$			
	$c = 10^{-3}/\text{mol dm}^{-3}$	$c = 10^{-2}/\text{mol dm}^{-3}$	$c = 10^{-1}/\text{mol dm}^{-3}$	$c = 1.0/\text{mol dm}^{-3}$
HCl (1:1)	0.996	0.904	0.796	0.809
KOH (1:1)	–	0.90	0.80	0.76
CaCl$_2$ (1:2)	–	0.903	0.741	0.608
CuSO$_4$ (2:2)	0.74	0.41	0.16	0.047
In$_2$(SO$_4$)$_3$ (2:3)	–	0.142	0.035	–

$$\log_{10} \gamma_{\pm} = -\frac{0.4072}{1 + 0.2}$$

$$\log_{10} \gamma_{\pm} = -0.3393$$

$$\gamma_{\pm} = 10^{-0.3383}$$

so

$$\gamma_{\pm} = 0.458$$

Table 7.6 cites a few sample values of $\gamma_{\pm}$ as a function of concentration. Note how multi-valent anions and cations cause $\gamma_{\pm}$ to vary more greatly than do mono-valent ions. The implications are vast: if an indium electrode were to be immersed in a solution of In$_2$(SO$_4$)$_3$ of concentration 0.1 mol dm^{-3}, for example, then a value of $\gamma_{\pm} = 0.035$ means that the activity (the perceived concentration) would be about 30 times smaller!

SAQ 7.15 From Worked Example 7.15, the mean ionic activity coefficient $\gamma_{\pm}$ is 0.911 for CuSO$_4$ at a concentration of 10^{-4} mol dm^{-3}. Show that adding MgSO$_4$ (of concentration 0.5 mol dm^{-3}) causes $\gamma_{\pm}$ of the CuSO$_4$ to drop to 0.06. [Hint: first calculate the ionic strength. [MgSO$_4$] is high, so ignore the CuSO$_4$ when calculating the ionic strength I.]

.4 Half-cells and the Nernst equation

Why does sodium react with water yet copper doesn't?

Standard electrode potentials and the $E^{\ominus}$ scale

Sodium reacts with in water almost explosively to effect the reaction

$$\text{Na}_{(s)} + \text{H}^{+}_{(aq)} \longrightarrow \text{Na}^{+}_{(aq)} + \tfrac{1}{2}\text{H}_{2(g)} \tag{7.35}$$

The protons on the left-hand side come from the water. Being spontaneous, the value of ΔG_r for Equation (7.25) is negative. The value of ΔG_r comprises two components:

(1) ΔG for the oxidation reaction $Na \longrightarrow Na^+ + e^-$; and

(2) ΔG for the reduction reaction $H^+ + 2e^- \longrightarrow \frac{1}{2}H_2$.

Since these two equations represent redox reactions, we have effectively separated a cell into its constituent half-cells, each of which is a single redox couple.

By contrast, copper metal does not react with water to liberate hydrogen in a reaction like Equation (7.35); on the contrary, black copper(II) oxide reacts with hydrogen gas to form copper metal:

$$CuO_{(s)} + H_{2(g)} \longrightarrow Cu_{(s)} + 2H^+ + O^{2-} \qquad (7.36)$$

The protons and oxide ions combine to form water. Again, the value of ΔG_r for Equation (7.36) is negative, because the reaction is spontaneous. ΔG would be positive if we wrote Equation (7.36) in reverse. The change in sign follows because the Gibbs function is a function of state (see p. 83).

The reaction in Equation (7.36) can be split into its two constituent half-cells:

(1) ΔG for the reaction $Cu^{2+} + 2e^- \longrightarrow Cu$; and

(2) ΔG for the reaction $H_2 \longrightarrow 2H^+ + 2e^-$.

Let us look at Equation (7.36) more closely. The value of ΔG_r comprises two components, according to Equation (7.36):

$$\Delta G_r = (\Delta G_{Cu^{2+} \rightarrow Cu}) + (\Delta G_{H_2 \rightarrow 2H^+}) \qquad (7.37)$$

> This change of sign follows from the change in direction of the second reaction.

If we wished to be wholly consistent, we could write both reactions as reduction processes. Reversing the direction of reaction (2) means that we need to change the sign of its contribution toward the overall value of ΔG_r, so

$$\Delta G_r = (\Delta G_{Cu^{2+} \rightarrow Cu}) - (\Delta G_{2H^+ \rightarrow H_2}) \qquad (7.38)$$

We remember from Equation (7.15) how $\Delta G_{(cell)} = -nF \times emf$. We will now invent a similar equation, Equation (7.39), which relates ΔG for a *half-cell* and its respective electrode potential $E_{O,R}$, saying:

$$\Delta G_{O,R} = -nFE_{O,R} \qquad (7.39)$$

Substituting for $\Delta G_{O,R}$ in Equation (7.38) with the invented expression in Equation (7.39) gives

$$\Delta G_r = (-nFE_{Cu^{2+},Cu}) - (-nFE_{H^+,H_2}) \qquad (7.40)$$

This expression does not relate to a true cell because the two electrode potentials are not measured with electrodes, nor can we relate ΔG_r to the *emf*, because electrons do not flow from one half-cell via an external circuit to the other. Nevertheless, Equation (7.40) is a kind of proof that the overall value of ΔG_r relates to the constituent half-cells.

If we write a similar expression to that in Equation (7.40) for the reaction between sodium metal and water in Equation (7.35), then we would have to write the term for the hydrogen couple first rather than second, because the direction of change within the couple is reversed. In fact, any couple that caused hydrogen gas to form protons would be written with the hydrogen couple first, and any couple that formed hydrogen gas from protons (the reverse reaction) would be written with the hydrogen term second.

This observation led the pioneers of electrochemical thermodynamics to construct a series of cells, each with the $H^+|H_2$ couple as one half-cell. The *emf* of each was measured. Unfortunately, there were always more couples than measurements, so they could never determine values for either $E_{Cu^{2+},Cu}$ or E_{H^+,H_2} (nor, indeed, for *any* electrode potential), so they commented on their *relative* magnitudes, and compiled a form of ranking order.

These scientists then suggested that the value of E_{H^+,H_2} should be *defined*, saying that at a temperature of 298 K, pumping the hydrogen gas at a pressure of hydrogen of 1 atm through a solution of protons at unit activity generates a value of E_{H^+,H_2} that is always zero. They called the half-cell '$H_{2(g)}(p = 1 \text{ atm})|H^+(a = 1)$' the *standard hydrogen electrode* (SHE), and gave it the symbol $E^{\ominus}_{H^+,H_2}$. The '$\ominus$' symbol indicates standard conditions.

> The 'standard electrode potential' $E^{\ominus}_{O,R}$ is the value of $E_{O,R}$ obtained at standard conditions.

Then, knowing $E^{\ominus}_{H^+,H_2}$, it was relatively easy to determine values of electrode potentials for any other couple. With this methodology, they devised the 'standard electrode potentials' $E^{\ominus}$ scale (often called the 'E nought scale', or the 'hydrogen scale').

> A pressure of $p = 1$ atm is *not* the same as $p^{\ominus}$, but its use is a permissible deviation within the SI scheme.

Table 7.7 contains a few such values of $E^{\ominus}$, each of which was determined with the same standard conditions as for the hydrogen couple, i.e. at $T = 298$ K, all activities being unity and $p = 1$ atm (the pressure is not, therefore, $p^{\ominus}$).

Negative values of $E^{\ominus}$ (such as $E^{\ominus}_{Na^+,Na} = -2.71$ V) indicate that the reduced form of the couple will react with protons to form hydrogen gas, as in Equation (7.35). The more negative the value of $E^{\ominus}$, the more potent the reducing power of the redox state, so $E^{\ominus}$ for the magnesium couple is -2.36 V, and $E^{\ominus}_{K^+,K} = -2.93$. Zinc is a less powerful reducing agent, so $E^{\ominus}_{Zn^{2+},Zn} = -0.76$ V, and a feeble reducing agent like iron yields a value of $E^{\ominus}_{Fe^{2+},Fe}$ of only -0.44 V.

> *Negative* values of $E^{\ominus}_{O,R}$ indicate that the reduced form of the couple will react with protons to form hydrogen gas.

And positive values of $E^{\ominus}$ indicate that the oxidized form of the redox couple will oxidize hydrogen gas to form protons, again

> *Positive* values of $E^{\ominus}_{O,R}$ indicate that the oxidized form of the redox couple will oxidize hydrogen gas to form protons.

Table 7.7   The electrode potential series (against the SHE). The electrode potential series is an arrangement of reduction systems in ascending order of their standard electrode potential $E^{\ominus}$

Couple[a,b,c]	$E^{\ominus}/V$	Couple[a,b,c]	$E^{\ominus}/V$
$Sm^{2+} + 2e^- = Sm$	−3.12	$Pb^{2+} + 2e^- = Pb$	−0.13
$Li^+ + e^- = Li$	−3.05	$Fe^{3+} + 3e^- = Fe$	−0.04
$K^+ + e^- = K$	−2.93	$Ti^{4+} + e^- = Ti^{3+}$	0.00
$Rb^+ + e^- = Rb$	−2.93	$2H^+ + 2e^- = H_2$ *(by definition)*	*0.000*
$Cs^+ + e^- = Cs$	−2.92	$AgBr + e^- = Ag + Br^-$	0.07
$Ra^{2+} + 2e^- = Ra$	−2.92	$Sn^{4+} + 2e^- = Sn^{2+}$	0.15
$Ba^{2+} + 2e^- = Ba$	−2.91	$Cu^{2+} + e^- = Cu^+$	0.16
$Sr^{2+} + 2e^- = Sr$	−2.89	$Bi^{3+} + 3e^- = Bi$	0.20
$Ca^{2+} + 2e^- = Ca$	−2.87	$AgCl + e^- = Ag + Cl^-$	0.2223
$Na^+ + e^- = Na$	−2.71	$Hg_2Cl_2 + 2e^- = 2Hg + 2Cl^-$	0.27
$Ce^{3+} + 3e^- = Ce$	−2.48	$Cu^{2+} + 2e^- = Cu$	0.34
$Mg^{2+} + 2e^- = Mg$	−2.36	$O_2 + 2H_2O + 4e^- = 4OH^-$	0.40
$Be^{2+} + 2e^- = Be$	−1.85	$NiOOH + H_2O + e^- = Ni(OH)_2 + OH^-$	0.49
$U^{3+} + 3e^- = U$	−1.79	$Cu^+ + e^- = Cu$	0.52
$Al^{3+} + 3e^- = Al$	−1.66	$I_3^- + 2e^- = 3I^-$	0.53
$Ti^{2+} + 2e^- = Ti$	−1.63	$I_2 + 2e^- = 2I^-$	0.54
$V^{2+} + 2e^- = V$	−1.19	$MnO_4^- + 3e^- = MnO_2$	0.58
$Mn^{2+} + 2e^- = Mn$	−1.18	$Hg_2SO_4 + 2e^- = 2Hg + SO_4^{2-}$	0.62
$Cr^{2+} + 2e^- = Cr$	−0.91	$Fe^{3+} + e^- = Fe^{2+}$	0.77
$2H_2O + 2e^- = H_2 + 2OH^-$	−0.83	$AgF + e^- = Ag + F$	0.78
$Cd(OH)_2 + 2e^- = Cd + 2OH^-$	−0.81	$Hg_2^{2+} + 2e^- = 2Hg$	0.79
$Zn^{2+} + 2e^- = Zn$	−0.76	$Ag^+ + e^- = Ag$	0.80
$Cr^{3+} + 3e^- = Cr$	−0.74	$2Hg^{2+} + 2e^- = Hg_2^{2+}$	0.92
$O_2 + e^- = O_2^-$	−0.56	$Pu^{4+} + e^- = Pu^{3+}$	0.97
$In^{3+} + e^- = In^{2+}$	−0.49	$Br_2 + 2e^- = 2Br^-$	1.09
$S + 2e^- = S^{2-}$	−0.48	$Pr^{2+} + 2e^- = Pr$	1.20
$In^{3+} + 2e^- = In^+$	−0.44	$MnO_2 + 4H^+ + 2e^- = Mn^{2+} + 2H_2O$	1.23
$Fe^{2+} + 2e^- = Fe$	−0.44	$O_2 + 4H^+ + 4e^- = 2H_2O$	1.23
$Cr^{3+} + e^- = Cr^{2+}$	−0.41	$Cl_2 + 2e^- = 2Cl^-$	1.36
$Cd^{2+} + 2e^- = Cd$	−0.40	$Au^{3+} + 3e^- = Au$	1.50
$In^{2+} + e^- = In^+$	−0.40	$Mn^{3+} + e^- = Mn^{2+}$	1.51
$Ti^{3+} + e^- = Ti^{2+}$	−0.37	$MnO_4^- + 8H^+ + 5e^- = Mn^{2+} + 4H_2O$	1.51
$PbSO_4 + 2e^- = Pb + SO_4^{2-}$	−0.36	$Ce^{4+} + e^- = Ce^{3+}$	1.61
$In^{3+} + 3e^- = In$	−0.34	$Pb^{4+} + 2e^- = Pb^{2+}$	1.67
$Co^{2+} + 2e^- = Co$	−0.28	$Au^+ + e^- = Au$	1.69
$Ni^{2+} + 2e^- = Ni$	−0.23	$Co^{3+} + e^- = Co^{2+}$	1.81
$AgI + e^- = Ag + I^-$	−0.15	$Ag^{2+} + e^- = Ag^+$	1.98
$Sn^{2+} + 2e^- = Sn$	−0.14	$S_2O_8^{2-} + 2e^- = 2SO_4^{2-}$	2.05
$In^+ + e^- = In$	−0.14	$F_2 + 2e^- = 2F^-$	2.87

[a]The more positive the value of $E^{\ominus}$, the more readily the half-reaction occurs in the direction left to right; the more negative the value, the more readily the reaction occurs in the direction right to left.

[b]Elemental fluorine is the strongest oxidizing agent and Sm^{2+} is the weakest. Oxidizing power increases from Sm^{2+} to F_2.

[c]Samarium is the strongest reducing agent and F^- is the weakest. Reducing power increases from F^- to Sm.

with the magnitude of $E^{\ominus}$ indicating the oxidizing power: $E^{\ominus}_{Cu^{2+},Cu} = +0.34$ V, but a powerful oxidizing agent such as bromine has a value of $E^{\ominus}_{Br_2,Br^-} = +1.09$ V.

In summary, sodium reacts with water and copper does not in consequence of their relative electrode potentials.

Why does a torch battery eventually 'go flat'?

The Nernst equation

A new torch battery has a voltage of about 1.5 V, but the *emf* decreases with usage until it becomes too small to operate the torch for which we bought it. We say the battery has 'gone flat', and throw it away.

We need to realize from Faraday's laws that chemicals within a battery are consumed every time the torch is switched on, and others are generated, causing the composition within the torch to change with use. Specifically, we alter the relative amounts of oxidized and reduced forms within each half-cell, causing the electrode potential to change.

The relationship between composition and electrode potential is given by the *Nernst equation*

$$E_{O,R} = E^{\ominus}_{O,R} + \frac{RT}{nF} \ln\left(\frac{a_{(O)}}{a_{(R)}}\right) \qquad (7.41)$$

> Though it is relatively easy to formulate relations like the Nernst equation here for a cell, Equation (7.41) properly relates to a half-cell.

where $E^{\ominus}_{O,R}$ is the standard electrode potential determined at s.t.p. and is a constant, $E_{O,R}$ is the electrode potential determined at non s.t.p. conditions. R, T, n and F have their usual definitions.

The bracket on the right of Equation (7.41) describes the relative activities of oxidized and reduced forms of the redox couple within a half-cell. The battery goes flat because the ratio $a_{(O)}/a_{(R)}$ alters with battery usage, so the value of $E_{O,R}$ changes until the *emf* is too low for the battery to be useful.

Worked Example 7.17 A silver electrode is immersed into a dilute solution of silver nitrate, $[AgNO_3] = 10^{-3}$ mol^{-3}. What is the electrode potential $E_{Ag^+,Ag}$ at 298 K? Take $E^{\ominus}_{Ag^+,Ag} = 0.799$ V.

The Nernst equation, Equation (7.41), for the silver couple is

$$E_{Ag^+,Ag} = E^{\ominus}_{Ag^+,Ag} + \frac{RT}{F} \ln\left(\frac{a_{(Ag^+)}}{a_{(Ag)}}\right)$$

> We use the approximation 'concentration = activity' because the solution is very dilute.

For simplicity, we assume that the concentration and activity of silver nitrate are the same, i.e. $a_{(Ag^+)} = 10^{-3}$. We also assume that the silver is pure, so its activity is unity.

The value of RT/F is 0.0257 V at 298 K.

Note how, as a consequence of the laws of arithmetic, we multiply the RT/F term with the logarithm term *before* adding the value of $E_{O,R}^{\ominus}$.

Inserting values into Equation (7.41):

$$E_{Ag^+,Ag} = 0.799 \text{ V} + 0.0257 \text{ V} \ln\left(\frac{0.001}{1}\right)$$

so

$$E_{Ag^+,Ag} = 0.799 \text{ V} + (0.0257 \text{ V} \times -6.91)$$

and

$$E_{Ag^+,Ag} = 0.799 \text{ V} - 0.178 \text{ V}$$

$$E_{Ag^+,Ag} = 0.621 \text{ V}$$

Note how the difference between E and $E^{\ominus}$ is normally quite small.

SAQ 7.16 A wire of pure copper is immersed into a solution of copper nitrate. If $E_{Cu^{2+},Cu}^{\ominus} = 0.34$ V and $E_{Cu^{2+},Cu} = 0.24$ V, what is the concentration of Cu^{2+}? Assume that $a_{(Cu^{2+})}$ is the same as $[Cu^{2+}]$.

The Nernst equation cannot adequately describe the relationship between an electrode potential $E_{O,R}$ and the concentration c of the redox couple it represents, unless we substitute for the activity, saying from Equation (7.28), $a = c \times \gamma_{\pm}$.

The form of Equation (7.41) will remind us of the equation of a straight line, so a plot of $E_{O,R}$ as the observed variable (as 'y'), against $\ln(a_{(O)} \div a_{(R)})$ (as 'x') should be linear with a gradient of $RT \div nF$ and with $E_{O,R}^{\ominus}$ as the intercept on the y-axis.

Worked Example 7.18   Determine a value for the standard electrode potential $E_{Ag^+,Ag}^{\ominus}$ with the data below. Assume that $\gamma_{\pm} = 1$ throughout.

$[AgNO_3]$/mol dm^{-3}	0.001	0.002	0.005	0.01	0.02	0.05	0.1
$E_{Ag^+,Ag}$/V	0.563	0.640	0.664	0.682	0.699	0.723	0.741

Figure 7.11 shows a *Nernst graph* drawn with the data in the table. The intercept of the graph is clearly 0.8 V.

Why does $E_{AgCl,Ag}$ change after immersing an SSCE in a solution of salt?

Further calculations with the Nernst equation

Care: In some books, SSCE is taken to mean a sodium chloride saturated calomel electrode.

Take a rod of silver, and immerse it in a solution of potassium chloride. A thin layer of silver chloride forms on its surface when the rod is made positive, generating a redox couple of AgCl|Ag. We have made a *silver–silver chloride electrode* (SSCE).

Now take this electrode together with a second redox couple (i.e. half-cell) of constant composition, and dip them together in a series

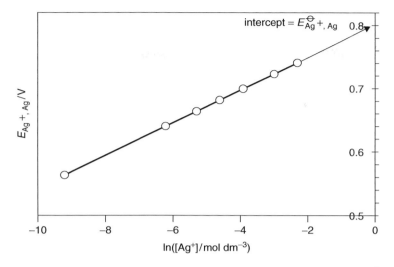

Figure 7.11   Nernst graph of the electrode potential $E_{Ag^+,Ag}$ as 'y' against $\ln([Ag^+]/mol\,dm^{-3})$ as 'x'. A value of $E^{\ominus}_{Ag^+,Ag} = 0.8$ V is obtained as the intercept on the y-axis

of salt solutions, and measure the *emf*. The magnitude of the *emf* will depend on the concentration of the salt. The silver rod and its outer layer of silver chloride do not alter, so why does the *emf* change?

The electrode potential $E_{AgCl,Ag}$ relates to the following redox reaction:

$$AgCl_{(s)} + e^- \longrightarrow Ag^0_{(s)} + Cl^-_{(aq)} \tag{7.42}$$

This redox couple is more complicated than any we have encountered yet, so the Nernst equation will appear to be a little more involved than those above:

$$E_{AgCl,Ag} = E^{\ominus}_{AgCl,Ag} + \frac{RT}{F} \quad \ln\left(\frac{a_{(AgCl)}}{a_{(Ag^0)}a_{(Cl^-)}}\right) \tag{7.43}$$

Silver chloride is the oxidized form, so we write it on top of the bracketed fraction, and silver metal is the reduced form, so we write it beneath. But we must also write a term for the chloride ion, because $Cl^-_{(aq)}$ appears in the balanced reduction reaction in Equation (7.42).

If we immerse a silver electrode bearing layer of AgCl in a concentrated solution of table salt, then the activity $a_{(Cl^-)}$ will be high; if the solution of salt is dilute, then $E_{AgCl,Ag}$ will change in the opposite direction, according to Equation (7.43).

SAQ 7.17 An SSCE electrode is immersed in a solution of $[NaCl] = 0.1\ mol\,dm^{-3}$. What is the value of $E_{AgCl,Ag}$? Take $E^{\ominus}_{AgCl,Ag} = 0.222$ V. Take all $\gamma_\pm = 1$. [Hint: the activities $a_{(AgCl)}$ and $a_{(Ag^0)}$ are both unity because both are *pure solids*.]

Why 'earth' a plug?

Reference electrodes

An electrical plug has three connections (or 'pins'): 'live', 'neutral' and 'earth'. The earth pin is necessary for safety considerations. The potential of the earth pin is the same as that of the ground, so there is no potential difference if we stand on the ground and accidentally touch the earth pin in a plug or electrical appliance. We will not be electrocuted. Conversely, the potentials of the other pins are different from that of the earth – in fact, we sometimes cite their potentials *with respect to* the earth pin, effectively defining the potential of the ground as being zero.

The incorporation of an earth pin is not only desirable for safety, it also enables us to know the potential of the other pins, because we cite them with respect to the earth pin.

A reaction in an electrochemical cell comprises two half-cell reactions. Even when we want to focus on a single half-cell, we must construct a whole cell and determine its cell *emf*, which is defined as '$E_{\text{(positive electrode)}} - E_{\text{(negative electrode)}}$'. Only when we know both the *emf* and the value of one of the two electrode potentials can we calculate the unknown electrode potential.

A *reference electrode* is a constant-potential device. We need such a reference to determine an unknown electrode potential.

A device or instrument having a known, predetermined electrode potential is called a *reference electrode*. A reference electrode is always necessary when working with a redox couple of unknown $E_{O,R}$. A reference electrode acts in a similar manner to the earth pin in a plug, allowing us to know the potential of any electrode with respect to it. And having defined the potential of the reference electrode – like saying the potential of earth is zero – we then know the potential of our second electrode.

Aside

At the heart of any reference electrode lies a redox couple of *known* composition: any passage of current through the reference electrode will change its composition (we argue this in terms of Faraday's laws in Table 7.1). This explains why we must never allow a current to flow through a reference electrode, *because a current will alter its potential.*

The standard hydrogen electrode – the primary reference

We *define* the value of $E_{\text{(SHE)}}$ as zero at all temperatures.

The internationally accepted primary reference is the *standard hydrogen electrode* (SHE). The potential of the SHE half-cell is defined as 0.000 V at all temperatures. We say the schematic for the half-cell is

$$Pt|H_2(a = 1)|H^+(a = 1, aq)|$$

The SHE is depicted in Figure 7.12, and shows the electrode immersed in a solution of hydrogen ions at unit activity (corresponding to 1.228 mol dm^{-3} HCl at 20 °C). Pure hydrogen gas at a pressure of 1 atm is passed over the electrode. The electrode itself consists of platinum covered with a thin layer of 'platinum black', i.e. finely divided platinum, electrodeposited onto the platinum metal. This additional layer thereby catalyses the electrode reaction by promoting cleavage of the H–H bonds.

Table 7.8 lists the advantages and disadvantages of the SHE.

> We employ hydrochloric acid of concentration 1.228 mol dm^{-3} at 20 °C because the activity of H$^+$ is less than its concentration, i.e. $\gamma_\pm < 1$.

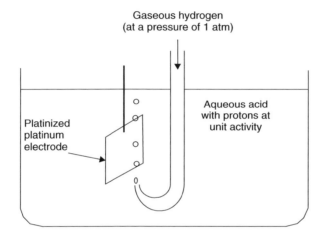

Figure 7.12   Schematic depiction of the standard hydrogen electrode (SHE). The half-cell schematic is therefore Pt|H$_2$($a = 1$)|H$^+$($a = 1$)

Table 7.8   Advantages and disadvantages of using the standard hydrogen electrode (SHE)

Advantage of the SHE

The SHE is the international standard

Disadvantages of the SHE

Safety	The SHE is intrinsically dangerous because H$_2$ gas is involved
Size	The SHE requires cumbersome apparatus, including a heavy cylinder of hydrogen
Cost	The SHE can be expensive because of using H$_2$ gas
Accuracy	With the SHE it is difficult to ensure that the activity of the protons is exactly unity
Precision	The SHE is prone to systematic errors, e.g. cyclic fluctuations in the H$_2$ pressure

Measurement with the hydrogen electrode   The SHE is the *primary* reference electrode, so other half-cell potentials are measured relative to its potential. In practice, if we wish to determine the value of $E^{\ominus}_{M^{n+},M}$ then we construct a cell of the type

$$Pt|H_2(a = 1)|H^+(a = 1)\|M^{n+}(a = 1)|M$$

> Remember that the proton is always solvated to form a hydroxonium ion, so $a_{(H_3O^+)} = a_{(H^+)}$.

Worked Example 7.19   We immerse a piece of silver metal into a solution of silver ions at unit activity and at s.t.p. The potential across the cell is 0.799 V when the SHE is the negative pole. What is the standard electrode potential $E^{\ominus}$ of the Ag^+, Ag couple?

By definition

$$emf = E_{(positive\ electrode)} - E_{(negative\ electrode)}$$

Inserting values gives

$$0.799\ V = E^{\ominus}_{Ag^+,Ag} - E_{(SHE)}$$

$$0.799\ V = E^{\ominus}_{Ag^+,Ag}$$

> The words *unit activity* here mean that the activity of silver ions is one (so the system *perceives* the concentration to be 1 mol dm^{-3}).

because $E_{(SHE)} = 0$.

The value of $E_{Ag^+,Ag}$ in this example is the *standard* electrode potential because $a_{(Ag^+)} = 1$, and s.t.p. conditions apply. We say that $E_{Ag^+,Ag} = E^{\ominus}_{Ag^+,Ag} = 0.799\ V$ versus the SHE. Most of the values of $E^{\ominus}$ in Table 7.7 were obtained in a similar way, although some were *calculated*.

> Just because a half-reaction appears in a table is no guarantee that it will actually work; such potentials are often calculated.

We should be aware from Table 7.8 that the SHE is an *ideal* device, and the electrode potential will not be exactly 0 V with non-standard usage.

> The SHE is sometimes erroneously called a *normal* hydrogen electrode (NHE).

Secondary reference electrodes

The SHE is experimentally inconvenient, so potentials are often measured and quoted with respect to reference electrodes other than the SHE. By far the most common reference is the saturated calomel electrode (SCE). We will usually make our choice of reference on the basis of experimental convenience.

> Note that the 'S' of 'SCE' here does NOT mean 'standard', but 'saturated'.

The SCE

By far the most common secondary reference electrode is the SCE:

$$Hg|Hg_2Cl_2|KCl(sat'd)|$$

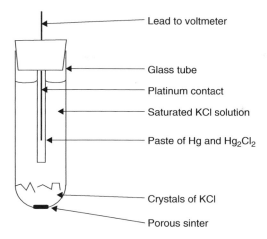

Figure 7.13   Schematic representation of the saturated calomel electrode (SCE)

The potential of the SCE is 0.242 V at 298 K relative to the SHE.

At the 'heart' of the SCE is a paste of liquid mercury and mercurous chloride (Hg_2Cl_2), which has the old-fashioned name 'calomel'. Figure 7.13 depicts a simple representation of the SCE.

The half-cell reaction in the SCE is

$$Hg_2Cl_2 + 2e^- \longrightarrow 2Cl^- + 2Hg \qquad (7.44)$$

So $E_{(SCE)} = E_{Hg_2Cl_2,Hg}$. From this redox reaction, the Nernst equation for the SCE is

$$E_{Hg_2Cl_2,Hg} = E^{\ominus}_{Hg_2Cl_2,Hg} + \frac{RT}{2F} \ln \left(\frac{a_{(Hg_2Cl_2)}}{a^2_{(Hg)} a^2_{(Cl^-)}} \right) \qquad (7.45)$$

the square terms for mercury and the chloride ion are needed in response to the stoichiometric numbers in Equation (7.44). Both mercury and calomel are pure substances, so their activities are unity. If the activity of the chloride ion is maintained at a constant level, then $E_{(SCE)}$ will have a constant value, which explains why the couple forms the basis of a reference electrode.

Changing $a_{(Cl^-)}$ must alter $E_{Hg_2Cl_2,Hg}$, since these two variables are interconnected. In practice, we maintain the activity of the chloride ions by placing surplus KCl crystals at the foot of the tube. The KCl solution is *saturated* – hence the 'S' in SCE. For this reason, we should avoid any SCE not showing a crust of crystals at its bottom, because its potential will be unknown. Also, currents must never be allowed to pass through an SCE, because charge will cause a redox change in $E_{(SCE)}$.

Table 7.9 lists the advantages and disadvantages of the SCE reference. Despite these flaws, the SCE is the favourite secondary reference in most laboratories.

> *Calomel* is the old-fashioned name for mercurous chloride, Hg_2Cl_2. Calomel was a vital commodity in the Middle Ages because it yields elemental mercury ('quick silver') when roasted; the mercury was required by alchemists.

> Oxidative currents reverse the reaction in Equation (7.44). Cl^- and Hg are consumed and Hg_2Cl_2 forms. The denominator of Equation (7.45) decreases, and $E_{(SCE)}$ increases.

Table 7.9   The advantages and disadvantages of using the saturated calomel electrode (SCE)

Advantages of the SCE	
Cost	SCEs are easy to make, and hence they are cheap
Size	SCEs can be made quite small (say, 2 cm long and 0.5 cm in diameter)
Safety	Unlike the SHE, the SCE is non-flammable
Disadvantages of the SCE	
Contamination	Chloride ions can leach out through the SCE sinter
Temperature effects	The value of dE/dT is quite large at 0.7 mV K^{-1}
Solvent	The SCE should not be used with non-aqueous solutions

The silver–silver chloride electrode

> We make the best films of AgCl by anodizing a silver wire in aqueous KCl, not in HCl. The reasons for the differences in morphology are not clear.

The silver–silver chloride electrode (SSCE) is another secondary reference electrode. A schematic of its half-cell is:

$$Ag|AgCl|KCl(aq, sat'd)$$

The value of $E^{\ominus}_{AgCl,Ag} = 0.222$ V.

The AgCl layer has a pale beige colour immediately it is made, but soon afterwards it assumes a pale mauve and then a dark purple aspect. The colour changes reflect chemical changes within the film, caused as a result of photolytic breakdown:

$$AgCl + h\nu + \text{electron donor} \longrightarrow Ag^0 + Cl^- \qquad (7.46)$$

> Photolytic breakdown is a big problem when we light the laboratory with fluorescent strips.

The purple colour is caused by colloidal silver, formed in a similar manner to the image on a black-and-white photograph after exposure to light. For this reason, an SSCE should be remade fairly frequently. (We introduce colloids in Chapter 10.)

Table 7.10 lists the advantages and disadvantages of SSCEs.

Table 7.10   The advantages and disadvantages of using the silver–silver chloride electrode (SSCE)

Advantages of the SSCE	
Cost	The SSCE is easy and extremely cheap to make
Stability	The SSCE has the smallest temperature voltage coefficient of any common reference electrode
Size	An SSCE can be as large or small as desired; it can even be microscopic if the silver is thin enough
Disadvantages of the SSCE	
Photochemical stability	The layer of AgCl must be remade often
Contamination	The solid AgCl loses Cl$^-$ ions during photolytic breakdown

.5 Concentration cells

Why does steel rust fast while iron is more passive?

Concentration cells

Steel is an impure form of iron, the most common contaminants being carbon (from the coke that fuels the smelting process) and sulphur from the iron oxide ore.

> Many iron ores also contain iron sulphide, which is commonly called *fool's gold*.

Pure iron is relatively reactive, so, given time and suitable conditions of water and oxygen, it forms a layer of red hydrated iron oxide ('rust'):

$$4Fe_{(s)} + 3O_2 + nH_2O \longrightarrow 2Fe_2O_3 \cdot (H_2O)_{n(s)} \qquad (7.47)$$

By contrast, steel is considerably more reactive, and rusts faster and to a greater extent.

The mole fraction x of Fe in *pure* iron is unity, so the activity of the metallic iron is also unity. The mole fraction x of iron in steel will be less than unity because it is impure. The carbon is evenly distributed throughout the steel, so its mole fraction $x_{(C)}$ is constant, itself ensuring that the activity is also constant. Conversely, the sulphur in steel is *not* evenly distributed, but resides in small (microscopic) 'pockets'. In consequence, the mole fraction of the

> We define a *concentration cell* as a cell in which the two half-cells are identical except for their relative concentrations.

iron host $x_{(Fe)}$ fluctuates, with x being higher where the steel is more pure, and lower in those pockets having a high sulphur content. To summarize, there are differences in the activity of the iron, so a *concentration cell* forms.

The *emf* of a concentration cell (in this case, on the surface of the steel where the rusting reaction actually occurs) is given by

> The electrolyte on the surface of the iron comprises water containing dissolved oxygen (e.g. rain water).

$$emf = \frac{RT}{nF} \quad \ln\left(\frac{a_2}{a_1}\right) \qquad (7.48)$$

Notice how this *emf* has no standard electrode potential $E^\ominus$ terms (unlike the Nernst equation from which it derives; see Justification Box 7.2).

A voltage forms between regions of higher irons activity a_1 and regions of lower iron activity a_2 (i.e. between regions of high purity and low iron purity); see Figure 7.14. We can write a schematic for a microscopic portion of the iron surface as:

> The commas in this schematic indicate that the carbon and sulphur impurities reside within the *same* phase as the iron.

$$Fe(a_2), S,C|O_2, H_2O|Fe(a_1), S,C$$

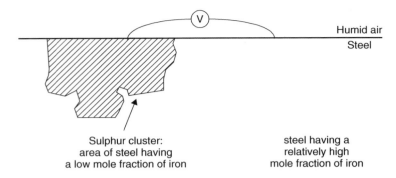

Figure 7.14 Concentration cells: a voltage forms between regions of higher iron activity a_1 and regions of lower iron activity a_2 (i.e. between regions of high purity and low iron purity). The reaction at the positive 'anode' is $4Fe + 3O_2 \rightarrow 2Fe_2O_3$, and the reaction at the negative 'cathode' is $S + 2e^- \rightarrow S^{2-}$

There is no salt bridge or any other means of stopping current flow in the microscopic 'circuit' on the iron surface, so electrochemical reduction occurs at the right-hand side of the cell, and oxidation occurs at the left:

> *at the LHS*, the oxidation reaction is formation of rust (Equation (7.47));

> *at the RHS*, the reduction reaction is usually formation of sulphide, via $S + 2e^- \rightarrow S^{2-}$.

We can draw several important conclusions from the example of rusting steel. Firstly, if the impurities of carbon and sulphur are evenly distributed throughout the steel then, whatever their concentrations, the extent of rusting will be less than if the impurities cluster, because the *emf* of a concentration cell is zero when the ratio of activities is unity.

Secondly, it is worth emphasizing that while oxide formation would have occurred on the surface of the iron whether it was pure or not, the steel containing impurities rusts *faster* as a consequence of the *emf*, and also more extensively than pure iron alone.

Thermodynamics of concentration cells

If the two half-cells were shorted then reduction would occur at the right-hand half-cell, $Cu^{2+}_{(aq)} + 2e^- \rightarrow Cu_{(s)}$, and oxidation would proceed at the left-hand side, $Cu_{(s)} \rightarrow Cu^{2+}_{(aq)} + 2e^-$.

A concentration cell contains the same electroactive material in both half-cells, but in different concentration (strictly, with different *activities*). The *emf* forms in response to differences in chemical potential μ between the two half-cells. Note that such a concentration cell does not usually involve different electrode reactions (other than, of course, that shorting causes one half-cell to undergo reduction while the other undergoes oxidation).

Worked Example 7.20 Consider the simple cell $Cu|Cu^{2+}(a = 0.002)\|Cu^{2+}(a = 0.02)|Cu$. What is its *emf*?

Inserting values into Equation (7.48):

$$emf = \frac{0.0257 \text{ V}}{2} \quad \ln\left(\frac{0.02}{0.002}\right)$$

$$emf = 0.0129 \text{ V} \ln(10)$$

$$emf = 0.0129 \text{ V} \times 2.303$$

$$emf = 29.7 \text{ mV}$$

so the *emf* for this two-electron concentration cell is about 30 mV. For an analogous one-electron concentration cell, the *emf* would be 59 mV.

SAQ 7.18 The concentration cell $Zn|[Zn^{2+}](c = 0.0112 \text{ mol dm}^{-3})|[Zn^{2+}]$ $(c = 0.2 \text{ mol dm}^{-3})|Zn$ is made. Calculate its *emf*, assuming all activity coefficients are unity.

Justification Box 7.2

Let the redox couple in the two half-cells be $O + ne^- = R$. An expression for the *emf* of the cell is

$$emf = E_{(RHS)} \, E_{(LHS)}$$

The Nernst equation for the O,R couple on the RHS of the cell is:

$$E_{O,R} = E_{O,R}^{\ominus} + \frac{RT}{nF} \quad \ln\left(\frac{a_{(O)RHS}}{a_{(R)RHS}}\right)$$

and the Nernst equation for the same O,R couple on the LHS of the cell is:

$$E_{O,R} = E_{O,R}^{\ominus} + \frac{RT}{nF} \ln\left(\frac{a_{(O)LHS}}{a_{(R)LHS}}\right)$$

Substituting for the two electrode potentials yields an *emf* of the cell of

$$emf = E_{O,R}^{\ominus} + \frac{RT}{nF} \ln\left(\frac{a_{(O)RHS}}{a_{(R)RHS}}\right) - E_{O,R}^{\ominus} - \frac{RT}{nF} \ln\left(\frac{a_{(O)LHS}}{a_{(R)LHS}}\right)$$

It will be seen straightaway that the two $E^{\ominus}$ terms cancel to leave

$$emf = \frac{RT}{nF} \ln\left(\frac{a_{(O)RHS}}{a_{(R)RHS}}\right) - \frac{RT}{nF} \ln\left(\frac{a_{(O)LHS}}{a_{(R)LHS}}\right)$$

A good example of a concentration cell would be the iron system in the worked example above, in which $a_{(R)} = a_{(Cu)} = 1$; accordingly, for simplicity here, we will assume that the reduced form of the couple is a pure solid.

The *emf* of the concentration cell, therefore, becomes

$$emf = \frac{RT}{nF} \ln[a_{(O)RHS}] - \frac{RT}{nF} \ln[a_{(O)LHS}]$$

which, through the laws of logarithms, simplifies readily to yield Equation (7.48).

If we assume that the activity coefficients in the left- and right-hand half-cells are the same (which would certainly be a very reasonable assumption if a swamping electrolyte was also in solution), then the activity coefficients would cancel to yield

$$emf = \frac{RT}{nF} \ln \left(\frac{[O]_{RHS}}{[O]_{LHS}} \right)$$

How do pH electrodes work?

The pH–glass electrode

A pH electrode is sometimes also called a 'membrane' electrode. Figure 7.15 shows how its structure consists of a glass tube culminating with a bulb of glass. This bulb is filled with a solution of chloride ions, buffered to about pH 7. A slim silver wire runs down the tube centre and is immersed in the chloride solution. It bears a thin layer of silver chloride, so the solution in the bulb is saturated with AgCl.

The bulb is usually fabricated with common soda glass, i.e. glass containing a high concentration of sodium ions. Finally, a small reference electrode, such as an SCE, is positioned beside the bulb. For this reason, the pH electrode ought properly to be called a *pH combination electrode*, because it is combined with a reference electrode. If the pH electrode does not have an SCE, it is termed a *glass electrode* (GE). The operation of a glass electrode is identical to that of a combination pH electrode, except that an *external* reference electrode is required.

Empirical means found from experiment, rather than from theory.

To determine a pH with a pH electrode, the bulb is fully immersed in a solution of unknown acidity. The electrode has fast response because a potential develops rapidly across the layer of glass

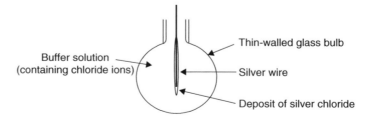

Figure 7.15   Schematic representation of a pH electrode (also called a 'glass electrode')

between the inner chloride solution and the outer, unknown acid. Empirically, we find the best response when the glass is extremely thin: the optimum seems to be 50 μm or so (50 μm = 0.05 mm = 50×10^{-6} m). Unfortunately, such thin glass is particularly fragile. The glass is not so thin that it is porous, so we do not need to worry about junction potentials E_j (see Section 7.6). The non-porous nature of the glass does imply, however, that the cell resistance is extremely large, so the circuitry of a pH meter has to operate with minute currents.

The magnitude of the potential developing across the glass depends on the difference between the concentration of acid inside the bulb (which we know) and the concentration of the acid outside the bulb (the analyte, whose pH is to be determined). In fact, the *emf* generated across the glass depends in a linear fashion on the

A *pH meter* is essentially a precalibrated voltmeter.

pH of the analyte solution provided that the internal pH does not alter, which is why we buffer it. This pH dependence shows why a pH meter is really just a pre-calibrated voltmeter, which converts the measured *emf* into a pH. It uses the following formula:

$$emf = K + \frac{2.303RT}{F}\text{pH} \tag{7.49}$$

SAQ 7.19 An *emf* of 0.2532 V was obtained by immersing a glass electrode in a solution of pH 4 at 25 °C. Taking $E_{(SCE)} = 0.242$ V, calculate the 'electrode constant' K.

SAQ 7.20 Following from SAQ 7.19, the same electrode was then immersed in a solution of anilinium hydrochloride of pH = 2.3. What will be the new *emf*?

In practice, we do not know the electrode constant of a pH electrode.

Electrode 'slope'

We can readily calculate from Equation (7.49) that the *emf* of a pH electrode should change by 59 mV per pH unit. It is common to see this stated as 'the electrode has a *slope* of 59 mV *per decade*'. A moment's pause shows how this is a simple statement of the obvious: a graph of *emf* (as '*y*') against $[H^+]$ (as '*x*') will have a gradient of 59 mV (hence 'slope'). The words 'per decade' point to the way that each pH unit represents a concentration change of 10 times, so a pH of 3 means that $[H^+] = 10^{-3}$ mol dm^{-3}, a pH of 4 means $[H^+] = 10^{-4}$ mol dm^{-3} and a pH of 5 means $[H^+] = 10^{-5}$ mol dm^{-3}, and so on. If the glass electrode does have a slope of 59 mV, its response is said to be *Nernstian*, i.e. it obeys the Nernst equation. The discussion of pH in Chapter 6 makes this same point in terms of Figure 6.1.

Table 7.11 lists the principle advantages and disadvantages encountered with the pH electrode.

SAQ 7.21 Effectively, it says above: 'From this equation, it can be readily calculated that the *emf* changes by 59 mV per pH unit'. Starting with the Nernst equation (Equation (7.41)), show this statement to be true.

Table 7.11 Advantages and disadvantages of the pH electrode

Advantages

1. If recently calibrated, the GE and pH electrodes give an accurate response
2. The response is rapid (possibly millisecond)
3. The electrodes are relatively cheap
4. Junction potentials are absent or minimal, depending on the choice of reference electrode
5. The electrode draws a minimal current
6. The glass is chemically robust, so the GE can be used in oxidizing or reducing conditions; and the internal acid solution cannot contaminate the analyte
7. The pH electrode has a very high *selectivity* – perhaps as high as $10^5 : 1$ at room temperature, so only one foreign ion is detected per 100 000 protons (although see disadvantage 6 below). The selectivity does decrease a lot above ca 35 °C

Disadvantages

Both the glass and pH electrodes alike have many disadvantages

1. To some extent, the constant K is a function of the area of glass in contact with the acid analyte. For this reason, no two glass electrodes will have the same value of K
2. Also, for the same reason, K contains contributions from the strains and stresses experienced at the glass.
3. (Following from 2): the electrode should be recalibrated often
4. In fact, the value of K may itself be slightly pH dependent, since the strains and stresses themselves depend on the amount of charge incorporated into the surfaces of the glass
5. The glass is very fragile and, if possible, should not be rested against the hard walls or floor of a beaker or container
6. Finally, the measured *emf* contains a response from ions other than the proton. Of these other ions, the only one that is commonly present is sodium. This error is magnified at very high pH (>11) when very few protons are in solution, and is known as the 'alkaline error'

Justification Box 7.3

At heart, the pH electrode operates as a simple concentration cell. Consider the schematic $H^+(a_2)\|H^+(a_1)$, then the Nernst equation can be written as Equation (7.48):

$$emf = \frac{RT}{F} \ln\left(\frac{a_2}{a_1}\right)$$

which, if written in terms of logarithms in base 10, becomes

$$emf = \frac{2.303RT}{F} \log_{10}\left(\frac{a_2}{a_1}\right) \qquad (7.50)$$

Subsequent splitting of the logarithm terms gives

$$emf = \frac{2.303RT}{F} \log_{10} a_2 - \frac{2.303RT}{F} \log_{10} a_1 \qquad (7.51)$$

If we say that a_1 is the analyte of known concentration (i.e. on the *inside* of the bulb), then the last term in the equation is a constant. If we call the term associated with a_2 'K', then we obtain

> *Care:* we have assumed here that the activities and concentrations of the solvated protons are the same.

$$emf = K + \frac{2.303\,RT}{F} \log_{10} a_2$$

If a_2 relates to the acidic solution of unknown concentration then we can substitute for '$\log_{10} a_2$', by saying that $pH = -\log_{10}[H^+]$, so:

> This derivation is based on the Nernst equation written in terms of ionic activities, but pH is usually discussed in terms of concentration.

$$emf = K + \frac{2.303\,RT}{F} \times -pH \qquad (7.52)$$

which is the same as Equation (7.49)

.6   Transport phenomena

How do nerve cells work?

Ionic transport across membranes

The brain relays information around the body by means of nerves, allowing us to register pain, to think, or to instruct the legs to walk and hands to grip. Although the way nerves operate is far from straightforward, it is nevertheless clear that the nerve pathways conduct charge around the body, with the charged particles (electrons and ions) acting as the brain's principal messengers between the brain and body.

The brain does not send a continuous current through the nerve, but short 'spurts'. We call them *impulses*, which transfer between nerve fibres within the *synapses* of cells (see Figure 7.16). The cell floats within an ionic solution called plasma. The membrane separating the synapse from the solution with which the nerve fibre is in contact surrounding the cell is the *axon*, and is essential to the nerve's operation.

The charge on the inside of a cell is negative with respect to the surrounding solution. A potential difference of about -70 mV forms across the axon (cell membrane) when the cell is 'at rest', i.e. before passing an impulse – we sometimes call it a *rest potential*, which is caused ultimately by differences in concentration either side of the axon (membrane).

> No potential difference forms *along* the membrane surface, only *across* it.

Movement of charge across the membrane causes the potential to change. A huge difference in concentration is seen in composition between the inside of the axon and the remainder of the nerve structure. For example, consider the compositional

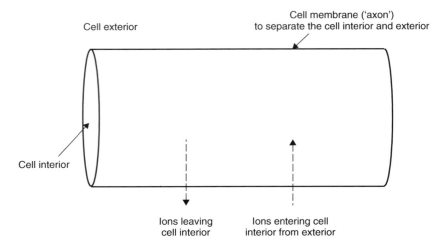

Cell exterior

Cell membrane ('axon')
to separate the cell interior and exterior

Cell interior

Ions leaving
cell interior

Ions entering cell
interior from exterior

Figure 7.16 Schematic diagram showing a portion of a cell, the membrane ('axon') and the way ions diffuse across the axon

Table 7.12 Concentrations of ions inside and outside nerve components

	$[Na^+]/mol\,dm^{-3}$	$[K^+]/mol\,dm^{-3}$	$[Cl^-]/mol\,dm^{-3}$
Inside the axon	0.05	0.40	0.04–0.1
Outside the axon	0.46	0.01	0.054

Source: J. Koryta, *Ions, Electrodes and Membranes*, Wiley, Chichester, 1991, p. 172.

> In this context, *permeable* indicates that ions or molecules can pass through the membrane. The mode of movement is probably diffusion or migration.

> Some texts give the name 'diffusion potential' to E_j.

differences in Table 7.12. The data in Table 7.12 refer to the nerves of a squid (a member of the *cephalopod* family) data for other species show a similar trend, with massive differences in ionic concentrations between the inside and outside of the axon. These differences, together with the exact extent to which the axon membrane is selectively *permeable* to ions, determines the magnitude of the potential at the cell surface.

The membrane encapsulating the axon is *semi-permeable*, thereby allowing the transfer of ionic material into and out from the axon. Since the cell encapsulates fluid and also floats in a fluid, we say the membrane represents a 'liquid junction'. A potential forms across the membrane in response to this movement of ions across the membrane, which we call a 'junction potential' E_j. If left unchecked, ionic movement across the membrane would occur until mixing was intimate and the two solutions were identical.

For a nerve to transmit a 'message' along a nerve fibre, ions traverse the axons and transiently changing the sign of the potential across the membrane, as represented schematically in Figure 7.17. We call this new voltage an *action potential*, to differentiate it from the rest potential. To effect this change in potential, potassium cations

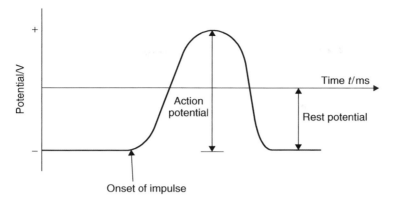

Figure 7.17   The potential across the axon–cell membrane changes in response to a stimulus, causing the potential to increase from its rest potential to its action potential

move from inside the axon concurrently with sodium ions moving in from outside. With a smaller difference in composition either side of the membrane, the junction potential decreases.

A nerve consists of an immense chain of these axons. Impulses 'conduct' along their length as each in turn registers an action potential, with the net result that messages transmit to and from the brain.

> To achieve this otherwise difficult process, chemical 'triggers' promote the transfer of ions.

Liquid junction potentials

A *liquid junction potential* E_j forms when the two half-cells of a cell contain different electrolyte solutions. The magnitude of E_j depends on the concentrations (strictly, the *activities*) of the constituent ions in the cell, the charges of each moving ion, and on the relative rates of ionic movement across the membrane. We record a constant value of E_j because equilibrium forms within a few milliseconds of the two half-cells adjoining across the membrane.

Liquid junction potentials are rarely large, so a value of E_j as large as 0.1 V should be regarded as exceptional. Nevertheless, junction potentials of 30 mV are common and a major cause of experimental error, in part because they are difficult to quantify, but also because they can be quite irreproducible.

> In most texts, the liquid junction potential is given the symbol E_j. In some books it is written as $E_{(lj)}$ or even $E_{(ljp)}$.

We have already encountered expressions that describe the *emf* of a cell in terms of the potentials of its constituent half-cells, e.g. Equation (7.23). When a junction potential is also involved – and it usually is – the *emf* increases according to

$$emf = E_{(\text{positive half–cell})} - E_{(\text{negative half–cell})} + E_j \qquad (7.53)$$

which explains why we occasionally describe E_j as 'an *additional* source of potential'.

While it is easy to measure a value of *emf*, we do not know the magnitude of E_j. SAQ 7.21 illustrates why we need to minimize E_j.

SAQ 7.22 The *emf* of the cell SHE $|Ag^+|Ag$, is 0.621 V. Use the Nernst equation to show that $a_{(Ag^+)} = 10^{-3}$ if $E_j = 0$ V, but only 4.6×10^{-4} if $E_j = 20$ mV. $E^{\ominus}_{Ag^+,Ag} = 0.799$ V. [Hint: to compensate for E_j in the second calculation, say that only 0.601 V of the *emf* derives from the $Ag^+|Ag$ half-cell, i.e. $E_{Ag^+,Ag} = 0.601$ V.]

What is a 'salt bridge'?

Minimizing junction potentials

> It's called a *bridge* because it connects the two half-cells, and *salt* because we saturate it with a strong ionic electrolyte.

In normal electrochemical usage, the best defence against a junction potential E_j is a *salt bridge*. In practice, the salt bridge is typically a thin strip of filter paper soaked in electrolyte, or a U-tube containing an electrolyte. The electrolyte is usually KCl or KNO₃ in relatively high concentration; the U-tube contains the salt, perhaps dissolved in a gelling agent such as agar or gelatine.

We connect the two half-cells by dipping either end of the salt bridge in a half-cell solution. A typical cell might be written in schematic form as:

$$Zn_{(s)}|Zn^{2+}_{(aq)}|S|Cu^{2+}_{(aq)}|Cu_{(s)}$$

We write the salt bridge as '$|S|$', where the S is the electrolyte within the salt bridge.

But how does the salt bridge minimize E_j? We recognize first how the electrolyte in the bridge is viscous and gel-like, so ionic motion *through* the bridge is slow. Secondly, the ionic diffusional processes of interest involve only the two *ends* of the salt bridge. Thirdly, and more importantly, the concentration of the salt in the bridge should greatly exceed the concentrations of electrolyte within either half-cell (exceed, if possible, by a factor of between 10–100 times).

The experimental use of a salt bridge is depicted in Figure 7.18. The extent of diffusion *from* the bridge, as represented by the large arrows in the diagram, is seen to be much greater than diffusion *into* the bridge, as represented by the smaller of the two arrows. A liquid junction forms at both ends of the bridge, each generating its own value of E_j. If the electrolyte in the bridge is concentrated, then the diffusion of ions moving *from* the bridge will dominate both of these two E_j. Furthermore, these E_j will be almost equal and opposite in magnitude, causing them to cancel each other out.

Table 7.13 shows how the concentration of the salt in the bridge has a large effect on E_j: it is seen that we achieve a lower value of E_j when the bridge is constructed with larger concentrations of salt. A junction potential E_j of as little as 1–2 mV can be achieved with a salt bridge if the electrolyte is concentrated.

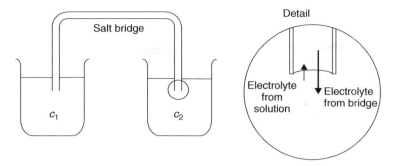

Figure 7.18   The two half-cells in a cell are joined with a salt bridge. *Inset*: more ions leave the bridge ends than enter it; the relative sizes of the arrows indicate the relative extents of diffusion

Table 7.13   Values of junction potential in aqueous cells as a function of the concentration of inert KCl within a salt bridge

[KCl]/$c^{\ominus}$	E_j/mV
0.1	27
1.0	8.4
2.5	3.4
4.2 (sat'd)	<1

Minimizing junction potentials with a swamping electrolyte

The second method of minimizing the junction potential is to employ a 'swamping electrolyte' *S*. We saw in Section 4.1 how diffusion occurs in response to entropy effects, themselves due to differences in activity. Diffusion may be minimized by decreasing the differences in activity, achieved by adding a high concentration of ionic electrolyte to both half-cells. Such an addition increases their ionic strengths *I*, and decreases all activity coefficients $\gamma_{\pm}$ to quite a small value.

If all values of $\gamma_{\pm}$ are small, then the differences between activities also decrease. Accordingly, after adding a swamping electrolyte, fewer ions diffuse and a smaller junction potential forms.

7.7   Batteries

How does an electric eel produce a current?

Introduction to batteries

The electric eel (*Electrophorus electricus*) is a thin fish of length 3–5 feet; see Figure 7.19. It is capable of delivering an electric shock of about 600 V as a means

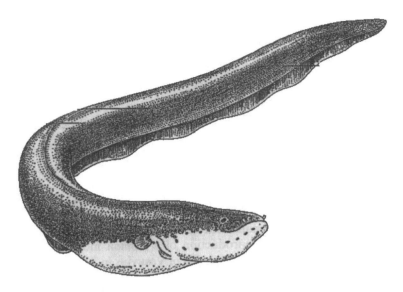

Figure 7.19   The electric eel (*Electrophorus electricus*) is a long, thin fish (3–5 feet) capable of delivering an electric shock of about 600 V. (Figure reprinted from *Ions, Electrodes and Membranes* by Jiří Koryta. Reproduced by permission of John Wiley and Sons Ltd)

of self-protection or for hunting. The eel either stuns a possible aggressor, or becomes an aggressor itself by stunning its prey, prior to eating it.

Fundamentally, the eel is simply a living *battery*. The tips of its head and tail represent the poles of the eel's 'battery'. As much as 80 per cent of its body is an electric organ, made up of many thousands of small platelets, which are alternately super-abundant in potassium or sodium ions, in a similar manner to the potentials formed across axon membranes in nerve cells (see p. 339). In effect, the voltage comprises thousands of concentration cells, each cell contributing a potential of about 160 mV. It is probable that the overall eel potential is augmented with junction potentials between the mini-cells.

The eel produces its electric shock when frightened, hungry or when it encounters its prey. The shock is formed when the eel causes the ionic charges on the surfaces of its voltage cells to redistribute (thereby reversing their cell polarities), and has the effect of summing the *emf*s of the mini-cells, in just the same way as we sum the voltages of small batteries incorporated within a *series circuit*. The ionic strength of seawater is very high, so conduction of the current from the eel to its prey is both swift and efficient.

Battery terminology

A *battery* is defined as a device for converting chemical energy into electrical energy. A battery is therefore an electrochemical cell that spontaneously produces a current when the two electrodes are connected externally by a conductor. The conductor will be the sea in the example of the eel above, or will more typically be a conductive metal such as a piece of copper wire, e.g. in a bicycle headlamp. The battery produces electrons as a by-product of the redox reaction occurring at the cathode. These electrons pass through the *load* (a bulb, motor, etc.) and do *work*, before re-entering the battery where the anode consumes them. Electrochemical reduction occurs at the positive pole (the anode) of the battery simultaneously with electrochemical oxidation at the negative pole (the cathode).

There are several types of battery we can envisage. A majority of the batteries we meet are classed as *primary* batteries, i.e. a chemical reaction occurs in both compartments to produce current, but when all the chemicals have been consumed, the battery becomes useless, so we throw it away. In other words, the electrochemical reactions inside the battery are not *reversible*. The most common primary batteries are the Leclanché cell, as described below, and the silver-oxide battery, found inside most watches and slim-line calculators.

> A battery is sometimes called a *galvanic* cell.

By contrast, *secondary* batteries may be reused after regenerating their original redox chemicals. This is achieved by passing a current through the battery in the opposite direction to that during normal battery usage. The most common examples of secondary batteries are the lead–acid cell (there is one inside most cars) and nickel–cadmium batteries (commonly called 'NiCad' batteries).

> In the shops, secondary batteries are usually called *rechargeable* batteries.

What is the earliest known battery?

Battery types

We have evidence that batteries were not unknown in the ancient world. The Parthians were a race living in the Mediterranean about 2000 years ago, from ca 300 BC until AD 224, when they were wiped out by the Romans. They are mentioned in the Bible, e.g. see *The Acts of the Apostles*, Chapter 2.

A device was found in 1936 near what is now modern Baghdad, the capital of Iraq. It consisted of a copper cylinder housing a central iron rod. The identity of the ionic electrolyte is now wholly unknown. The device was held together with asphalt as glue.

If the copper was tarnished and the iron was rusty (i.e. each was covered with a layer of oxide), then an approximate *emf* for this 2000-year-old battery would probably be in the range 0.6–0.7 V. We do not know what the battery was used for.

The Daniell cell

One of the first batteries in recent times was the *Daniell* cell, $Zn|Zn^{2+}:Cu^{2+}|Cu$. This battery comprised two concentric terracotta pots, the outer pot containing a zinc solution and the inner pot containing a copper solution. Metallic rods of copper and

> The vertical *dotted* line in this schematic indicates a *porous* membrane.

zinc were then immersed in their respective solutions. The electrode reaction at the zinc anode is $Zn \rightarrow Zn^{2+} + 2e^-$, while reduction occurs at the positive electrode, $Cu^{2+} + 2e^- \rightarrow Cu$.

Although this battery was efficient, it was never popular because it required aqueous solutions, which can be a danger if they slopped about. Its market share also suffered when better batteries were introduced to the market.

The Leclanché 'dry-cell' battery

The Leclanché cell was first sold in 1880, and is still probably the most popular battery in the world today, being needed for everyday applications such as torches, radios, etc. It delivers an *emf* of ca 1.6 V.

Figure 7.20 depicts the Leclanché cell in schematic form. The zinc can is generally coated with plastic for encapsulation (i.e. to prevent it from splitting) and to stop the intrusion of moisture. Plastic is an insulator, and so we place a conductive cap of stainless steel at the base of the cell to conduct away the electrons originating from the dissolution of the zinc from the inside of the can. A carbon rod then acts as an inert electrode to conduct electrons away from the reduction of MnO_2 at the cathode.

The reaction at the cathode is given by

$$2MnO_{2(s)} + 2H_2O + 2e^- \longrightarrow 2MnO(OH)_{(s)} + 2OH^-{}_{(aq)} \tag{7.54}$$

and the reaction at the zinc anode is: $Zn \rightarrow Zn^{2+} + 2e^-$.

We incorporate an ammonium salt to immobilize the Zn^{2+} ions: NH_4Cl is prepared as a paste, and forms a partially soluble complex with zinc cations produced at the

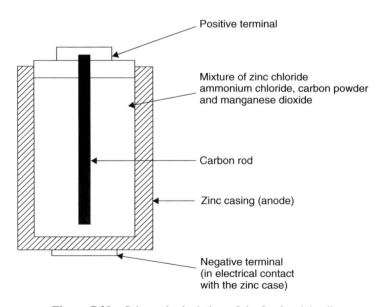

Figure 7.20 Schematic depiction of the Leclanché cell

Table 7.14   Advantages and disadvantages of the Leclanché cell

---

Advantages

It is cheap to make
It has a high energy density
It is not toxic
It contains no liquid electrolytes

Disadvantages

Its *emf* decreases during use as the material is consumed
It cannot readily deliver a high current

---

anode. We sometimes add starch to the paste to provide *additional* stiffness. The juxtaposition of the zinc ions with the zinc of the casing forms a redox buffer, thereby decreasing the extent to which the potential of the zinc half-cell wanders while drawing current.

Table 7.14 lists the advantages and disadvantages of the Leclanché cell.

> *Alkaline* manganese cells are broadly similar in design to the Leclanché cell, but they contain concentrated KOH as the electrolyte instead of NH_4Cl.

The lead–acid battery

The lead–acid cell was invented by Planté in 1859, and has remained more-or-less unchanged since Fauré updated it in 1881. The lead–acid cell is the world's most popular choice of *secondary battery*, meaning it is rechargeable. It delivers an *emf* of about 2.0 V. Six lead–acid batteries in series produce an *emf* of 12 V.

Table 7.15   Advantages and disadvantages of the lead–acid battery

---

Advantages

It is relatively easy to make, and so can be quite cheap
It has a high energy density, producing much electrical energy per unit mass
It can readily deliver a very high current

Disadvantages

It contains toxic lead
Also, since it contains lead, its power density is low
The acid is corrosive
Furthermore, the acid is a liquid electrolyte
Given time, lead sulphate (which is non-conductive) covers the electrode. Having 'sulphated up', the energy density of the battery is greatly impaired. To avoid sulphating up, it ought to be recharged often

---

Plates of lead, each coated with lead dioxide, are immersed in fairly concentrated sulphuric acid. Lead is oxidized at the lead anode during discharge:

$$Pb_{(s)} + HSO_{4(aq)}^- \longrightarrow PbSO_{4(s)} + H^+_{(aq)} + 2e^- \tag{7.55}$$

The reaction at the cathode during discharge is

$$PbO_{2(s)} + 2H^+_{(aq)} + H_2SO_{4(aq)} + 2e^- \longrightarrow PbSO_{4(s)} + 2H_2O \tag{7.56}$$

Both half-cell reactions are fully electro-reversible. In practice, there are two types of lead: the 'collector' electrode is made of lead alloy 'mesh' in order to give it greater structural strength, and is made with about 5 per cent antimony. 'Spongy lead' $(Pb + PbO_2)$ is introduced into the holes of the mesh.

> *Spongy lead* has a higher surface area than normal lead.

Table 7.15 lists the advantages and disadvantages of the lead–acid battery.

8

Chemical kinetics

Introduction

In previous chapters, we considered questions like: 'How much energy does a reaction liberate or consume?' and 'In which direction will a reaction proceed?' We then asked questions like: 'To what extent will a reaction proceed in that direction, before it stops?' and even 'Why do reactions occur at all?' In this chapter, we look at a different question: 'How *fast* does a reaction proceed?' Straightaway, we make assumptions. Firstly, we need to know whether the reaction under study *can* occur: there is no point in looking at how fast it is not going if a reaction is not thermodynamically feasible! So we first assume the reaction can and does occur.

Secondly, we assume that reactions can be treated according to their *type*, so 'reaction order' is introduced and discussed in terms of the way in which concentrations vary with time in a manner that characterizes that order.

Finally, the associated energy changes of reaction are discussed in terms of the thermodynamic laws learnt from previous chapters. Catalysis is discussed briefly from within this latter context.

8.1 Kinetic definitions

Why does a 'strong' bleach clean faster than a weaker one does?

Introduction to kinetics: rate laws

We often clean away the grime and dirt in a kitchen with bleach, the active ingredient of which is the hypochlorite ion ClO^-. The cleaning process we see by eye ('the bleaching reaction') occurs between an aqueous solution of ClO^- ion and coloured species stuck to the kitchen surfaces, which explains why the dirt or grease, etc., appears to vanish during the reaction. The reaction proceeds concurrently with colour loss in this example.

Care: a supermarket uses the word 'strong' in a different way from chemists: remember from Chapter 6 that the everyday word 'strong' has the specific chemical meaning 'a large equilibrium constant of dissociation'.

> We ignore the complication here that solution-phase ClO^- is in equilibrium with chlorine.

> This reaction could be one of the steps in a more complicated series of reactions, in a so-called *multi-step reaction*. If this reaction is the *rate-determining step* of the overall complicated series, then this rate law still holds; see p. 357.

> We define the rate of reaction as the speed at which a chemical conversion proceeds from start to its position of equilibrium, which explains why the rate is sometimes written as $d\xi/dt$, where ξ is the extent of reaction.

> We formulate the rate of a reaction by multiplying the rate constant of the reaction by the concentration of each reactant, i.e. by each species appearing at the *tail end* of the arrow. We can only do this if the reaction is elementary (proceeds in a single step)

We soon discover that a 'strong' bleach cleans the surfaces *faster* than a more dilute bleach. The reason is that 'strong' bleaches are in fact more concentrated, since they contain more ClO^- ions per unit volume than do 'weaker' bleaches.

We will consider the chemical reaction between the hypochlorite ion and coloured grease to form a colourless product P (the 'bleaching' reaction) as having the following stoichiometry:

$$ClO^- + grease \longrightarrow P \qquad (8.1)$$

We wish to know the *rate* at which this reaction occurs. The rate is defined as the number of moles of product formed per unit time. We define this rate according to

$$rate = \frac{[product]}{dt} \qquad (8.2)$$

As far as equations like Equation (8.2) are concerned, we tend to think of a chemical reaction occurring in a forward direction, so the product in Equation (8.2) is the chemical at the *head* of the arrow in Equation (8.1). Consequently, the concentration of product will always increase with time until the reaction reaches its position of equilibrium (when the rate will equal zero). This explains why the rate of reaction always has a positive value. The rate is generally cited with the units of $mol\ dm^{-3}\ s^{-1}$, i.e. concentration change per second.

The numerical value of the rate of reaction is obtained from a *rate equation*, which is obtained by first multiplying together the concentrations of *each* reactant involved in the reaction. (Before we do this, we must be sure of the identities of each reactant – in a complicated multi-step reaction, the reacting species might differ from those mentioned in the stoichiometric equation.) The following simple equation defines exactly the rate at which the reaction in Equation (8.1) occurs:

$$rate = k[ClO^-][grease] \qquad (8.3)$$

where the constant of proportionality k is termed the *rate constant*. The value of k is generally constant provided that the reaction is performed at constant temperature T. Values of rate constant are always positive, although they may appear to be negative in some of the more complicated mathematical expressions. Table 8.1 contains a few representative values of k.

We see from Equation (8.3) that the reaction proceeds faster (has a faster rate) when performed with a more concentrated ('strong')

Table 8.1   Selection of rate constants k

Reaction	Phase	Temperature/$^{\circ}$C	$k/$ units[a]
First-order reactions			
$SO_2Cl_2 \rightarrow SO_2 + Cl_2$	Gas	320	2×10^{-5}
Cyclopropane $\rightarrow$ propene	Gas	500	6.71×10^{-5}
$C_2H_6 \rightarrow 2CH_3^{\bullet}$	Gas	700	5.36×10^{-4}
$ClH_2C\text{–}CH_2Cl \rightarrow CH_2{=}CHCl + HCl$	Gas	780	4.4×10^{-3}
Second-order reactions			
$ClO^- + Br^- \rightarrow BrO^- + Cl^-$	Aqueous	25	4.2×10^{-7}
$CH_3COOC_2H_5 + NaOH \rightarrow CH_3CO_2Na + C_2H_5OH$	Aqueous	30	1.07×10^{-2}
$H_2 + I_2 \rightarrow 2HI$	Gas	400	2.42×10^{-2}
$2NO_2 \rightarrow 2NO + O_2$	Gas	300	0.54
$H_2 + 2NO \rightarrow N_2 + 2H_2O$	Gas	700	145.5
$2I \rightarrow I_2$	Gas	23	7×10^9
$H^+ + OH^- \rightarrow H_2O$	Aqueous	25	1.35×10^{11}

[a]For first-order reactions, k has the units of s^{-1}. For second-order reactions, k has units of $dm^3 mol^{-1} s^{-1}$.

bleach because the concentration term '$[ClO^-]$' in Equation (8.3) has increased.

> *Care*: A 'rate constant' is written as a lower case k in contrast to the more familiar 'equilibrium constant', which is written as an upper case K.

SAQ 8.1 Consider the reaction between ethanoic acid and ethanol to form the pungent ester ethyl ethanoate and water:

$$CH_3COOH + CH_3CH_2OH \longrightarrow CH_3COOC_2H_5 + H_2O$$

Write an expression for the rate of this reaction in a similar form to that in Equation (8.3), assuming the reaction proceeds in a single step as written.

SAQ 8.2 Write an expression for the rate of the reaction $Cu^{2+}_{(aq)} + 4NH_{3(aq)} \rightarrow [Cu(NH_3)_4]^{2+}_{(aq)}$, assuming that the reaction proceeds in a single step as written.

Why does the bleaching reaction eventually stop?

Calculating rates and rate constants

When cleaning in the kitchen with a pool of bleach on tables and surfaces, there comes a time when the bleaching action seems to stop. We might say that the bleach is 'exhausted', and so pour out some more bleach from the bottle.

When thinking about reaction kinetics, we need to appreciate that reactions involve chemical changes, with reactants being *consumed* during a reaction, and products being *formed*. After a time of reacting, the concentration of one or more of the chemicals will have decreased to zero. We generally say the chemical is 'used up'. Now look again at Equation (8.3). The rate of reaction depends on the concentrations of both the grease and the ClO^- ion from the bleach. If the concentration $[ClO^-]$ has decreased to zero, then the rate will also be zero, whatever the value of k or the concentration of grease. And if the rate is zero, then the reaction stops.

> The value of the rate constant is important because it tells us how fast a reaction occurs.

Although we appreciate from Equation (8.3) that the reaction will stop when one or both of the concentration terms reaches zero, we should also appreciate that the concentration terms reach zero faster if the value of k is large, and the concentrations deplete more slowly if k is smaller. We see how the value of the rate constant is important, because it tells us how fast a reaction occurs.

Worked Example 8.1   In solution, the cerium(IV) ion reacts with aqueous hydrogen peroxide with a 1:1 stoichiometry. The reaction has a rate constant of 1.09×10^6 dm^3 mol^{-1} s^{-1}. How fast is the reaction that occurs between Ce^{IV} and H_2O_2, if $[Ce^{IV}] = 10^{-4}$ mol dm^{-3} and $[H_2O_2] = 10^{-3}$ mol dm^{-3}?

By 'how fast', we are in effect asking 'What is the value of the rate of this reaction?' The reaction has a 1:1 stoichiometry; so, following the model in Equation (8.3), the rate equation of reaction is

$$\text{rate} = k[Ce^{IV}][H_2O_2] \tag{8.4}$$

> Placing the concentration brackets together – without a mathematical sign between them – implies that the concentrations are to be *multiplied*.

where k is the rate constant. Inserting values for k and for the two concentrations:

$$\text{rate} = 1.09 \times 10^6 \text{ dm}^3 \text{ mol}^{-1} \text{ s}^{-1} \times 10^{-4} \text{ mol dm}^{-3}$$
$$\times 10^{-3} \text{ mol dm}^{-3}$$
$$\text{rate} = 0.109 \text{ mol dm}^{-3} \text{ s}^{-1}$$

or, stated another way, the concentration of product changes (increases) by the amount 0.109 mol dm^{-3} per second, or just over 1 mol is formed during 10 s. This is quite a fast reaction.

SAQ 8.3   Show that the rate of reaction in Worked Example 8.1 *quadruples* if both $[Ce^{IV}]$ and $[H_2O_2]$ are doubled.

> The rate constant k is truly constant and only varies with temperature and (sometimes) with ionic strength I.

Worked Example 8.1 shows a calculation of a reaction rate from a rate constant k of known value, but it is much more common to know the reaction rate but be ignorant of the rate constant. A rate equation such as Equation (8.3) allows us to obtain a numerical value for k. And if we know the value of k, we can calculate from the rate equation exactly what length of time is required for the reaction to proceed when performed under specific reaction conditions.

Worked Example 8.2   Ethyl ethanoate (0.02 mol dm^{-3}) hydrolyses during reaction with aqueous sodium hydroxide (0.1 mol dm^{-3}). If the rate of reaction is $3 \times 10^2 \text{ mol dm}^{-3} \text{ s}^{-1}$, calculate the rate constant k.

The rate equation of reaction is given by

$$\text{rate} = k[CH_3COOCH_2CH_3][NaOH] \tag{8.5}$$

Rearranging Equation (8.5) to make k the subject gives

$$k = \frac{\text{rate}}{[CH_3COOCH_2CH_3][NaOH]}$$

Inserting values yields

$$k = \frac{3 \times 10^2 \text{ mol dm}^{-3} \text{ s}^{-1}}{0.02 \text{ mol dm}^{-3} \times 0.1 \text{ mol dm}^{-3}}$$

and

$$k = 1.5 \times 10^5 \text{ mol}^{-1} \text{ dm}^3 \text{ s}^{-1}$$

This is quite a large value of k.

Worked Example 8.2 yields a value for the rate constant k, but an alternative and usually more accurate way of obtaining k is to prepare a *series* of solutions, and to measure the rate of each reaction. A graph is then plotted of 'reaction rate' (as 'y') against 'concentration(s) of reactants' (as 'x') to yield a linear graph of gradient equal to k.

Worked Example 8.3   We continue with the reaction between Ce^{IV} and H_2O_2 from Worked Example 8.1. Consider the following data:

$[Ce^{IV}]/\text{mol dm}^{-3}$	3×10^{-5}	5×10^{-5}	8×10^{-5}	1×10^{-4}
$[H_2O_2]/\text{mol dm}^{-3}$	8×10^{-6}	2×10^{-5}	3×10^{-5}	5×10^{-5}
$\text{rate/mol dm}^{-3} \text{ s}^{-1}$	0.0009	0.0011	0.0026	0.0055

Equation (8.4) above can be seen to have the form '$y = mx$', in which 'rate' is 'y' and the mathematical product '$[Ce^{IV}] \times [H_2O_2]$' is '$x$'. The gradient m will be equal to k.

Figure 8.1 shows such a graph. The gradient of the graph is $1.09 \times 10^6 \text{ dm}^3 \text{ s}^{-1}$ (which is the same value as that cited in Worked Example 8.1). Notice how the intercept is zero, which confirms the obvious result that the rate of reaction is zero (i.e. no reaction can occur) when no reactants are present.

> We are saying here that the rate is of the form $y = mx$ (straight line). The intercept is zero.

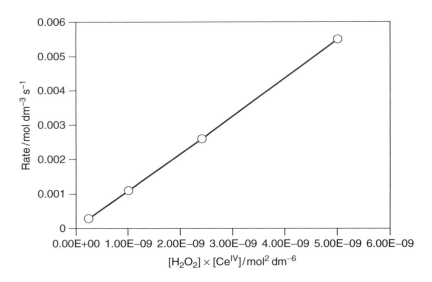

Figure 8.1   Graph of reaction rate against the product '[Ce^{IV}] × [H$_2$O$_2$]'. The numerical value of the gradient of the graph is the rate constant k

Why does bleach work faster on some greases than on others?

Rates expressions, reaction stoichiometry and reaction order

Each reaction has a unique value of rate constant k. For example, the value of k in Worked Example 8.2 would have been different if we had chosen ethyl formate, or ethyl propanoate, or ethyl butanoate, etc., rather than ethyl ethanoate. The value of k depends ultimately on the Gibbs function of forming reaction intermediates, as discussed below.

Most forms of grease in the kitchen derive from organic materials in the home – some derive from meat, some from vegetable oils and some from pets in the home, or even human tissue such as oily fingerprints. (Kitchen grease is, in fact, a complicated mixture of chemicals, each of which reacts with bleach at a different rate and, therefore, with a different value of k.)

But the reaction conditions are still more complicated because the stoichiometry of reaction might alter. Many greases and oils comprise the esters of long-chain fatty acids. The hydrolysing reaction between NaOH and an ester such as ethyl ethanoate proceeds with a stoichiometry of 1:1, but a tri-ester, such as most natural oils (e.g. olive oil or sunflower oil), occurs with a 1:3 stoichiometry, consuming one hydroxide ion per ester bond. Clearly, the hydroxide will be consumed more quickly when hydrolysing a tri-ester than a mono-ester. The rate depends on the *stoichiometry* of reaction.

> The rate of reaction depends on the stoichiometry.

Worked Example 8.4   The active ingredient within many weedkillers is methyl violo-gen, $MV^{+\bullet}$ (**I**).

$$Cl^-$$
(I)

Being a weedkiller, $MV^{+\bullet}$ is said to be *phytotoxic*. As a complication, two radicals of $MV^{+\bullet}$ will dimerize in solution to form the non-toxic dimer species $(MV)_2^{2+}$:

$$2\,MV^{+\bullet} \longrightarrow (MV)_2^{2+} \tag{8.6}$$

Why does the rate of the reaction in Equation (8.6) quadruple when we double the con-centration of the radical cation, $MV^{+\bullet}$?

The rate of reaction is written as the rate constant of reaction multiplied by the concen-tration of each reacting species on the tail end of the arrow. Accordingly, we write

$$\text{rate} = k[MV^{+\bullet}]^2 \tag{8.7}$$

Why does the rate quadruple if we double $[MV^{+\bullet}]$? We start by saying that $[MV^{+\bullet}]_{(new)} = 2 \times [MV^{+\bullet}]_{(old)}$. By doubling the concentration of $MV^{+\bullet}$ and (from Equation (8.7)) squaring its new value, we see that the new rate $= k \times \{2 \times [MV^{+\bullet}]_{(old)}\}^2$. This new rate is the same as $4 \times k[MV^{+\bullet}]_{(old)}^2$, which explains why the rate quadruples when the concentration of $MV^{+\bullet}$ doubles.

The reason why the rate equation includes the square term $[MV^{+\bullet}]^2$ rather than just $[MV^{+\bullet}]$ should not surprise us. Notice that we could have written the equation for the chemical reaction in a slightly different way, as

$$MV^{+\bullet} + MV^{+\bullet} \longrightarrow (MV)_2^{2+} \tag{8.8}$$

which is clearly the same equation as Equation (8.6). If we derive a rate equation for this alternative way of writing the reaction, with a concentration term for each participating reactant, then the rate equation of the reaction is

$$\text{rate} = k[MV^{+\bullet}][MV^{+\bullet}] \tag{8.9}$$

and multiplying the two $MV^{+\bullet}$ terms together yields $[MV^{+\bullet}]^2$. We see that Equation (8.9) is the same rate equation as that obtained in Equation (8.7).

SAQ 8.4   Write the rate equations for the following reactions. In each case, assume the reaction proceeds according to its stoichiometric equation.

(1) $2NO_{2(g)} \rightarrow N_2O_{4(g)}$

(2) $2SO_{2(g)} + O_{2(g)} \rightarrow 2SO_{3(g)}$

(3) $Ag^+_{(aq)} + Cl^-_{(aq)} \rightarrow AgCl_{(s)}$

(4) $HCl_{(aq)} + NaOH_{(aq)} \rightarrow NaCl_{(aq)} + H_2O$

Reaction order

> The *order* of a reaction is the same as the number of concentration terms in the rate expression.

The *order* of a reaction is the same as the number of concentration terms in the rate expression. Consider the general rate equation:

$$\text{rate} = k[A]^a[B]^b \ldots \qquad (8.10)$$

The sum of the powers is equal to the order of the reaction.

We see that Equation (8.9) involves two concentration terms, so it is said to be a *second-order* reaction. The rate expression for SAQ 8.4 (1) involves two concentration terms, so it is also a second-order reaction. Although each of the two concentrations in Equation (8.7) is the same, there are nevertheless two concentrations, and Equation (8.7) also represents a second-order reaction. In fact, the majority of reactions are second order.

> Third-order reactions are very rare.

The rate expression for SAQ 8.4 (2) involves three concentrations, so it is a *third-order* reaction. Third-order reactions are very rare, and there are no fourth-order or higher reactions.

SAQ 8.5 Analyse the following rate equations, and determine the orders of reaction:

(1) $\text{rate} = k[Cu^{2+}][NH_3]$

(2) $\text{rate} = k[OH^-]$

(3) $\text{rate} = k[NO]^2[O_2]$

It is increasingly common to see the rate constant given a subscripted descriptor indicating the order. The rate equation in SAQ 8.5 (1) would therefore be written as k_2.

Why do copper ions amminate so slowly?

The kinetic treatment of multi-step reactions, and the rate-determining step

Addition of concentrated ammonia to a solution of copper(II) yields a deep-blue solution of $[Cu(NH_3)_4]^{2+}$. The balanced reaction is given by

$$Cu^{2+}_{(aq)} + 4NH_{3(aq)} \longrightarrow [Cu(NH_3)_4]^{2+}_{(aq)} \qquad (8.11)$$

A quick look at the reaction suggests that the rate of this ammination reaction should be $k[Cu^{2+}{}_{(aq)}][NH_{3(aq)}]^4$, where the power of '4' derives from the stoichiometry (provided that the reaction as written was the rate-determining step). It would be a fifth-order reaction, and we would expect that doubling the concentration of ammonia would cause the rate to increase 16-fold (because $2^4 = 16$). But the increase in rate is not 16-fold; and, as we have just seen, a fifth-order reaction is not likely.

In fact, the ammination reaction forming $[Cu(NH_3)_4]^{2+}$ occurs stepwise, with first one ammonia ligand bonding to the copper ion, then a second, and so forth until the tetra-amminated complex is formed. And if there are four separate reaction steps, then there are four separate kinetic steps – one for each ammination step, with each reaction having its own rate constant k – we call them $k_{(1)}, k_{(2)}, k_{(3)}$ and $k_{(4)}$. This observation helps explain why the increase in reaction rate is not 16-fold when we double the concentration of ammonia.

> Each step in a multi-step reaction sequence proceeds at a different rate.

Aside

When we write $k_{(1)}$ with the subscripted '1' in brackets, we mean the rate constant of the first step in a multi-step reaction. When we write k_1 with a subscripted '1' but no brackets, we mean the rate constant of a first-order reaction. We adopt this convention to avoid confusion.

Most organic reactions occur in multi-step reactions, with only a small minority of organic reactions proceeding with a single step. We find, experimentally, that it is extremely unlikely for any two steps to proceed with the same rate constant, which means that we can only follow one reaction at a time. And the reaction that *can* be followed is always the *slowest* reaction step, which we call the *rate-determining step* – a term we often abbreviate to RDS.

A simple analogy from everyday life illustrates the reasonableness of this assumption. Imagine driving north from Italy to Norway in a journey involving travel along a fast motorway, along a moderately fast main road, and crawling through a 'contraflow' system, e.g. caused by road works. No matter how fast we travel along the main roads or the motorway, the overall time required for the journey depends crucially on the slowest bit, the tedious stop–start journey through the 'contraflow'. In a similar way, the only step in a multi-step reaction that we are able to follow experimentally is the slowest. We call it the rate-determining step because it 'determines' the rate.

> It is not possible to follow the rates in a multi-step reaction sequence: only the slowest step can be followed.

> We call the *slowest* step in a multi-step process its *rate-determining step*, often abbreviated to 'RDS'. The overall (observed) rate of a multi-step reaction is equal to the rate of the rate-determining reaction.

The separate reaction steps *will* proceed with the same rates in the unlikely event that all steps are *diffusion controlled*; see p. 416.

We will discuss multiple-step reactions in much greater detail in Section 8.4.

SAQ 8.6   Consider the oxidation of aqueous ethanol (e.g. in wine) to form vinegar:

$$\text{ethanol} + O_2 \xrightarrow[\text{slow}]{\text{first reaction}} \text{ethanal} + H_2O \tag{8.12}$$

$$2\,\text{ethanal} + O_2 \xrightarrow[\text{fast}]{\text{second reaction}} 2\,\text{ethanoic acid} \tag{8.13}$$

Decide which step is rate determining, and then write an expression for its rate (assuming that each reaction proceeds according to the stoichiometric equation written).

The pair of reactions in SAQ 8.6 explains how old wine 'goes off', i.e. becomes too acidic to drink when left too long.

How fast is the reaction that depletes the ozone layer?

Use of pressures rather than concentrations in a rate equation

Many substances react in the gas phase rather than in solution. An important example is the process thought to deplete the ozone layer: the reaction between gaseous ozone, O_3, and chlorine radicals, high up in the stratosphere. Ultimately, the chlorine derives from volatile halocarbon compounds, such as the refrigerant Freon-12 or the methyl chloroform thinner in correction fluid.

The ozone-depleting reaction involves a rather complicated series of reactions, all of which occur in the gas phase. Equation (8.14) describes the rate-determining step:

$$Cl^{\bullet} + O_3 \longrightarrow ClO^{\bullet} + O_2 \tag{8.14}$$

The $Cl^{\bullet}$ radical is retrieved quantitatively, and is therefore a *catalyst*.

The $ClO^{\bullet}$ radical product of Equation (8.14) then reacts further, yielding the overall reaction:

$$2O_{3(g)} + Cl^{\bullet}_{\text{(as a catalyst)}} \longrightarrow 3O_{2(g)} + Cl^{\bullet}_{\text{(retrieved catalyst)}} \tag{8.15}$$

The Earth is constantly irradiated with UV light, much of which is harmful to our skin. Ozone absorbs the harmful UV frequencies and thereby filters the light before it reaches the Earth's surface. Normal diatomic oxygen, O_2, does not absorb UV in this way, so any reaction that removes ozone has the effect of allowing more harmful UV light to reach us. The implications for skin health are outlined in Chapter 9.

The rate equation for Equation (8.14) is expressed in terms of 'pressures' p rather than *concentrations*, such as $[Cl^{\bullet}]$. We write the rate as

$$\text{rate} = k\,p(Cl^{\bullet})\,p(O_3) \tag{8.16}$$

where k here is a gas-phase rate constant. This rate expression has a similar form to the rate expressions above, except that the now-familiar concentration terms are each replaced with a pressure term, causing the units of k to differ.

The kinetics of reactions such as those leading to ozone depletion are treated in greater depth in subsequent sections.

Why is it more difficult to breathe when up a mountain than at ground level?

The dependence of rate on reactant pressure

We all breathe oxygen from the air to maintain life. Although we require oxygen, the air contains other gases. Normal air contains about 21 per cent of oxygen, the remainder being about 1 per cent argon and 78 per cent nitrogen.

We term the component of the total air pressure due to oxygen its 'partial pressure', $p(O_2)$. From Dalton's law (see Section 5.6), the partial pressure of oxygen $p(O_2)$ is obtained by multiplying the mole fraction of the oxygen by the overall pressure of the gases in air.

SAQ 8.7 The air we breathe at sea level has a pressure of 100 kPa. Show that the partial pressure of oxygen is 21 kPa.

> 'Standard pressure' $p^{\ominus}$ equals 10^5 Pa, and is sometimes called 1 bar.

We start to feel a bit breathless if the partial pressure of oxygen decreases below about 15 kPa; and we will feel quite ill (light-headed, breathless and maybe nauseous) if $p(O_2)$ drops below 10 kPa.

Air is taken into the lungs when we breathe. There, it is transported through the maze of progressively smaller bronchial tubes until it reaches the tiny sacs of delicate tissue called *alveoli*. Each sac look like bunches of grapes. The alveoli are the sites where oxygen from the air enters the blood, and the carbon dioxide from the body passes into the air (Figure 8.2). Oxygenated blood then flows around the body. Each *alveolus* is tiny, but there are 300 000 000 in each lung.

> As the molar masses of oxygen, nitrogen and argon are so similar, we can approximate the mole fractions of the gases to their percentage compositions.

We can think of the oxygen transfer from the lung to the blood as a simple chemical reaction: molecules of gas strike the alveoli. By analogy with simple solution-phase reactions, the rate equation describing the rate at which oxygen enters the blood is formulated according to

$$\text{rate} = k \times f(O_2) \times f(\text{alveoli}) \tag{8.17}$$

where each f simply means 'function of'. The thermodynamic function chosen to represent the oxygen is its partial pressure, $p(O_2)$. The alveoli are solid, so we omit

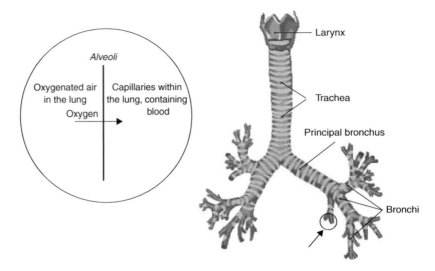

Figure 8.2 Oxygen from the air is taken into the lungs and is transported across the delicate *alveoli* tissue covering the inside of the lung and into the blood

them from the rate equation, Equation (8.17). Accordingly, Equation (8.17) becomes

$$\text{rate} = k \times p(O_2) \qquad (8.18)$$

which is seen to be very similar to the rate equations we formulated earlier for a reaction between solution-phase chemicals, with partial pressures replacing concentrations.

The overall pressure of the air decreases as we journey upwards, away from sea level, but the *proportions* of the gases in the air remains constant. At a height of 2 miles above sea level, the air pressure has dropped to about two-thirds of $p^{\ominus}$.

Worked Example 8.5 If the overall air pressure at a height of about 2 miles has dropped to 67 kPa, what is the partial pressure of oxygen?

Strategy. we calculate the partial pressure of oxygen $p(O_2)$ from Dalton's law (see p. 221), saying $p(O_2)$ = total pressure of the air × mole fraction of oxygen in the air.
Substituting values into the above equation:

$$p(O_2) = 67 \text{ kPa} \times 0.21$$

So the partial pressure of oxygen is 14 kPa.

This simple calculation shows why it is more difficult to breathe when up a mountain than at ground level: the pressure term in Equation (8.18) decreases, so the rate at which oxygen enters the blood decreases in proportion to the decrease in the oxygen partial pressure. And the partial pressure is smaller at high altitudes than at sea level.

By corollary, more oxygen can enter the blood if the oxygen partial pressure is increased. Two simple methods are available to increase $p(O_2)$:

(1) Breathe air of normal composition, but at a greater overall pressure. An example of this approach is the diver who breathes underwater while fitted out with SCUBA gear.

> 'SCUBA' is an acronym for 'self-contained underwater breathing apparatus'.

(2) In cases where the lungs are damaged, a doctor will place a patient in an 'oxygen tent'. The overall pressure of gas is the same as air pressure (otherwise the tent would explode) but the percentage of oxygen in the air is much greater than in normal air. For example, if the overall air pressure is the same as $p^{\ominus}$, but the air comprises 63% oxygen rather than the 21% in normal air, then the partial pressure of oxygen will treble.

SAQ 8.8 Consider the equilibrium reaction between hydrogen and chlorine to form HCl:

$$H_{2(g)} + Cl_{2(g)} \underset{\text{back}}{\overset{\text{forward}}{\longrightarrow}} 2HCl_{(g)}$$

Write two separate rate laws, one for the forward reaction and one for back reactions.

.2 Qualitative discussion of concentration changes

> ### Why does a full tank of petrol allow a car to travel over a constant distance?

Qualitative reaction kinetics: molecularity

Cars and buses are fuelled by a volatile mixture of hydrocarbons. The mixture is called 'petrol' in the UK, and 'gas' (short for gasoline) in the USA. One of the main chemicals in petrol is octane, albeit in several isomeric forms. In the internal combustion engine, the carburettor first vaporizes the petrol to form an aerosol (see Section 10.2) comprising tiny droplets of petrol suspended in air (Figure 8.3). This vaporization process is similar to that which converts liquid perfume into a fine spray.

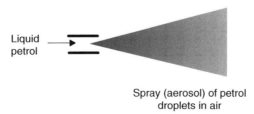

Figure 8.3 A carburettor in a car engine vaporizes the petrol to form an aerosol comprising tiny droplets of petrol suspended in air

These droplets burn in a controlled manner inside the car engine to release their chemical energy while the petrol combines chemically with oxygen:

$$C_8H_{18(\text{both g and l})} + 12.5O_{2(g)} \longrightarrow 8CO_{2(g)} + 9H_2O_{(g)} \qquad (8.19)$$

> The word 'stoichiometry' comes from the Greek word *stoic*, meaning 'indifference'. A stoichiometric reaction is, therefore, indifferent to all external conditions, and proceeds with a predetermined ratio of products and reactants.

This equation relates the overall ratios of reactants and products. It may be termed a *fully balanced* equation, or is more commonly termed a *stoichiometric* equation.

We employ octane as the fuel rather than, say, paper because burning octane releases such a large amount of heat energy (which is converted into kinetic energy). Inside the cylinders of the engine, this energy is released quickly, causing the gaseous products of combustion to heat up rapidly, which causes the pressure inside the cylinder to increase greatly. An additional cause of increasing pressure is the change in the number of moles of gas formed during combustion, since 13.5 mol of reactant form 17 mol of gaseous product. The ultimate cause of the car's motion is the large amount of pressure work performed during burning.

A constant amount of energy is released per mole of petrol. Furthermore, the size of the car's petrol tank predetermines the car's capacity to contain the alkane fuel; so we see how the overall amount of energy available to the car between refuelling stops cannot alter much. As the amount of energy is constant, the amount of work that the car can perform is also constant.

A well-tuned car will, therefore, travel essentially the same number of miles per tank of petrol because the amount of energy released for work is simply a function of the tank's capacity.

Why do we add a drop of bromine water to a solution of an alkene?

Reaction stoichiometry and molecularity

Bromine readily adds across an alkenic double bond by electrophilic addition (Figure 8.4). The brominated compound is usually colourless, but bromine in solution ('bromine water') has a red colour. Addition of bromine water to an alkene is accompanied by a loss of the red colour as reaction proceeds. The stoichiometry of reaction is almost always 1:1, with one molecule of bromine reacting per double bond.

Elemental analysis can be employed to show that the reaction between Br_2 and a C=C double bond always occurs with this stoichiometry of 1:1. It is a law of nature.

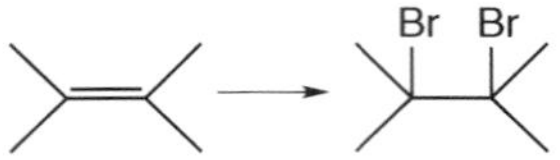

Figure 8.4   The red colour of elemental bromine is lost during addition across an alkenic double bond; the brominated compound is usually colourless

But sometimes we find that the kinetic data obtained experimentally bear little resemblance to the fully balanced reaction. In other words, the reaction proceeds by a mechanism that is different from the fully balanced equation.

The argument (above) concerning petrol centres on the way that chemicals always react in fixed proportions. The reaction above (Equation (8.19)) is the *stoichiometric* reaction because it cites the *overall* amounts of reaction occurring. But we should appreciate that the reaction *as written* will not occur in a single step: it is impossible even to imagine one molecule of octane colliding simultaneously with 12.5 molecules of oxygen. The probability is simply too vast. Even if there is a probability of getting the molecules together, how do we conceive of one and a half molecules of oxygen? How can we have *half* a molecule? Even if we could, is it possible to arrange these 12.5 molecules of oxygen physically around a single molecule of octane?

Similarly, we have also seen already how the copper(II) *tetrakis* (amine) complex forms in a step manner with four separate stages, rather than in a single step, forming the mono-ammine complex, then the *bis*-ammine, the *tris*-ammine and finally the *tetrakis*-ammine complex. So we start to appreciate that the actual reaction occurring during the burning of octane is more complicated than it first appears to be: the ratios in the stoichiometric equation are *not* useful in determining the reaction mechanism.

> The stoichiometric ratios in a fully balanced equation are usually not useful for determining the reaction mechanism.

In the case of burning a hydrocarbon, such as octane, the first step of the reaction usually occurs between a peroxide radical $O_2^\bullet$ generated by the spark of the sparking plug. A radical inserts into a C–H bond of a hydrocarbon molecule with the likely mechanism:

$$\sim C\text{–}H + O_2^\bullet \longrightarrow \sim C\text{–}O^\bullet\text{–}O\text{–}H \qquad (8.20)$$

The peroxide bond in the product is weak and readily cleaves to form additional radicals. Because more radicals are formed, any further reaction proceeds by a chain reaction, termed *radical propagation*, until all the petrol has been consumed.

The rate-limiting process in Equation (8.20) involves the two species (peroxide and octane) colliding within the car cylinder and combining chemically. Because *two* species react in the rate-limiting reaction step, we say that the reaction step represents a *bi*molecular reaction. In alternative phraseology, we say 'the *molecularity* of the reaction is two'.

> We describe the *molecularity* with the familiar Latin-based descriptors 'uni' = 1, 'bi' = 2 and 'tri' = 3.

The overwhelming majority of reactions are bimolecular. Some reactions are unimolecular and a mere handful of processes proceed as a trimolecular reactions. No quadrimolecular (or higher order) reactions are known.

We must appreciate the essential truth that the *molecularity* of a reaction and the *stoichiometric equation* are two separate things, and do not necessarily coincide. Luckily, we find that reactions are quite often 'simple' (or 'elementary'), by which we mean that they involve a single reaction step. The molecularity and the reaction order are the same if the reaction

> *Care*: the molecularity and the order of a reaction need not be the same.

> Kinetically, a reaction is *simple* if the molecularity and the reaction order are the same, usually implying that the reaction proceeds with a single step.

involves a single step, so we say that many inorganic reactions are simple because they are *both* second order and involve a bimolecular reaction mechanism.

In SAQ 8.2, we considered the case of forming $[Cu(NH_3)_4]^{2+}$ from copper(II) and ammonia. We have already seen that a reaction cannot be *quintimolecular* – five species colliding simultaneously. In fact, involves a sequence of a bimolecular reactions.

When magnesium dissolves in aqueous acid, why does the amount of fizzing decrease with time?

Reaction profiles

Magnesium ribbon reacts with sulphuric acid to cause a vigorous reaction, as demonstrated by the large volume of hydrogen gas evolved, according to

$$Mg_{(s)} + H_2SO_{4(aq)} \longrightarrow MgSO_{4(aq)} + H_{2(g)} \uparrow \qquad (8.21)$$

We know the reaction is complete when no more grey metal remains and the solution is clear. In fact, we often say the magnesium dissolves, although such dissolution is in fact a redox reaction (see Chapter 7).

But observant chemists will notice that the rate at which the gas is formed decreases even before the reaction has stopped. Stated another way, the rate at which H_2 is formed will decrease *during* the course of the reaction.

We can explain this in terms of a rate expression, as follows. First, consider the case where 1 mol of magnesium is reacted with 1 mol of sulphuric acid (we will also say that the overall volume of the solution is 1 dm^3). Initially there is no product, but at the end of reaction there will be 1 mol of $MgSO_4$ and 0 mol of magnesium metal or sulphuric acid. Therefore, the amount of product *in*creases while the amounts of the two reactants will both *de*crease as the reaction proceeds. The amounts of material change and, as the volume of solution is constant, the concentrations change.

> The concentrations of all reactants and products change during the course of a reaction.

We have already seen that the concentrations of reactants dictate the rate of reaction. For the consumption of magnesium in acid, the rate of reaction is given by

$$\text{rate} = k[Mg][H_2SO_4] \qquad (8.22)$$

This result was obtained by recalling how the rate of a reaction is equal to the rate constant of the process, multiplied by the concentration of each species at the *tail* end of the arrow in Equation (8.21).

There is 1 mol of magnesium at the start of reaction, and the concentration of sulphuric acid was 1 mol dm^{-3} (because there was 1 mol of sulphuric acid in 1 dm^3).

We next consider the situation after a period of time has elapsed such that half of the magnesium has been consumed. By taking proportions, 0.5 mol of the acid has also reacted, so its new concentration is 0.5 mol dm^{-3}. Accordingly, the value of 'rate' from Equation (8.22) is smaller, meaning that the rate has slowed down. And so the rate at which hydrogen gas is produced will also decrease.

Further reflection on Equation (8.22) shows that the concentrations of the two reactants will always alter with time, since, by the very nature of a chemical reaction, reactants are consumed. Accordingly, the rate of reaction will decrease continually throughout the reaction. The rate will reach zero (i.e. the reaction will stop) when one or both of the concentrations reaches zero, i.e. when one or all of the reactants have been consumed completely. The rate at which hydrogen gas is formed will reach zero when there is no more magnesium to react.

Figure 8.5 is a graph of the amount of sulphuric acid remaining as a function of time. The trace commences with 1 mol of H_2SO_4 and none is left at the end of the reaction, so $[H_2SO_4] = 0$. The rate of reaction is zero at all times after the sulphuric acid is consumed because the rate equation, Equation (8.22), involves multiplying the k and [Mg] terms by $[H_2SO_4]$, which is zero. Note that the abscissa (y-axis) could have been written as 'concentration', because the volume of the solution does not alter during the course of reaction, in which case the trace is called a *concentration profile*.

> We call a graph of concentration of reactant or product (as 'y') against time (as 'x') a *concentration profile*.

Also depicted on the graph in Figure 8.5 is the number of moles of magnesium sulphate produced. It should be apparent that the two concentration profiles (for reactant and product) are symmetrical, with one being the mirror image of the other. This symmetry is a by-product of the reaction stoichiometry, with 1 mol of sulphuric acid forming 1 mol of magnesium sulphate product.

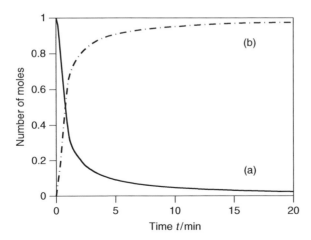

Figure 8.5 Concentration profiles for the reaction between sulphuric acid and magnesium to form magnesium sulphate. (a) Profile for the *consumption* of sulphuric acid, and (b) profile for the *formation* of magnesium sulphate. The initial concentration of H_2SO_4 was 1.0 mol dm^{-3}

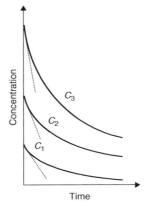

Figure 8.6 Concentration profile for the reaction between sulphuric acid and magnesium to form magnesium sulphate: reaction with three concentrations of $[H_2SO_4]_{t=0}$. The same amount of magnesium reacted in each case, and $c_3 > c_2 > c_1$

Figure 8.6 is a similar graph to Figure 8.5, showing the concentration profiles of sulphuric acid as a function of its initial concentration. The gradients of each trace are different, with the more concentrated solutions generating the steepest traces. In fact, we have merely rediscovered the concept of rate, because the gradient of the concentration profile is reaction rate, being the rate at which a compound or chemical reacts as a function of time. We say the rate of reaction is the gradient, after Equation (8.22).

> We obtain the rate of reaction as the gradient of a concentration profile.

Graphs such as those in Figures 8.5 and 8.6 are an ideal means of determining the rates of reaction. To obtain the rate, we plot the concentration of a reactant or product as a function of time, and measure the slope. (Strictly, since the slopes are negative for reactants, so the rate is 'slope $\times -1$'.)

Aside

Because product forms at the expense of reactant, the *magnitudes* of the rates of forming product and consuming reactant are the same. Although the magnitudes of the rates are the same, their *signs* are not: the rate of forming product is positive because the concentration increases with time.

For example, consider the reaction, $a\text{A} + b\text{B} = c\text{C} + d\text{D}$:

$$\text{rate} = \frac{d[\text{Product}]}{dt} = -\frac{d[\text{reactant}]}{dt}$$

For the individual chemical components, we say,

$$\text{reactants} \quad \text{rate} = -1 \times \frac{1}{a} \times \frac{d[\text{A}]}{dt} \quad \text{rate} = -1 \times \frac{1}{b} \times \frac{d[\text{B}]}{dt}$$

$$\text{products} \quad \text{rate} = \frac{1}{c} \times \frac{d[\text{C}]}{dt} \quad \text{rate} = \frac{1}{d} \times \frac{d[\text{D}]}{dt}$$

In each case, the minus sign indicates a decreasing concentration with time.

The rate of loss of reactant is negative because the concentration decreases with time The gradient at the start of the reaction is called the *initial rate*. Analysing the initial rates method is an extremely powerful way of determining the order of a reaction.

Worked Example 8.6   The following kinetic data were obtained for the reaction between nitric oxide and hydrogen at 700 °C. Determine: (1) the order of the reaction of the reaction at this temperature; (2) the rate constant of reaction.

	Experiment A	Experiment B	Experiment C
Initial concentration of NO/mol dm^{-3}	0.025	0.025	0.0125
Initial concentration of H$_2$/mol dm^{-3}	0.01	0.005	0.01
Initial rate/mol dm^{-3} s^{-1}	2.4×10^{-6}	1.2×10^{-6}	0.6×10^{-6}

The rates listed in the table were obtained as the gradients of graphs like those in Figure 8.6.

Answer Strategy

(1) *To determine the order of reaction.* It is always good research strategy to change only one variable at a time. That way, the measured response (if any) can be attributed unambiguously to the change in that variable. And the variable of choice in this example will be one or other of the two initial concentrations.

The rate equation for the reaction will have the following form:

$$\text{rate} = k[\text{NO}]^x[\text{H}_2]^y \tag{8.23}$$

We determine values for the exponents x and y varying [NO] and [H$_2$]. First, consider experiments A and B. In going from experiment B to experiment A, the concentration [NO] remains constant but we double [H$_2$], as a consequence of which the rate doubles. There is, therefore, a *linear* relationship between [H$_2$] and rate, so the value of y is '1'.

Second, consider experiments A and C. In going from C to A, we double the concentration [NO] and the rate increases by a factor of four. Accordingly, the value of x cannot be '1' because the increase in rate is not linear. In fact, as $2^2 = 4$, we see how the exponent x has a value of '2'.

The reaction is first order in H$_2$, second order in NO and, therefore, third order overall. Inserting values into Equation (8.23), we say, rate $= k[\text{NO}]^2[\text{H}_2]$.

(2) *To determine the rate constant of reaction.* We know the rate and concentrations for several sets of experimental conditions, so we rearrange Equation (8.23) to make k the subject and insert the concentrations.

$$k = \frac{\text{rate}}{[\text{NO}]^2[\text{H}_2]}$$

The question cites data from three separate experiments, each of which will give the same answer. We will insert data from 'experiment A':

$$k = \frac{2.4 \times 10^{-6} \ \text{mol dm}^{-3} \ \text{s}^{-1}}{[0.25 \ \text{mol dm}^{-3}]^2 [0.01 \ \text{mol dm}^{-3}]}$$

so

$$k = 3.84 \times 10^3 \ \text{dm}^6 \ \text{mol}^{-2} \ \text{s}^{-1}$$

SAQ 8.9 Iodide reacts with thiosulphate to form elemental iodine. If the reaction solution contains a tiny amount of starch solution, then this I_2 is seen by eye as a blue complex. The data below were obtained at 298 K. Determine the order of reaction, and hence its rate constant k.

	Experiment A	Experiment B	Experiment C
Initial concentration of I^-/mol dm^{-3}	0.1	0.1	0.2
Initial concentration of $S_2O_8{}^{2-}$/mol dm^{-3}	0.05	0.025	0.05
Initial rate/mol dm^{-3} s^{-1}	1.5×10^{-4}	7.5×10^{-5}	3.0×10^{-4}

8.3 Quantitative concentration changes: integrated rate equations

Why do some photographs develop so slowly?

Extent of reaction and integrated rate equations

A common problem for amateur photographers who develop their own photographs is gauging the speed necessary for development. When the solution of thiosulphate is first prepared, the photographs develop very fast, but this speed decreases quite rapidly as the solution 'ages', i.e. the concentration of $S_2O_8{}^{2-}$ decreases because the thiosulphate is consumed. So most inexperienced photographers have, at some time, ruined a film by developing with an 'old' solution: they wait what seems like forever without realizing that the concentration of thiosulphate is simply too low, meaning that development will never occur. They ruin the partially processed film by illuminating it after removal from the developing bath. These amateurs need to answer the question, 'How much thiosulphate remains in solution *as a function of time*?' We need a mathematical equation to relate concentrations and time.

The concentrations of each reactant and product will vary during the course of a chemical reaction. The so-called *integrated rate equation* relates the amounts of reactant remaining in solution during a reaction with the time elapsing since the reaction started. The integrated rate equation has a different form according to the order of reaction.

Let us start by considering a *first-order reaction*. Because the reactant concentration depends on time t, we write such concentrations with a subscript, as $[A]_t$. The initial reactant concentration (i.e. at time $t = 0$) is then written as $[A]_0$. The constant of proportionality in these equations will be the now-familiar rate constant k_1 (where the subscripted '1' indicates the order).

The relationship between the two concentrations $[A]_0$, $[A]_t$ and t is given by

$$\ln\left(\frac{[A]_0}{[A]_t}\right) = k_1 t \qquad (8.24)$$

Equation (8.24) is the *integrated first-order rate equation*. Being a logarithm, the left-hand side of Equation (8.24) is dimensionless, so the right-hand side must also be dimensionless. Accordingly, the rate constant k will have the units of s^{-1} when the time is expressed in terms of the SI unit of time, the second.

> The first-order rate constant k will have the units of s^{-1}.

Worked Example 8.7   Methyl ethanoate is hydrolysed when dissolved in excess hydrochloric acid at 298 K. The ester's concentration was 0.01 mol dm^{-3} at the start of the reaction, but 8.09×10^{-2} after 21 min. What is the value of the first-order rate constant k_1?

Answer Strategy. (1) we convert the time into SI units of seconds; (2) we insert values into Equation (8.24).

(1)   Convert the time to SI units: 21 min $= 21$ min $\times$ 60 s min$^{-1} = 1260$ s

(2)   Next, inserting values into Equation (8.24)

$$\ln\left(\frac{0.01 \text{ mol dm}^{-3}}{0.008\,09 \text{ mol dm}^{-3}}\right) = 1260 \text{ s} \times k_1$$

> Notice how the units of concentration will cancel here.

so

$$\ln(1.236) = k_1 \times 1260 \text{ s}$$

Taking the logarithm yields

$$0.211 = k_1 \times 1260 \text{ s}$$

and rearranging to make k_1 the subject gives

$$k_1 = \frac{0.211}{1260\ \text{s}} = 1.68 \times 10^{-4}\ \text{s}^{-1}$$

This value of k_1 is relatively small, indicating that the reaction is rather slow.

SAQ 8.10 (Continuing with the same chemical example): what is the concentration of the methyl ethanoate after a time of 30 min? Keep the same value of k_1 – it's a constant.

> We can calculate a rate constant k without knowing an *absolute* value for $[A]_0$ by following the *fractional* changes in the time-dependent concentration $[A]_t$.

We note, when looking at the form of Equation (8.24), how the bracket on the left-hand side contains a *ratio* of concentrations. This ratio implies that we do not need to know the actual concentrations of the reagent $[A]_0$ when the reaction starts and $[A]_t$ after a time t has elapsed since the reaction commenced; all we need to know is the fractional decrease in concentration. Incidentally, this aspect of the equation also explains why we could perform the calculation in terms of a *percentage* (i.e. a form of ratio) rather than a 'proper' concentration.

In fact, because the integrated first-order rate equation (Equation (8.24)) is written in terms of a *ratio* of concentrations, we do not need actual concentrations in moles per litre, but can employ any physicochemical parameter that is proportional to concentration. Obvious parameters include conductance, optical absorbance, the angle through which a beam of plane-polarized light is rotated (polarimetry), titre from a titration and even mass, e.g. if a gas is evolved.

Worked Example 8.8   Consider the simple reaction 'A → product'. After 3 min, 20 per cent of A has been consumed when the reaction occurs at 298 K. What is the rate constant of reaction k_1?

> If we know the amount of reactant consumed, then we will need to calculate how much remains.

We start by inserting values into the integrated Equation (8.24), noting that if 20 per cent has been consumed then 80 per cent remains, so:

$$\ln\left(\frac{100\% \text{ of } [A]_0}{80\% \text{ of } [A]_0}\right) = k_1 \times (3 \times 60\ \text{s})$$

The logarithmic term on the left-hand side is $\ln(1.25) = 0.223$, so

$$k_1 = \frac{0.223}{180\ \text{s}}$$

> Remember that we should always cite a rate constant at the temperature of measurement, because k itself depends on temperature.

where the term '180 s' comes from the seconds within the 3 min of observation time. We see that $k_1 = 1.24 \times 10^{-3}\ \text{s}^{-1}$ at 298 K.

SAQ 8.11   If the molecule A reacts by a first-order mechanism such that 15% is consumed after 1276 s, what is the rate constant k?

Justification Box 8.1

Integrated rate equations for a first-order reaction

The rate law of a first-order reaction has the form 'rate $= k_1[A]$'. And, by 'rate' we mean the rate of change of the concentration of reactant A, so

$$\text{rate} = -\frac{d[A]}{dt} = k_1[A] \qquad (8.25)$$

> The minus sign in Equation (8.25) is essential to show that the concentration of A *decreases* with time.

We will start at $t = 0$ with a concentration $[A]_0$, with the concentration decreasing with time t as $[A]_t$. The inclusion of a minus sign is crucial, and shows that species A is a reactant and thus the amount of it *decreases* with time.

Separating the variables (i.e. rearranging the equation) and indicating the limits, we obtain

$$-\int_{[A]_0}^{[A]_t} \frac{1}{[A]_t} \, d[A] = k_1 \int_0^t dt$$

Note how we can place the rate constant outside the integral, because it does not change with time. Integration then yields

$$-[\ln[A]]_{[A]_0}^{[A]_t} = k_1[t]_0^t$$

And, after inserting the limits, we obtain

$$-(\ln[A]_t - \ln[A]_0) = k_1 t$$

Using the laws of logarithms, the equation may be tidied further to yield Equation (8.24):

$$\ln\left(\frac{[A]_0}{[A]_t}\right) = k_1 t$$

which is the *integrated first-order rate equation*.

Note that if a *multiple-step* reaction is occurring, then this equation relates only to the case where the slowest (i.e. rate-limiting) step is kinetically first order. We will return to this idea when we consider pseudo reactions in Section 8.4.

Graphical forms of the rate equations

Similar to the integrated first-order rate equation is the *linear first-order rate equation*:

$$\underset{y}{\ln[A]_t} = \underset{m}{\Big|{-k_1}} \; \underset{x}{t} + \underset{c}{\Big|\ln[A]_0} \qquad (8.26)$$

We obtain the first-order rate constant k by drawing a graph with the integrated first-order rate equation, and multiplying its slope by -1.

which we recognize as having the form of the equation for a straight line: plotting $\ln[A]_t$ (as 'y') against time (as 'x') will be linear for reactions that are first order.

The rate constant k_1 is obtained from such a first-order rate graph as ($-1 \times$ gradient) if the time axis is given with units of seconds. Accordingly, the units of the first-order rate constant are s^{-1}.

Worked Example 8.9   Consider the following reaction: hydrogen peroxide decomposes in the presence of excess cerous ion Ce^{III} (which reacts to form ceric ion Ce^{IV}) according to a first-order rate law. The following data were obtained at 298 K:

Time t/s	2	4	6	8	10
$[H_2O_2]_t$/mol dm^{-3}	6.23	4.84	3.76	3.20	2.60

Time t/s	12	14	16	18	20
$[H_2O_2]_t$/mol dm^{-3}	2.16	1.85	1.49	1.27	1.01

This reaction *appears* to be first order because the cerium ions are in excess, so the concentration does not really change. We look at pseudo-order reactions on p. 387.

Figure 8.7 shows the way the concentration of hydrogen peroxide decreases with time. The trace is clearly curved, and Figure 8.8 shows a graph constructed with the linear form of the first-order integrated rate equation, Equation (8.26). This latter graph is clearly linear.

The rate constant is obtained from the figure as ($-1 \times$ 'gradient'), so $k = 0.11$ s^{-1} at 298 K.

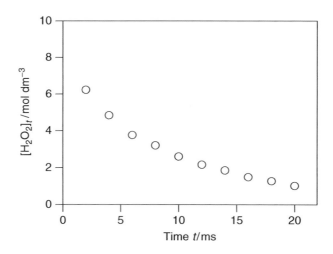

Figure 8.7   Plot of $[H_2O_2]_t$ against time. Notice the pronounced plot curvature

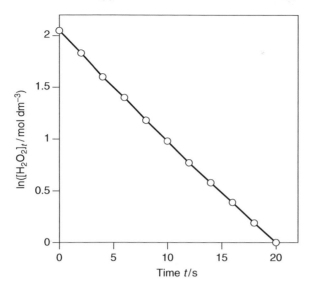

Figure 8.8   Graph constructed by drawing $\ln[H_2O_2]_t$ (as 'y') against time t (as 'x'), i.e. with the axes of a linear first-order rate law. Notice the linearity of the trace

---

Justification Box 8.2

The linear form of the integrated first-order rate equation

If we start with the now-familiar integrated rate equation of Equation (8.24):

$$\ln\left(\frac{[A]_0}{[A]_t}\right) = k_1 t$$

Using the laws of logarithms, we can split the left-hand side:

$$\ln[A]_0 - \ln[A]_t = k_1 t$$

Next, we multiply by -1:

$$\ln[A]_t - \ln[A]_0 = -k_1 t$$

Then, by adding the $[A]_0$ term to the right-hand side we obtain Equation (8.26):

$$\ln[A]_t = -k_1 t + \ln[A]_0$$

which is the linear form of the equation, as desired. We note that the intercept in Equation (8.26) is $\ln[A]_0$.

We could have obtained Equation (8.26) alternatively by integrating without limits during the derivation in Justification Box 8.1.

The SI unit of time is the second (see p. 15), but it is sometimes more convenient to cite k in terms of non-SI units, such as min^{-1} or even $year^{-1}$.

Worked Example 8.10   What is the relationship between the values of rate constant expressed in units of s^{-1}, and expressed in units of min^{-1} and $year^{-1}$?

By Analogy. Let the rate constant be k. Conversion into SI is easy: k in $min^{-1} = 60 \times (k$ in $s^{-1})$ because there are 60 s per minute, so 60 times as much reaction can occur during a minute.

Via Dimensional Analysis. Again, let the rate constant be k. In this example, imagine the value of k is 3.2 $year^{-1}$.

The number of seconds in a year is $(60 \times 60 \times 24 \times 365.25) = 3.16 \times 10^7$ s. We can write this result as 3.16×10^7 s yr^{-1}, which (by taking reciprocals) means that 2.37×10^{-8} yr s^{-1}.

To obtain k with units of s^{-1}, we say

To convert k from SI units to the time unit of choice, just multiply the value of k by the fraction of the time interval occurring during a single second.

$$k = \boxed{(3.2 \text{ yr}^{-1})} \times \boxed{(2.37 \times 10^{-8} \text{ yr s}^{-1})}$$

original value of k   **conversion factor**

so $k = 1.01 \times 10^{-7}$ s^{-1}.

SAQ 8.12   Show that the rate constants 1.244×10^4 yr^{-1} and 3.94×10^{-4} s^{-1} are the same.

Integrated rate equations: second-order reactions

For a second-order reaction, the form of the integrated rate equation is different:

Notice that the units of the second-order rate constant k_2 are $dm^3 \, mol^{-1} \, s^{-1}$ which are, in effect, (concentration)$^{-1}$ s^{-1}.

$$\frac{1}{[A]_t} - \frac{1}{[A]_0} = k_2 t \qquad (8.27)$$

where the subscripted '2' on k reminds us that it represents a *second*-order rate constant. The other subscripts and terms retain their previous meanings.

SAQ 8.13   Show that the units of the second-order rate constant are $dm^3 \, mol^{-1} \, s^{-1}$. [Hint: you will need to perform a simple dimensional analysis of Equation 8.27.]

We do not need to know the temperature in order to answer this question; but we *do* need to know that T remained constant, i.e. that the reaction was thermostatted.

Worked Example 8.11   We encountered the dimerization of methyl viologen radical cation $MV^{+\bullet}$ in Equation (8.6) and Worked Example 8.4. Calculate the value of the second-order rate constant k_2 if the initial concentration of $MV^{+\bullet}$ was 0.001 $mol \, dm^{-3}$ and the concentration dropped to 4×10^{-4} $mol \, dm^{-3}$ after 0.02 s. (The temperature was 298 K.)

Inserting values into Equation (8.27):

$$\frac{1}{4 \times 10^{-4} \text{ mol dm}^{-3}} - \frac{1}{10^{-3} \text{ mol dm}^{-3}} = k_2 \times 0.02 \text{ s}$$

so

$$2500 \text{ (mol dm}^{-3})^{-1} - 1000 \text{ (mol dm}^{-3})^{-1} = k_2 \times 0.02 \text{ s}$$

and

$$1500 \text{ (mol dm}^{-3})^{-1} = k_2 \times 0.02 \text{ s}$$

Rearranging, we say

$$k_2 = \frac{1500 \text{ (mol dm}^{-3})^{-1}}{0.2 \text{ s}}$$

so

$$k_2 = 7.5 \times 10^4 \text{ dm}^3 \text{ mol}^{-1} \text{ s}^{-1} \text{ at 298 K}$$

which is relatively fast.

SAQ 8.14 Remaining with the same system from Worked Example 8.11, having calculated k_2 (i.e. having 'calibrated' the reaction), how much $MV^{+\bullet}$ remains after 40 ms (0.04 s)?

SAQ 8.15 Consider a second-order reaction which consumes 15 per cent of the initial material after 12 min and 23 s. If $[A]_0$ was 1×10^{-3} mol dm^{-3}, calculate k_2. [Hint: first calculate how much material remains.]

An alternative form of the integrated rate equation is the so-called *linear* form

$$\underset{y}{\frac{1}{[A]_t}} = \underset{m}{k_2} \, \underset{x}{t} + \underset{c}{\frac{1}{[A]_0}} \qquad (8.28)$$

which we recognize as relating to the equation of a straight line, so plotting a graph of $1/[A]_t$ (as 'y') against time (as 'x') will be linear for a reaction that is second order. The rate constant k_2 is obtained directly as the gradient of the graph.

> We obtain the second-order rate constant k_2 as the slope of a graph drawn according to the integrated second-order rate equation.

Worked Example 8.12 Consider the data below, which relate to the second-order racemization of a glucose in aqueous hydrochloric acid at 17 °C. The concentrations of glucose and hydrochloric acid are the same, '$[A]$'.

Time t/s	0	600	1200	1800	2400
$[A]$/mol dm^{-3}	0.400	0.350	0.311	0.279	0.254

Care: The gradient is only truly k if the time axis is given with the SI units of time (the second).

A graph of the concentration [A] (as 'y') against time (as 'x') is clearly not linear; see Figure 8.9(a). Conversely, a different, linear, graph is obtained by plotting $1/[A]_t$ (as 'y') against time (as 'x'); see Figure 8.9(b). This follows the integrated second-order rate equation. The gradient of Figure 8.9(b) is the second-order rate constant k_2, and has a value of 6.00×10^{-4} dm^3 mol^{-1} s^{-1}.

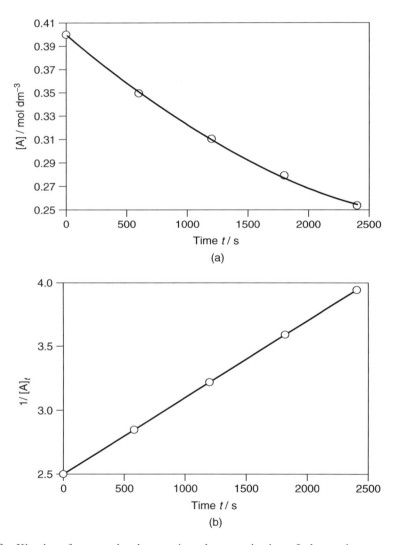

Figure 8.9   Kinetics of a second-order reaction: the racemization of glucose in aqueous mineral acid at 17 °C: (a) graph of concentration (as 'y') against time (as 'x'); (b) graph drawn according to the linear form of the integrated second-order rate equation, obtained by plotting $1/[A]_t$ (as 'y') against time (as 'x'). The gradient of trace (b) equals the second-order rate constant k_2, and has a value of 6.00×10^{-4} dm^{3}mol^{-1}s^{-1}

SAQ 8.16 Consider the following data concerning the reaction between triethylamine and methyl iodide at $20\,°C$ in an inert solvent of CCl_4. The initial concentrations of $[CH_3I]_0$ and $[N(CH_3)_3]_0$ are the same. Draw a suitable graph to demonstrate that the reaction is second order, and hence determine the value of the second-order rate constant k_2.

> Remember that when plotting kinetic graphs the concentration at $t = 0$ is also a valid data point.

Time t/s	0	2400	5400	9000	18 000
$[CH_3I]_0 = [N(CH_3)_3]_0/mol\ dm^{-3}$	2.112	0.813	0.149	0.122	0.084

Warning: if a chemical process comprises several reaction steps, only the progress of the *slowest* step can be followed kinetically. These graphical methods of determining k are only useful for obtaining the *rate-determining step* (RDS) of such reactions. Although the reaction may appear kinetically simple, it is wisest to assume otherwise.

Justification Box 8.3

Integrated rate equations for a second-order reaction

We will only consider the derivation of the simplest case of a second-order reaction, where the concentrations of the two reacting species are the same. Being second order, the rate law has the form rate $= k_2[A]^2$. The subscript '2' on k indicates a second-order process. Again, by 'rate' we mean the rate of change of the concentration of reactant A.

> The minus sign in Equation (8.29) is essential to show that the concentration of A *decreases* with time.

$$\text{rate} = -\frac{d[A]}{dt} = k_2[A]^2 \qquad (8.29)$$

As before, we shall start at $t = 0$ with a concentration $[A]_0$. The value of $[A]_t$ decreases with time t, hence a minus sign is inserted.

Rearranging the equation, and indicating the limits yields

$$-\int_{[A]_0}^{[A]_t} \frac{1}{[A]_t^2}\,d[A] = k_2 \int_0^t dt$$

Integrating gives

$$-\left[\frac{-1}{[A]_t}\right]_{[A]_0}^{[A]_t} = k_2[t]_0^t$$

The two minus signs on the left will cancel. Inserting limits, and rearranging slightly gives Equation (8.27):

$$\frac{1}{[A]_t} - \frac{1}{[A]_0} = k_2 t$$

This equation is known as the *integrated second-order rate equation*.

Second-order reactions of unequal concentration

We will start with the reactants A and B having the concentrations $[A]_0$ and $[B]_0$ respectively. If the rate constant of reaction is k_2, and if the concentrations at time t are $[A]_t$ and $[B]_t$ respectively, then it is readily shown that

$$\frac{1}{[B]_0 - [A]_t} \times \ln\left(\frac{[A]_0 \times [B]_t}{[B]_0 \times [A]_t}\right) = k_2 t \qquad (8.30)$$

We will need to look further at this equation when thinking about kinetic situations in which one of the reactants is in great excess (the so-called 'pseudo order' reactions described in Section 8.4).

SAQ 8.17 A 1:1 reaction occurs between A and B, and is second order. The initial concentrations of A and B are $[A]_0 = 0.1 \text{ mol dm}^{-3}$ and $[B]_0 = 0.2 \text{ mol dm}^{-3}$. What is k_2 if $[A]_t = 0.05 \text{ mol dm}^{-3}$ after 0.5 h? Remember to work out a value for $[B]_t$ as well.

Why do we often refer to a 'half-life' when speaking about radioactivity?

Half-lives

A radioactive substance is one in which the atomic nuclei are unstable and spontaneously decay to form other elements. Because the nuclei decay, the amount of the radioactive material decreases with time. Such decreases follow the straightforward kinetic rate laws we discussed above.

> We shall look at why radioactive materials are toxic in Section 8.3, below.

But many people talk emotionally of radioactivity 'because radioactive materials are so poisonous', and one of the clinching arguments given to explain why radioactivity is undesirable is that radioactive materials have long 'half-lives'. What is a half-life? And why is this facet of their behaviour important? And, for that matter, is it true that radioactive materials are poisonous?

Table 8.2 Half-lives of radioactive isotopes (listed in order of increasing atomic number)

Isotope	Half-life	Source of radioactive isotope
^{12}B	0.02 s	Unnatural (manmade)
^{14}C	5570 years	Natural
^{40}K	1.3×10^9 years	Natural: 0.011% of all natural potassium
^{60}Co	10.5 min	Unnatural: made for medicinal uses
^{129}I	1.6×10^7 years	Unnatural: fallout from nuclear weapons
^{238}U	4.5×10^8 years	Natural: 99.27% of all uranium
^{239}Pu	2.4×10^4 years	Unnatural: by-product of nuclear energy

The *half-life* of radioactive decay or of a chemical reaction is the length of time required for exactly half the material under study to be consumed, e.g. by chemical reaction or radioactive decay. We often give the half-life the symbol $t_{1/2}$, and call it 'tee half'.

> A 'half-life' $t_{1/2}$ is the time required for the amount of material to halve.

The only difference between a chemical and a radioactive half-life is that the former reflects the rate of a *chemical* reaction and the latter reflects the rate of radioactive (i.e. *nuclear*) decay. Some values of radioactive half-lives are given in the Table 8.2 to demonstrate the huge range of values $t_{1/2}$ can take. The difference between chemical and radioactive toxicity is mentioned in the Aside box on p. 382. A *chemical* half-life is the time required for half the material to have been consumed *chemically*, and a *radioactive* half-life is the time required for half of a *radioactive* substance to disappear by nuclear disintegration. Since most chemicals react while dissolved in a constant volume of solvent, the half-life of a chemical reaction equates to the time required for the *concentration* to halve.

> ^{60}Co is a favourite radionuclide within the medical profession, because its half life is conveniently short.

Worked Example 8.13 The half-life of ^{60}Co is 10.5 min. If we start with 100 g of ^{60}Co, how much remains after 42 min?

Answer Strategy

1. We determine how many of the half-lives have occurred during the time interval.

2. We then successively halve the amount of ^{60}Co, once per half-life.

(1) The number of half-lives is obtained by dividing 10.5 min into 42 min; so *four* half-lives elapse during 42 min.

(2) If four half-lives have elapsed, then the original amount of ^{60}Co has halved, then halved again, then halved once more and then halved a fourth time:

$$100 \text{ g} \xrightarrow[\text{half-life}]{\text{1st}} 50 \text{ g} \xrightarrow[\text{half-life}]{\text{2nd}} 25 \text{ g} \xrightarrow[\text{half-life}]{\text{3rd}} 12.5 \text{ g} \xrightarrow[\text{half-life}]{\text{4th}} 6.25 \text{ g}$$

We see that one-sixteenth of the ^{60}Co remains after four half-lives, because $(\frac{1}{2})^4 = \frac{1}{16}$. In fact, a general way of looking at the amount remaining after a few half-lives is to say that

$$\text{fraction remaining} = (\tfrac{1}{2})^n \tag{8.31}$$

where n is the number of half-lives.

SAQ 8.18 The half-life of radioactive ^{14}C is 5570 years. If we start with 10 g of ^{14}C, show that the amount of ^{14}C remaining after 11 140 years is 2.5 g.

Often, though, we don't know the half-life. One of the easier ways to determine a value of $t_{1/2}$ is to draw a graph of amount of substance (as 'y') – or, if in solution, of concentration – against time (as 'x').

Worked Example 8.14 The table below shows the amount of a biological metabolite T-IDA as a function of time.

1.   What is the half-life of T-IDA?

2.   Show that the data follow the integrated first-order rate equation.

Time t/min	0	10	20	30	40	50	60	70
[T-IDA]/μmol dm^{-3}	100	50	25	12.5	6.25	3.13	1.56	0.781

Figure 8.10 shows a plot of the amount of material (as 'y') as a function of time t (as 'x'), which is exponential. This shape should not surprise us, because Equation (8.31) is also exponential in form.

To obtain the half-life, we first choose a concentration – any concentration will do, but we will choose [T-IDA] = 50 μmol dm^{-3}. We then draw a horizontal arrow from this concentration on the y-axis, note the time where this arrow strikes the curve, and then read off the time on the x-axis, and call it t_1. Next, we repeat the process, drawing an arrow from half this original concentration, in this case from [T-IDA] = 25 μmol dm^{-3}. We note this new time, and call it t_2. The half-life is simply the difference in time between t_1 and t_2. It should be clear that the half-life in this example is 10 min.

But then we notice that the time needed to decrease from 60 to 30 μmol dm^{-3}, or from 2 to 1 μmol dm^{-3} will also be 10 min each. In fact, we deduce the important conclusion that the half-life of a first-order reaction is independent of the initial concentration of material.

As long as a reaction is first order, the duration of a half-life will be the same length of time *regardless of the initial amount of material present.*

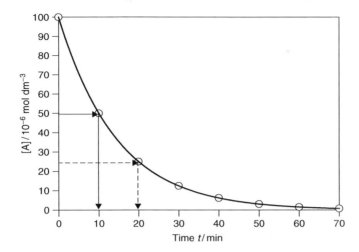

Figure 8.10   Kinetic trace concerning the change in concentration as a function of time: graph of [T-IDA] (as 'y') against time (as 'x') to show the way half-life is independent of the initial concentration

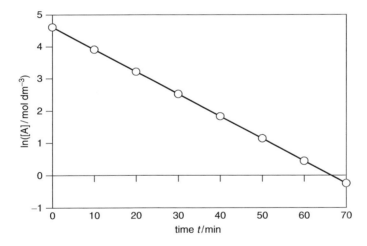

Figure 8.11   Graph plotted with data from Figure 8.10, plotted with the axes of the linear form of the integrated first-order rate equation, with ln[A] as 'y' against time t as 'x'

To show that this reaction is kinetically first order, we take the logarithm of the concentration, and plot $\ln[A]t$ (as 'y') against time t (as 'x'); see Figure 8.11. That the graph in Figure 8.11 is linear with this set of axes demonstrates its first-order character.

The half-life of second- or third-order reactions is *not* independent of the initial concentration in this way (see p. 387).

Aside

The difference between chemical and radiochemical toxicity

It is good that we should be concerned about the environmental impact of what we, as chemists, do to our planet. But many environmental campaigners too easily confuse *radioactive* toxicity and *chemical* toxicity. For example, the radon gas emanating from naturally occurring granite rocks is chemically inert, because it is a rare gas, but it is toxic to humans because of its radioactivity. Conversely, sodium cyanide contains no radioactive constituents yet is chemically toxic.

The conceptual problems start when considering materials such as plutonium, which is a by-product of the nuclear electricity industry. Plutonium is one of the most chemically toxic materials known to humanity, and it is also radioactive. The half-life of ^{238}Pu is so long at 4.5×10^8 years (see Table 8.2) that we say with some certainty that effectively none of it will disappear from the environment by radioactive decay; and if none of it decays, then it cannot have emitted ionizing α and β particles, etc. and, therefore, cannot really be said to be a radioactive hazard. Unfortunately, the long half-life also means that the ^{238}Pu remains more-or-less for ever to pollute the environment with its lethal *chemistry*.

> If we accept that plutonium is *chemically* toxic, then we also need to recognize that the extent of its toxicity will depend on how the plutonium is bonded *chemically* (see p. 59).

But if we accept that plutonium is *chemically* toxic, then we must also recognize that the extent of its toxicity will depend on how the plutonium is bonded chemically, i.e. in what redox and chemical form it is present. As an example, note how soldiers were poisoned with chlorine gas during the First World War (when it was called Mustard Gas), but chloride in table salt is vital for life. Some plutonium compounds are more toxic than others.

At the other extreme are materials with very short radioactive half-lives, such as ^{12}B (which has a relatively short $t_{1/2}$ of 0.02 h). ^{12}B is less likely to cause *chemical* poisoning than ^{238}Pu simply because its residence time is so short that it will transmute to become a different element and, therefore, have little time to interact in a chemical sense with anything in the environment (such as us). On the other hand, its short half-life means that the speed of its radioactive decay will generate many subatomic particles (α, β and γ particles) responsible for radioactive poisoning per unit time, causing a larger dose of radioactive poisoning.

This Aside is not intended to suggest that the threat of radioactivity is to be ignored or marginalized; but we should always aim to be well informed when confronting an environmental problem.

How was the Turin Shroud 'carbon dated'?

Quantitative studies with half-lives

The Turin Shroud is a long linen sheet housed in Italy's Turin Cathedral. Many people believe that the surface of the cloth bears the image of Jesus Christ (see

Figure 8.12   Many people believe the Turin Shroud bears an image of Jesus Christ, imprinted soon after his crucifixion. Radiocarbon dating suggests that the flax of the shroud dates from 1345 AD

Figure 8.12), imprinted soon after he died by crucifixion at the hands of the Roman authorities in about AD 33. There has been constant debate about the Turin Shroud and its authenticity since it first came to public notice in AD 1345: some devout people want it to be genuine, perhaps so that they can know what Jesus actually looked like, while others (many of whom are equally devout) believe it to be a fake dating from the Middle Ages.

An accurate knowledge of the Turin Shroud's age would allow us to differentiate between these two simplistic extremes of AD 33 and AD 1345, effectively distinguishing between a certain fake and a possible relic of enormous value.

The age of the cloth was ascertained in 1988 when the Vatican (which has jurisdiction over Turin Cathedral) allowed a small piece of the cloth to be analysed by *radiocarbon dating*. By this means, the shroud was found to date from AD 1320 ± 65. Even after taking account of the uncertainty of ± 65 years, the age of the shroud is consistent with the idea of a medieval forgery. It cannot be genuine.

But the discussion about the shroud continues, so many people now assert that the results of the test itself are part of a 'cover up', or that the moment of Jesus's resurrection occurred with a burst of high-energy sub-atomic particles, which upset the delicate ratio of carbon isotopes.

Radiocarbon dating

In 1946, Frank Libby of the Institute of Nuclear Sciences in Chicago initiated the dating of carbon-based artifacts by analysing the extent of radioactive decay.

The physicochemical basis behind the technique of radiocarbon dating is the isotopic abundances of carbon's three isotopes: ^{12}C is the 'normal' form and constitutes 98.9 per cent of all naturally occurring carbon. ^{13}C is the other naturally occurring isotope, with an abundance of about 1 per cent. ^{14}C does not occur naturally, but tiny amounts of it are formed when high-energy particles from space collide with gases in the upper atmosphere, thus causing radiochemical modification.

All living matter is organic and, therefore, contains carbon; and since all living material must breathe CO_2, all carbon-based life forms ingest ^{14}C. Additionally, living matter contains ^{14}C deriving from the food chain.

The *beta particle* emitted during radioactive decay is an energetic electron.

But ^{14}C is radioactive, meaning that atoms of ^{14}C occasionally self-destruct to form a beta particle, β^-, and an atom of ^{14}N:

$$^{14}C \longrightarrow {}^{14}N + \beta^- \tag{8.32}$$

The half-life of the process in Equation (8.32) is 5570 years.

Following death, flora and fauna alike cease to breathe and eat, so the only ^{14}C in a dead body will be the ^{14}C it died with. And because the amounts of ^{14}C decrease owing to radioactive decay, the amount of the ^{14}C in a dead plant or person decreases whereas the amounts of the ^{12}C and ^{13}C isotopes do not. We see why the *proportion* of ^{14}C decreases steadily as a function of time following the instant of death.

Radiocarbon dating is also called 'radiometric dating' or 'radiochemical dating'.

By corollary, if we could measure accurately the ratio of ^{12}C to ^{14}C in a once-living sample, we could then determine roughly how long since it was last breathing. This explains why we can 'date' a sample by analysing the residual amounts of ^{14}C.

In the carbon-dating experiment, a sample is burnt in pure oxygen and converted into water and carbon dioxide. Both gases are fed into a specially designed mass spectrometer, and the relative abundances of $^{12}CO_2$ and $^{14}CO_2$ determined. The proportion of $^{14}CO_2$ formed from burning an older sample will be smaller.

Oil and oil-based products contain no ^{14}C, because the creatures from which the oil was formed died so many millions of years ago. Accordingly, Equation (8.32) has proceeded to its completion.

Knowing this ratio, it is a simple matter to back calculate to ascertain the length of time since the sample was last alive. For example, we know that a time of one half-life $t_{1/2}$ has elapsed if a sample contains exactly half the expected amount of ^{14}C, so the sample died 5570 years ago.

Great care is needed during the preparation of the sample before dating to eliminate the possibility of contamination with additional sources of carbon.

Great care is needed during the preparation of the sample, since dirt, adsorbed CO_2 and other impurities can all contain additional sources of carbon. The dirt may come from the sample, or it could have been adsorbed during sample collection or even contamination during the dating procedure. The more recent the contamination, the higher the proportion of carbon that is radioactive ^{14}C that has not yet decayed, causing the artifact to appear younger. It has been suggested, for example, that the Turin Shroud was covered with much 'modern' pollen and dust at the time of its radiocarbon dating, so the date of AD 1320 refers to the age of the modern pollen rather than that of the underlying cloth itself.

How old is Ötzi the iceman?

Calculations with half-lives

Approximately 5000 years ago, a man set out to climb the Tyrolean Alps on the Austrian–Italian border. At death, he was between 40 and 50 years old and suffered from several medical ailments. Some scientists believe he was caught in a heavy snowfall, fell asleep, and froze to death. Others suppose he was murdered during his journey. Either way, his body was covered with snow almost immediately and, due to the freezing weather, rapidly became a mummy – 'The Iceman'. In 1991, his body was re-exposed and discovered by climbers in the Ötzal Alps, explaining why the 'Iceman', as he was called, was given the nickname 'Ötzi' (or, more commonly, as just Otzi).

His body (see Figure 8.13) was retrieved and taken to the Department of Forensic Medicine at the University of Innsbruck. Their analytical tests – principally radiocarbon dating – suggest that Ötzi died between 3360 and 3100 BC. Additional radiocarbon dating of wooden artifacts found near his body show how the site of his death was used as a mountain pass for millennia before and after his lifetime.

But how, having defined the half-life $t_{1/2}$ as the time necessary for half of a substance to decay or disappear, can we quantitatively determine the time elapsing since Ötzi died? In Justification Box 8.4, we show how the half-life and the rate constant of decay k are related according to

> Scientific techniques, such as radiocarbon dating, applied to archaeology is sometimes termed *archaeometry*.

> Equation (8.33) suggests the half-life is independent of the amount of material initially present, so radioactive decay follows the mathematics of first-order kinetics.

$$t_{1/2} = \frac{\ln(2)}{k} \tag{8.33}$$

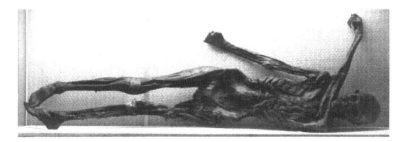

Figure 8.13 The body of 'Ötzi the Iceman' was preserved in the freezing depths of a glacier. Radiocarbon dating suggests that Özti froze to death in about BC 3360–3100

Worked Example 8.15   A small portion of Ötzi's clothing was removed and burnt carefully in pure oxygen. The amount of ^{14}C was found to be 50.93 per cent of the amount expected if the naturally occurring fabric precursors had been freshly picked. How long is it since the crop of flax was picked, i.e. what is its age?

We start by inserting the known half-life $t_{1/2}$ into Equation (8.33) to obtain a 'rate constant of radioactive decay'.

$$k = \frac{\ln(2)}{5570 \text{ yr}}$$

By calculating k, we are in effect *calibrating* the experiment. We only need to do this calibration once.

By this means, we calculate the rate constant as $k = 1.244 \times 10^{-4} \text{ yr}^{-1}$. (Alternatively, we could have calculated k in terms of the SI unit of time (the second), in which case k has the value $3.94 \times 10^{-12} \text{ s}^{-1}$.)

To calculate t, the length of time since the radioactive decay commenced (i.e. since the fabric precursor died), we again insert values into the integrated form of the first-order rate equation, Equation (8.33). We then insert our previously calculated value of k:

Using our calculators, we need to type 'ln(0.5093)' as the numerator rather than a percentage. The minus sign comes from the laws of logarithms.

$$t = \frac{\ln(50.93\%)}{-1.244 \times 10^{-4} \text{ yr}^{-1}} = 5420 \text{ yr}$$

So the interval of time t since the flax was picked and thence woven into cloth is 5420 yr, so the cloth dates from about 3420 BC.

SAQ 8.19   To return to the example of the Turin Shroud. Suppose a sloppy technique caused the precision of the ^{14}C measurement to decrease from 92.23 per cent to 90 ± 2 per cent. Calculate the range of ages for the shroud. [Hint: perform two calculations, one for either of the extreme values of percentage.]

Justification Box 8.4

Justification Box 8.1 shows how the integrated first-order rate equation is given by Equation (8.24):

$$\ln\left(\frac{[A]_0}{[A]_t}\right) = kt$$

After a length of time equal to one half-life $t_{1/2}$, the concentration of A will be $[A]_t$, which, from the definition of half-life, has a value of $\frac{1}{2}[A]_0$.

Inserting these two concentrations into Equation (8.24) gives

$$\ln\left(\frac{[A]_0}{\frac{1}{2}[A]_0}\right) = kt_{1/2}$$

The two $[A]_0$ terms cancel, causing the bracket on the left to simplify to just '2'. Accordingly, Equation (8.24) becomes

$$\ln(2) = kt_{1/2}$$

So, after rearranging to make $t_{1/2}$ the subject, the half-life is given by Equation (8.33):

$$t_{1/2} = \frac{\ln(2)}{k}$$

SAQ 8.20 Show from Equation (8.27) that the half-life of a *second*-order equation is given by the expression:

$$t_{1/2} = \frac{1}{[A]_0 k_2}$$

> The half-life of a second-order reaction is *not* independent of the initial concentrations used.

Why does the metabolism of a hormone not cause a large chemical change in the body?

'Pseudo-order' reactions

A *hormone* is a chemical that transfers information and instructions between cells in animals and plants. They are often described as the body's 'chemical messengers', but they also regulate growth and development, control the function of various tissues, support reproductive functions, and regulate metabolism (i.e. the process used to break down food to create energy).

Hormones are generally quite small molecules, and are chemically uncomplicated. Examples include adrenaline (**II**) and β-phenylethylamine (**III**).

(II)   (III)

Most hormones are produced naturally in the body (e.g. adrenaline (**II**) is formed in the adrenal glands). From there, the hormone enters the bloodstream and is consumed chemically (a physiologist would say 'metabolized') at the relevant sites in the body – in fact, adrenaline accumulates and is then broken down chemically in the muscles and lungs. Adrenaline is generated in equal amounts in men and women, and promotes a stronger, faster heartbeat in times of crisis or panic. The body is thus made ready for aggression ('fight') or necessary feats of endurance to escape ('flight'). Adrenaline is also administered artificially in medical emergencies, e.g. immediately following an anaphylactic shock, in order to give the heart a 'kick start' after a heart attack.

β-Phenylethylamine (**III**) is a different type of hormone, and is metabolized in women's bodies to a far higher extent than in a man's. The mechanism of metabolism is still a mystery, but it appears to cause feelings of excitement, alert feelings and in terms of mood, perhaps a bit of a 'high'. Unfortunately, the bodies of most men do not metabolize this hormone, so they do not feel the same 'high'.

> We say a reaction is *pseudo* first-order if it is second- or third-order, but behaves mathematically as though it were first order.

Hormones are potent and are produced in tiny concentrations (generally with a concentration of nanomoles per litre). By contrast, the chemicals in the body with which the hormone reacts have a sizeable concentration. For example, reaction with oxygen in the blood is one of the first processes to occur during the metabolism of adrenaline. The approximate range of $[O_2]_{(blood)}$ is $0.02–0.05 \ mol \ dm^{-3}$, so the change in oxygen concentration is virtually imperceptible even if *all* the adrenaline in the blood is consumed. As a good corollary, then, the only concentration to change is that of the hormone, because it is consumed.

> A *pseudo-order reaction* proceeds with all but one of the reactants in excess. This ensures that the only concentration to change appreciably is that of the *minority* reactant.

Although such a reaction is clearly second order, it behaves like a *first*-order reaction because only one of the concentrations changes. We say it is a *pseudo* first-order reaction.

Why do we not see radicals forming in the skin while sunbathing?

Pseudo-order rate constants

One of the most common causes of skin cancer is excessive sunbathing. Radicals are generated in the skin during irradiation with high-intensity UV-light, e.g. while lying on a beach. These radicals react with other compounds in the skin, which may ultimately cause skin cancer. But we never see these radicals by eye, because their concentration is so minuscule. And the concentration is so small because the radicals react so fast. (Photo-ionization is discussed in Chapter 9.)

> The accumulated concentration of such a fast-reacting intermediate will always be extremely low: perhaps as low as $10^{-10} \ mol \ dm^{-3}$.

Reaction *intermediates* are common in mechanistic studies of organic reactions. They are called 'intermediates' because they react as soon as they form. Such intermediates are sometimes described as 'reactive' because they react so fast they disappear more or less 'immediately'. Indeed, these intermediates are so reactive, they may react with solvent or even dissolved air in solution, i.e. with other chemicals in high concentration.

Since the reaction of intermediates is so fast, the concentration of the radical intermediate changes dramatically, yet the concentrations of the natural compounds in the skin with which the intermediate reacts (via second-order processes) do not change perceptibly. How in practice, then, do we determine kinetic parameters for pseudo-order reactions such as these?

> Intermediates react fast because their activation energy is small – see Section 8.5.

We will call the intermediate 'A' and the other reagent, which is in excess, will be called 'B'. In the example here, B will be the skin, but is more generally the solvent in which the reaction is performed, or an additional chemical in excess.

Because the concentration of B, [B], does not alter, this reaction will obey a *first-order* kinetic rate law, because only one of the concentrations changes with time. Because the reaction is first order (albeit in a pseudo sense), a plot of $\ln[A]_t$ (as 'y') against time (as 'x') will be linear for all but the longest times (e.g. see Figure 8.14). In an analogous manner to a straightforward first-order reaction, the gradient of such a plot has a value of 'rate constant $\times -1$' (see Worked Example 8.9). We generally call the rate constant k', where the *prime* symbol indicates that the rate constant is not a *true* rate constant, but is pseudo.

> Pseudo-order rate constants are generally indicated with a prime, e.g. k'.

The proper rate law for reaction for the second-order reaction between A and B is:

$$\text{rate} = k_2[A][B] \tag{8.34}$$

as for any normal second-order reaction. By contrast, the *perceived* rate law we measure is first order, as given by

$$\text{rate} = k'[A] \tag{8.35}$$

> Writing a pseudo rate constant k' without an order implies that it is pseudo *first* order.

where k' is the *perceived* rate constant.

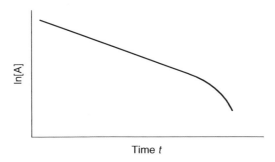

Time t

Figure 8.14 The reaction of A and B, with B greatly in excess is a second-order reaction, but it follows a kinetic rate law for a first-order reaction. We say it is *pseudo first-order* reaction. The deviation from linearity at longer times occurs because the concentration of B (which we assume is constant) does actually change during reaction, so the reaction no longer behaves as a first-order reaction

Comparing Equations (8.34) and (8.35), we obtain

$$k' = k_2[B] \qquad (8.36)$$

> The concentration $[B]_t$ barely changes, so we can write $[B]_0$ in Equations (8.34) and (8.36).

This little relationship shows that a pseudo-order rate constant k' is not a genuine rate constant, because its value changes in proportion to the concentration of the reactant in excess (in this case, with $[B]$).

Worked Example 8.16   The reaction of the ester ethyl methanoate and sodium hydroxide in water is performed with NaOH in great excess ($[NaOH]_0 = 0.23 \text{ mol dm}^{-3}$). The reaction has a half-life that is independent of the initial concentration of ester present. 13.2 per cent of the ester remains after 14 min and 12 s. What is the second-order rate constant of reaction k_2?

Strategy. (1) We ascertain the order of reaction, (2) we determine the pseudo rate constant k', (3) from k', we determine the value of the second-order rate constant k_2.

(1) Is the reaction a pseudo first-order process? The question says that the half-life $t_{1/2}$ of reaction is independent of initial concentration of ester, so the reaction must behave as though it was a *first-order* reaction in terms of [ester]. In other words, NaOH is in excess and its concentration does not vary with time.

(2) What is the value of the pseudo first-order rate constant k'? We calculate the pseudo first-order rate constant k' by assuming that the reaction obeys first-order kinetics. Accordingly, we write from Equation (8.24):

$$\ln\left(\frac{[\text{ester}]_0}{[\text{ester}]_t}\right) = k't$$

so

$$\ln\left(\frac{100\%}{13.2\%}\right) = k' \times [(14 \times 60) + 12] \text{ s}$$

and

$$\ln(7.57) = k' \times 852 \text{ s}$$

Because $\ln(7.57) = 2.02$ we say

$$k' = \frac{2.02}{852 \text{ s}} = 2.37 \times 10^{-3} \text{ s}^{-1}$$

> Note how manipulating the units in the fraction yields the correct units for k_2.

(3) What is the value of the second-order rate constant k_2? The value of rate constant k_2 can be determined from Equation (8.36), as $k' \div [NaOH]$, so

$$k_2 = \frac{2.37 \times 10^{-3} \text{ s}^{-1}}{0.23 \text{ mol dm}^{-3}} = 1.03 \times 10^{-3} \text{ dm}^3 \text{ mol}^{-1} \text{ s}^{-1}$$

SAQ 8.21 Potassium hexacyanoferrate(III) in excess oxidizes an alcohol at a temperature of 298 K. The concentration of $K_3[Fe(CN)_6]$ is $0.05 \, mol \, dm^{-3}$. The concentration of the alcohol drops to 45 per cent of its $t = 0$ value after 20 min. Calculate first the pseudo first-order rate constant k', and thence the second-order rate constant k_2.

Justification Box 8.5

When we first looked at the derivation of integrated rate equations, we looked briefly at the case where two species A and B were reacting but $[A]_0$ does not equal $[B]_0$. The integrated rate equation for such a case is Equation (8.30):

$$\frac{1}{[B]_0 - [A]_t} \times \ln\left(\frac{[A]_0 \times [B]_t}{[B]_0 \times [A]_t}\right) = k_2 t$$

Though we do not need to remember this fearsome-looking equation, we notice a few things about it. First, we assume that $[B]_0 \gg [A]_0$, causing the first term on the left-hand side to behave as $1/[B]_0$.

Also, the major change in the logarithm bracket is the change in [A], since the difference between $[B]_0$ and $[B]_t$ will be negligible in comparison, causing a cancellation of the [B] terms on top and bottom.

Accordingly, Equation (8.30) simplifies to

$$\frac{1}{[B]_0} \times \ln\left(\frac{[A]_0}{[A]_t}\right) = k_2 t$$

> We argue this statement by saying that if [B] is, say, 100 times larger than [A], then a complete consumption of [A] (i.e. a 100 per cent change in its concentration) will be associated with only a 1 per cent change in [B] – which is tiny.

Since $\ln([A]_0/[A]_t) = k't$ for a pseudo first-order reaction (by analogy with Equation (8.24)), we say that

$$\frac{1}{[B]_0} \times (k't) = k_2 t$$

and cancelling the two t terms, and rearranging yields

$$k' = k_2[B]_0$$

so we re-obtain Equation (8.36).

Alternatively, the value of the true second-order rate constant may be obtained by treating Equation (8.36) as the equation of a straight line, with the form $y = mx$:

$$
\begin{array}{ccc}
k' = & \left| k_2 \right| & [B]_0 \\
y & \left| m \right| & x
\end{array}
$$

Accordingly, we perform the kinetic experiment with a series of concentrations $[B]_0$, the reactant in excess, and then plot a graph of k' (as 'y') against $[B]_0$ (as 'x'). The gradient will have a value of k_2.

A graphical method such as this is usually superior to a simplistic calculation of $k_2 = k' \div [NaOH]$ (e.g. in the Worked Example 8.16), because scatter and/or chemical back or side reactions will not be detected by a single calculation. Also, the involvement of a back reaction (see next section) would be seen most straightforwardly as a non-zero intercept in a plot of k' (as 'y') against [reagent in excess] (as 'x').

Worked Example 8.17   The following kinetic data were obtained for the second-order reaction between osmium tetroxide and an alkene, to yield a 1,2-diol. Values of k' are pseudo-order rate constants because the OsO_4 was always in a tiny minority. Determine the second-order rate constant k_2 from the data in the following table:

$[alkene]_0/$ $mol\,dm^{-3}$	0.01	0.02	0.03	0.04	0.05	0.06	0.07	0.08
$k'/10^{-4}\,s^{-1}$	3.2	6.4	9.6	12.8	16.0	19.2	22.4	25.6

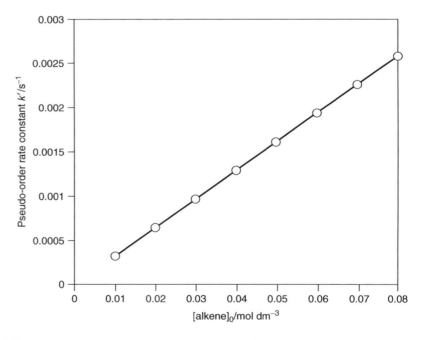

Figure 8.15   The rate constant of a pseudo-order reaction varies with the concentration of the reactant in excess: graph of k' (as 'y') against $[alkene]_0$ (as 'x'). The data refer to the formation of a 1,2-diol by the dihydrolysis of an alkene with osmium tetroxide. The gradient of the graph yields k_2, with a value of 3.2×10^{-2} $dm^3\,mol^{-1}\,s^{-1}$

The data clearly show that k' is not a true rate constant, because its value varies as the concentration of the alkene increases. That k' increases linearly with increased [alkene] suggests a straightforward pseudo first-order reaction. Figure 8.15 shows a graph of k' (as 'y') against $[\text{alkene}]_0$ (as 'x'). The graph is linear, and has a gradient of 3.2×10^{-2} dm^3 mol^{-1} s^{-1}, which is also the value of k_2.

> Graphs such as that in Figure 8.15 generally pass through the origin.

8.4 Kinetic treatment of complicated reactions

Why is arsenic poisonous?

'Concurrent' or 'competing' reactions

Arsenic is one of the oldest and best known of poisons. It is so well known, in fact, that when the wonderful Frank Capra comedy *Arsenic and Old Lace* was released, everyone knew that it was going to be a murder mystery in which someone would be poisoned. In fact, it has even been rumoured that Napoleon died from arsenic poisoning, the arsenic coming from the green dye on his wallpaper. We deduce that even a small amount of arsenic will cause death, or at least an unpleasant and lingering illness.

> *Care*: it is unwise to call a complicated reaction such as these a 'complex reaction', since the word 'complex' implies an association compound.

Arsenic exists in several different redox states. The characteristic energy at which one redox state converts to the other depends on its electrode potential E (see Chapter 7). The nervous system in a human body is 'instructed' by the brain much like a microprocessor, and regulated by electron 'relay cycles' as the circuitry, which consume or eject electrons. Unfortunately, the electrons acquired or released by arsenic in the blood interfere with these naturally occurring electron relay cycles, so, following arsenic poisoning, the numbers of electrons in these relay cycles is wrong – drastically so, if a large amount of arsenic is ingested.

Improper numbers of electrons in the relay cycles cause them not to work properly, causing a breakdown of those bodily functions, which require exact amounts of charge to flow. If the nervous system fails, then the lungs are not 'instructed' how to work, the heart is not told to beat, etc., at which point death is not too far away.

But arsenic is more subtle a poison than simply a reducing or oxidizing agent. Arsenic is a metalloid from Group V(B) of the periodic table, immediately below the elements nitrogen and phosphorus, both of which are vital for health. Unfortunately, arsenic is chemically similar to both nitrogen and phosphorus, and is readily incorporated into body tissues following ingestion. Arsenic effectively tricks the body into supposing that straightforward incorporation of nitrogen or phosphorus has occurred.

> Arsenic and nitrogen *compete* for the electrons participating in the natural electron-relay cycles in the body. The number of electrons transferred by nitrogen depends on the number of arsenic atoms competing for them.

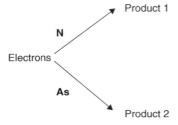

Figure 8.16 Arsenic and nitrogen compete for electrons both for and from the natural relay cycles in the body. The overall rate at which electrons are transferred by the nitrogen will alter when arsenic competes for them. The arsenic is poisonous, since these two pathways yield different products

---

'Sequester' means to confiscate, seize or take control of something to prevent its further use. The word comes from the Latin *sequester*, meaning a 'trustee' or 'agent' whose job was to seize property.

Concurrent (or *competing*) reactions are so called because two reactions occur at the same time. We occasionally call them *simultaneous* reactions, although this terminology can be confusing.

A beam of plane-polarized light is caused to rotate by an angle θ as it passes through a solution of a chiral compound. The magnitude of θ depends on the concentration of the chiral compound.

The first time the body realizes that arsenic has been incorporated is when the redox activity (as above) proceeds at potentials when nitrogen or phosphorus are inert. By the time we detect the arsenic poisoning (i.e. we feel unwell), it is generally too late, since atoms of arsenic are covalently *bound* within body tissue and cannot just be flushed out or treated with an antidote. The arsenic sequesters electrons that might otherwise be involved in other relay cycles, which is a *concurrent* kinetic process; see Figure 8.16.

So arsenic is toxic because it has 'fooled' the body into thinking it is something else.

Why is the extent of Walden inversion smaller when a secondary alkyl halide reacts than with a primary halide?

Reaction profiles for complicated reactions

Alkyl halides react by a substitution reaction with hydroxide ions to yield an alcohol. A primary halide, such as 1-bromopentane, reacts by a simple bimolecular S_N2 mechanism, where the 'S' stands for substitution, the 'N' for nucleophilic (because the hydroxide ion is a *nucleophile*) and the '2' reminds us that the reaction is bimolecular. Being a single-step reaction, the substitution reaction is necessarily the rate-determining step. The reaction is accompanied by stereochemical inversion about the central tetrahedral carbon atom to which the halide is attached – we call it *Walden inversion*; see Figure 8.17(a). Each molecule of primary halide inverts during the S_N2 reaction. For this reason, we could monitor the rate of the S_N2 reaction by following changes in the angle of rotation θ of plane polarized light. This angle θ changes as a function of the extent of reaction ξ, so we know the reaction is complete when θ remains constant at a new value that we call $\theta_{(final)}$.

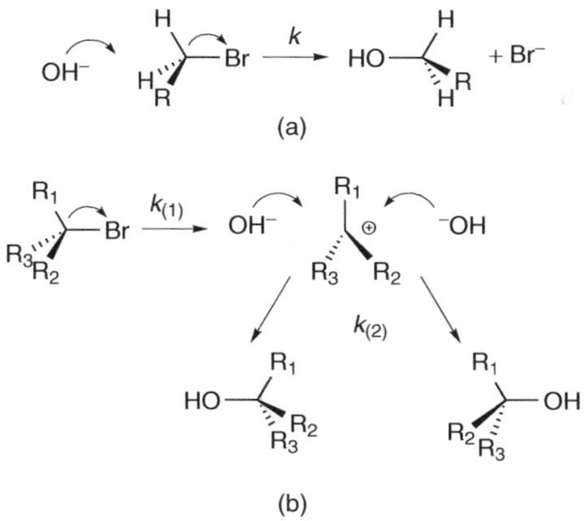

(a)

(b)

Figure 8.17   Reaction of an alkyl halide with hydroxide ion. (a) A primary halide reacts by an S_N2 mechanism, causing Walden inversion about the central, chiral carbon. (b) A tertiary halide reacts by an S_N1 mechanism (the rate-determining step of which is unimolecular dissociation, minimizing the extent of Walden inversion and maximizing the extent of racemization). Secondary alcohols often react with both S_N1 and S_N2 mechanistic pathways proceeding concurrently

By contrast, tertiary halides, such as 2-bromo-2,2-dimethylpropane, cannot participate in an S_N2 mechanism because it would be impossible to fit two methyl groups, one bromine and a hydroxide around a single carbon. The steric congestion would be too great. So the tertiary halide reacts by a different mechanism, which we call S_N1. It's still a nucleophilic substitution reaction (hence the 'S' and the 'N') but this time it is a unimolecular reaction, hence the '1'. The rate-determining step during reaction is the slow unimolecular dissociation of the alkyl halide to form a bromide ion and a carbocation that is planar around the reacting carbon.

Addition of hydroxide occurs as a rapid follow-up reaction. Even if the alkyl halide was chiral before the carbocation formed, racemization occurs about the central carbon atom because the hydroxide can bond to the planar central carbon from either side (see Figure 8.17(b)). Statistically, equal numbers of each racemate are formed, so the angle through which the plane polarized light rotated during reaction will, therefore, decrease toward $0°$, when reaction is complete.

> Polarimetry is the technique of following the rotation of plane-polarized light.

> Genuinely first-order reactions are unusual. It is likely that the alkyl halide collides with another body (such as solvent) with sufficient energy to cause bond cleavage.

In summary, primary halides react almost wholly by a bimolecular process and tertiary halides react by a unimolecular process. Secondary halides are structurally between these two extreme structural examples, since reaction occurs by *both* S_N2 and S_N1 routes. These two mechanisms proceed in *competition*, and occur *concurrently*.

When following the (dual-route) reaction of a secondary halide with hydroxide ion, we find that the angle θ through which plane polarized light is rotated will decrease, as for primary and tertiary halides, but will not reach zero at completion. In fact, the final angle will have a value between $0°$ and θ_{final} because of the mixtures of products, itself a function of the mixture of S_N1 and S_N2 reaction pathways.

We looked briefly at *reaction profiles* in Section 8.2. Before we look at the reaction profile for the concurrent reactions of hydrolysing a secondary alkyl halide, we will look briefly at the simpler reaction of a primary alkyl halide, which proceeds via a single reaction path. And for additional simplicity, we also assume that the reaction goes to completion. We will look not only at the rate of change of the reactants' concentration but also at the rate at which product forms.

> We include the minus sign in Equation (8.37) to show how the product concentration INcreases while the reactant concentration DEcreases.

Consider the graph in Figure 8.18, which we construct with the data in Worked Example 8.12. We have seen the upper, lighter line before: it represents the rate of a second-order decay of molecule A with time as it reacts with the stoichiometry A + B → product. The bold line in Figure 8.18 represents the concentration of the product. It is a 1:1 reaction, so each molecule of A *consumed* by the reaction will *generate* one molecule of product, with the consequence that the two traces are mirror images. Stated another way, the rate of forming product is the same as the rate of consuming A:

$$\text{rate} = -\frac{d[A]}{dt} = \frac{d[\text{product}]}{dt} \qquad (8.37)$$

where the minus sign is introduced because one concentration *increases* while the other *decreases*.

Now, to return to the hydrolysis of the secondary alkyl halides, we will call the reactions (1) and (2), where the '1' relates to the S_N1 reaction and the '2' relates to the S_N2 reactions. (And we write the numbers with brackets to avoid any confusion, i.e. to prevent us from thinking that the '1' and '2' indicate first- and second-order reactions respectively.) We next say that the rate constants of the two concurrent reactions are $k_{(1)}$ and $k_{(2)}$ respectively. As the two reactions proceed with the same 1:1

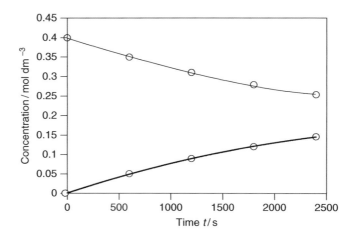

Figure 8.18 Concentration profile for a simple reaction of a primary alkyl halide + OH⁻ → alcohol. The bolder, lower line represents the concentration of product as a function of time, and the fainter, upper line represents the concentration of reactant

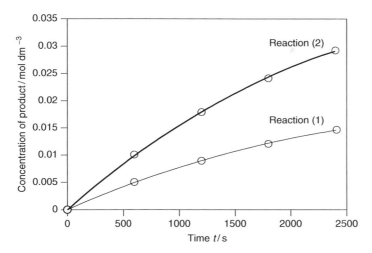

Figure 8.19 Concentration profiles for a concurrent reaction, e.g. of a secondary alkyl halide + OH⁻ → alcohol: reaction (2) is twice as fast as reaction (1) in this example

stoichiometry, the ratio of the products will relate to the respective rate constants in a very simple way, according to

$$\frac{k_{(1)}}{k_{(2)}} = \frac{\text{moles of product formed via reaction (1)}}{\text{moles of product formed via reaction (2)}} \quad (8.38)$$

so we see that the amounts of product depend crucially on the *relative* magnitudes of the two rate constants. (We shall return to this theme when we look at the way the human body generates a high temperature to cure a fever in Section 8.5).

We now look at the concentration profiles for reactions (1) and (2). For simplicity, we shall say that reaction (2) is twice as fast as reaction (1), which is likely – unimolecular reactions are often quite slow. We should note how, at the end of the reaction at the far right-hand side of the profile in Figure 8.19, the *total* sum of product will be the same as the initial concentration of reactant. Also note how, at all stages during the course of the reaction, the ratio of products will be 2:1, since that was the ratio of the two rate constants.

> *Care*: do not confuse k_1 and $k_{(1)}$. k_1 is the rate constant of a first-order process, and $k_{(1)}$ is the rate constant of the first process in a multi-step reaction.

> We can inter-convert between 'number of moles' and 'concentration' here because the reactions are performed within a constant volume of solvent.

Why does 'standing' a bottle of wine cause it to smell and taste better?

Consecutive reactions

It is often said that a good wine, particularly if red, should 'stand' before serving. By 'standing', we mean that the bottle should be uncorked some time before consumption,

A wine is also said to *age* and *breathe*, which means the same thing.

The enzymes in wine are killed if the percentage of alcohol exceeds about 13 per cent. Fermentation alone cannot make a stronger wine, so spirits are prepared by *distilling* a wine. Adding brandy to wine makes a fortified wine, such as sherry.

to allow air into the wine. After a period of about an hour or so, the wine should taste and smell better.

Wines contain a complicated mixture of natural products, many of which are alcohols. Ethanol is the most abundant alcohol, at a concentration of 3–11 per cent by volume. The amounts of the other alcohols generally total no more than about 0.1–1 per cent.

The majority of the smells and flavours found in nature comprise esters, which are often covalent liquids with low boiling points and high vapour pressures. For that reason, even a very small amount of an ester can be readily detected on the palate – after all, think how much ester is generated within a single rose and yet how overwhelmingly strong its lovely smell can be!

Esters are the product of reaction between an alcohol and a carboxylic acid. Although the reaction can be slow – particularly at lower temperatures – the equilibrium constant is sufficiently high for the eventual yield of ester to be significant. We see how even a small amount of carboxylic acid and alcohol can generate a sufficient amount of ester to be detected by smell or taste.

This helps explain why a wine should be left to stand: some of the natural alcohols in the wine are oxidized by oxygen dissolved from the air to form carboxylic acids. These acids then react with the natural alcohols to generate a wide range of esters, which connoisseurs of wines will recognize as a superior taste and 'bouquet'.

The 'mash' that ferments to form wine sometimes includes grape skins; which are rich in enzymes. It is likely, therefore, that the oxidation in Equation (8.39) is mediated (i.e. *catalysed*) enzymatically.

And the reason why the wine must 'stand' (rather than the reaction occurring 'immediately' the oxygen enters the bottle on opening) is that the reaction to form the ester is not a straightforward one-step reaction: the first step (Equation (8.39)) is quite slow and occurs in low yield:

$$\text{alcohol}_{(aq)} + O_{2(aq)} \xrightarrow{\text{first reaction}} \text{carboxylic acid}_{(aq)} + H_2O_{(l)}$$

$$(8.39)$$

where the rate constant of Equation (8.39) is that of a second-order reaction. Once formed, the acid reacts more rapidly to form the respective ester:

$$\text{carboxylic acid}_{(aq)} + \text{alcohol}_{(aq)} \xrightarrow{\text{second reaction}} \text{ester}_{(aq)} \qquad (8.40)$$

where the rate constant for this second step is larger than the respective rate constant for the first step, implying that the second step is faster.

In summary, we see that esterification is a two-step process. The first step – production of a carboxylic acid – is relatively slow because its rate is proportional to the concentration of dissolved oxygen; and the $[O_2]_{(soln)}$ is low. Only after the wine bottle has been open for some time (it has had sufficient time to 'breathe') will $[O_2]_{(soln)}$ be higher, meaning that, after 'standing', the esters that lend additional flavour and aroma are formed in higher yield.

The case of esterification is an example of a whole class of reactions, in which the product of an initial reaction will itself undergo a further reaction. We say that there is a *sequence* of reaction steps, so the reaction as a whole is *sequential* or *consecutive*.

> To remember this latter terminology, we note that for a *consecutive* reaction to occur, the second step proceeds as a *consequ*ence of the first.

Although this example comprises two reactions in a sequence, many reactions involve a vast series of steps. Some radioactive decay routes, for example, have as many as a dozen species involved in a sequence before terminating with an eventual product.

For simplicity, we will denote the reaction sequence by

$$A \xrightarrow{k_{(1)}} B \xrightarrow{k_{(2)}} C \qquad (8.41)$$

All the other reactants will be ignored here to make the analysis more straightforward, even if steps (1) and (2) are, in fact, bimolecular. We again write the reaction number within brackets to avoid confusion: we do not want to mistake the subscripted *number* for the *order* of reaction. We call the rate constant of the first reaction $k_{(1)}$ and the rate constant of the second will be $k_{(2)}$.

The material A is the *initial* reactant or *precursor*. It is usually stable and only reacts under the necessary conditions, e.g. when mixed and/or refluxed with other reactants. We know its concentration because we made or bought it before the reaction commenced. Similarly, the material C (the *product*) is usually easy to handle, so we can weigh or analyse it when the reaction is complete, and examine or use it. By contrast, the material B is not stable – if it

> The maximum rate of reaction occurs when the concentration of the intermediate B is highest.

was, then it would not react further to form C as a second reaction step. Accordingly, it is rare that we can isolate B. We call B an *intermediate*, because it forms during the consumption of A but before the formation of C.

The concentration profile of a simple 1:1 reaction is always easy to draw because product is formed at the expense of the reactant, so the rate at which reactant is consumed is the same as the rate of product formation. No such simple relation holds for a consecutive reaction, because two distinct rate constants are involved. Two extreme cases need to be considered when dealing with a consecutive reaction: when $k_{(1)} > k_{(2)}$ and when $k_{(1)} < k_{(2)}$. We shall treat each in turn.

Why fit a catalytic converter to a car exhaust?

Consecutive reactions in which the second reaction is slow

First, we consider the case where the first reaction is very fast compared with the second, so $k_{(1)} > k_{(2)}$. This situation is the less common of the two extremes. When all the A has been consumed (to form B), the second reaction is only just starting to convert the B into the final product C. A graph of concentration against time will show a rapid decrease in [A] but a slow increase in the concentration of the eventual product, C. More specifically, the concentration of intermediate B will initially increase and only at later times will its concentration decrease once more as the slower reaction (the one with rate constant $k_{(2)}$) has time to occur to any significant extent. The concentration profile of B, therefore, has a maximum; see Figure 8.20.

> Most modern cars are fitted with a 'catalytic converter', one purpose of which is to speed up this second (slower) process in the reaction sequence.

A good example of a consecutive process in which the second reaction is much slower than the first is the reaction occurring in a car exhaust. The engine forms carbon monoxide (CO) as its initial product, and only at later times will $CO_{(g)}$ oxidize to form $CO_{2(g)}$. In fact, the second reaction ($CO_{(g)} + \frac{1}{2}O_{2(g)} \rightarrow CO_{2(g)}$) is so slow that the concentration of $CO_{(g)}$ is often high enough to cause serious damage to health.

Why do some people not burn when sunbathing?

Consecutive reactions in which the second reaction is fast

> If $k_{(1)} < k_{(2)}$, an intermediate is generally termed a *reactive intermediate* to emphasize how soon it reacts.

We now look at the second situation, i.e. when $k_{(1)} < k_{(2)}$. The first reaction produces the intermediate B very slowly. B is consumed immediately – as soon as each molecule of intermediate is formed – by the second reaction, which forms the final product C. Accordingly, we call it a *reactive intermediate*, to emphasize how rapidly it is consumed.

The rate of the first reaction in the sequence is slow because $k_{(1)}$ is small, so the rate of decrease of [A] is not steep. Conversely, since $k_{(2)}$ is fast, we would at first expect the rate d[C]/dt to be quite high. A moment's thought shows

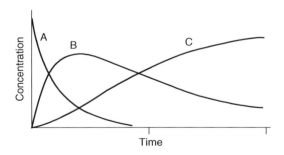

Figure 8.20 Schematic graph of concentration against time ('a concentration profile') for a consecutive reaction for which $k_{(1)} > k_{(2)}$. Note the maximum in the concentration of the intermediate, B. This graph was computed with $k_{(2)}$ being five times slower than $k_{(1)}$

that this is not necessarily the case:

$$\text{rate of forming the final product, C} = \frac{d[C]}{dt} = k_{(2)}[B] \tag{8.42}$$

We see from Equation (8.42) that the rate of forming C is quite slow because [B] is tiny no matter how fast the rate constant $k_{(2)}$.

We now consider the concentration profile of the reactive intermediate B. Because the second reaction is so much faster than the first, the concentration of B is never more than minimal. Its concentration profile is virtually horizontal, although it will show a very small maximum (it must, because [B] = 0 at the start of the reaction at $t = 0$; and all the intermediate B has been consumed at the end of reaction, when [B] is again zero). At all times between these two extremes, the concentration of B at time t, $[B]_t$, is not zero. The concentration profile is shown in Figure 8.21.

Sunbathing to obtain a tan, or simply to soak up the heat, is an inadvertent means of studying photochemical reactions in the skin. It is also a good example of a consecutive reaction for which $k_{(1)} < k_{(2)}$.

Small amounts of organic radicals are formed continually in the skin during photolysis (in a process with rate constant $k_{(1)}$). The radicals are consumed 'immediately' by natural substances in the skin, termed *antioxidants* (in a different process with rate constant $k_{(2)}$). Vitamin C (L-(+)-ascorbic acid, **IV**) is one of the best naturally occurring antioxidants. Red wine and tea also contain efficient antioxidants.

> An *antioxidant* is a chemical that prevents oxidation reactions.

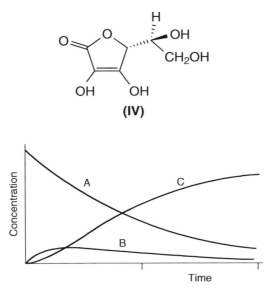

(IV)

Figure 8.21 Schematic graph of concentration against time ('a concentration profile') for a consecutive reaction for which $k_{(2)} > k_{(1)}$. This graph was computed with $k_{(2)}$ being five times faster than $k_{(1)}$

The vast majority of organic radical reactions involve the radical as a reactive intermediate, since their values of $k_{(2)}$ are so high, although we need to note that the second reaction need not be particularly fast: it only has to be fast *in relation to the first reaction*. As a good generalization, the intermediate may be treated as a *reactive* intermediate if $k_{(2)}/k_{(1)} > 10^{-3}$.

Integrated rate equations for consecutive reactions

Our consecutive reaction here has the general form $A \rightarrow B \rightarrow C$. Deriving an integrated rate equation for a consecutive reaction is performed in much the same way as for a simple one-step reaction (see Section 8.2), although its complexity will prevent us from attempting a full derivation for ourselves here.

For the precursor A, the rate of change of [A] is given by

$$\text{rate} = -\frac{d[A]}{dt} = k_{(1)}[A] \tag{8.43}$$

where the minus sign reflects the way that [A] is *consumed*, meaning its concentration decreases with time.

> Writing an integrated rate expression for a complicated reaction is difficult because we don't readily know the time-dependent concentration of the intermediate, $[B]_t$.

The rate of change of C has been given already as Equation (8.42). Equations (8.42) and (8.43) show why the derivation of integrated rate equations can be difficult for consecutive reactions: while we can readily *write* an expression for the rate of forming C, the rate expression requires a knowledge of [B], which first increases, then decreases. The problem is that [B] is itself *a function of time*.

The rate of change of [B] has two components. Firstly, we form B from A with a rate of '$+k_{(1)} \times [A]$'. The second part of the rate equation concerns the subsequent removal of B, which occurs with a rate of '$-k_{(2)} \times [B]$'. The minus sign here reminds us that [B] *decreases* in consequence of this latter process.

> The rate expression of a complicated reaction comprises *one term* for *each reaction step*. In this case, species B is involved in two reactions, so the rate equation comprises two terms.

Combining the two rate terms in Equations (8.42) and (8.43) yields

$$\frac{d[B]}{dt} = \boxed{k_{(1)}[A]} - \boxed{k_{(2)}[B]} \tag{8.44}$$

rate of the process *forming* the B   rate of the process *removing* the B

Equation (8.44) helps explain why the concentration profile in Figure 8.20 contains a maximum. Before the peak, and at short times, the second term on the right-hand side of Equation (8.44) is tiny because [B] is small. Therefore, the net rate $d[B]/dt$ is positive, meaning that the concentration of B *increases*. Later – after the peak maximum – much of the [A] has been consumed, meaning more [B] resides in solution. At this stage in the reaction, the first term on the right-hand side of Equation (8.44) is relatively small, because [A] is substantially depleted, yet the second term is now quite large in response to a higher value of [B]. The second term in Equation (8.44) is consequently larger than the first, causing the overall rate to be negative, which means that [B] *decreases* with time.

Table 8.3 lists the concentrations $[A]_t$, $[B]_t$ and $[C]_t$, which are exact, and will give the correct concentrations of A, B and C at any stages of a reaction. We will not derive them here, nor will we need to learn them. But it is a good idea to recognize the interdependence between $[A]_t$, $[B]_t$ and $[C]_t$.

How do Reactolite® sunglasses work?

The kinetics of reversible reactions

Reactolite® sunglasses are photochromic: they are colourless in the dark, but become dark grey–black when strongly illuminated, e.g. on a bright summer morning. The reaction is fully reversible (in the thermodynamic sense), so when energy is removed from the system, e.g. by allowing the lenses to cool in the dark, the photochemical reaction reverses, causing the lenses to become uncoloured and fully transparent again.

The photochromic lenses contain a thin layer of a silver-containing glass, the silver being in its $+1$ oxidation state. Absorption of a photon supplies the energy for an electron-transfer reaction in the glass, the product of which is atomic silver:

> The word 'photochromic' comes from the Greek words *photos*, meaning 'light' and *khromos* meaning 'colour'. A photochromic substance acquires colour when illuminated.

$$Ag^+ + e^- + h\nu \longrightarrow Ag^0 \qquad (8.45)$$

where the electron comes from some other component within the glass. It is the silver that we perceive as colour. In effect, the colour

> Excited states are defined in Chapter 9.

Table 8.3 Mathematical equations to describe the concentrations[a] of the three species A, B and C involved in the consecutive reaction, $A \rightarrow B \rightarrow C$

$[A]_t = [A]_0 \exp(-k_{(1)}t)$

$[B]_t = \dfrac{k_{(1)}}{k_{(2)} - k_{(1)}}[A]_0\{\exp(k_{(1)}t) - \exp(k_{(2)}t)\}$

$[C]_t = [A]_0 \left\{ 1 - \dfrac{k_{(2)}\exp(-k_{(1)}t) + k_{(1)}t}{k_{(2)} - k_{(1)}} \right\}$

---

[a]Rate constants are denoted as k, where the subscripts indicate either the first or the second reaction in the sequence. The subscript '0' indicates the concentration at the commencement of the reaction. Concentrations at other times are denoted with a subscript t.

indicates a very long-lived excited state, the unusually long time is achieved because colouration occurs in the solid state, so the reaction medium is extremely viscous.

Atomic silver is constantly being formed during illumination, but, at the same time, the reverse reaction also occurs during 'cooling' (also termed 'relaxation' – see Chapter 9) in a process called charge *recombination*. Such recombination is only seen when removing the bright light that caused the initial coloration reaction, so the reaction proceeds in the opposite direction.

> *Reversible* reactions are also termed *equilibrium* or *opposing* reactions.

> The reaction might be so slow that the reaction never actually *reaches* the RHS during a realistic time scale, but the *direction* of the reaction is still the same.

> For a thermodynamically reversible reaction, the rate constants of the forward and reverse reactions are k_n and k_{-n} respectively.

Rate laws for reversible reactions

All the reactions considered so far in this chapter have been irreversible reactions, i.e. they only go in one direction, from fully reactants on the left-hand side to fully products on the right-hand side. (They might stop before the reaction is complete.)

We now consider the case of a reversible reaction:

$$A \; \underset{k_{-1}}{\overset{k}{\underset{\longleftarrow}{\longrightarrow}}} \; B \qquad (8.46)$$

where k_1 is the rate constant for the forward reaction and k_{-1} is the rate constant for the reverse reaction. The minus sign is inserted to tell us that the reaction concerned is that of the reverse reaction.

When writing a rate expression for such a reaction, we note that two arrows involve the species B, so, straightaway, we know that the rate expression for species B has two terms. This follows since the rate of change of the concentration of B involves two separate processes: one reaction forms the B (causing [B] to increase with time) while the other reaction is consuming the B (causing [B] to decrease). The rate is given by

$$\text{rate of change of } [B] = \frac{d[B]}{dt} = k_1[A] - k_{-1}[B] \qquad (8.47)$$

where the minus and (implicit) plus signs indicate that the concentrations decrease or increase with time respectively. Note that a new situation has arisen whereby the expression to describe the rate of change of [B] itself involves [B] – this is a general feature of rate expressions for simple reversible reactions. Similarly:

$$\text{rate of change of } [A] = \frac{d[A]}{dt} = -k_1[A] + k_{-1}[B] \qquad (8.48)$$

> The concentrations stop changing when the reaction reaches equilibrium.

Note that the rate of change of [A] is equal but opposite to the rate of change of [B], which is one way of saying that A is consumed at the expense of B; and B is formed at the expense of A.

When the reaction has reached equilibrium, the rate of change of both species must be zero, since the concentrations do not alter any more – that is what we mean by true 'equilibrium'. From Equation (8.48):

$$0 = k_1[A]_{(eq)} - k_{-1}[B]_{(eq)}$$

so

$$k_1[A]_{(eq)} = k_{-1}[B]_{(eq)}$$

which, after a little algebraic rearranging, gives a rather surprising result:

$$\frac{k_1}{k_{-1}} = \frac{[B]_{(eq)}}{[A]_{(eq)}} \qquad (8.49)$$

We recognize the right-hand side of the equation as the equilibrium constant K. We give the term *microscopic reversibility* to the idea that the ratio of rate constants equals the equilibrium constant K.

> The principle of 'microscopic reversibility' demonstrates how the ratio of rate constants (forward to back) for a *reversible* reaction equals the reaction's equilibrium constant K.

Worked Example 8.18 Consider the reaction between pyridine and heptyl bromide, to make 1-heptylpyridinium bromide. It is an equilibrium reaction with an equilibrium constant $K = 40$. What is the rate constant of back reaction k_{-1} if the value of the forward rate constant $k_1 = 2.4 \times 10^3$ dm^3 mol^{-1} s^{-1}?

We start with Equation (8.49):

$$K = \frac{k_1}{k_{-1}}$$

and rearrange it to make k_{-1} the subject, to yield

$$k_{-1} = \frac{k_1}{K}$$

We then insert values:

$$\frac{2.4 \times 10^3 \text{ dm}^3 \text{ mol}^{-1} \text{ s}^{-1}}{40}$$

so $k_{-1} = 60$ dm^3 mol^{-1} s^{-1}.

SAQ 8.22 A simple first-order reaction has a forward rate constant of 120 s^{-1} while the rate constant for the back reaction is 0.1 s^{-1}. Calculate the equilibrium constant K of this reversible reaction by invoking the principle of microscopic reversibility.

Integrated rate equations for reversible reactions

In kinetics, we often term the concentrations at equilibrium 'the infinity concentration'. The Reactolite® glasses do not become

> In kinetics, the equilibrium concentration $[A]_{(eq)}$ is often termed the *infinity concentration*, and cited with an infinity sign as $[A]_\infty$.

We should never throw away a reaction solution without measuring a value of $[A]_{(eq)}$.

progressively darker with time because the concentration of atomic silver reaches its infinity value $[Ag]_{(eq)}$.

We will consider the case of a first-order reaction, $A \rightleftarrows$ product. Following integration of an expression similar to Equation (8.48), we arrive at

$$\ln\left(\frac{[A]_0 - [A]_{(eq)}}{[A]_t - [A]_{(eq)}}\right) = (k_1 + k_{-1})t \tag{8.50}$$

which is very similar to the equation we had earlier for a simple reaction (i.e. one proceeding in a single direction), Equation (8.24). There are two simple differences. Firstly, within the logarithmic bracket on the left-hand side, each term on top and bottom has the infinity reading subtracted from it. Secondly, the time t is not multiplied by a single rate constant term, but by the *sum* of both rate constants, forward and back.

This equation can be rearranged slightly, by separating the logarithm:

$$\underset{y}{\ln([A]_t - [A]_{(eq)})} \quad \Bigg| \quad \underset{m}{-(k_1 + k_{-1})} \quad \Bigg| \quad \underset{x}{t} \quad \Bigg| \quad \underset{c}{+ \ln([A]_0 - [A]_{(eq)})} \tag{8.51}$$

Note that the final term on the right-hand side is a constant. Accordingly, a plot of $\ln([A]_t - [A]_{(eq)})$ (as 'y') against time (as 'x') will yield a straight line of gradient $-(k_1 + k_{-1})$.

Worked Example 8.19 The data below relate to the first-order isomerization of 2-hexene at 340 K, a reaction for which the equilibrium constant is known from other studies to be 10.0. What are the rate constants k_1 and k_{-1}?

Time t/min	0	20	47	80	107	140
$([A]_t - [A]_{(eq)})$/mol dm^{-3}	0.114	0.103	0.091	0.076	0.066	0.055

Strategy. (1) we need to plot a graph of $\ln([A]_t - [A]_{(eq)})$ (as 'y') against time t (as 'x'); (2) determine its gradient; (3) then, knowing the equilibrium constant K, we will be able to determine the two rate constants algebraically.

(1) Figure 8.22 shows a graph for a reversible first-order reaction with the axes for an integrated rate equation $\ln([A]_t - [A]_{(eq)})$ (as 'y') against time (as 'x'). The gradient is -5.26×10^{-3} min^{-1}.

The *microscopic-reversibility* relationship $K = k_1 \div k_{-1}$ cannot be applied unless we know we have a simple *reversible* reaction.

(2) The gradient is $-(k_1 + k_{-1})$, so $(k_1 + k_{-1}) = +5.26 \times 10^{-3}$ min^{-1}.

(3) Next, we perform a little algebra, and start by saying that $K = k_1 \div k_{-1}$, i.e. we invoke the principle of *microscopic reversibility*. Multiplying the bracket $(k_1 + k_{-1})$ by $k_{-1} \div k_{-1}$, i.e. by 1, yields

$$\text{gradient} = k_{-1}(K + 1) \tag{8.52}$$

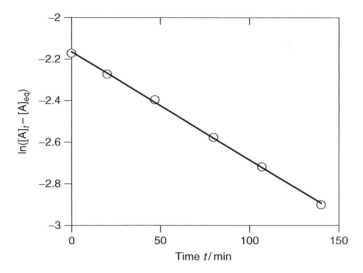

Figure 8.22   Kinetic graph for a reversible first-order reaction with the axes for an integrated rate equation $\ln([A]_t - [A]_{(eq)})$ (as 'y') against time (as 'x'). The gradient is -5.26×10^{-3} min^{-1}

Substituting values into Equation (8.52) gives

$$5.26 \times 10^{-3} \text{ min}^{-1} = k_{-1}(10.0 + 1)$$

so

$$k_{-1} = \frac{5.26 \times 10^{-3} \text{ min}^{-1}}{11} = 4.78 \times 10^{-4} \text{ min}^{-1}$$

Next, since we know both K and k_{-1}, we can calculate k_1. We know from the gradient of the graph that

$$(k_1 + k_{-1}) = 5.26 \times 10^{-3} \text{ min}^{-1}$$

which, after a little rearranging, gives

$$k_1 = 5.26 \times 10^{-3} \text{ min}^{-1} - k_{-1}$$

and, after inserting values

$$k_1 = (5.26 \times 10^{-3} - 4.78 \times 10^{-4}) \text{ min}^{-1} = 4.78 \times 10^{-3} \text{ min}^{-1}.$$

In summary, $k_1 = 4.78 \times 10^{-3}$ min^{-1} and $k_{-1} = 4.78 \times 10^{-4}$ min^{-1}.

> Notice how the ratio '$k_1 \div k_{-1}$' yields the value of K.

It is wrong (but common) to see a reversible reaction written with a double-headed arrow, as 'A ↔ B'. Such an arrow implies *resonance*, e.g. between the two extreme valence-bond structures of Kekulé benzene.

SAQ 8.23 Consider a reversible first-order reaction. Its integrated rate equation is given by Equation (8.50). People with poor mathematical skills often say (erroneously!) that taking away the infinity reading from both top and bottom is a waste of time because the two infinity concentration terms will cancel. Show that the infinity terms cannot be cancelled in this way; take $[A]_{(eq)} = 0.4 \text{ mol dm}^{-3}$, $[A]_0 = 1 \text{ mol dm}^{-3}$ and $[A]_t = 0.7 \text{ mol dm}^{-3}$.

8.5 Thermodynamic considerations: activation energy, absolute reaction rates and catalysis

Why prepare a cup of tea with boiling water?

The temperature dependence of reaction rate

The instructions printed on the side of a tea packet say, 'To make a perfect cup of tea, add boiling water to the tea bag and leave for a minute'. The stipulation for 'one minute' suggests the criterion for brewing tea in water at 100 °C is a *kinetic* requirement. In fact, it reflects the *rate* at which flavour is extracted from the tea bag and enters the water.

Remember that the temperature of boiling water $T_{(boil)}$ is itself a function of the external pressure, according to the Clausius–Clapeyron equation (see Section 5.3).

Pure water boils at 100 °C (273.15 K). If the tea is prepared with cooler water, then the time taken to achieve a good cup of tea is longer (and by 'good', here, we mean a solution of tea having a sufficiently high concentration). If the water is merely tepid, then a duration as long as 10 min might be required to make a satisfactory cup of tea; and if the water is cold, then it is possible that the tea will never brew, and will always remain dilute. In summary, the rate of flavour extraction depends on the temperature because the rate constant of flavour extraction increases with increasing temperature.

Why store food in a fridge?

The temperature dependence of rate constants

Food is stored in a fridge to prevent (or slow down) the rate at which it perishes. Foods such as milk or butter will remain fresh for longer if stored in a fridge, but they decompose or otherwise 'go off' more quickly if stored in a warmer environment.

The natural processes that cause food to go bad occur because of enzymes and microbes, which react with the natural constituents of the food, and multiply. When these biological materials have reached a certain concentration, the food smells and tastes bad, and is also likely to be toxic.

The growth of each microbe and enzyme occurs with its own unique rate. A fridge acts by cooling the food in order to slow these rates to a more manageable level. At constant temperature, the rate of each reaction equals the respective rate constant k multiplied by the concentrations of all reacting species. For example, the rate of the reaction causing milk to 'go off' occurs between lactic acid and an enzyme. The rate of the process is written formally as

> The rate constant k_2 is truly a constant at a fixed temperature, but can vary significantly.

$$\text{rate} = k_2[\text{lactose}][\text{enzyme}] \qquad (8.53)$$

where, as usual, the subscripted '2' indicates that the reaction is second order. Neither [lactose] nor [enzyme] will vary with temperature, so any variations in rate caused by cooling must, therefore, arise from changes in k_2 as the temperature alters.

> The rate constant increases as the temperature increases and decreases as the temperature decreases.

The rate constant k_2 is truly a constant at a fixed temperature, but can vary significantly: increasing as the temperature increases and decreasing as the temperature decreases. This result explains why rate of reaction depends so strongly on temperature.

Why do the chemical reactions involved in cooking require heating?

Activation energy E_a and the Arrhenius theory

Cooking is an applied form of organic chemistry, since the molecules in the food occur naturally. We heat the food because the reactions occurring in, say, a pie dish require energy; and an oven is simply an excellent means of supplying large amounts of energy over extended periods of time.

> 'Naturally occurring' was the old-fashioned definition of 'organic chemistry', and persisted until nearly the end of the 19th century.

The natural ingredients in food are all organic chemicals, and it is rare for organic reactions to proceed without an additional means of energy, which explains why we usually need to reflux a reaction mixture.

It is easy to see why an endothermic reaction requires energy to react – the energy to replace the bonds, etc. must be supplied from the surroundings. But why does an *exothermic* reaction require additional energy? Why do we need to *add* any energy, since it surely seeks to lose energy?

At the heart of this form of kinetic theory is the *activated complex*. In this context, the word 'activated' simply means a species brimming with energy, and which will react as soon as possible in order to decrease that energy content.

A reaction can be thought of as a multi-step process: first the reactants approach and then they collide. Only after touching do they react. One of the more useful definitions of reaction is 'a rearrangement of bonds'. We are saying that, as a good generalization, the atomic nuclei remain stationary during the reaction while the electrons move. This idea is important, since it is the electrons that act as the 'glue' between the nuclei. Such movement occurs in such a way that the bonds between the atoms are different in the product than in the reactant.

The *Franck–Condon* principle states that atomic nuclei are stationary during a reaction, with only electrons moving – see p. 451.

This simple yet profound notion, that atomic nuclei are stationary during the reaction and that only electrons have time to move, is called the *Franck–Condon principle*. We shall see its important consequences later, in Chapter 9.

We now move on slightly, conceptually. Consider a single pair of reactant molecules combining to form a product. As electrons rearrange as the reaction commences, we pass smoothly from a structure that is purely reactant to one of pure product. The *transition* from one to the other is seamless; see Figure 8.23.

There will soon come a point where some bonds are almost broken and others almost formed. We have neither reactant nor product: it is a hybrid, being a mixture of both reactant and product. It is extremely unstable, and hence of extremely high energy (i.e. with respect to initial reactants or the eventual products). We call it the *transition-state complex*, and often give it the initials TS. To a first approximation, the character of the complex is predominantly reactant before the TS is formed, and predominantly product afterwards.

The transition-state complex TS is only ever formed in minute concentrations and for a mere fraction of a second, e.g. 10^{-12} s, so we do not expect to 'see' it except by the most sophisticated of spectroscopic techniques, such as laser flash photolysis.

Forming the TS is like pushing a marble over a large termite hill: most of the marbles cannot ascend the slope and, however high they rise up the hill's slope, they do not ascend as far as the summit. Those rare marbles that do reach the summit appear to stay immobile for a mere moment in time, and are then propelled by their own momentum (and their own instability, in terms of potential energy) down over the termite hill and onto the other side. A chemical reaction is energetically similar: the reaction commences when molecules of reactant collide with 'sufficient energy'. If sufficient energy *is* available, then the two or more reactants join to form the transition-state complex TS, i.e. the electrons rearrange with the net results that, in effect, atoms or groups of atoms move their positions (the bonds change).

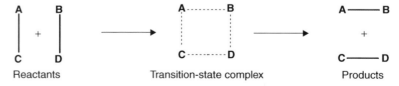

Figure 8.23 During a reaction, the participating species approach, collide and then interact. A seamless transition exists between pure reactants and pure products. The rearrangement of electrons requires large amounts of energy, which is lost as product forms. The highest energy on the activation energy graph corresponds to the formation of the transition-state complex. The relative magnitudes of the bond orders are indicated by the heaviness of the lines

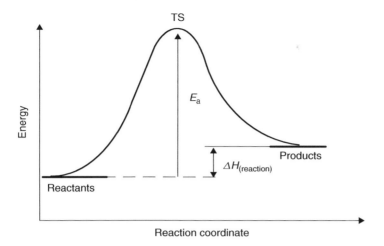

Figure 8.24   Reaction profile of energy (as 'y') against reaction coordinate (as 'x'). The activation energy E_a is obtained as the vertical difference between the reactants and the peak of the graph, at 'TS'

Figure 8.24 shows a graph of energy as a function of reaction progress. The transition complex is formed at the energy maximum. The figure will remind us of Figures 3.1 and 3.2, except with the peak on top. It is similar. The enthalpy of reaction is obtained as the vertical difference between 'REACTANTS' and 'PRODUCTS' on the graph.

The 'sufficient energy' we mentioned as needed to form the transition-state complex is termed the *activation energy*, which is given the symbol E_a (with the E denoting 'energy' and the subscripted 'a' for 'activation'). The word 'activation' ought to suggest additional energy is required; in fact, the activation energy is always positive because the TS is always higher in energy than the reactants. Stated another way, its formation is *always* endothermic.

> The activation energy E_a is always *positive*, so the formation of a transition-state complex is always *endothermic*.

Why does a reaction speed up at higher temperature?

The Arrhenius equation

In Chapters 1 and 2, we met the idea that the simplest way to increase the energy of a chemical, material or body is to raise its temperature. So heating a reaction mixture gives more energy to its molecules. Although only a tiny proportion of these molecules will ever have sufficient energy to collide successfully and form an activated complex TS, even a small increase in the amounts of energy possessed by a molecule will increase the *proportion* that

> The simplest way to increase the energy of a chemical, material or body is to raise its temperature – see p. 34.

Heating a reaction mixture increases the number of *successful* collisions between the reactant species, increasing the amount of product formed per unit time.

have sufficient energy to form the TS. Therefore, heating a reaction mixture, e.g. by reflux, increases the number of *successful* collisions between reactant species, increasing the amount of product formed per unit time.

By increasing the temperature T, we have not changed the magnitude of the activation energy, nor have we changed the value of ΔH of reaction. The increased rate is a kinetic result: we have enhanced the number of successful reaction collisions per unit time.

The simplest relationship between temperature T and rate constant k is given by the *Arrhenius* equation (Equation (8.54)), which relates the rate constant of reaction k with the thermodynamic temperature T at which the reaction is performed:

$$k = A \exp\left(-\frac{E_a}{RT}\right) \tag{8.54}$$

The Arrhenius equation is written in terms of *thermodynamic* temperature.

where R is the gas constant and E_a is the *activation energy* (above), which is a constant for any particular reaction. T is the thermodynamic temperature (in kelvin), and A is called usually called the Arrhenius 'pre-exponential' factor. The value of E_a depends on the reaction being studied.

The logarithmic form of Equation (8.54)

$$\ln k = \ln A - \frac{E_a}{RT}$$

The activation energy is obtained as '$-1 \times R \times$ gradient', where 'gradient' refers to the slope of the Arrhenius plot.

reminds us of the equation of a straight line, '$y = c + mx$', so a plot of ln(rate constant) (as 'y') against $1/T$ (as 'x') will yield a straight-line graph of gradient '$-E_a \div R$'. In practice, we repeatedly perform the experiment to determine a value of its rate constant k, each determined at a single value of T.

We should not attempt to memorize the Arrhenius equation until we can 'read' it, and have satisfied ourselves that it is reasonable. Firstly, we note that R, E_a and the 'constant' term will not vary. We are, therefore, looking at the effect of T on k. Next, we see that the first term on the right-hand side of the logarithmic form of Equation (8.54) decreases as the temperature increases and so the logarithmic term on the left must also decrease. However, since there is a minus sign on the right-hand side of the equation, we are saying that as T increases, so the right-hand side becomes less negative (more positive). In other words, as the temperature increases, so the logarithm of the rate constant also increases, and hence k gets larger.

Worked Example 8.20   Consider the following data that relate to the rate of removing a naturally occurring protein with bleach on a kitchen surface. What is the activation

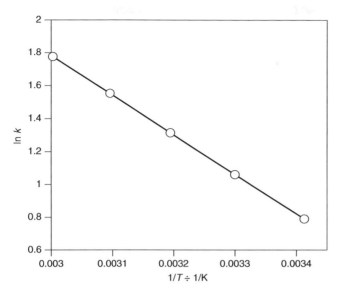

Figure 8.25   An Arrhenius plot of ln(rate constant) (as 'y') against $1/T$ (as 'x'). The data relate to the rate of removing a naturally occurring protein with bleach on a kitchen surface

energy of reaction?

Temperature $T/^\circ C$	20	30	40	50	60
Rate constant k/s^{-1}	2.20	2.89	3.72	4.72	5.91

Answer Strategy: before we can plot anything, we first convert the temperatures from Celsius to thermodynamic temperatures in kelvin. Then we plot $\ln(k/s^{-1})$ (as 'y') against $1/T$ (as 'x').

Such a plot is seen in Figure 8.25. Its gradient is equal to $-E_a/R$. This graph is seen to be linear, with a gradient of -2400 K. From the Arrhenius equation, the value of activation energy is obtained as '$-$gradient $\times R$'. Therefore:

> We take the logarithm of $k \div s^{-1}$ because it is mathematically impossible to take the logarithm of anything expect a *number*.

$$E_a = -2400 \text{ K} \times 8.314 \text{ J K}^{-1} \text{mol}^{-1} = 20\,000 \text{ J mol}^{-1}$$

The activation energy is 20 kJ mol^{-1}.

Aside

Activation in biological systems

Many biological systems show an activated behaviour, since they appear to follow the Arrhenius equation (Equation (8.45)). In fact, *all* biological processes are activated, but the complicated array of biological rates means that we only see the rate of the slowest, rate-limiting step. Nevertheless, look at the following examples:

(1)   The rate of development of the water flea (*Alona affinis*) from egg to adult is activated, since a plot of ln(development rate) (as '*y*') against $1/T$ (as '*x*') is linear.

(2)   The walking speed of an ant (*Liometopum apiculetum*) follows an Arrhenius plot over a limited temperature range. Over a wider temperature range, the plot is curved, indicating the involvement of different, finely balanced processes, each being rate limiting over a different temperature range.

(3)   Figure 8.26 shows a plot of ln(heart beat) (as '*y*') against $1/T$ (as '*x*') for the diamond-backed terrapin (*Malaclemys macrospilota*).

(4)   The Mediterranean cicadas (*Homoptera, Auchenorrhyncha*) chirp at a frequency so closely related to the Arrhenius equation that we can 'hear' the temperature by measuring its rate.

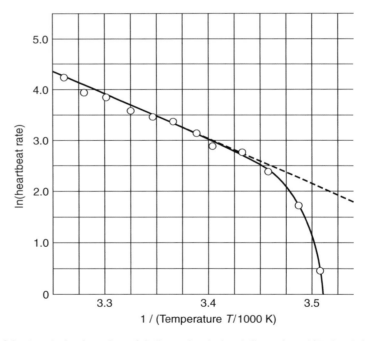

Figure 8.26 An Arrhenius plot of ln(heart beat) (as '*y*') against $1/T$ (as '*x*') for the diamond-backed terrapin (*Malaclemys macrospilota*) is linear over a limited range of temperatures, showing that the rate-limiting process dictating its heart rate is activated. (Figure reproduced from *Chemical Kinetics and Mechanism*, Michael Mortimer and Peter Taylor (eds), Royal Society of Chemistry, Cambridge, 2002, p. 77. Reproduced with permission)

Why does the body become hotter when ill, and get 'a temperature'?

The thermodynamics of competing reactions

One of the worst aspects of being ill is the way the body develops a 'temperature'. A healthy human should have a more-or-less constant body temperature of 37 °C (which, in the old-fashioned temperature scale of Fahrenheit, is 98.6 °F). A body temperature of above 100 °F is best avoided, and temperatures above about 104 °F are often lethal. As a temperature is so harmful to the body, why does the body itself generate the extra heat?

All reactions proceed via a transition-state complex, and with an activation energy E_a. The values of E_a vary tremendously, from effectively zero (for a so-called *diffusion-controlled* reaction, as below) to several hundreds of kilojoules per mole (for reactions that do not proceed at all at room temperature). The rate constant of a reaction is relatively insensitive to temperature if E_a is small.

> If E_a is small, then the rate constant of reaction is relatively insensitive to temperature; if E_a is large, then k is more sensitive to temperature fluctuations.

An alternative form of the Arrhenius equation is the *integrated form*:

$$\ln\left(\frac{k \text{ at } T_2}{k \text{ at } T_1}\right) = -\frac{E_a}{R}\left(\frac{1}{T_2} - \frac{1}{T_1}\right) \qquad (8.55)$$

which is similar to Equation (8.54), but applies to two temperatures and two rate constants. If E_a is small, then the left-hand side of the equation will have to be small, implying that the ratio of rate constants must be close to unity. Conversely, a large value of E_a causes the rate constants k to be more sensitive to changes in temperature.

> Equation (8.55) comes ultimately from the Maxwell–Boltzmann distribution in Equation (1.16).

Antibodies are naturally occurring substances in the blood that fight an infection or illness. Like all reactions, the rate at which antibodies fight an illness has an activation energy, which is quite high. Whenever the illness appears to be 'winning' the battle, the body raises its temperature to increase the rate at which antibodies react. A higher temperature means that the body's antibodies operate faster, i.e. with an increased rate constant. A body, therefore, has 'a temperature' in order to fight an infection more rapidly.

SAQ 8.24 A person is ill. The rate at which their antibodies react needs to increase twofold. This rate increase is achieved by raising the body's temperature from 37 °C to 40 °C. What is the activation energy of the reaction? [Hint: first convert from Celsius to kelvin.]

Aside

Another reason for the body's temperature rising during illness is an increase in the rate of metabolism of white blood cells. White blood cells are an essential part of the body's defence mechanism, and 'attack' any foreign bodies or pathogens in the body, such as bacteria or viruses. Pathogens are engulfed or consumed by white blood cells when we are ill by a process known as *phagocytosis*.

Why are the rates of some reactions insensitive to temperature?

Diffusion-controlled reactions

Some rare reactions occur at a rate that appears to be insensitive to temperature. Such reactions are extremely rapid, and are termed *diffusion-controlled* reactions.

> The value of E_a is effectively zero for *diffusion-controlled* reactions.

If the activation energy is extremely small – of the order of 1 kJ mol^{-1} or so – then all the energy necessary to overcome the activation energy is available from the solvent, etc., so reaction occurs 'immediately' the reactants collide.

In fact, the only kinetic limitation to such a reaction is the speed at which they move through solution before the collision that forms product. This rate is itself dictated by the speed of *diffusion* (which is not generally an efficient form of transport). The rate of reactants colliding is, therefore, said to be 'diffusion controlled'. Typically, diffusion-controlled processes in which E_a is tiny involve radical intermediates.

If a second-order reaction is diffusion controlled, then its rate constant has a magnitude of about 10^{10} or even $10^{11} \text{ dm}^3 \text{ mol}^{-1} \text{ s}^{-1}$.

SAQ 8.25 The rate of a reaction is said to be 'diffusion controlled' because its activation energy is 1.4 kJ mol^{-1}. The rate constant of reaction is 4.00×10^{10} dm^3 mol^{-1}s^{-1} at 298 K. Show that the rate constant is effectively the same at 330 K.

The Eyring approach to kinetic theory

The Arrhenius theory (above) was wholly empirical in terms of it derivation. A more rigorous, but related, form of the theory is that of Eyring (also called the theory of *absolute reaction rates*). The *Eyring equation* is

$$\ln \left(\frac{k}{T} \right) = \frac{-\Delta H^{\ddagger}}{RT} + \frac{\Delta S^{\ddagger}}{R} + \ln \left(\frac{k_B}{h} \right) \tag{8.56}$$

where k_B is the Boltzmann constant and h is the Planck constant. $\Delta H^{\ddagger}$ is the enthalpy change associated with forming the activated complex and $\Delta S^{\ddagger}$ is the change in entropy.

A plot of $\ln(k \div T)$ (as 'y') against $1/T$ (as 'x') should be linear, of gradient $-\Delta H^{\ddagger} \div R$ and intercept '$\Delta S^{\ddagger} \div R + \ln(k_B \div h)$'.

Worked Example 8.21   The rate of hydrolysis for a biological molecule was studied over the temperature range 300–500 K, and the rates found to be as follows. Use a suitable graphical method to determine $\Delta H^{\ddagger}$, $\Delta S^{\ddagger}$ and $\Delta G^{\ddagger}$, for the reaction at 298 K.

T/K	300	350	400	450	500
$k/\mathrm{dm^3 mol^{-1} s^{-1}}$	7.9×10^6	3.0×10^7	7.9×10^7	1.7×10^8	3.2×10^8

Figure 8.27 shows a graph of $\ln(k \div T)$ (as 'y') against '$1 \div T$' (as 'x'). Its gradient of $-2386 \mathrm{~K^{-1}}$ is equal to '$-\Delta H^{\ddagger} \div R$', so $\Delta H^{\ddagger} = 2386 \mathrm{~K^{-1}} \times 8.314 \mathrm{~J~K^{-1}~mol^{-1}} = 19.8 \mathrm{~kJ~mol^{-1}}$.

The intercept on Figure 8.27 is 18.15, and is equal to '$\Delta S^{\ddagger}/R + \ln(k_B/h)$'. The logarithmic term $\ln(k_B/h)$ has a value of 23.76, so

$$\frac{\Delta S^{\ddagger}}{R} = 18.15 - 23.76 = -5.61$$

so

$$\Delta S^{\ddagger} = -5.61 \times 8.314 \mathrm{~J~K^{-1}~mol^{-1}} = 46.6 \mathrm{~J~K^{-1}~mol^{-1}}$$

Finally, the value of $\Delta G^{\ddagger}$ is obtained via the equation $\Delta G^{\ddagger} = \Delta H^{\ddagger} - T\Delta S^{\ddagger}$ (cf. Equation (4.21)). Inserting values and saying

> The value of $\Delta G^{\ddagger}$ is positive, implying that the equilibrium constant of forming the transition-state complex is minuscule, as expected.

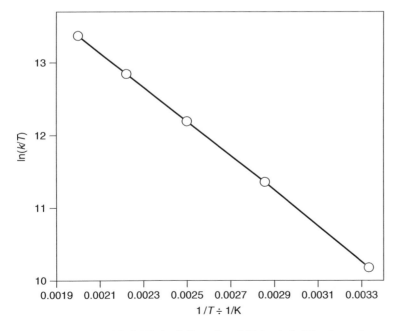

Figure 8.27   An Eyring plot of $\ln(k/T)$ (as 'y') against $1/T$ (as 'x'). The data relate to the rate of hydrolysing a biological molecule in the temperature range 300–500 K

$T = 298$ K:

$$\Delta G^{\ddagger} = 19\,800 \text{ J mol}^{-1} - (298 \text{ K} \times -46.6 \text{ J K}^{-1} \text{ mol}^{-1})$$

$$\Delta G^{\ddagger} = 5.9 \text{ kJ mol}^{-1}$$

Aside

Comparing the Arrhenius and Eyring equations

It is extremely common, but *wrong*, to see the Eyring equation written in a similar form, with the Arrhenius ordinate of $\ln(k)$ (as 'y') and the final logarithmic term written as $\ln(k_B T \div h)$. Although such an equation might be readily achieved from the brief derivation in Justification Box 8.6, it is seen straightaway to be nonsensical, for the following three reasons:

(1) If the intercept contains a temperature term T, it is nonsense to have an intercept that contains one of the equation's principal variables.

(2) As an intercept, $1 \div T = 0$, so the only sensible temperature to include as T in the intercept term would be $T = \infty$, which means that $\Delta S^{\ddagger} = -\infty$. Again, this is not realistic.

(3) More importantly, however, is a physicochemical concept behind the equations: if both equations are written as $\ln(k)$ (as 'y') as a function of $1 \div T$ (as 'x'), then it is dishonest to suppose that the gradients of the respective graphs can be different, one a function of E_a and the other a function of $\Delta H^{\ddagger}$.

As a further implication, $\Delta H^{\ddagger}$ cannot be the same as E_a. In fact, from the Eyring theory, we can show readily that

> Eyring theory says that $\Delta H^{\ddagger} = E_a + RT$, explaining why values of $\Delta H^{\ddagger}$ are not constant, but depend on temperature.

$$\Delta H^{\ddagger} = E_a + RT \tag{8.57}$$

This equation explains why values of $\Delta H^{\ddagger}$ are not constant, but depend on temperature. Conversely, E_a is a true constant.

We will employ the form of the Eyring equation written as Equation (8.56).

SAQ 8.26 The following table contains the rate constant k for the demethylation reaction of *N*-methyl pyridinium bromide by aqueous sodium hydroxide as a function of temperature:

T/K	298	313	333	353
$k/10^2$ dm^3 mol^{-1} s^{-1}	8.39	21.0	77.2	238

(1)   Calculate the activation energy E_a and pre-exponential factor A by plotting an Arrhenius graph.

(2)   Calculate $\Delta G^{\ddagger}$, $\Delta H^{\ddagger}$ and $\Delta S^{\ddagger}$ for the reaction at 298 K by plotting an Eyring graph.

(3)   What is the relationship between $\Delta H^{\ddagger}$ and E_a?

Justification Box 8.6

From Eyring, the rate constant of reaction k depends on a *pseudo* equilibrium constant $K^{\ddagger}$, relating to the formation of a transition-state complex, TS. Clearly, $K^{\ddagger}$ will always be virtually infinitesimal.

The values of k and $K^{\ddagger}$ are related as

$$k = \frac{k_B T}{h} K^{\ddagger} \qquad (8.58)$$

where T is the thermodynamic temperature. From an equation which will remind us of the van't Hoff isotherm (cf. Equation (4.59)), we relate the equilibrium constant with the change in Gibbs function:

> *Care*: there are three different types of 'k' in Equation (8.58), so we must be careful about the choice of big or small characters, and subscripts.

$$\Delta G^{\ddagger} = -RT \ln K^{\ddagger} \qquad (8.59)$$

where $\Delta G^{\ddagger}$ here is specifically the change in Gibbs function associated with forming the transition-state complex. After rearrangement, we obtain

$$K^{\ddagger} = \exp\left(\frac{-\Delta G^{\ddagger}}{RT}\right) \qquad (8.60)$$

and by substituting for $K^{\ddagger}$ we obtain

$$k = \frac{k_B T}{h} \exp\left(\frac{-\Delta G^{\ddagger}}{RT}\right) \qquad (8.61)$$

Next, we recall, from the second law of thermodynamics, that $\Delta G^{\ominus} = \Delta H^{\ominus} - T\Delta S^{\ominus}$. By direct analogy, $\Delta G^{\ddagger} = \Delta H^{\ddagger} - T\Delta S^{\ddagger}$, where $\Delta H^{\ddagger}$ is the enthalpy of forming the transition-state complex (akin to the activation energy E_a). $\Delta S^{\ddagger}$ is the entropy of forming the transition-state complex.

By inspection alone, we can guess that $\Delta S^{\ddagger}$ will be negative for bimolecular reactions, since two components associate to form one (the TS). The value of $\Delta S^{\ddagger}$ is positive for unimolecular processes, such as gas-phase dissociation.

Substitution for $\Delta G^{\ddagger}$ into Equation (8.61) gives

$$k = \frac{k_B T}{h} \exp\left(\frac{-\Delta H^{\ddagger}}{RT}\right) \exp\left(\frac{\Delta S^{\ddagger}}{R}\right) \qquad (8.62)$$

The Arrhenius pre-exponential factor A is $\dfrac{k_B T}{h} \exp\left(\dfrac{\Delta S^{\ddagger}}{R}\right)$.

where, by dividing both sides by T, we obtain

$$\frac{k}{T} = \frac{k_B}{h} \exp\left(\frac{-\Delta H^{\ddagger}}{RT}\right) \exp\left(\frac{\Delta S^{\ddagger}}{R}\right) \qquad (8.63)$$

Taking logarithms of both sides yields the *Eyring equation*, Equation (8.56):

$$\ln\left(\frac{k}{T}\right) = \frac{-\Delta H^{\ddagger}}{RT} + \frac{\Delta S^{\ddagger}}{R} + \ln\left(\frac{k_B}{h}\right)$$

This equation can be thought of as a more quantitative form of the Arrhenius equation.

What are catalytic converters?

Catalysis

A *catalytic converter* is a part of a car exhaust. It is sometimes called a 'CAT' or, worse, given the wholly non-descriptive abbreviation 'CC'.

It has been appreciated for many years how the exhaust gases from a car are hazardous to the health of most mammals. The most harmful gases are carbon monoxide (CO) and low-valence of oxides of sulphur (SO_2) and nitrogen (NO_x). CO is doubly undesirable in being toxic (it causes asphyxiation) and being a greenhouse gas. SO_2 or NO_x are not particularly toxic – in fact, minute amounts of NO_x may even be beneficial to health. But both gases hydrolyse in water to form acid rain, with attendant environmental damage. The source of the sulphur and nitrogen, before combustion, are natural compounds in the petrol or diesel fuels.

A catalytic converter is a stainless steel tube located near the exhaust manifold, lined with finely divided metal salts, e.g. of platinum and palladium.

The purpose of the catalytic converter is to oxidize some of the oxides in the gas phase. NO_x is reduced to elemental nitrogen. The principal reaction at the catalytic converter is oxidation of CO:

$$CO_{(g)} + \tfrac{1}{2}O_{2(g)} \longrightarrow CO_{2(g)} \qquad (8.64)$$

This reaction is energetically spontaneous, but it occurs quite slowly if the gases are just mixed because the activation energy to reaction is too high at 80 kJ mol^{-1}. The reaction is much faster if the CO is burnt, but a naked flame is considered unsafe in a car exhaust.

Although the reaction in Equation (8.64) is slow under normal conditions, it can be accelerated if performed with a *catalyst*; hence the incorporation of a catalytic converter in the car exhaust system.

Catalysis theory

All reactions occur by molecules surmounting an energetic activation barrier, as described above.

Consider the diagram of reaction profile in Figure 8.28: the reaction is clearly endothermic, since the energy of the final state is higher than the energy of the initial state i.e. ΔH is positive.

We recall that enthalpy H is a *state function* (see Section 3.1), so the overall enthalpy change of the reaction is independent of the chemical route taken in going from start to finish. It is clear from Figure 8.28 that the initial and final energies, of the reactants and products respectively, are wholly unaffected by the presence or otherwise of a catalyst: we deduce that a catalyst changes the *mechanism* of a reaction but does not change the *enthalpy change* of reaction.

It is quite common to see reaction profiles that are more complicated than that depicted in Figure 8.28, e.g. with two or three separate maxima corresponding to two or three activation energies. The lower reaction profile in Figure 8.29 has two 'peaks' and is included to show how the second activation energy is obtained. The second activation energy is usually smaller than the first. (The words 'first' and 'second' in this context are meant to imply the sequence during the course of a concurrent reaction from reactant to product).

> 'Catalysis' comes from the Greek *cata*, meaning 'away from' or 'alongside' and *lusis*, meaning 'dissolution' or 'cleavage'; so a catalyst promotes bond cleavage *away from* the usual mechanistic route.

> The principal advantage of catalysis is to make a reaction proceed more quickly because the activation energy decreases.

> A catalyst changes the *mechanism* of a reaction, but not its *enthalpy change*.

> The energies of the reactants and products remain unchanged following catalysis, so ΔH_r is unaffected.

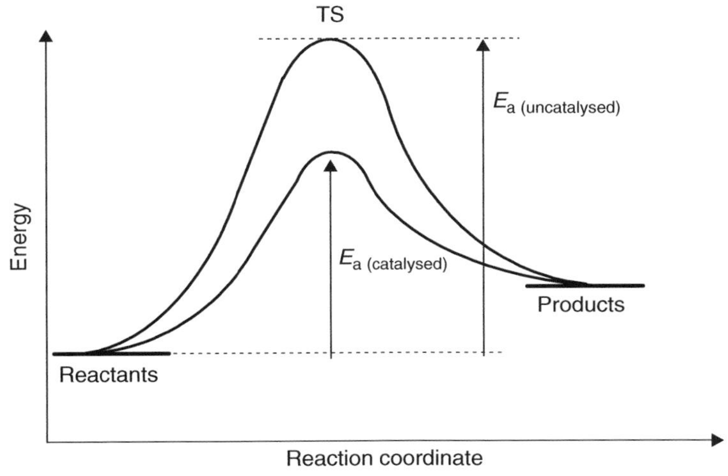

Figure 8.28 Reaction profiles for an uncatalysed reaction (upper curve) and catalysed (lower curve). Note that the reaction energies start and finish at the same energies, so the magnitude of ΔH_r is not affected by the catalyst

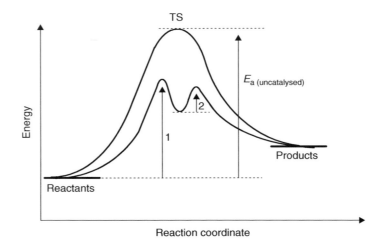

Figure 8.29 Reaction profiles for an uncatalysed reaction (upper single curve) and catalysed (curve with a double maximum). The two solid arrows represents E_a for the catalysed reactions: note how '1' (E_a for the first reaction step) is so much larger than the activation energy for the second step, '2'

Though a catalyst alters the *mechanism* of a reaction, the reaction's *position* of equilibrium is unaltered, meaning its chemical yield remains the same.

And we see a further point: the equilibrium constant K of a reaction is a direct function of ΔG_r according to the van't Hoff isotherm (Equation (4.55)). If the overall energy of reaction remains unaltered by the catalyst, then the *position* of equilibrium will also remain unaltered.

9

Physical chemistry involving light: spectroscopy and photochemistry

Introduction

We look in this chapter at the interactions of light with matter. All photochemical processes occur following the absorption of a particle of light called a photon, causing excitation from the ground state to one of several higher energy states. The physico-chemical properties of the molecule in the excited material will often differ markedly from that of the ground state.

Spectroscopic techniques look at the way photons of light are absorbed quantum mechanically. X-ray photons excite inner-shell electrons, ultra-violet and visible-light photons excite outer-shell (valence) electrons. Infrared photons are less energetic, and induce bond vibrations. Microwaves are less energetic still, and induce molecular rotation. Spectroscopic selection rules are analysed from within the context of optical transitions, including charge-transfer interactions

The absorbed photon may be subsequently emitted through one of several different pathways, such as fluorescence or phosphorescence. Other photon emission processes, such as incandescence, are also discussed.

Finally, we look at other spectroscopic phenomena, such as light scattering.

9.1 Introduction to photochemistry

Why is ink coloured?

Chromophores

We know that we have filled our pens with blue ink because the colour *looks* blue. Black ink is clearly black because it *looks* black. We know a colour because we see it.

The word 'chromophore' comes from two Greek words: *khroma*, meaning colour, and *phoro*, which means to give or to impart. A chromophore is therefore a species that imparts colour.

After a further moment's thought, we realize that the only reason why we see a colour at all is because there is *something* in the solution that gives it its colour. There is an innate redness to chemicals dissolved in red ink, which interact with the light, and allow us to recognize its colour. There are different chemicals in blue ink, and altogether different chemicals in green ink. Each of these chemicals interacts in a characteristic way, which is why we see different colours.

A chemical that imparts a colour is called a *chromophore*. We could see no colour in a solution without a chromophore.

Why do neon streetlights glow?

Emission and absorption

Neon lights are commonly seen on advertising hoardings, outside shops, restaurants, and cinemas. Their colour is so distinctive that to most people the word 'neon' has come to mean a dark pink glow, although neon is merely an elemental rare gas (element number 10).

Neon is wholly colourless if placed in a normal flask, yet the glow of a neon light is visible even at night, so its colour cannot be due to the way it interacts with light – there is no ambient light to interact with. Neon lights glow because they *emit* light under suitable conditions; see p. 480.

The principal difference between neon lights and the colours seen in daylight is that a neon light *emits* light, whereas wine, paint and chromophores, in general, *absorb* light. The difference between absorbance and emission is illustrated schematically in Figure 9.1.

The word 'light' here does not just mean visible light, but all wavelengths.

This difference between absorption and emission is crucial, and underlies many of pivotal aspects of photochemistry. We will look in depth at the underlying *photochemistry* of neon in Section 9.3.

At its simplest, photochemical absorbances involve the uptake of photons, so the number of photons leaving a sample will be less

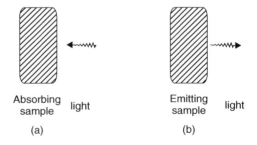

Absorbing sample   light   Emitting sample   light

(a)   (b)

Figure 9.1 The difference between absorption and emission: (a) light enters a sample during absorption and (b) leaves a sample following emission

than the number entering it. Some photons are consumed. The remainder are absorbed. By contrast, a photochemical emission occurs when there are more photons leaving then entering it, i.e. more after the interaction than before. It is the phenomenon of absorbance that we study in most forms of spectroscopy: a sample

> The light striking the sample is called the *incident* light.

is illuminated with photons of various energies, and a detector of some sort analyses the decrease in the number of photons. Usually, the detector is placed behind the sample and the light source is in front of it. The light striking the sample is called the *incident* light.

In studies involving emission, it is usual for the sample to be irradiated while the number of photons emitted is measured as a function of the incident light. To prevent the incident and emitted

> To *irradiate* means to shine light upon.

beams getting confused, the emitted light is generally analysed with the detector placed at right angles to the light source. The topic of emission underlies fluorescence, luminescence and phosphorescence, which are discussed in Section 9.3.

Why do we get hot when lying in the sun?

Introduction to photochemistry

People are invisible to radio waves. In other words, there should be no effect when standing between a radio-wave transmitter and its receiver. Conversely, humans are not invisible to light in the wavelength range 100–1000 nm – we say they are *opaque*, which is most easily proved by asking a person to stand in the path of a light source such as a torch, and seeing the shadow cast.

> *Opaque* is the opposite extreme to *invisible*.

We see such a shadow because the light not blocked by a person travels from the torch and impinges on the ground behind them, where it interacts, i.e. looks brighter. Light cannot pass through the person, who is opaque, so no light is available to interact with those portions of the ground behind them.

> The crackle we sometimes hear when walking round a radio set is because of metals and other conductors in our clothing; human flesh is 'invisible' to radio waves.

A more subtle way of showing that we are opaque (wholly or partially) to some wavelengths of light is to stand in front of an oven, or in the sun. If we were invisible, the energy from the heat source would pass straight through us without our noticing, just as we do not 'feel' anything when bathed in radio waves. But we do get hot when lying in the sun because an interaction occurs between the infrared radiation (the heat) and us, as we are exposed. We say this infrared light is *absorbed*, and its energy is transferred in some way to us; as we saw in Chapter 2, the simplest way to tell that something has gained energy is that its temperature goes up.

> The science of the way light interacts with a species, such as a chromophore in red wine or a body in the sun, is called *photochemistry*.

The *first law of photochemistry* states that only light that is absorbed can have any photochemical effect.

This principle is so simple that it has been given the title the *first law of photochemistry*, and was first expressed by Grotthus and Draper in the early 19th century. They stated it as the (hopefully) obvious truth: 'Only light that is absorbed can have any photochemical effect'.

The primary interaction occurring during a photochemical process is between light and an analyte, such as a molecule, ion, atom, etc. The reaction is often written in the generalized form

$$M + light \longrightarrow M^* \tag{9.1}$$

The light contains energy, so Equation (9.1) alerts us to the truth that the body M has *acquired* energy during the photochemical reaction. We sometimes say that the product is a 'hot' atom, ion, etc. for this reason. This energy cannot just be 'tacked on' to the body M, but is stored within it in some way, so the asterisk against the product shows that it is formed in an *excited state*.

These simple concepts underlie the whole of photochemistry.

Why is red wine so red?

Photon absorption: the excited state

The colour of a chromophore is seen only after light passes *through* it, or through a solution containing it.

The first thing we see when pouring a glass of red wine is its beautiful colour. It seems almost to *glow*. The colour is more impressive still if we hold the glass up to the light and see the way light streams through the wine it contains. Yet we could see no colour if the room was dark: the wine does not emit light. Indeed, we only see the colour after the light has passed though the solution. We see the colour following *transmission* of the light.

We see a colour because a chromophore interacts with light.

The colours we see around us are each a consequence of light. Particles of light – we call them *photons* – enter the wine, pass through it, and are detected by the eye only after transmission. We deduce that the action of the chromophore in imparting its colour is due to the light interacting with the chromophore during the transmission of light through the wine.

A particle of electromagnetic radiation (light) is called a *photon*.

The fundamental feature of photochemistry that separates it from other branches of chemistry is its emphasis on an excited state. A coloured chromophore, be it a molecule ion or atom in solution, has electrons within the bonds which characterize it. These electrons usually reside in the orbitals of lowest energy. We say the molecule is in its *ground state* if it is in its lowest, unexcited state.

The only orbitals we are interested in are the *highest occupied molecular orbital* (HOMO) and the *lowest unoccupied molecular orbital* (LUMO). Collectively, they are often referred to as the *frontier orbitals*. We can usually ignore all the other orbitals.

We have already seen that photons of light interact with the chromophore in solution. In fact, the energy locked up within the photon is wholly taken up into an electron within the chromophore. In its place, there exists an electron of higher energy. We say the photon is *absorbed* by the chromophore, and its energy enables the *excitation* of an electron from a ground-state orbital to an orbital of higher energy, i.e. from the HOMO to the LUMO, as represented in the schematic drawing in Figure 9.2.

We say the molecule is in its first excited state. It is possible to excite into the second excited state, which is of an even higher energy.

> The term *ground state* is easy to understand and then remember: a ball thrown into the air will descend, and come to rest at a position of lowest potential energy. Inevitably, this usual position of lowest energy is resting on the ground.

Why are some paints red, some blue and others black?

Absorption at different wavelengths: spectra

The light we see by eye is termed *visible* light. Normal, everyday light is also termed 'white' light, and is a mixture of different colours. We can readily demonstrate this composite nature of light by 'splitting' it as it passes through a prism to generate a *spectrum*. If white light (e.g. from the sun) passes through a bottle of blue ink, then some of the photons are absorbed by the chromophore that makes it appear blue, so the light transmitted (i.e. the remainder that is not absorbed) only contains the colours that are not absorbed.

> Some light is *absorbed*, and the remainder is either *transmitted* or *scattered*; see later.

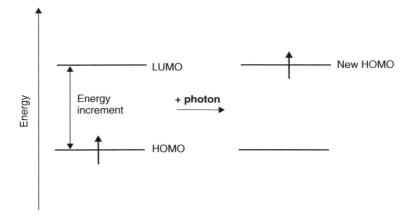

Figure 9.2 Schematic representation of a photon being absorbed by a single molecule of chromophore. The photon causes excitation of an electron (depicted by a vertical arrow) from the HOMO to the LUMO

> 'Spectrum' is a Latin word, so its plural is *spectra*, never 'spectrums'.

We should be aware from elementary physics that there exists a continuum of photon energies, with radio waves at one extreme, which have a tiny energy per photon, through to gamma rays with a massive energy per photon. We term this continuum of energies a *spectrum*. Figure 9.3 is a reminder of the different photon energies expressed as wavelengths in a spectrum. Because a particle can also behave as a wave (see p. 431 below), Figure 9.3 also indicates the wavelength λ of each photon.

> Whatever its acquired meaning, to a scientist the word *spectrum* means 'a range of photon energies'.

The word 'spectrum' is generally employed in a clumsy way in everyday life – we might talk about a 'spectrum of opinions', meaning a range or spread. To a scientist, the word means a spread of photon energy. As an excellent example, consider the colours formed when white light passes through a prism: the range of colours indicates a range of photon energies from 171 kJ mol^{-1} (red light at 700 nm) to 342 kJ mol^{-1} (violet light of λ = 350 nm). The frequencies on a radio-tuning dial are another example.

The word 'spectrum' also means a graph depicting photon intensity (as 'y') against photon energy (as 'x'). The discipline of obtaining a spectrum is termed *spectroscopy*, and will be discussed in more detail in Sections 9.2 and 9.3.

> Graphs such as Figure 9.4 are often called *electronic spectra* because the optical absorption is caused by the excitation of *electrons* between various molecular orbitals.

Figure 9.4 shows a spectrum of methyl viologen in water. The x-axis of the spectrum covers the wavelength range 220–700 nm. The range 190–350 nm is the *ultraviolet* (UV) region (see Figure 9.3), and 350–700 nm is the region *visible* to the human eye. Accordingly, Figure 9.4 is most usually described as a UV–visible spectrum.

The function on the y-axis of the spectrum in Figure 9.4 is the *absorbance* (as defined on p. 441). Absorbance is also called *optical density* or *optical absorbance* in older books; these three terms each mean the same thing. We can see from the spectrum that more light is absorbed at 300 nm (in the near infrared) than at 500 nm

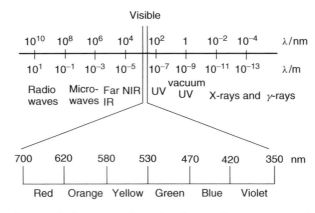

Figure 9.3 A continuum of photon energies exists from radio waves through to γ-rays. We call it a 'spectrum'. The visible region extends from 350 to 700 nm

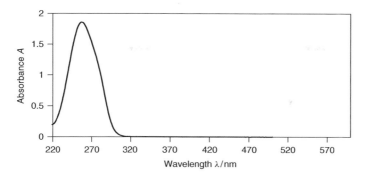

Figure 9.4 UV–visible spectrum of methyl viologen in water at a concentration of 10^{-3} mol dm^{-3}. Methyl viologen compound is the active ingredient in many weedkillers, and has the IUPAC name 1,1'-dimethyl-4,4'-bipyridilium dichloride

(in the visible region). The maximum amount of light absorbed is light of wavelength 260 nm. We often call the peak in this spectrum, a spectroscopic *band*, and the wavelength at which it appears is termed the *wavelength maximum*, which we give the symbol $\lambda_{(max)}$. The value of $\lambda_{(max)}$ in a spectrum corresponds to the very tip of the peak in the spectrum.

> We often call the peak in this spectrum, a spectroscopic *band*.

The spectrum in Figure 9.4 shows that only light within the range 200–300 nm is absorbed. It follows, therefore, that if light of these wavelengths is absorbed, then such light cannot also be available to be 'seen', i.e. to enter the eye. In fact, we only see the remainder of the light – that which is *not* absorbed. We say that the colour observed by the eye is the *complementary* colour to that absorbed. For example, removal of red light makes a thing appear blue.

> The word 'spectroscopy' comes ultimately from the Greek *skopein*, which means 'to watch'. The word 'spectrum' derives from the Greek *metrein*, meaning to measure.

Worked Example 9.1   The permanganate ion absorbs green light ($\lambda_{(max)} = 500$ nm). Why does it appear to the eye to be purple?

Light that is green occurs at the centre of the visible region of the spectrum. If photons of green light are removed by absorption, then the major colours remaining will be at the two extremes of the spectrum, at red and blue. The colour we see following removal of green light will, therefore, be a mixture of blue and red, which the eye perceives as purple.

Why can't we see infrared light with our eyes?

'Types' of photon

When sitting on a beach in the sun, or standing in front of an oven, we often refer to 'waves of heat' hitting us. A thermometer would tell us that our temperature was increasing. In fact, this heat comprises photons of infrared light, which is why we

Red light has the longest wavelengths in the visible portion of the electromagnetic spectrum. Longer wavelengths are, therefore, beyond the red, which, from Latin is *infrared*.

We experience photons of infrared light as heat.

The visible light having the shortest wavelength is violet in colour. Light having an even shorter wavelength is therefore 'more than' or 'beyond' violet which, from Latin, is *ultraviolet*.

sometimes see a heater advertised as an 'infrared lamp'. Although we cannot see the heat with our eyes, nevertheless the photons of infrared radiation strike our bodies and cause us to feel a sensation that we call 'heat'.

The reason why we can see the sun when we look up at the sky on a clear day is because it emits photons of visible light. Furthermore, photons of UV light may be 'seen' by the way they cause our skin to acquire a darker colour following irradiation, because a tan forms when UV light reacts with *melanin* in the skin (see further discussion, later). Again, we notice photons of radiowaves because the radio beside us on the beach plays music and relays the latest news to us.

Infrared, visible, UV and radio-wave forms of light each comprise photons, yet of different types, which explains why each is experienced in a different way. Table 9.1 lists the various types of photon, together with typical wavelengths and applications.

We 'experience' heat, visible light, UV and radio waves by the way they interact with our thermometers, our eyes, skin and our radio sets respectively. This is a tremendously important concept. Photons of infrared light are experienced as heat. The photons that cause photochemical changes in the retina at the back of the eye are termed 'visible'. These photochemical reactions in the eye generate electrical signals which the brain encodes to allow the reconstruction of the image in our mind: this is why we *see* a scene only with visible light – indeed this is why we call it 'visible'.

Just as a golf ball will go down a golf hole and a football cannot, and cotton can go through the eye of a needle yet string cannot, so each mode of 'experiencing' these types of light depends on the

Table 9.1 Summary of the different types of photon, and ways in which each is experienced in everyday life

Name	Source	Typical λ/m	Application
γ-rays	Nuclear decay	$<10^{-10}$	Some forms of medicine (radiotherapy)
X-rays	High-energy electron collisions	$10^{-8}-10^{-10}$	X-rays for detecting broken limbs
Ultraviolet	The sun, mercury vapour lamps	$4 \times 10^{-7}-10^{-8}$	Forming a sun tan
Visible	The sun, incandescence	$4 \times 10^{-7}-7.5 \times 10^{-7}$	All forms and applications of vision
Infrared	The sun, combustion	$7.5 \times 10^{-7}-10^{-4}$	Warmth; remote-control handsets
Microwaves	Electron excitation	$10^{-4}-10^{-1}$	Mobile phone receivers; microwave ovens
Radio waves	Electron excitation	$10^{-1}->10^2$	Radar; radio transmission

energy of the respective process. Each type of detection has been tailored within the natural world to respond to only one type of photon.

So we cannot see infrared light even when looking at photons coming off a heat source because photons of infrared light do not interact with the chemicals at the back of the eye (see later example), unlike photons of visible light.

> The amount of energy per particle of light is fixed. We say it is *quantized*.

But we must be careful with our words: when we say these are different 'types' of photon, we do not mean that they are different in *kind*, only in *magnitude*. In kind, all photons are identical: each comprises a 'packet' of energy, which is termed a *quantum*, because the amount of energy in each photon is quantized (or 'fixed'). Whereas the amount of energy per UV photon is fixed, the amount of energy per type of photon can vary enormously.

> 'Quantum' is another Latin word, so its plural is *quanta*, never 'quantums'.

Wave–particle duality

There is a mind-blowing paradox at the heart of all discussions about light. Light is a form of energy that exhibits both wave-like and particle-like properties. In other words, a photon is simultaneously both a wave and a particle. We can never fully understand this paradox, but will merely say that extremely small particles exhibit a *duality of matter*.

Figure 9.5 shows a *wave of light*. It is a form of electromagnetic radiation, with properties that arise from the electric and magnetic waves oscillating sinusoidally at right angles to each other. The waves clearly have a 'wavelength' λ. Also, when we think of light as a wave, we must recognize that it travels (*propagates*) through space or another medium in a straight line, so it has a direction. The speed of the light is very fast: all light from γ-rays through to radio waves travels at 'the speed of light', which is usually given the symbol c. Light travelling through a vacuum has a speed $c = 3 \times 10^8 \text{ m s}^{-1}$.

Light in the visible region of the spectrum is also characterized by its colour, which is a straightforward function of its wavelength λ (see Figure 9.6), so λ for green light is 510 nm, and λ for orange light is 590 nm.

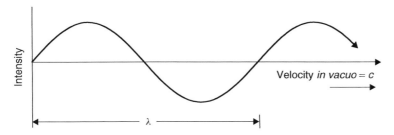

Figure 9.5 Light can be thought of as waves of frequency ν, speed c and wavelength λ. The frequency is $c \div \lambda$

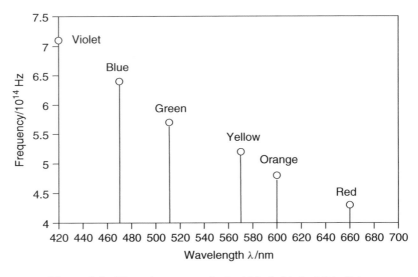

Figure 9.6   The colours comprised within 'white' visible light

Worked Example 9.2   How many waves of microwave radiation pass through food per second in a microwave oven?

In a microwave oven, the microwave 'light' typically has a wavelength of ~12 cm. From simple mechanics, speed c, wavelength λ and frequency ν are related as

$$c = \lambda \times \nu \tag{9.2}$$

Frequency and wavelength are inversely proportional, so a longer wavelength propagates at a lower frequency. If we assume that the speed of light is the same through food as through a vacuum, then c has the value 3×10^8 m s^{-1}.

First, we rearrange Equation (9.2) to make frequency ν the subject: $\nu = c/\lambda$.

Second, we convert the wavelength to the SI unit of length, the metre. The necessary relationship is 1 m $= 10^2$ cm, so 1 cm $= 10^{-2}$ m. Accordingly, 12 cm $= 12 \times 10^{-2}$ m.

Third, inserting values into the rearranged equation yields

$$\nu = \frac{3 \times 10^8 \text{ m s}^{-1}}{12 \times 10^{-2} \text{ m}} = 2.5 \times 10^9 \text{ s}^{-1}$$

> Note that the wavelength is first converted to metres (which is SI).

> One cycle per second is termed one *hertz* (H$_3$), so 2.5×10^9 Hz is 2.5 gigahertz (GHz).

We see how 2500 million waves of microwave radiation pass through the food each second.

SAQ 9.1   What is the wavelength of a wave of light having a frequency of 10^{12} Hz?

But light is also a particle. Some properties of light cannot be explained by the wave-like nature of light, such as the photoelectric effect and blackbody radiation (see Section 9.4), so we also need to think of light comprising particles, i.e. photons. Each photon has a *direction* as it travels. A photon moves in a straight line, just like a tennis ball would in the absence of gravity, until it interacts in some way (either it reflects or is absorbed).

And each photon has a fixed energy E. We say the energy is *quantized*. The intensity of a beam of light is merely a function of the number of photons within it per unit time; see below.

Only one photon at a time may interact with matter. This means that the energy available to each recipient atom or molecule is the same as the energy possessed by the single photon with which it interacts. This truth was refined by Stark and Einstein, who called it the *second law of photochemistry*: 'If a species absorbs radiation, then one particle (molecule, ion, atom, etc.) is excited for each quantum of radiation (photon) that is absorbed'.

We will modify and expand this idea when we discuss quantum yields, in Section 9.2.

> The intensity of light is often expressed as the number of photons per second, which is termed the *flux*.

> The *second law of photochemistry* says that if a species absorbs radiation, then only one particle is excited for each photon absorbed.

> The photon must have sufficient energy to cause the photochemical changes in a single molecule, ion, atom, etc.

How does a dimmer switch work?

Photon flux

The lights in a baby's bedroom, or in a sophisticated restaurant, are dimmed in the evening, off during the day and bright when visibility needs to be maximized, e.g. during cleaning. A *dimmer switch* modifies the brightness of a light bulb, i.e. dictates the amount of light it emits.

A dimmer switch is merely a device that alters the voltage applied to the filament in a light bulb. A higher voltage allows more energy to enter the bulb, to be emitted as a greater number of photons of visible light. A lower voltage means that less energy can be emitted, and the light emitted is feeble, with fewer photons emitted.

> In practice, the dimmer switch incorporates a small variable resistor into the switch box, which allows differing amounts of voltage to be 'tapped off'.

A dim light emits fewer photons than does a bright light. So a dimmer switch is merely a device for varying the *flux* of photons emitted by a light bulb. The energy per photon is not significantly altered in this way.

A laser, with light of wavelength 1000 nm at a power of 100 W (i.e. 10^2 J s^{-1}), emits a flux of 5×10^{21} photons per second.

Why does UV-b cause sunburn yet UV-a does not?

Energy per photon

It is easy to get burned by the sun while out sunbathing, because the second law of photochemistry shows how each UV photon from the sun releases its energy as it impinges on the skin. This energy is not readily dissipated because skin is an insulator, so the energy remains in the skin, causes photo-excitation, which is experienced as damage in the form of sunburn.

The best and simplest way to avoid burning is merely to avoid lying in the sun; but if we insist on sunbathing, then we must stop the UV light from releasing its energy into the delicate molecules within the skin. We can reflect the light – which is why some sunglasses have a mirror finish – or we stop the light from reaching the skin, by absorbing it first. For this reason, sun creams (see below) are employed to absorb the UV light before photo-excitation of molecules within the skin can occur.

The meaning of 'photolytic' may be deduced from its Greek roots: *photo* relates to light, and the ending '–lytic' comes from *lysis*, which means to cleave or split, so a process is 'photolytic' if splitting or scission occurs during irradiation.

In recent years, advertisements have tended to sound more technological, and many now proudly claim protection against 'harmful UV-b light'. What does this mean?

In practice, the term 'UV light' represents a fairly wide range of wavelengths from 100 nm through to about 400 nm. Although there is a continuum of UV energies, it is often convenient to subdivide it into three 'bands'; see Table 9.2. UV-a comprises wavelengths in the range 320–400 nm. Such light penetrates the top, outermost layer of skin, causing photolytic damage to the layers beneath, as seen by premature aging, sagging, wrinkles and coloured blotches. UV-a can contribute to the onset of skin cancer (*melanoma*) but is otherwise relatively safe provided that suitable protection is taken; see below.

UV-b radiation has wavelengths in the range 290–320 nm. It is much more dangerous to the skin than UV-a because each photon possesses more energy. In consequence, the photolytic processes caused by UV-b are more extreme than those caused by UV-a. For example, UV-b causes thermal degradation of the skin (we call it 'sunburn') but, additionally, it inhibits DNA and RNA replication, which is why over-exposure to UV-b will ultimately lead to skin cancer.

We call the extreme end of the UV-c range (of $\lambda \lesssim 190$ nm) the *vacuum UV*. It's the 'vacuum' UV because the O=O double bonds in oxygen absorb UV-a, -b or -c, causing air to become opaque. Although absorbed by air, such photons readily pass through a vacuum (or gaseous atmospheres containing only monatomic gases).

UV light in the wavelength range 100–290 nm is called UV-c. Such UV light can generally be filtered out by ozone (O_3) in the

Table 9.2 Classification of types of UV light

UV type	Wavelength range/nm
a	320–400
b	290–320
c	100–290
Vacuum	100–190

upper atmosphere before it reaches the Earth's surface. Exposure to UV-c is more hazardous than UV-a or -b because of the greater energy possessed by each photon. Incidentally, it also explains the current concern about ozone depletion, which allows more UV-c light to reach the Earth's surface.

So, in summary, the extent of the damage caused by these different types of UV light depend on the amount of energy per photon of light, itself a function of its wavelength.

The Planck–Einstein relationship

At a quantum-mechanical level, there is a simple relationship that ties together the twin modes by which we visualize photons: we say that the energy of a photon particle is E and the frequency of a light wave is v. The Planck–Einstein equation, Equation (9.3), says

$$E = hv \tag{9.3}$$

where h is the Planck constant. Within the SI system, h has the value 6.626×10^{-34} Js if v is expressed as a frequency in cycles per second. The energy E is expressed in joules.

Equation (9.3) is valid for all photons of all types. We see from Equation (9.3) how the only difference between different types of photon is the variation in the energy each comprises, causing the frequency to vary proportionately. Photons of high energy, such as γ-rays and X-rays, are characterized by very high frequencies and high energy, whereas radio waves are characterized by low frequencies and low energy.

Worked Example 9.3   What is the energy per photon of a γ-ray that has a frequency of 3×10^{18} Hz?

Inserting values into Equation (9.3):

> We recall that 1 Hz = 1 cycle per second.

$$E = h \times v = 6.626 \times 10^{-34} \text{ J s} \times 3 \times 10^{18} \text{ s}^{-1}$$

$$E = 1.99 \times 10^{-15} \text{ J per photon}$$

SAQ 9.2   What is the frequency of a photon having an energy per photon of 10^{-18} J?

Worked Example 9.4   What is the *molar* energy of the γ-particle photons in Worked Example 9.3?

In Worked Example 9.3, we calculated that each photon has an energy of 1.99×10^{-15} J. To convert from the energy *per particle* to the energy *per mole*, we multiply the answer by the Avogadro number, $L = 6.022 \times 10^{23}$ mol^{-1}:

$$\text{Energy per mole} = (hv) \times L \tag{9.4}$$

$$\text{Energy per mole} = (1.99 \times 10^{-15} \text{ J}) \times 6.022 \times 10^{23} \text{ mol}^{-1} = 1.2 \times 10^{9} \text{ J mol}^{-1}$$

This energy is so large that we suddenly realize why γ-rays are dangerous.

Furthermore, the magnitude of the energy in this answer helps explains why we sometimes wish to employ light to effect a reaction, because a chemical reagent simply does not possess enough energy (see later examples).

SAQ 9.3   What is the frequency ν of a photon having a molar energy of 1.0 MJ mol^{-1}?

In practice, most chemists prefer to talk in terms of *wavelength* rather than frequency. For this reason, we often combine Equations (9.2) and (9.3), and so obtain a modified form of the Planck–Einstein equation:

$$E = \frac{hc}{\lambda} \tag{9.5}$$

Equation (9.5) shows that a spectrum plotted with wavelength λ as 'x' has its highest energies on the left-hand side and the lowest energies on the right.

Equation (9.5) is interesting, because it shows that the energy per photon is *inversely* proportional to wavelength. The equation also shows how our usual way of representing a spectrum, with wavelength λ along the x-axis, is illogical: it is more usual, in physical chemistry, to have numbers increasing from left to right along the x-axis. We see from Equation (9.5) that if we plot wavelength λ (as 'x'), then in fact the highest energies are on the left-hand side of the spectrum and the lowest energies are on the right.

Worked Example 9.5   What is the energy per photon of an X-ray of wavelength 5×10^{-10} m?

Many textbooks prefer to cite 10^{-10} m as 1 Ångstrom (symbol Å and pronounced as 'ang-strom'). Others will cite it as 100 pm, i.e. 100×10^{-12} m.

Inserting values into Equation (9.5):

$$E = \frac{6.626 \times 10^{-34} \text{ J s} \times 3 \times 10^{8} \text{ m s}^{-1}}{5 \times 10^{-10} \text{ m}} = 3.98 \times 10^{-16} \text{ J}$$

This energy equates to a molar energy of 0.24 GJ mol^{-1}, which is simply massive, and helps explain why X-rays can also cause much damage to the body. The energy per X-ray photon is so large that radiographers who work with X-ray machines require complete body protection.

How does a suntan protect against sunlight?

Absorbance as a function of wavelength

Melanin is a complicated mixture of optically absorbing materials.

The colour of our skin is caused by the presence of pigments (i.e. chromophores), of which *melanin* is the most important. Melanin is a complicated mixture of optically absorbing materials, formed as an end product during the photo-assisted metabolism of the amino acid tyrosine. Both are bound covalently to the surrounding proteins within the skin, and other pigmented regions of the body.

Melanin from natural sources falls into two general classes. The first component is *pheomelanin* (**I**), which has a yellow-to-reddish brown colour, and is found in red feathers and red hair. The other component is *eumelanin* (which has two principal components, **II** and **III**). Eumelanin is a dark brown–black compound, and is found in skin, hair, eyes, and some internal membranes, and in the feathers of birds and scales of fish. Melanin is particularly conspicuous in the black dermal *melanocytes* (pigment cells) of dark-skinned peoples and in dark hair; and is conspicuous in the freckles, and moles of people with lighter skins.

(I)   **(II)**

(III)

Melanin is formed by a photochemical reaction, so the concentration of melanin within the human *epidermis* (the outer layer of the skin) increases following exposure to photochemical reactions with tyrosine in the skin. This increase is seen readily by the formation of a suntan. This increase in the concentration of melanin following exposure to the sun is more obvious for fair-skinned people, although darker people usually tan more quickly because melanin is produced more efficiently in their skins.

The enhanced pigmentation engendered by a suntan is advantageous because the brown melanin pigment absorbs the UV rays in sunlight, effectively filtering them out, and thereby generates a barrier between the sensitive inner layers of skin and the harmful effects of UV. It also explains why the indigenous people who live at or near the equator need to have more melanin in their skins than people living nearer the poles.

Unfortunately, melanin does not protect against all UV light, only that of wavelengths longer than 370 nm. We can readily show this

> The word 'epidermis' is defined on p. 254.

> A suntan is merely an increase in the concentration of *melanin* in the skin.

> A suntan forms as nature's way of protecting the skin against energetic photons.

A spectrum is obtained with a device called a *spectrometer*.

to be the case by placing a sample of melanin in the beam of a *spectrometer*, i.e. a device that obtains a *spectrum*. We have already seen how a 'spectrum' is defined as a graph of intensity (as '*y*') against photon energy (as '*x*'); see Figure 9.4. The spectrometer irradiates the sample with light of a single frequency. The light beam passes through the sample of melanin, during which some of the light is absorbed. The spectrometer determines the exact decrease in intensity that is caused by transmission through the sample at this wavelength, and then repeats the measurement with as many other wavelengths as are needed.

The most common form of spectrum required by a chemist is a graph of the absorbance A (see p. 441), plotted on the y-axis, as a function of wavelength λ (as '*x*'). Such a spectrum obtained for melanin is shown in Figure 9.7. It shows how melanin is not particularly effective for UV-c, thereby allowing some of the harmful effects of the UV-c light to occur. But the large values of absorbance A in the wavelength range 200–350 nm clearly show where melanin absorbs most strongly, and confirms that the more harmful wavelengths of UV light are absorbed before they can reach the sensitive tissues beneath the melanin-containing layer in the skin.

The spectrum of melanin has minimal optical absorbances at wavelengths greater than about 750 nm, suggesting that it does not protect us against the heat from the sun (i.e. infrared light).

The spectrum of melanin components in Figure 9.7 is different from that of methyl viologen in Figure 9.4. The latter spectrum has a clear, symmetrical peak. We call it an optical *band*. By contrast, the spectrum of melanin components increases but never quite reaches a peak. In fact, the spectrum only shows the *side* of a huge optical band. We sometimes say that spectra like Figure 9.7 contain a *band edge*.

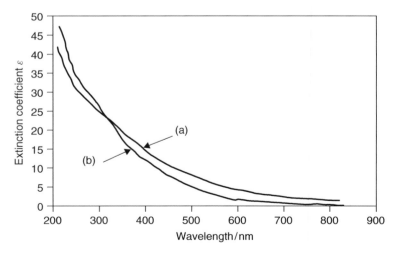

Figure 9.7 UV–visible spectrum of the skin pigment melanin. The spectrum contains two traces: (a) eumelanin and (b) pheomelanin. Both component compounds protect the skin by absorbing harmful UV light. All pigment concentrations were 1 mg dm^{-3}

How does sun cream block sunlight?

Absorbing light

We have just seen that melanin is quite efficient at stopping UV light below about 370 nm. If the skin is damaged or the body cannot generate its own melanin – which is the case for an *albino* – then the UV will not be blocked. It is also common for young children and babies to have insufficient amounts of melanin in their skin to block the UV effectively. For this reason, babies should be protected from direct sunlight.

> An albino does not have the ability to generate melanin, so their hair is snow white, and their skin looks pale pink in colour because the blood corpuscles beneath its surface are the only chromophore they have.

Furthermore, following concerns over the way the ozone layer is being damaged by man-made chemicals, some UV-c now reaches the Earth's surface, particularly in Australasia and some parts of South America. People living in such places are not adequately protected against the harmful effects of the sun's rays by the melanin in their skin.

If insufficient natural melanin is available in the skin, then it is advisable to protect ourselves artificially; the best form of protection is to apply chemicals to the skin, such as sun cream (or 'sunscreen'). The best sun creams are emulsions containing chemicals that absorb light in a similar manner to naturally occurring melanin. The organic compounds benzophenones, anthranilates and dibenzoyl methanes are good at blocking UV-a, and microfine zinc oxide can also act as a 'physical blocker'. To block UV-b, the recommended compounds are the salicylate or cinnamate derivatives of camphor or *p*-aminobenzoic acid, together with microfine titanium dioxide (TiO_2). TiO_2 is the only permitted block against UV-c. A good-quality sun cream will contain some of each compound to protect against all UV-a, -b and -c.

> Emulsions are discussed in Section 10.2.

> There are now some health concerns concerning the safety of these compounds, since irradiation of TiO_2 or ZnO causes the formation of radicals.

Figure 9.8 depicts the spectra of a few of the more popular sunblocking compounds contained within sun creams, and shows just how effective they are. Remember, the UV region of the spectrum ends at about 400 nm.

Absorbance and transmittance

The 'transmittance' T of a sample is a quantitative measure of how much of the light entering a sample is absorbed. (Transmittance is also called *optical transmittance*, which means the same thing.) The amount of light entering a sample is called the *incident* light.

> We say that the beam of light striking a sample is the incident light.

A simple measure of the number of photons entering the sample is the intensity I. The intensity of the incident light is conveniently symbolized as I_0. Similarly, the number of photons leaving the sample after some is absorbed is

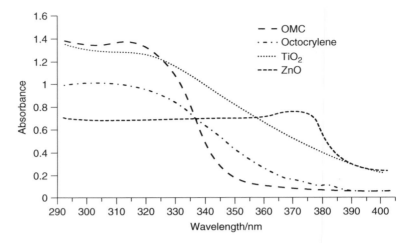

Figure 9.8 UV–visible spectra of some of the active (i.e. sun blocking) components within commercially available sun creams. OMC = ethylhexyl methoxycinnamate

An optical measurement determined at a single, invariant wavelength is said to be *monochromatic*, from the Greek roots *mono* (meaning one) and *khroma* (meaning colour).

A transmittance of 100 per cent implies no absorbance at all. A sample that is wholly opaque has a transmittance of 0 per cent because $I = 0$.

best gauged by a new intensity, which we will symbolize as just I, i.e. without a subscript.

The transmittance T of the sample is given by the ratio of the intensities of light entering the sample (I_o) and leaving it (I):

$$T = \frac{I}{I_o} \qquad (9.6)$$

The measurement of an optical transmittance is only accurate if I and I_o are both determined at the *same*, invariant wavelength. We say the light source is *monochromatic*.

If no light is absorbed at all, then $I = I_o$ and the transmittance is unity. A large number of spectroscopists cite transmittance data as a percentage. A transmittance of unity represents a transmittance of 100 per cent, and a value of T of, say, 0.67 is 67 per cent. A large transmittance indicates that few photons are absorbed, so the amount of colour will be slight.

Worked Example 9.6 A sample of blackcurrant juice is strongly coloured to the extent that six photons are absorbed out of every seven in the incident light. What is the transmittance T of the sample?

If six photons in seven are absorbed, then only one photon in seven is transmitted. Inserting values into Equation (9.6):

$$T = \frac{1}{7} = 0.14$$

So the transmittance T of the blackcurrent juice sample is 14 per cent.

SAQ 9.4 What is the transmittance T when 12 photons are transmitted for every 19?

The transmittance T is a useful measure of how many photons are absorbed; nevertheless, most spectroscopists prefer to work in terms of the 'absorbance' A. (Absorbance is also called *optical absorbance* and *optical density* – each of these three terms means the same thing.)

The absorbance of a sample is defined according to

$$A = \log_{10}\left(\frac{I_{\mathrm{o}}}{I}\right) \tag{9.7}$$

Values of absorbance obtained from Equation (9.7) are only valid if I and I_{o} are determined at the same, single wavelength, so the absorbance should only be measured with monochromatic light.

SAQ 9.5 A solution of beetroot juice is strongly coloured. The incident beam shining through it I_{o} emits 4×10^{10} photons per second, and the emergent beam emits 10^9 photons per second. Calculate the absorbance of the solution.

Comparing Equations (9.6) and (9.7), the relationship between absorbance A and transmittance T is given by

$$A = \log_{10}\left(\frac{1}{T}\right) = -\log_{10} T \tag{9.8}$$

If less light remains following transmission through a sample then, from Equation (9.7), $I < I_{\mathrm{o}}$, and $A > 1$. It should be clear that the absorbance is zero when the same number of photons enter the sample as leave it. A highly coloured sample is characterized by a high absorption, so A is large.

On the other hand, the value of A cannot be negative because it would be bizarre if more light came out of the sample than entered it!

> An absorbance A cannot be negative.

We should appreciate the consequences of the logarithm term in Equation (9.7): an absorbance of $A = 1$ means that one in ten photons is transmitted and an absorbance of $A = 2$ indicates a solution in which ten times more are absorbed, so only one photon in 100 is transmitted. Similarly, an absorbance of $A = 3$ follows if a mere one photon per 1000 is transmitted.

At this point we ought to hold our breath: if 999 photons per 1000 are absorbed (with only one being transmitted), then the detector measuring the intensity of the transmitted light I needs to be exceptionally sensitive. Stated another way, an absorbance of $A = 3$ is unlikely to be particularly accurate.

> The best-quality absorbance data are obtained when the absorbance A lies in the range 0.75–1.

At the opposite extreme, if only one photon per 100 is absorbed (a transmittance of 99 per cent), then the logarithm term in

Care: the words
'strong' and 'weak'
refer to the extent
to which a weak
acid dissociates – see
Section 3.1 – and not
concentrated and dilute
respectively.

Beer's law says the
absorbance *A* of a
chromophore in solu-
tion increases in direct
proportion to its con-
centration.

The relationship
summarized by
Equation (9.9) is only
a law if absorbances
are determined with
monochromatic light
(i.e. a single value
of λ), and if the
chromophore does not
associate or dissociate.

Beer's law requires
monochromatic light
because the value of *k*
depends on λ.

Equation (9.7) will magnify any errors. For these reasons, the best-quality absorbance data are obtained when the absorbance *A* lies in the range 0.75–1.

Why does tea have a darker colour if brewed for longer?

Beer's Law

The best way to tell how strong we have made a cup of tea is to look at its colour. A strong cup of tea has a darker, more intense colour than does a weak one. In fact, some people, when asked, 'How strong would you like your tea?' will respond by saying simply, 'Dark!'

Similarly, a glass of fruit cordial or even gravy looks darker if the coloured compounds in it are more concentrated. This is a general observation, and has been codified as *Beer's law*: the absorbance *A* of a solution of chromophore increases in direction proportion to its concentration *c*, according to

$$A = kc \qquad (9.9)$$

where *k* is merely an empirical constant.

An alternative way of presenting this law is to note that a graph of absorbance *A* (as '*y*') against concentration *c* (as '*x*') will be linear. Figure 9.9 shows such a graph for the permanganate ion in water as a function of concentration.

Equation 9.9 also suggests the (hopefully obvious) corollary that an absorbing sample has no absorbance if its concentration is zero. In other words, we have no absorbance if there is no sample in solution to absorb, which is why the extrapolation of the line in Figure 9.9 passes through the origin.

Why does a glass of apple juice appear darker when viewed against a white card?

The Lambert law

Apple juice usually has a pleasant pale orange–yellow colour. But, if the glass is held against a white card or curtain, the intensity of the colour appears to increase significantly – it may look brown. It appears darker when held against a white card because light travels through the glass and juice *twice*: once before striking the white

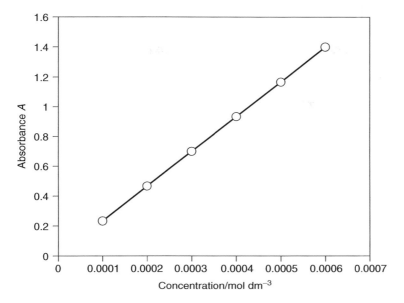

Figure 9.9   An illustration of Beer's law: absorbance of solutions of permanganate ion MnO_4^- as a function of concentration. The optical path length l was 1 cm, and the wavelength of observation was 523 nm

back card, and a second time after light reflects off the card, as shown schematically in Figure 9.10. We say that the *optical path length* has doubled. This length is usually symbolized as l.

A similar optical effect is seen at the swimming baths. The water at the deep end appears to have a more intense blue colour than at the shallow end. Again, this intensification of colour arises because the path length l alters. Again, the photograph in Figure 9.11 shows the view around the coast of Hawaii. From the varying intensity of the colour of the sea, it is easy to see which sections of the water are deeper. Shallow parts of the sea appear as a lighter hue.

The absorbance A has long been known to be a function of path length l, according to the Lambert law:

$$A = k'l \qquad (9.10)$$

where k' is merely another empirical constant (different from that in Equation (9.9)). We see that the absorbance should increase with l while going from the shallow to the deep end of the pool. Also note the (hopefully obvious) corollary, that an absorbing sample has no absorbance if its path length is zero. In other words, we can have no absorbance if there is no absorbing sample to pass through. Stated another way, Equation (9.10) tells us that a graph of absorbance A (as 'y') against path length l (as 'x') will be linear, and pass through the origin.

> The law summarized by Equation (9.10) only holds for absorbances determined with monochromatic light.

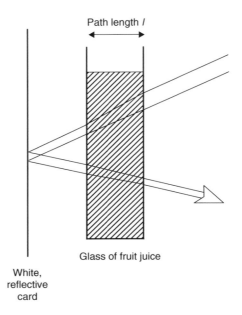

Figure 9.10   Illustration of Lambert's law: the absorbance A of a glass of juice is proportional to the optical path length l, so holding the glass against a white card makes its colour appear twice as intense because the path length has been doubled. The width of the beam here is proportional to its intensity

Figure 9.11   Part of the Hawaii coastline. The lighter areas indicate areas where the sea is shallower

Why are some paints darker than others?

The molar absorption coefficient ε

To the eye, cobalt blue and woad have a very similar colour. We say they have the same *hue*. But woad is considerably paler in intensity, even when the concentration c

and path length l are the same. Cobalt blue is simply more intensely blue than woad. We say that it has a larger 'molar absorption coefficient' ε.

At heart, this greater intensity may be explained as follows. The ease with which an electron may be photo-excited depends on the probability of successful excitation, which itself depends on the likelihood of photon absorption. If the probability of excitation in the woad was 20 per cent, then 20 from every 100 incident photons are absorbed (assuming each absorption results in a successful electron excitation). By contrast, cobalt blue is more intense because it has a higher probability of photon uptake, so fewer photons remain to be seen, and the absorbance increases.

> Older books often give the name *extinction coefficient* to the molar absorption coefficient ε, or 'molar extinction coefficient'.

A few molar absorption coefficients are listed in Table 9.3. The higher the value of ε, the more intense the colour we see.

What is ink?

The Beer–Lambert law

Ink is a fluid or paste having an intense colour. Inks comprise a pigment or dye as the chromophore dissolved or dispersed in a liquid called the *vehicle*.

The first inks for writing comprised soot suspended in water. 'India ink' was the first truly waterproof ink and is a dispersion of carbon black, stabilized to stop sedimentation. India ink is still a favourite drawing medium with draughtsmen.

> India ink is stabilized to prevent sedimentation by adding substances such as shellac in borax solution, soap, gelatine, glue, gum arabic, and dextrin.

The first patent for making coloured inks was issued in England in 1772. In the 19th century, ink compositions became more complicated, as drying agents were added to allow the production of a wide variety of synthetic pigments for coloured inks. It was not until the beginning of the early 20th century that ink-making became a complicated chemical–industrial process.

Modern writing inks usually contain ferrous sulphate, together with a small amount of mineral organic acid. The resulting solution is light bluish black, and results in

Table 9.3 Typical values of molar extinction coefficient ε

Species	$\varepsilon^{a}/dm^3\ mol^{-1}\ cm^{-1}$
$[Mn(H_2O)_6]^{2+}$	0.04 (312) (a spin and Laporte forbidden transition)
$[Cr(H_2O)_6]^{2+}$	10 (610) (a Laporte forbidden d–d band on the chromium)
$[Cr(NCS)_6]^{3-}$	160 (565) (a Laporte forbidden d–d band on the chromium)
Trytophan	540 (280)
Tyrosine triphosphate	1028 (430)
MnO_4^{-}	2334 (523) (a charge-transfer band)
$H_{0.04}WO_3$	5220 (950) (an intervalence charge-transfer band)
Methyl viologen[b]	16 900 (600) (an intervalence charge-transfer band)

[a]Wavelength of observation in nanometres indicated in parentheses.
[b]The radical cation of 1,1'-dimethyl-4,4'-bipyridilium.

writing that is only faint coloured on paper. The molar absorption coefficient ε is small, so a more intense image requires writing over the same part of the paper a few times, thereby increasing the path length l of chromophore.

To make the written image darker and more legible at the outset, additional dyes are added to the mixture. Modern coloured inks, and washable inks, contain a soluble synthetic dye as the sole chromophore, ensuring ε is large and obviating the requirement for overwriting.

Modern inks for printing are usually less fluid than writing inks. The composition, viscosity, density, volatility, and diffusibility of ink may vary, but adding less solvent has the effect of increasing the intensity of the colour, i.e. the absorbance is proportional to the concentration c of the chromophore.

So a broad overview of ink technology shows how the intensity of colour (its absorbance) is a function of three variables: molar absorption coefficient ε, path length l and chromophone concentration c. We have already met Beer's and Lambert's laws. We now combine the two to yield the Beer–Lambert Law:

$$A = \varepsilon l c \qquad (9.11)$$

We see how the proportionality factor in Lambert's law (Equation (9.9)) is εc, and the proportionality factor in Beer's law (Equation (9.10)) is εl.

9.2 Photon absorptions and the effect of wavelength

Most of the previous section concerned UV and visible light. In this section we will look in greater depth at the other common forms of light. From previous chapters, we are now familiar with the concept that different physical and chemical processes require differing amounts of energy. More specifically, it was shown in the previous section how the energies of photons can also vary. In this section, we see how the energies of different types of photon are manifested, and how their interactions may be followed.

A bit of theory introduces this section. We then look first at infrared light and microwave light. We will also need to look a bit at the way bonds break following absorption of light, i.e. photolysis. These ideas are summarized as Table 9.4.

> *Care*: although we give different names to these photons, in reality there exists a continuum of wavelengths from 10^{-11} m (for γ-rays) through to hundreds of metres (for radio waves). The titles we use, such as 'microwave' or 'X-ray' are just historical artifacts, and merely describe the way scientists first encountered them.

Why do radical reactions usually require UV light?

Photo-excitation and bond length

We have seen already how photons of UV light from the sun can cause burning of the skin. UV photons often cause bonds to cleave,

Table 9.4 Types of photon, and their uses in spectroscopy

Photon	Parameter observed	Spectroscopic technique
γ-rays	Excitation of the atomic nucleus	Mössbauer
X-rays	Structure determination	X-ray diffraction
	Excitation of inner-shell electrons	X-ray photoelectron spectroscopy, XPS
UV	Excitation of inner and valence electrons	UV–visible spectroscopy
Visible	Excitation of valence and outer-shell electrons	UV–visible spectroscopy
Infrared	Excitation between vibrational quantum states	Infrared spectroscopy
Microwaves	Excitation between rotational quantum states	Rotational spectroscopy
	Excitation between electronic spin states	Electron paramagnetic resonance, EPR
Radio waves	Excitation between nuclear spin states	Nuclear magnetic resonance, NMR

thereby explaining why cancers of the skin occur if we are over-irradiated with UV – particularly energetic UV-b or -c.

A majority of radical reactions require irradiation with UV light. A simple example is the chlorination of methane (CH_4), in which CH_4 and elemental chlorine are mixed and irradiated to yield a mixture of chlorohydrocarbons, such as CH_3Cl and CCl_4. The energy for reaction comes from the UV photons. Diels–Alder and other pericyclic reactions also require photons of light.

> UV photons are energetic enough to break bonds. Photons of visible light cannot break any but the weakest of bonds.

A typical value for bond energy in organic chemistry is 150 kJ mol^{-1}.

Worked Example 9.7 What wavelength of light corresponds to an energy of 150 kJ mol^{-1}?

We will perform a simple calculation in two parts. First, we will divide by the Avogadro number to obtain the energy per bond. Second, we convert the energy to a wavelength λ with the Planck–Einstein relation (Equation (9.4)), $E = hc/\lambda$.

To obtain the energy per bond, we divide 150 kJ mol^{-1} by L:

> Remember to convert from kilojoules to joules.

$$E = \frac{150\,000 \text{ J mol}^{-1}}{6.022 \times 10^{23} \text{ mol}^{-1}} = 2.49 \times 10^{-19} \text{ J}$$

Next, we determine the wavelength, and start by rearranging the Planck–Einstein relation to make wavelength λ the subject:

$$\lambda = \frac{hc}{E}$$

> The units involving J and s cancel, leaving a bond length with the unit of metres.

Finally, we insert values:

$$\lambda = \frac{6.626 \times 10^{-34} \text{ J s} \times 3 \times 10^{8} \text{ m s}^{-1}}{2.49 \times 10^{-19} \text{ J}} = 800 \text{ nm}$$

We find that photons of wavelength 800 nm are capable (in principle) of breaking a bond of energy 150 kJ mol^{-1}. Photons of lower energy (which have a longer wavelength) will not have sufficient energy. In practice, many chemical bonds are considerably stronger than 150 kJ mol^{-1} (e.g. see Table 2.4), so bond cleavage requires light of shorter wavelengths (and high energy).

As experimental chemists, we do not effect a photochemical reaction by irradiating a reaction with photons having exactly the right amount of energy, but prefer to supply an excess by irradiating with UV light.

Morse curves

Imagine we have two atoms of the element X – they could be bromine or chlorine. Each is infinitely far from the other, so there can be no interaction between them. Now imagine that they slowly approach from infinity, and we monitor their energy all the time. If we were to draw a graph of their energy as a function of distance r during their approach, we would generate something like Figure 9.12. We call the graph a *Morse curve*.

Although the two atoms of X do not interact at extreme separations r, they *do* interact when closer than a minimum distance, which is generally about twice the normal bond length. At these closer distances, a result of the interaction between the atoms of X is a decrease in the energy of the two atoms. We recall from Chapter 2 that all matter seeks to decrease its energy, so the lowering of the energy of our two atoms suggests a willingness to get closer. The interaction causing the decrease in energy arises from an overlap of the orbitals on the approaching atoms. In other

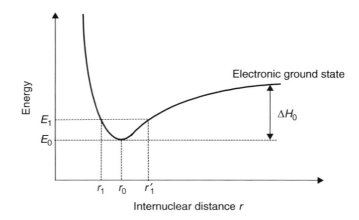

Figure 9.12 Morse curve of the diatomic molecule X$_2$ in the ground state

words, a bond starts to form. The decrease in energy on overlap is none other than the first part of the bond energy being released.

The two atoms of X are brought yet closer in our thought experiment, causing the energy to decrease until a minimum is reached. The energy then rises steeply at short distances r, so a minimum forms in the graph. The separation at the minimum represents the preferred separation between the two atoms of X. A molecule has been formed, which we will call X_2.

> The deep energy minimum in a Morse curve is often called an energy 'well'.

The rise in energy at very short distances tells us that the two atoms resist any further approach. They do not want to get closer than the minimum separation.

Another way of looking at the Morse curve in Figure 9.12 is to say it represents the energy E (as 'y') of the two atoms of X as a function of their bond length r (as 'x'). The two atoms of X form a simple diatomic molecule in its ground state, i.e. before it absorbs a photon of light.

The preferred bond length is r_o, because the value on the energy axis is lowest when the internuclear separation corresponds to this value of bond length. The energy of a bond having exactly this bond length is E_o.

Not all molecules of X_2 will have the same energy, because molecules are always swapping energy as a result of inelastic collisions. Accordingly, a graph of energy against number of molecules of X_2 having that energy follows a straightforward Boltzmann type of distribution, (see Section 1.4), so many molecules will have

> Remember that it is *electrons* that hold a bond together.

more energy than E_o. Let us concentrate on molecules having a greater energy, which we will call E_1. From the graph it should be clear that a molecule having the energy E_1 will have a bond length between the two extremes of r_1 and r_1'. In practice, the two atoms in a molecule of X_2 having the energy E_1 will vibrate such that the bond length will vary between these two extremes of r_1 and r_1', as though connected by a spring.

And a molecule of X_2 having even more energy in its bond will vibrate more strongly, causing more extreme variations in the bond length. If the energy in the molecule's bond is enormous, then we can envisage the (simplistic) situation in which the two atoms of X are so far apart that the bond has effectively 'snapped', i.e. causing bond cleavage. The energy of the bond when cleavage occurs will necessarily correspond to the bond energy ΔH_o, so if we shine light of sufficient energy on the molecule X_2, then the molecule will cleave owing to excessive vibration.

> The 'o' in ΔH_o relates to the bond dissociation energy of the molecule in the *ground* state.

We used the word 'simplistic' in the previous paragraph because we described the vibrations getting more and more violent, as though there was no alternative to eventual bond cleavage. In fact, there is a very straightforward alternative: absorption of a photon (i.e. energy) to X_2 will also cause an electron to *photo-excite*, as follows.

Look at Figure 9.13, which now shows *two* Morse curves. The lower curve is that of the molecule in the ground state (in fact, it

> We call it the 'first' excited state to emphasize that an electron can be excited further to the second excited state; if the photon energies are vast, then excitation is to the third state, etc.

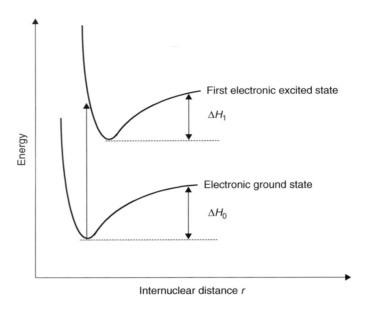

Figure 9.13   Morse curves of the diatomic molecule X_2 in its ground and first excited states

is the same as that in Figure 9.12), and the upper Morse curve shows the relationship between energy and bond length in the first excited state of the molecule, i.e. in X_2^*.

Following photon absorption, an electron from the HOMO of X_2 is excited from the ground to the first excited state. The electronic excitation that occurs on photon absorption is represented on the figure by an arrow from the lower (ground state) Morse curve to the higher (excited state) curve. The time required for excitation of the electron is very short, at about 10^{-16} s. By contrast, because the atomic nuclei are so much more massive than the electron, any movement of the nuclei occurs only some time after photo-excitation of the electron – a safe estimate is that nuclear motion occurs only after about 10^{-8} s, which is 10^8 times slower.

> The atomic nuclei are so much heavier than the electrons that we approximate and say the nuclei are *stationary* during the electronic excitation. We call this idea the *Born–Oppenheimer approximation.*

The motion of the electrons is so much faster than the motion of the atomic nuclei that, in practice, we can make an important approximation: we say the nuclei are *stationary* during the electronic excitation. This idea is known as the *Born–Oppenheimer approximation.*

The Born–Oppenheimer approximation has a further serious consequence. At the instant of excitation, the length of the bond in the excited state species is the *same* as that within the ground state. While close inspection readily convinces us that the two Morse curves are very similar and have the same shape, it is important to recognize that the *equilibrium bond lengths differ.*

The Born–Oppenheimer principle says that the atomic nuclei do not move during the electronic excitation; only later will the excited state structure 'relax' to minimize its conformational energy. An arrow represents the photo-excitation from the ground state to the excited state structures. The requirement for the excited-state structure to move only *after* the absorption explains why the arrow strikes the upper curve at a different position than its minimum energy.

This idea may be summarized from within the *Franck–Condon principle*. Because the atomic nuclei are relatively massive and effectively immobile, the transition is from the ground state to the excited state lying *vertically* above it. We say that the electronic excitation is *vertical*, which explains why the arrow drawn on Figure 9.13 is vertical.

The molecule in its excited state rearranges *after* the photon is absorbed, and rearranges its bond lengths and angles until it has reached its new minimum energy, i.e. attained a structure corresponding to the minimum in the upper Morse curve.

We see following consideration of Figure 9.13 the common situation whereby the bond length in the excited state is longer than that in the ground-state structure. For example, the C=C bond length in ethene increases from 0.134 nm to 0.169 nm following photo-excitation from the ground to first excited state.

The second difference caused by photo-excitation is to decrease the bond dissociation energy (see Figure 9.13). We will call the new value of bond dissociation energy ΔH_1 ('1' here because we refer to the *first* excited state.) This decrease arises from differences in the localization of electrons within the molecular orbitals in the ground and excited states.

Because the value of ΔH_1 is smaller than ΔH_0, it becomes more likely that photo-dissociation of molecular X_2 occurs when the molecule is in the excited state than in its ground state. This decrease in ΔH also explains why photo-dissociation can sometimes be achieved following irradiation with light of longer wavelength (lower energy) than calculations such as that in Worked Example 9.7 first suggest. Nevertheless, tables of bond dissociation energy generally only contain data for molecules in the ground state (i.e. values of ΔH_0 rather than ΔH_1).

So, in summary, photo-initiated chlorination of methane requires UV light to generate 'hot' atoms of molecular chlorine, which subsequently dissociate:

> The *Franck–Condon principle* states that the excited state is formed with the same geometry as that of the ground state from which it derived. The transition is from the ground state to the excited state lying *vertically* above it.

> More subtle change can occur to a molecule's structure following photo-excitation. For example, the bond angle in methanal (formaldehyde) increases, so the molecule is flat in the ground state but bent by 30° in the first excited state; see Figure 9.14.

$$Cl_2 + h\nu \longrightarrow Cl_2^* \longrightarrow 2Cl^\bullet \qquad (9.12)$$

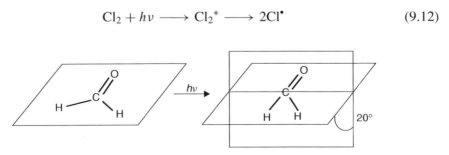

Figure 9.14 The structure of methanal changes following photo-excitation, from a flat ground-state molecule to a bent structure in the first photo-excited state

Why does photolysis require a powerful lamp?

Quantum yield

It is advisable to employ a *high-power* lamp when performing a photochemical reaction because it produces more photons than a low-power lamp. Its flux is greater. When we looked at the laws of photochemistry, we saw how the second law stated the idea that when a species absorbs radiation, one particle is excited for each quantum of radiation absorbed. This (hopefully) obvious truth now needs to be investigated further.

The 'quantum yield' Φ is a useful concept for quantifying the number of molecules of reactant consumed per photon of light. It may be defined mathematically by

$$\Phi = \frac{\text{number of molecules of reactant consumed}}{\text{number of photons consumed}} \qquad (9.13)$$

Its value can lie anywhere in the range 10^{-6} to 10^{6}. A value of 10^{-6} implies that the photon absorption process is very inefficient, with only one molecule absorbed per million photons. In other words, the energetic requirements for reaction are not being met.

Conversely, a quantum yield Φ of greater than unity cannot be achieved during a straightforward photochemical reaction, since the second law of photochemistry clearly says that one photon is consumed per species excited. In fact, values of $\Phi > 1$ indicate that a secondary reaction(s) has occurred. A value of $\Phi > 2$ implies that the product of the photochemical reaction is consumed by another molecule of reactant, e.g. during a chain reaction, with one photon generating a simple molecule of, say, excited chlorine, which cleaves in the excited state to generate two radicals. Each radical then reacts in propagation reactions until the reaction mixture is exhausted of reactant.

Note that the overall quantum yield is Φ and the primary quantum yield is ϕ.

In fact, the term 'secondary quantum yield' is a misnomer, since it refers to chemical reactions rather than photochemical (and hence quantum-based) processes.

To help clarify the situation, we generally define two types of quantum yield: primary and secondary. The magnitude of the *primary quantum yield* refers solely to the photochemical formation of a product so, from the second law of Photochemistry, the value of $\phi_{\text{(primary)}}$ cannot be greater than unity.

The so-called *secondary quantum yield* refers to the total number of product molecules formed via secondary (chemical) reactions; its value is not limited.

The primary quantum yield ϕ should always be cited together with the photon pathway occurring: it is common for several possible pathways to coexist, with each characterized by a separate value of ϕ.

As a natural consequence of the second law of photochemistry, the sum of the primary quantum yields cannot be greater than unity.

Why are spectroscopic bands not sharp?

Vibrational and rotational energy

The energetic separation between the two Morse curves in Figure 9.13 is fairly well defined. We recall from Chapter 2 how each molecule vibrates somewhat if the temperature is greater than absolute zero, so the actual bond length for the molecule varies as the two X atoms vibrate about their mean positions, as described above. The extent of vibration depends on the temperature, and the range of vibrational energies will follow a straightforward Boltzmann-type distribution law, such as that seen in Figure 1.9. For these reasons, there will be a small range of energies between the two Morse curves in Figure 9.13, and we should, therefore, expect a spectroscopic band to be relatively fine and narrow, reflecting this small range of internuclear separations. In other words, a spectrum of X_2 should contain a narrow line, with a peak corresponding to the separations between the two Morse curves (i.e. the transition $X_2 + h\nu \rightarrow X_2^*$).

Each electronic state also has vibrational sub-states.

Even a cursory glance at a spectrum will show quite how wide a typical UV–visible spectroscopic band actually is. For example, the band for the methyl viologen dication shown in Figure 9.4 is about 100 nm wide at half peak height. There is clearly an additional variable to consider.

Figure 9.13 is not the whole story, since it only depicts the simplistic situation whereby an electron is excited from the lower Morse curve to a higher one. In reality, an electron in the ground state is also described by *vibrational energy levels* (we sometimes call them *sub-states*); see Figure 9.15. This finding applies to all ground and all excited states, so any photo-effected change in electronic state also occurs with a change in vibrational energy level: the electron excites from a vibrational energy level in the lower, ground-state Morse curve, to a vibrational level in the Morse curve for the first excited state.

An electron excites from a vibrational energy level in the lower, ground-state Morse curve, to a vibrational level in the excited-state Morse curve.

Instead of saying the vibrational level $v'' = 0$, we could say that the vibrational 'quantum number' $v'' = 0$.

The multiplicity of excitations possible are shown more clearly in Figure 9.16, in which the Morse curves have been omitted for clarity. Initially, the electron resides in a (quantized) vibrational energy level on the ground-state Morse curve. This is the case for electrons on the far left of Figure 9.16, where the initial vibrational level is $v'' = 0$. When the electron is photo-excited, it is excited vertically (because of the Franck–Condon principle) and enters one of the vibrational levels in the first excited state. The only vibrational level it cannot enter is the one with the same vibrational quantum number, so the electron cannot photo-excite from $v'' = 0$ to $v' = 0$, but must go to $v' = 1$ or, if the energy of the photon is sufficient, to $v' = 1$, $v' = 2$, or an even higher vibrational state.

We describe the vibrational levels in the lower (ground-state) Morse curve with the quantum number v'', the lowermost vibrational level being $v'' = 0$. The vibrational states in the upper (excited-state) Morse curve are described by the quantum number v'.

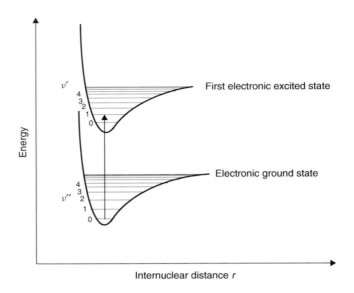

Figure 9.15 Morse curve of a diatomic molecule X_2 showing vibrational fine structure

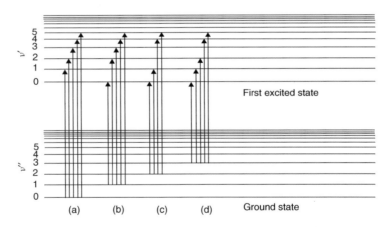

Figure 9.16 Schematic representation of the excitation of an electron from a vibrational level in a ground-state configuration to vibrational levels in the first excited state. The ground-state electron can have any vibrational sub-state: (a) $v'' = 0$; (b) $v'' = 1$; (c) $v'' = 2$; (d) $v'' = 3$

Alternatively, the ground-state electron could be in $v'' = 1$. In which case it can photo-excite to $v' = 0$, $v' = 2$, etc. but not $v' = 1$, because that would imply no change in v during the excitation process. Such a situation is also shown in Figure 9.16(b).

The simple quantum-mechanical rule governing transitions between vibrational levels is summarized by

$$\Delta v = \pm 1, \pm 2, \pm 3, \text{ etc.} \tag{9.14}$$

so we see how it is possible for molecules of sample in solution to absorb photons of subtly different energy.

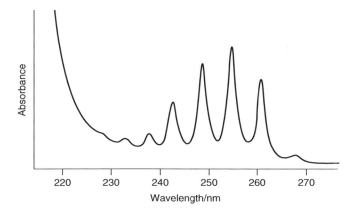

Figure 9.17 UV–visible spectrum of benzene $(6 \times 10^{-4} \text{ mol dm}^{-3})$ in cyclohexane at 298 K, showing vibrational fine structure

Figure 9.17 shows a spectrum of benzene (**IV**) in cyclohexane, which clearly shows small 'peaklets' superimposed on a broader band (or *envelope*). These peaklets are called *vibrational fine structure*. In benzene, they are caused by excitation from $v'' = 0$ to $v' = 1$, $v' = 2$, etc. Consideration of Figures 9.15 and 9.16 suggests that excitation from $v'' = 0$ to $v' = 1$ requires less energy than from $v'' = 0$ to $v' = 2$, so the excitation $v'' = 0 \rightarrow v' = 1$ occurs on the right-hand side of the figure, i.e. relates to processes of lower energy.

(IV)

Close inspection of Figure 9.15 shows how the separations between the vibrational levels become progressively smaller as we ascend the figure, going to higher energies. It becomes quite rare for the energetic separation between the vibrational levels to be quite as clear as that shown by benzene, so the peaklets are not often discernible individually, and we only see a broad *envelope* incorporating each of them. These spectroscopic peaks are quite broad, being hundreds of nanometres in width.

> It is quite rare for the energetic separation between the vibrational levels to be quite as clear as that shown by benzene.

Why does hydrogen look pink in a glow discharge?

Photo-excitation to anti-bonding orbitals

Consider a light bulb containing hydrogen gas at a low concentration of, say, 30 Pa (which is about 0.03 per cent of atmospheric pressure). A pale pink glow is seen when a voltage is applied between two electrodes. The hydrogen molecules are excited to form H_2^*, which immediately splits to form two hydrogen atoms. They contain excess energy, and must emit energy to return to the ground state, which is achieved by the emission of a photon. We see these photons as pink light.

We saw in Figure 9.15 how photon absorption leads to the excitation of an electron from the ground state to the first excited state. It is usual for the excited-state structure to form in a non-equilibrium state, so it must subsequently rearrange to achieve a lower energy.

> The first excited state in the hydrogen molecule is an anti-bonding orbital.

The hydrogen molecule comprises only two electrons and may be described by just two orbitals. The ground-state orbital is a bonding orbital, and the first excited state is an *anti-bonding* orbital. Figure 9.18 shows a Morse curve for the hydrogen molecule. An electron photo-excited to the first excited state is not in its minimum energy state, so it will seek to undergo whatever physicochemical processes are necessary to decrease the energy. The usual process is molecular rearrangement, such as lengthening of the bond.

> Photo-excitation of the hydrogen molecule leads to bond dissociation.

The upper Morse curve in Figure 9.18 has no minimum, but the energy decreases monotonically with increasing bond length. In other words, the lowest energy occurs at an internuclear separation of infinity. To decrease the energy of the first-excited-state H_2^* molecule, the bond length increases until the two hydrogen atoms are too far apart for there to be any electron density between the two nuclei: the bond breaks.

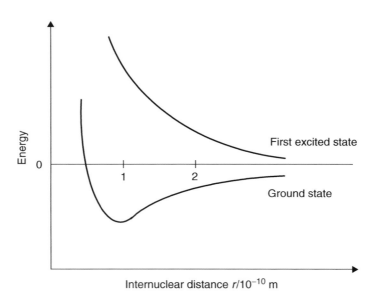

Figure 9.18   Morse curve of a diatomic H_2. Photo-ionization occurs following photo-excitation to a high-energy state that has no bonding character

Why do surfaces exposed to the sun get so dusty?

Photo-ionization

The top of a bookcase or tabletop exposed to the sun soon shows a thin surface of dust. It's not just that sunlight makes the dust more visible; rather, it is because light from the sun actually *promotes* the collection of the dust.

Look at Figure 9.19, which shows two Morse curves, one each for the ground and first excited states. It is a fairly typical situation, except that the upper curve is quite shallow (which is common). Excitation from the ground to the excited state proceeds following the absorption of a photon.

Because the upper curve is so shallow, the electron enters the Morse curve for the excited-state structure at an energy that is quite high, causing the bond length to vibrate over a very wide range of lengths. At its maximum, the bond length r will correspond to the horizontal plateau on the right of the Morse curve. Being horizontal, there is essentially no difference in energy as the bond increases in length until there is no electron density between the two atoms (or groups of atoms), at which point the bond has broken.

The simplest form of such *photo-dissociation* is that when one of the fragments is simply an electron:

$$Y + h\nu \longrightarrow Y^* \longrightarrow Y^+ + e^- \qquad (9.15)$$

in which case the dissociation process is more correctly termed *photo-ionization*.

> By convention, the groups of atoms formed during *photo-dissociation* are usually called *fragments*.

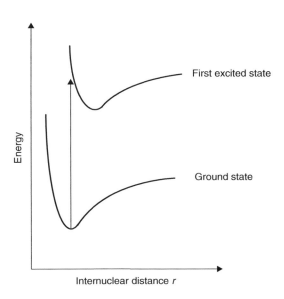

Figure 9.19   Morse curve of a diatomic molecule in which the first excited state has a relatively shallow well

It is usual to employ photons of UV radiation if photo-ionization is intended, because most ionization energies are relatively large. Irradiation will cause some ionization. For this reason, a surface left in the sun – like a tabletop placed in direct sunlight near a window – will soon acquire a small positive charge. Dust is a decomposition product from an organic-rich material (such as skin) and contains lone pairs and other electron-rich moieties. These lone pairs help the dust particle behave as though it had a slight negative charge.

We saw in Chapter 2 how *Coulomb's law* states that charges of the opposite sign attract each other. We call this an *electrostatic attraction*. In consequence, the charges on the dust and tabletop are mutually attractive, so the dust no longer floats past the table but forms a weak electrostatic *bond* with it. The amount of dust on the table will, therefore, be much greater than would be suggested by a mere statistical consideration of dust accumulation.

Photo-ionization of this sort is of profound importance in the upper atmosphere, where photons of vacuum UV light are absorbed and participate in reactions, e.g. with ozone and nitrogen oxides. These reactions help explain why the Earth's surface is relatively free of such harmful UV light, because the photons are absorbed *en route* through the Earth's atmosphere.

Experimentally, we often want to ionize a molecule, e.g. for photo-ionization spectroscopy. In which case, a sample is bombarded with energetic photons of UV (i.e. short wavelength) or low-energy X-rays.

> The table top acquires a positive charge because it was originally charge neutral, and an electron was lost following photo-dissociation.

> Photo-ionization of nitrous oxide in the upper atmosphere only occurs with light of wavelengths shorter than 134 nm, to cause the reaction $NO + h\nu \rightarrow NO^+ + e^-$.

> UV-photoelectron spectroscopy (UPS) and X-ray photoelectron spectroscopy (XPS) are powerful analytical tools: the energy of the photons required to eject the electron is characteristic of an element and of its oxidation state.

Why is microwave radiation invisible to the eye?

The photochemistry of vision

We have mentioned microwave ovens a few times already. Although we might employ such an oven regularly, no one has actually *seen* microwave radiation. The photons are wholly invisible. The phenomenon of 'sight' can be simplified to the photo-effected transformation of a pigment related to *retinal* within the retina at the back of the eye. Retinal is derived from vitamin A.

The outermost layer of the retina is the pigment *epithelium* (with supporting cells for the neural portion of the retina). The epithelium is darkened with melanin (as described above), thereby decreasing the extent of light scattering within the eye. Directly beneath this layer is the *bacillary* layer (also called the rod and cone layer), which comprises a layer of photoreceptor cells that respond to light. The rods are highly sensitive, and monitor the intensity of incident light; the cones are less sensitive, and relay information that is ultimately encoded within the brain as colour.

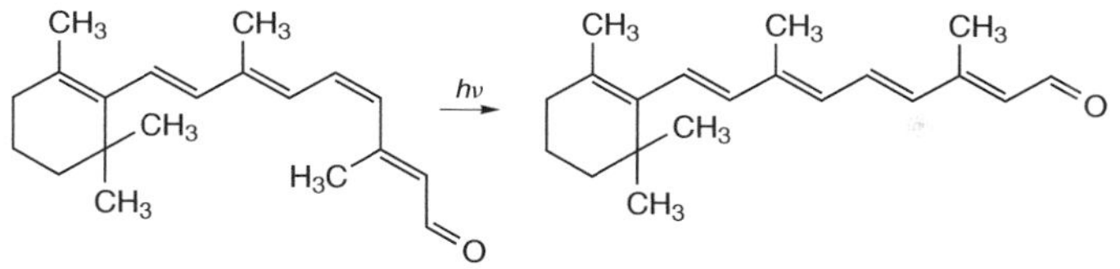

Figure 9.20 The photo-activated process that accounts for vision: the 11-cis → 11-trans photo-isomerism of retinal

The active compound within the bacillary layer is retinal. To simplify the photo-physics within the rods and cones hugely, absorption of a photon initiates a series of conformational changes that lead ultimately to photo-isomerization of retinal from the 11-cis isomer to the 11-trans isomer; see Figure 9.20. The uncoiling of the molecule following photo-excitation triggers a neural impulse, which is detected and deconvoluted by the brain. The photochemical reaction is breakage and, after rotation, re-formation of the C=C bond.

> The absorption of some photons does not lead to bond cleavage, but to molecular rearrangements.

Photons of microwave light do not possess sufficient energy to break the bond in retinal, so the photochemical reactions in Figure 9.20 do not occur in the eye when looking at microwaves. Without the photochemical reaction in the bacillary layer, no electrical impulse is formed, and the brain does not 'see' the microwave radiation, so it is invisible light.

9.3 Photochemical and spectroscopic selection rules

We have now looked at the way photons are absorbed. Photons of UV and visible light cause electrons to promote between orbitals. Infrared photons have less energy, and are incapable of exciting electrons between orbitals, but they do allow excitation between quantized vibrational levels. The absorption of microwaves, which are less energetic still, effects the excitation between quantized rotational levels.

In this section, we shall look at the way these various absorptions are analysed by spectroscopists. There are four kinds of quantized energy: translational, rotational, vibrational and electronic, so we anticipate four corresponding kinds of spectroscopy. When a photon is absorbed or generated, we must conserve the total angular momentum in the overall process. So we must start by looking at some of the 'rules' that allow for intense UV–visible bands (caused by electronic motion), then look at infrared spectroscopy (which follows vibrational motion) and finally microwave spectroscopy (which looks at rotation).

Why is the permanganate ion so intensely coloured?

Optical charge transfer

The permanganate ion, MnO_4^-, has a beautiful, intense purple colour, and is a popular choice for oxidation reactions. The colour is intense even if the solution is dilute.

Care: don't confuse Group VII(a) (manganese, technetium and rhenium) with Group VII(b), the halogens.

Care: we write a *formal charge* with Arabic numerals, and means that the full charge exists as indicated: Cu^{2+} means a copper atom with fully two electronic charges missing. We write an *oxidation number* with Roman numerals, and does not relate to any physical loss or gain of electrons: it is purely a 'book-keeping' exercise. Mn^{VII} does *not* mean that a manganese atom has lost seven electrons.

The small increment of charge δ might be as small as 0.1 electronic charges.

It is simple to determine the molar absorption coefficient ε via Beer's law as 2334 $dm^3\,mol^{-1}\,cm^{-1}$ at its $\lambda_{(max)}$ of 523 nm (see Table 9.3).

The valence of the central manganese atom in the permanganate anion is +VII. Manganese is in Group VII(a) of the periodic table, so a valence of VII is achieved by removal of *all* five d-electrons and both s-electrons. The manganese in permanganate, therefore, has an electronic configuration of d^0. But this fact ought to make us think: if a colour is seen because electrons are photo-excited, how can the MnO_4^- ion have a colour, since it has no (outer-shell) electrons? The answer is *charge transfer*.

A very naïve view of the permanganate ion suggests the central manganese has a +7 charge, and each of the four oxygen atoms is an oxide ion of charge −2, i.e. the ion is held together with ionic bonds. This view is wholly false, since the bonding is almost completely covalent, with overlap of orbitals. But because the central Mn atom is d^0, each of the bonds is dative, with both electrons per bond coming from the oxygen atoms. In fact, it is probably safest to say that the formal charge on each oxygen atom is $-(2 - \delta)$, and the charge on the manganese is therefore $+4\delta$. This, then, is the electronic structure in the ground state, as discussed from within a valence bond model.

We will discuss the situation in terms of molecular orbitals, which is a better model. We will, however, simplify it a bit by talking as though there was a single Mn–O bond. As usual, absorption of a photon photo-excites an electron from the ground state to the first excited state. The molecular orbital for the ground state has the central manganese with an effectively d^0 structure, but the molecular orbital for the first excited state has a different distribution of electronic charge. This higher energy molecular orbital has one electron on the central manganese, and so fewer electrons on the oxygen. This situation is represented by

$$[Mn^{+\delta}\text{–}O^{-(2-\delta)}] + h\nu \longrightarrow [Mn^{+1}\text{–}O^{-1}] \qquad (9.16)$$

ground state   first excited state

The system relaxes back to the starting materials (the reverse of Equation (9.16)) within a tiny fraction of a second.

In a normal photo-excitation process, e.g. on a chromophore, the electron does not move spatially, but stays in the same physical location. Only its energy changes. The situation for MnO_4^- here is different: the electron not only changes its energy when photo-exited but also its position, from the oxygen in the ground state to the central manganese in the first excited state. We often say the electronic charge has transferred or, in short, that there is a *charge-transfer* process. In summary, although the permanganate ion is d^0, electrons can still be photo-excited, and hence the ion can act as chromophore.

The reason why the colour of MnO_4^- is so intense follows from the unusual way in which the electron changes its position. There are no restrictions (on a quantum-mechanical level) to the photo-excitation of an electron, so the probability of excitation is high. In other words, a high proportion of the MnO_4^- ions undergo this photo-excitation process. Conversely, if a photo-excited charge does not move spatially, then there *are* quantum-mechanic inhibitions, and the probability is lower.

In fact, the charge-transfer process is wholly *allowed*, but the more conventional transitions are only partially allowed. Indeed, the probability of some photo-excitation processes is so low that we generally say they are *forbidden*.

Charge transfer also occurs between ions in solution. The classical test for the Fe^{3+} ion in solution is to mix solutions of Fe^{3+} and thiocyanate ion SCN^-, to form the $[FeSCN]^{2+}$ complex, and a deep blood-red colour forms. The colour originates from a charge-transfer transition between Fe^{3+} and SCN^-. There was no red colour before mixing, confirming that the optical transition responsible for the colour did not originate from either constituent but from the new 'compound' formed.

> To ascertain if an optical transition arises from a charge-transfer process: (i) prepare two solutions, each containing one half of the charge-transfer couple; (ii) mix the two solutions. Charge transfer is responsible if a *new* optical band (and hence a new colour) forms.

Why is chlorophyll green?

Metal-to-ligand charge transfer

Chlorophyll is a member of one of the most important classes of pigments involved in photosynthesis, i.e. the process by which light energy from the sun is converted into chemical energy through the synthesis of organic compounds. It is found in virtually all photosynthetic organisms, including green plants, certain *protists* and bacteria, and *cyanophytes* (algae). The light energy absorbed by chlorophyll is consumed in order to convert carbon dioxide into carbohydrates.

> The word root 'cyan' in *cyanophyte* means a 'blue–green' colour, rather than cyanide.

The chlorophyll molecule comprises a central magnesium ion surrounded by an organic nitrogen-containing cyclic structure called *haem* (or 'heme'), (**V**) which is based on a porphyrin ring.

(V)

Naturally occurring compounds incorporating haem include vitamin B_{12}, haemoglobin, and chlorophyll. The central metal is iron in haemoglobin, magnesium in chlorophyll and cobalt in B_{12}.

Both magnesium and porphyrin are essentially colourless, but mixing solutions of magnesium and haem together causes a colour to form. The magnesium ion resides at the centre of the haem ring, which acts as a polydentate ligand.

The colour formed on mixing the two solutions, of magnesium and haem, arises from a charge-transfer bond. The resultant spectrum contains a broad, intense band that we call a *metal-to-ligand charge transfer* transition, or MLCT for short.

Why does adding salt remove a blood stain?

A charge-transfer couple

Haemoglobin is an essential component of the blood. The central, active part of the molecule comprises a haem ring (as above) at the centre of which lies an iron ion. We call it a charge transfer *couple*, since it requires two constituents, both the iron and the haem.

An old-fashioned way of removing the stain caused by blood is to place the stained item in a fairly concentrated solution of salt, and is usually performed by sprinkling table salt on the bloodied cloth after dampening it under a tap.

After a while, sodium ions from the salt swap for the iron ion at the centre of the haem ring. There is no longer a couple (one component is lost), and consequently no scope for an MLCT transition, so the red colour of the blood fades.

It is likely that the haem remains incorporated in the cloth, but it is not involved in a charge-transfer type of process and is now invisible.

What is gold-free gold paint made of?

Mixed-valence charge transfer

The 'gold paint' with which lamp-posts and other ornamental metalwork are decorated contains no gold, because of its prohibitive expense. The reflective gold-coloured substitute is sodium tungsten bronze, $Na_{0.3}WO_3$. So why is $Na_{0.3}WO_3$ reflective?

Other well-known compounds that exist with a mixed valence are Prussian blue [Fe^{3+}[Fe^{II}(CN)$_6^{4-}$]], and *ferredoxin*, in which iron has the valence of +II and +III.

We make $Na_{0.3}WO_3$ by annealing solid WO_3 in sodium vapour. Metallic sodium is a strong reducing agent, and readily reduces tungsten from W^{VI} to W^V, with each sodium atom becoming a sodium ion. (The sodium ions are necessary to maintain overall charge neutrality.)

If all the WO_3 was reduced with sodium, the product would be $NaWO_3$, in which the oxidation state of each the tungsten is +V. But the stoichiometry of $Na_{0.3}WO_3$ implies that only 30 per cent of the tungsten atoms have been reduced. A simple calculation to determine the oxidation number of the tungsten atoms in $Na_{0.3}WO_3$ soon shows how the compound is not electronically simple. In fact, the tungsten atoms exist in two separate valences, with W^V and W^{VI} coexisting side by side. $Na_{0.3}WO_3$ is, therefore, termed a *mixed-valence compound*, and shows *intervalence* behaviour.

Mixed valency of this sort is the cause of the reflective, gold colour of $Na_{0.3}WO_3$. In this system, like the MnO_4^- ion described above, electrons are excited optically following photon absorption from a ground-state electronic configuration to a vacant electronic state on an adjacent ion or atom. The colour is caused by a photo-effected intervalence transition between adjacent W^{VI} and W^V valence sites:

> The labels 'A' and 'B' merely serve to identify two adjacent atoms. Notice how the atoms do not move spatially, but an electron does move in response to the photo-excitation, going *from* W_B *to* W_A.

$$W_A^{VI} + W_B^V + h\nu \longrightarrow W_A^V + W_B^{VI} \qquad (9.17)$$

The system relaxes back to the starting materials (the reverse of Equation (9.17)) within a fraction of a second.

Such intervalence transitions are characterized by broad, intense and relatively featureless absorption bands in the UV, visible or near infrared (NIR).

Following the photo-excitation process, the electron moves spatially as well as changing energy, so this is again a charge-transfer process. Being a wholly allowed transition, the molar absorption coefficient ε is relatively large at 5600 $dm^3 \, mol^{-1} \, cm^{-1}$ (which is nearly three times more intense than for permanganate).

> Many books speak of *near* (wavelengths of 7×10^{-7} m to 4×10^{-4} m) and *far* (wavelengths of 4×10^{-4} m to about 0.01 m) infrared.

What causes the blue colour of sapphire?

Intervalence charge transfer

Sapphires are naturally occurring gem stones, and are transparent to translucent. We know from ancient records that they have always been highly prized. Natural sapphires are found in many igneous rocks, especially *syenites* and *pegmatites*. Alternatively, synthetic sapphires may be made by doping aluminium oxide Al_2O_3 with a chromophore.

The principal source of the sapphire's colour is the presence of small amounts of iron and titanium. The relative ratio of these two contaminants helps explain why the colour of sapphire ranges from a very pale blue through to deep indigo. The most highly valued sapphires have a medium-deep cornflower blue colour. Colourless, grey, yellow, pale pink, orange, green, violet, and brown varieties of the semi-precious gem corundum are also known as sapphire, although red sapphire is more properly called ruby.

But why is sapphire blue? Alumina is colourless, yet neither iron nor titanium commonly form blue compounds. The colour arises from a charge-transfer type of interaction. In the ground state, Fe^{II} and Ti^{IV} redox states reside close together. Following photo-excitation, an electron transfers from the Fe^{2+} to the Ti^{IV}:

$$Fe^{II} + Ti^{IV} + h\nu \longrightarrow Fe^{III} + Ti^{III} \tag{9.18}$$

The frequency of the light absorbed to effect the photo-excitation is in the red and near-infrared parts of the visible region, so the complementary colour seen is blue. This explains why sapphire is blue. It is again a charge-transfer excitation, but not of the mixed-valence type. The optical band formed is intense, so a strong colour is seen even though the concentrations of iron and titanium are minuscule.

The system relaxes back to the starting materials (the reverse of Equation (9.18)) within a fraction of a second (probably 10^{-15} s).

Why do we get hot while lying in the sun?

Vibrational energy

> Photons of infrared light are not sufficiently energetic to excite electrons, but can excite between quantum-mechanical vibrational levels.

Irradiation with infrared light causes the sensation of warmth, which is why we sit next to a radiator when cold, and why our temperature increases when we sunbathe on a sunny beach. Photons of infrared light are absorbed. Whereas electrons are excited following irradiation with visible light, photons of infrared light do not possess sufficient energy because their wavelengths are longer (remember that $E = hc/\lambda$). They cannot photo-excite electrons, but they can excite the *bonds*, causing them to vibrate.

We have already seen how, on the microscopic level, the vibrational energies of bonds are quantized in a similar manner to the way the energies required for electronic excitation are quantized. For this reason, irradiation with an infrared light from the sun or a lamp results in a photon absorption, and the bonds vibrate, which we experience as the sensation of heat.

> Equation (9.19) is a chemical version of *Hooke's law*, and only applies where the Morse curve is *parabolic*, i.e. near the bottom of the curve where molecular vibrations are of low energy.

Mathematically, the movement of vibrating atoms at either end of a bond can be approximated to simple-harmonic motion (SHM), like two balls separated by a spring. From classical mechanics, the force necessary to shift an atom or group away from its equilibrium position is given by

$$\text{force} = -kx \tag{9.19}$$

where x is the displacement of the atom from the position of equilibrium and k is the force constant of the bond. The magnitude of k reflects the *elasticity* of the bond, with a large value of k relating directly to a strong bond.

Force constants for bond stretching are much larger than for bond-angle bending, which are themselves considerably larger than for torsional vibrations. In consequence, the energy needed for molecular distortion decreases in going from stretching to bending to torsion. We can approximate this statement, saying that bonds do not deform, angles are stiff and torsion angles are flexible.

Molecules translate, rotate and vibrate at any temperature (except absolute zero), jumping between the requisite quantum-mechanically allowed energy levels. We call the common 'pool' of energy enabling translation, rotation and vibration 'the thermal energy'. In fact, we can now rephrase the statement on p. 34, and say that temperature is a macroscopic manifestation of these motions. Energy can be readily distributed and redistributed at random between these different modes.

We describe as *rigid-body rotation* any molecular motion that leaves the centre of mass at rest, leaves the internal coordinates unaltered, but otherwise changes the positions of the atomic nuclei with respect to a reference frame. Whereas in a simple molecule, such as carbon monoxide, it is easy to visualize the two atoms vibrating about a mean position, i.e. with the bond length changing periodically, we may sometimes find it easier to 'see' the vibration in our mind's eye if we think of one atom being stationary while the other atom moves relative to it.

It is usually easier, mathematically, not to think in terms of wavelength λ (which is inversely proportional to energy) but to employ variables that are directly proportional to energy. Most spectroscopists use ω, which is the frequency of the vibration normalized to the speed of light c, so $\omega = \nu \div c$, where ν is the frequency. In the context of infrared spectroscopy, we usually call ω the *wavenumber* of the band vibration.

The wavenumber of a bond vibration is given by

$$\omega = \frac{1}{2\pi c}\sqrt{\frac{k}{\mu}} \tag{9.20}$$

where c is the speed of light *in vacuo*; k is the bond force constant with units of N m^{-1}. This way, the wavenumber has the SI units of m^{-1}.

> *In vacuo* is Latin for 'in or through a vacuum'.

The parameter μ in Equation (9.20) is called the *reduced mass*. For the carbon monoxide molecule CO, μ is defined by

$$\mu = \frac{M_C \times M_O}{M_C + M_O} \times 1.66 \times 10^{-27} \tag{9.21}$$

where M_C and M_O are the *relative* atomic masses of the carbon and oxygen atoms respectively. We need the additional factor of 1.66×10^{-27} kg since this is the mass of one atomic mass unit. With this conversion factor in place, we can express μ in the SI units of kg molecule^{-1}.

Worked Example 9.8   What is the reduced mass of carbon and oxygen in the C=O bond?

The relative atomic masses of oxygen and carbon are 12.00 and 16.01 respectively. Inserting values into Equation (9.21) yields

$$\mu = \frac{12.00 \times 16.01}{12.00 + 16.01} \times 1.66 \times 10^{-27} = 6.858 \times 10^{-27} \text{ kg}$$

Notice how μ is smaller than either M_O or M_C.

Because different bonds connect different atoms or groups of atoms, the energy needed to excite the bonds varies in a characteristic way, so a C–H bond vibrates following absorption of a specific energy, which is different from the energy of, say, an N≡N bond. Table 9.5 contains some force constants for a few diatomic molecules. Clearly, single bonds such as Cl–Cl are weaker than double bonds, e.g. C=O. From the magnitudes of the numbers in Table 9.5, the triple bond in nitrogen is the strongest, as expected.

SAQ 9.6 What is the reduced mass of (1) chlorine, (2) hydrogen, and (3) H–Cl in the HCl molecule?

If we know the masses of the atoms involved in a vibration, and the wavenumber can be determined from a spectrum, then we can readily calculate a value for the force constant k.

Worked Example 9.9 A spectrum of pipe smoke contains a band at 2180 cm^{-1}, which is attributable to carbon monoxide. What is the force constant k for the C=O bond?

We must first convert the wavenumber ω to SI. If there are 2180 vibrations per centimetre, and there are 100 cm per metre, then there must be 2180×100 vibrations per metre, so $\omega = 2.18 \times 10^5$ m^{-1}.

Second, we rearrange Equation (9.20) to make k the subject. After squaring both sides:

$$\omega^2 = \frac{1}{4\pi^2 c^2} \frac{k}{\mu}$$

Then, multiplying both sides by $4\pi^2 c^2 \mu$ yields

$$k = (4\pi^2 c^2 \mu) \times \omega^2$$

Finally, we insert values. (The value of μ for the C=O bond was calculated in Worked Example 9.8 as 6.858×10^{-27} kg.)

> From Newton's second law, the unit of kg s^{-2} equates to N m^{-1}.

$$k = 4 \times (3.142)^2 \times (3 \times 10^8 \text{ m s}^{-1})^2 \times (6.858 \times 10^{-27} \text{ kg})$$
$$\times (2.18 \times 10^5 \text{ m}^{-1})^2$$

so

$$k = 1158 \text{ kg s}^{-2} = 1158 \text{ N m}^{-1}$$

SAQ 9.7 From Equation (9.20), calculate the wavenumber ω of the vibrational band for the fluorine molecule. The force constant k for F_2 is listed in Table 9.5.

Table 9.5   Force constants for some diatomic molecules

Molecule	Force constant $k/\text{N m}^{-1}$
H–Cl	516
H–F	964
Cl–Cl	320
F–F	445
O–O	1141
N=O	1548
C=O	1855
N≡N	2241

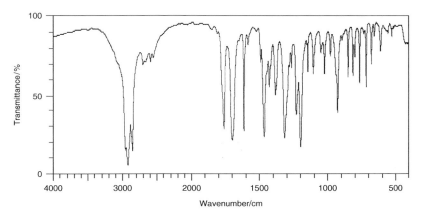

Figure 9.21   Infrared spectrum of aspirin, i.e. 2-acetoxybenzoic acid. (Solid aspirin was prepared within a Nujol mull)

What is an infrared spectrum?

Polyatomic molecules

Figure 9.21 depicts the infrared spectrum of aspirin (**VI**). The presence of several peaks demonstrates how several infrared photon energies are absorbed while others are not, so the spectrum shows how aspirin is transparent to photons of energy 2000 cm^{-1} (corresponding to a molar energy of 26.6 kJ mol^{-1}), but does absorb very strongly at 2900 cm^{-1}.

(VI)

A molecule containing more than one atom is termed *polyatomic*, since 'poly' is Greek for 'many'.

A *series* of vibrations is possible in a complicated molecule, by which we mean one containing three or more atoms. As a simple example, let us consider the molecule responsible for the bleaching action of common household bleach, hypochlorous acid, H–O–Cl. There are two bonds, H–O and O–Cl, and each can vibrate. Although the vibration of one bond is not wholly independent of the other, we can usually approximate and treat the vibrations as if they were independent. A single molecule can, therefore, have a wide range of frequencies and k values.

In fact, to complicate the situation further, a single bond can have several values of frequency because there are different *kinds* of vibration: the simplest bond vibration is movement along the direction of the bond axis. This is called a *stretch mode* vibration, since the bond *length* changes rhythmically while the bond *angle* remains unchanged. A stretch of this sort is depicted in Figure 9.22(a).

Next are the vibrations in which the bond angle varies periodically but the bond length remains constant. This is called a *scissor-mode* vibration, because it looks like the action of a pair of scissors, as depicted in Figure 9.22(b).

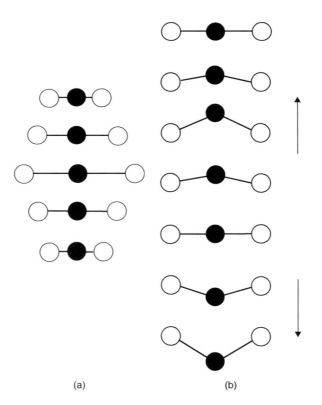

(a)   (b)

Figure 9.22   Simple vibration modes for carbon dioxide, O=C=O: (a) a symmetric stretching mode and (b) a 'scissor mode' vibration

The frequency of these two vibration modes will be different because the energy required for each is different.

An *infrared spectrum* is a different way of displaying such information, and represents a graph in which the function of energy chosen for the x-axis is the wavenumber ω (frequency). By convention, the x-axis is usually expressed in units of cm^{-1}, with values decreasing in going from left to right. Therefore, we say that the vibrations on the left-hand side (at the highest ω) represent vibrations of higher energy. As a good generalization, the peaks in the left-hand side of the spectrum represent vibrations of the stronger bonds. For example, a C=C double bond in a benzene ring will absorb photons of energy 20.3 kJ mol^{-1}, which corresponds to a wavenumber of 1700 cm^{-1}.

> *Remember*: an infrared spectroscopist calls these frequencies *wavenumbers*.

Other atoms and groups of atoms have different masses (and hence reduced masses μ), so they resonantly absorb infrared photons of different energies. Chemists usually want to know which bonds a molecule is made up of, so they will obtain an infrared spectrum by scanning over a wide range of energies, and look at which infrared energies are absorbed. The energy (wavenumber) of each peak is characteristic of a certain type of bond, so the chemist merely has to compare the peak energies from the spectrum with wavenumbers in data books.

Why does food get hot in a microwave oven?

Rotational energy

Not everyone cooks with a conventional oven: many people prefer to cook with a *microwave* oven, because it works faster and consumes less power (and, therefore, is more efficient) than a conventional oven. A microwave oven heats the food by absorption of electromagnetic radiation from the microwave region of the spectrum. Such microwaves can be viewed as either radio waves of very short wavelength or infrared light of longer wavelength. In the case of a typical microwave oven, the frequency is typically 2.45 GHz, and is usually generated by a *magnetron*. Radio waves in this frequency range are readily absorbed by water and by small molecules having the same mass as water.

> A microwave oven is more efficient than a conventional oven because it only heats the food without first heating the oven itself.

> Metals tend to reflect microwaves, hence the 'lightning' seen when we put a metallic item in a microwave oven. The microwaves' energy is not absorbed by the food, and is otherwise absorbed in order to ionize the (moist) air inside the oven.

SAQ 9.8 Show that 2.45 GHz (2450 MHz) corresponds to a wavelength of 12.2 cm.

Irradiation of food inside the microwave oven causes photon uptake. The energy liberated each time a photon is absorbed is not sufficient to cause bond breakage (as was the case with UV light); nor can these microwave photons cause excitation of electrons (which is why we see a colour during irradiation with visible light but not with microwaves). Again, the energy is insufficient to cause molecular vibrations, so there is no vibration of the atoms and groups within the molecules comprised by the food.

> The food in a micro-wave oven is effectively steam-cooked because the heating occurs when water molecules receive the energy of the microwaves.

> The overwhelming majority of plastics do not contain water, nor do glass or ceram-ics, which is why they do not get hot when placed within a micro-wave oven.

> Although 2.45 MHz does not coincide with the maximum micro-wave absorption for water, there is a huge *pressure broadening* effect at atmospheric pressure, so there is a sufficient overlap for the absorption process to be fairly efficient.

> A conventional oven cooks from the out-side first, explaining why the crust on the outside of a loaf of bread is crispy and brown yet the inside is soft and moist. Con-versely, bread made in a microwave oven has no crust.

But there is a type of excitation that requires less energy, i.e. *rotation*. Irradiation with microwave radiation causes small mole-cules such as water to absorb energy, and thence rotate at high speed. These absorptions are 'allowed', quantum mechanically.

Just as the absorption of UV or visible light causes electrons to excite between different electronic quantum states, so absorp-tion of infrared photons causes excitation between allowed vibra-tional states, and absorbing microwave radiation causes excitation between allowed rotational states in the absorbing molecule. As a crude physical representation, these quantum states correspond to different angular velocities of rotation, so absorption of two pho-tons of microwave radiation by a molecule results in a rotation that is twice as rapid as following absorption of one photon.

The molecules within the sample (e.g. the food in the microwave oven) cannot continue to absorb more and more energy indefinitely, so the energy must be dissipated somehow. In practice, the spinning water molecules collide with other molecular features within the topography of the food, resulting in inelastic collisions and the transfer of energy. We discern this energy transfer as an increase in heat (see Section 1.4), so the food gets hot.

The US Federal Communications Committee chose a frequency of 2.45 GHz for microwave ovens to ensure that they do not inter-fere with other equipment operated by microwaves, such as mobile phones. Therefore, we see that, contrary to popular belief, a fre-quency of 2.45 MHz is *not* a resonant frequency for molecules of water.

We see how the microwave oven is merely an extremely effi-cient means of warming anything containing water. Furthermore, the water is evenly distributed throughout the food, so heating is usually uniform: heating occurs everywhere all at once because the water molecules are all excited at the same time. The process of heating is different from that in a normal oven because we excite the atoms rather than 'conducting heat'.

Defrosting food in a microwave oven is not very efficient, because the molecules of water are incapable of rotation, and are 'locked' into position in the form of ice crystals.

A molecule must have a permanent dipole moment to be micro-wave active. As it rotates, the changing dipole moment interacts with the oscillating electric field of the electromagnetic radiation, resulting in absorption or emission of energy. This requirement means that homonuclear molecules such as H_2 are microwave inac-tive, but heteronuclear molecules such as SO_3, SO_2, NO and, of course, H_2O are active.

Rotational quantum numbers

Just as electronic excitations and vibrations are photo-induced, so rotational energy is also quantized. The energy of a photon of light is given by the Planck–Einstein equation, $E = h\nu$. When a rotating molecule absorbs a photon of light, it is excited from one quantum mechanically allowed rotational energy level to another one of higher energy.

We tend to give the letter J to the rotational quantum states. The rotational ground state has a rotational quantum number of J' and the excited rotational quantum number is J. To be *allowed* (in the quantum-mechanical sense), the excitation from J' to J must follow

$$\Delta J = \pm 1 \qquad (9.22)$$

Strictly, Equation (9.22) applies to diatomic molecules having a permanent dipole moment.

We call Equation (9.22) the *rotational selection rule*. Equation (9.22) assumes that the bond lengths do not alter during rotation. For this reason, we call the rotating body a *rigid rotor* (or 'rotator'). In fact, all the bonds will stretch to some slight extent as the speed of rotation increases, in consequence of centrifugal forces, just as a spring stretches when rotating a weight on its end.

The energies of rotation are quite small, so we require photons of relatively low energy to photo-excite between rotational quantum levels. For this reason, the spacings between rotational energy levels correspond to transitions in the far infrared and microwave regions of the electromagnetic spectrum.

> We say, in the quantum-mechanical sense, an excitation process is *allowed* if the probability of it occurring is very high. If the probability is low, we say the process is *forbidden*.

> Equation (9.22) assumes that the bond lengths do not alter during rotation. For this reason, we call the rotating body a *rigid rotor*.

> We follow the rotational behaviour of molecules with *microwave spectroscopy* because the spacings between each rotational energy level correspond to transitions in the far infrared and microwave regions of the spectrum.

Are mobile phones a risk to health?

Microwaves and human health

Microwave research started in earnest during World War II, when the need to detect and locate enemy aircraft at long distances, and at night, was crucial to the defence of Allied forces. The ability of microwaves to cook food was first noticed in a rather macabre way in 1945: at that time, army personnel operating the radar equipment were experimenting with high-power equipment to enable communication over longer distances. They noticed how the experiments occurred concurrently with dead birds dropping to the ground. Closer inspection of the bird carcasses showed that birds were cooked, or at least partially so. Microwave energy cooks flesh, even if accidentally!

The possibility of microwaves being a hazard to health has been investigated extensively, and much care has been spent on optimizing safety features, so it is now clear that microwave ovens are completely safe.

The new generation of cordless phones transmit their text and verbal messages via microwaves, prompting persistent rumours that mobile phone can cause a user's ear to 'cook', much like a pie in a microwave oven. Too few studies have been performed to clarify, so the situation if still inconclusive. Certainly, the energy consumption of a mobile phone is tiny compared with that in an oven (a microwave oven operates with a power of about 200–1000 W, whereas a phone operates at a maximum energy of about 1 W – and the regulatory authorities are decreasing this maximum quite rapidly).

But there are disturbing signs that mobile phone usage can be quite addictive. Some researchers think the onset of addiction follows from minute changes deep within the brain caused by 'cooking' tiny glands embedded there. Others prefer an explanation from psychology or sociology – such as peer group pressure.

While some studies suggest that these phones operating via microwaves are safe, others suggest that they are quite dangerous. In conclusion, whereas microwave ovens are definitively safe (if used correctly), many medical reports now suggest we try not to communicate via mobile phones too frequently.

9.4 Photophysics: emission and loss processes

How are X-rays made?

Generation of photons

The transfer or conversion of energy is always associated with the emission of electromagnetic waves. We met this concept in its simplest form in Chapter 2, when we looked at the transfer of infrared radiation (i.e. heat). This emission of photons occurs because all objects contain electrically charged particles; and, whenever an electrically charged particles accelerates, it emits electromagnetic waves.

All objects at temperatures above absolute zero contain some thermal energy, so electrically charged particles within them continually undergo thermal motion. If we could cool matter to 0 K, then the thermal motion would cease and the matter would not emit any radiation. Again, we saw this idea in Chapter 2.

But absolute zero is unattainable, so all particles move. Furthermore, the particles never retain an invariant speed because inelastic collisions cause some particles to decelerate and others to accelerate. As a result, everything emits some electromagnetic waves, even if merely in the context of a dynamic thermal equilibrium with the object exchanging energy with its surroundings.

The energy of an emitted photon depends on the magnitude of the change in velocity, so a steep velocity gradient generates (or requires) a more energetic photon. The intensity of the light emitted is simply a measure of how many particles change their velocity.

> The word 'gradient' here implies the slope of a graph of particle velocity (as 'y') against time (as 'x').

An electron is readily decelerated to a standstill simply by firing it through a vacuum toward a relatively massive object. The energy of the photons emitted relates to the difference between the electron's kinetic energy before and after the collision; so, if the electron is travelling extremely fast before the collision and stops dead (has a zero velocity), then the energy change is substantial, and the energy per photon is large.

We often go to hospital to obtain an X-ray photograph if we have a broken bone. At the heart of the machine generating the X-rays, electrons are accelerated to a very high energy, which corresponds to a fast velocity. These electrons are then smashed into a large metal target. Rapid deceleration of the electrons occurs simultaneously with X-rays emerging at an angle that is normal to the angle of impact.

> We simplify the argument here: in fact, the energy E of the photon relates to the frequency v of the photon and also the work function ϕ of the metal, according to the equation $E = hv - \phi$.

The frequency v of the X-rays relates to the energy change E on deceleration according to the Einstein equation $E = hv$ (Equation (9.3)). The actual speed at which the electrons decelerate on impact depends on the identity of metal with which the target is made: most X-ray generators employ copper as the target. For this reason, we often see the X-rays employed described as 'Cu Kα', where 'Kα' merely relates to electronic processes occurring within atoms of copper in the target.

Why does metal glow when hot?

Black body radiation

Warming iron or steel to a temperature of about $500\,°C$ causes it to glow a dull red colour, as seen on an electric cooker set at 'low'. The oven ring appears bright orange if the temperature increases further ($\sim 1000\,°C$). In these kitchen items, an electric current inductively heats a coil of wire.

Pure iron melts at $1532\,°C$, at which temperature the molten iron glows white–yellow. Further heating to about $2500\,°C$ causes the colour to change again to brilliant white. In short, all the colours of the visible spectrum are represented. Even before the iron begins to glow red, we can feel the emission of infrared light as the sensation of heat on our skin. A white-hot piece of iron also emits ultraviolet radiation, as detected by a photographic film.

But not all materials emit the same amount of light when heated to the same temperature: there is a spectral *distribution* of electromagnetic waves. For example, a piece of glass and a piece of iron when heated in the same furnace look different: the glass is nearly colourless yet feels hotter to the skin because it emits more infrared light; conversely, the iron glows because it emits visible as well as infrared light.

This observation illustrates the so-called *rule of reciprocity*: a body radiates strongly at those frequencies that it is able to absorb, and emits weakly at other frequencies.

The heated body emits light with a spectral composition that

> The correct explanation of black-body radiation was an early triumph of quantum theory.

The heated body emits light with a spectral composition that depends on the material's composition. That observation is not the case for an 'ideal' radiator or absorber: ideal objects will absorb and thence re-emit radiation of all frequencies equally and fully. A radiator/absorber of this kind is called a *black body*, and its radiation spectrum is referred to as *black-body radiation*, which depends on only one parameter, its temperature, so a hotter body absorbs more light and emits more light.

Strictly, a black body is defined as something that absorbs photons of all energies, and does not reflect light. Furthermore, a black body is also a perfect emitter of light. A black body is a *theoretical* object since, in practice, nothing behaves as a perfect black body. The best approximations are hot objects such as red- or white-hot metals.

How does a light bulb work?

The emission of light

At heart, a normal light bulb operates via the resistive heating of a metal filament. It's not particularly efficient, and a bulb only lasts about 1000 h at most. Its poor efficiency follows because the bulb simultaneously radiates a lot of infrared heat as well as light; in fact, only about 20 per cent of the energy emitted is light that is visible to the eye. Since the purpose of a bulb is to generate light, this heat represents a waste of energy. The bulb does not last long because the tungsten of the filament evaporates and deposits on the glass. The filament eventually breaks when it becomes too thin, and we say that it has 'burnt out'.

Why is a quartz–halogen bulb so bright?

Wien's law

> This type of bulb is called a quartz–halogen lamp because iodine is a halogen element (from Group VII of the periodic table).

Like a conventional light bulb, a quartz–halogen bulb contains a tungsten filament. The first major difference between the two bulbs is the encapsulation of the tungsten within a small quartz envelope. Because the envelope is so small, the distance between the quartz and the filament is tiny, so the quartz gets very hot. Glass would melt, so we need the higher melting temperature of quartz.

The second difference is the gas inside the bulb. Inside a normal light bulb, the gas is usually argon, but the gas inside the quartz halogen bulb is iodine vapour at low pressure, which has the ability to combine chemically with tungsten vapour. When the temperature is sufficiently high, the halogen gas combines with tungsten atoms as they evaporate from the filament, and redeposits them on the wire. This process of recycling enhances the durability of the filament.

More importantly, the addition of iodine makes it possible to operate the filament at a hotter temperature, with a higher proportion of the emitted light being visible light, although we still form much heat.

We start to discern a dim glow if a surface is heated to about 400 °C in a darkened room. We see the dull red of a heat lamp after heating to about 500 °C. A conventional light bulb emits light when a coiled tungsten wire is heated to a temperature of about 2500 °C, and incandesces. The yellow glow of the sun relates to a temperature of 5800 °C. A blue light requires an even hotter temperature. We deduce the generalization that the light emitted becomes progressively more blue light as the temperature increases.

Figure 9.23 depicts the distribution of the wavelengths emitted by the bulb. When heated electrically to about 2500 °C, the coil becomes 'white hot' and emits a great deal of visible light, although $\lambda_{(max)}$ is about 1000 nm. The wavelengths emitted lie in the range 500 to 6000 nm. Quartz–halogen bulbs operate at a temperature of 3000 °C, causing the maximum wavelength to shift to 800 nm. If the temperature were to increase to 4000 °C, the wavelength corresponding to the maximum intensity of photons would lie on the upper wavelength limit of the visible, at 700 nm.

Wien's law offers a simple relationship between the wavelength maximum $\lambda_{(max)}$ and the thermodynamic temperature T:

$$\lambda_{(max)}T = \text{constant} \qquad (9.23)$$

> Wien's law is also called Wien's *displacement* law.

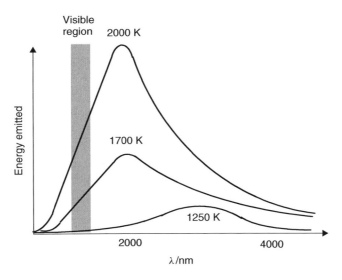

Figure 9.23 Wien's law and black-body radiation: as the temperature T of the black body is raised, so the wavelength maximum of the emitted radiation decreases. The area under the curve indicates the intensity of the energy emitted by the black body, and is proportional to T^4

The filament in a light bulb is made of tungsten because its melting point is so high. Operation of the bulb near its maximum temperature enables the value of $\lambda_{(max)}$ to shift to shorter wavelengths, i.e. closer to the sensitivity of the human eye.

The shape of each curve in Figure 9.23 is the same, but the maximum is higher as the temperature rises, indicating how more energy is releases at higher T. This shows a second reason why quartz–halogen bulbs are brighter: more energy is emitted (such a lamp generally operates at about 300 W, rather than the 100 W in a more conventional bulb).

What is 'limelight'?

Incandescence

'Limelight' is a term from the theatre of the 19th century. Thomas Drummond introduced the first theatrical spotlight in 1816. He was so successful that *limelight* soon became a popular term, meaning 'appearing on stage, and being in the public eye'. Drummond's light consisted of narrow rods of calcium oxide housed within small burners along the edge of the stage. Each rod was roasted strongly in jets of burning oxygen and hydrogen.

When something produces light on heating, we say it *incandesces*. In fact, anything will glow when heated sufficiently, but the amount of light emitted depends quite markedly on the material chosen. For example, steel is a fairly efficient producer of light, but glass is very poor. Heating calcium oxide to a temperature above about 1000 °C causes incandescence and a soft, yet brilliant white light is produced. With a suitable arrangement of reflectors around the hot lime, this light – the limelight – was projected onto the actors on the stage, and formed the first recorded type of spotlight.

Lime is the old-fashioned name for calcium oxide. As a root, it is also found in *lime*stone (calcium carbonate), *lime* pits (burial sites for the poor, which were lined with CaO) and *quick* lime (calcium hydroxide).

We choose lime because it has a high melting temperature, so it can be heated to a white glow without it melting. Pure iron melts at 1532 °C and lime melts at around 2500 °C. Heating glass causes it to glow, but it gives off much less light than the same volume of steel or lime.

Another everyday example of incandescence is the heating of a metal wire to about 1000 °C in a conventional light bulb. The light emitted from a candle or other form of fire is a further demonstration of incandescence.

The light emitted by a candle originates from hot particles of soot in the flame; these soot particles strongly absorb and thence re-emit it as visible light. By contrast, the gas flame of a kitchen oven is paler, despite being hotter than a candle flame, and does not emit much light owing to an absence of soot.

Why do TV screens emit light?

Luminescence

Luminescence is the emission of light by certain materials when they are relatively cool, in contrast to light emitted during incandescence, such as when a candle burns, when rock is molten, or when a wire is heated by an electric current within a conventional light bulb. Such luminescence is most commonly seen in neon and fluorescent light tubes, and in most television screens (though some are now based on liquid crystals).

The name 'luminescence' is generally accepted for all the light-emission phenomena not caused solely by a rise in temperature (as above), but in fact incorporates the different categories of phosphorescence and fluorescence, although the distinctions between the terms is still in some dispute.

Electrons are accelerated by a large electron gun positioned behind the screen. Since they are charged, the electrons are readily accelerated by means of a large voltage. The electron has a considerable momentum by the time it slams into the inside of the TV's front screen. In a black and white TV, the inside face is coated with a phosphor, i.e. a substance that emits light. Electromagnetic coils (or electronically charged metal plates) direct the beam of electrons from side to side and top to bottom, covering the entire screen every twentieth of a second or so. Every phosphor that is struck by an electron emits light; the light emitted by the screen is then perceived by the brain as an 'image' or 'picture'.

A colour TV screen operates in the same way as a black and white machine, except that the inside of the screen is coated with thousands of groups of dots: each group consists of three dots, one each for red, green and blue.

> The groups of dots are called 'picture elements' – or *pixels* for short.

The kinetic energy of the electron from the electron 'gun' is absorbed by the phosphor, and re-emitted as the visible light seen by the viewer.

Aside

Phosphors

The most important phosphors are sulphides and oxides of transition metals. The sulphides of zinc and of cadmium are the most important materials of the sulphide type. An important condition of achieving a highly efficient phosphor is to prepare a salt of the highest possible chemical purity. The emission of zinc sulphide can be shifted to longer wavelengths by increasingly replacing the zinc ions with cadmium.

Sulphide-type phosphors are produced from pure zinc or cadmium sulphide (or mixtures thereof) and heating them together at about $1000\,°C$ with small quantities (0.1–0.001 per cent) of copper, silver, gallium, or other salts, which are termed *activators*.

Some oxide-type minerals have been found to luminesce when irradiated. A simple example is ruby (aluminium oxide with chromium activator), which emits bright-red light. The phosphors are incorporated into colour television screens to emit the colours blue (silver-activated zinc sulphide), green (manganese-activated zinc orthosilicate), and red (europium-activated yttrium vanadate).

> Copper-activated zinc and cadmium sulphides exhibit a rather long afterglow when their irradiation has ceased, which is favourable for application in radar screens and self-luminous phosphors.

Why do some rotting fish glow in the dark?

Bioluminescence

In some parts of the world, one of the simplest tests of whether meat or fish has gone rotten is to look for the emission of light. The test for healthy flesh is to ask the question, 'Does it glow?' If the flesh does display a ghostly glow then it is *luminescent*, and contains biochemical organisms that are hazardous to health.

Bioluminescence results from a chemical reaction, so it is more strictly termed *chemiluminescence*. Biochemical energy is converted directly to radiant energy. The process is virtually 100 per cent efficient, so remarkably little heat is generated during emission. For this reason the emission is often called 'cold light', or *luminescence*.

This form of luminescence occurs sporadically in a wide range of natural organisms, such as protists (bacteria, fungi), animals, marine invertebrates and fish. It even exists naturally, albeit rarely, in plants or in amphibians, reptiles, birds, or mammals.

In most bioluminescent organisms, the essential light-emitting component is the oxidizable organic molecule *luciferin* and the enzyme *luciferase*. The luciferin–luciferase reaction is an enzyme–substrate reaction (i.e. as a catalyst; see Section 8.5) in which luciferin is catalytically oxidized by molecular oxygen. Luciferin is oxidized to form the ketone structure (**VII**), with emission of light occurring during the enzymic phase of the reaction.

(VII)

> One of the names of the devil in the Hebrew Bible is *Lucifer*, and comes from a Hebrew word meaning 'light bearer'.

Luciferin is one of the simplest examples of *chemiluminescence*, and it is remarkably efficient. The overall yield is one photon per molecule of luciferin.

The light emission continues until all of the luciferin has been oxidized. This type of reaction is found in fireflies, *Cypridina*, *Latia*, and many types of fish, such as lantern fish or hatchet fish.

How do 'see in the dark' watch hands work?

Phosphorescence

'See in the dark' watches operate with special material painted onto the watch hands, which allows them to be seen even in the dark. The paint usually looks pale green to the eye during daylight, and has a ghostly green–white colour at night. The watch hands can be seen many hours after sundown. The paint is *phosphorescent*, even though we often refer to it as 'fluorescent paint.'

Like fluorescence, phosphorescence is the emission of light by a substance previously irradiated with electromagnetic radiation. Unlike true fluorescence, however, the emission persists as an afterglow after the exciting radiation has been removed. The glow persists from about 10^{-3} s after the irradiation ceases, through to days or even weeks. In effect, we can say that the light is *stored* within the phosphorescent material. The length of time depends on the internal electronic levels involved.

Phosphorescence is different from the more straightforward process of fluorescence because an additional excitation occurs, which produces an emission of light in the visible region. Figure 9.24 highlights the difference between phosphorescence and fluorescence. Between the ground and excited electronic levels is a band of intermediate energy, called a *metastable level* (or *electron trap*). An electron in an excited level can demote to the metastable level by emission of radiation or by energy transfer (as below).

Because transitions between the metastable level and the ground or excited levels are so slow, we sometimes say they are *forbidden*. Once an electron enters the metastable level, it remains there until it can make a forbidden transition, in which case a photon is released. The time of residence for the electron in the metastable level determines the length of time that phosphorescence persists.

Glow-in-the-dark molecules possess such metastable quantum states. The reason why the light release persists is that demotion of an electron from the metastable state to the ground state can only

> The quantum-mechanical term *forbidden* indicates that the probability of the event or process is too tiny for it to occur to any significant extent.

> A *metastable* state in physics and chemistry is an energetically excited state in which an electron resides for an unusually long time. A metastable state, therefore, acts as a kind of temporary 'energy trap'.

> The relative distance of a line, or level, above a base line (the ground level) denotes the energy of an electron occupying that level.

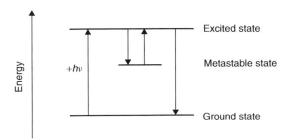

Figure 9.24 Energy-level diagram for a luminescent species, in which a metastable state slows the rate of emission. The metastable state is also termed an ion trap

occur if one of the electrons changes its direction of spin. This change of spin in these molecules is almost totally forbidden by the laws of physics, and proceeds with a very slow rate constant.

The visible consequence of such a metastable state is phosphorescence: straightforward irradiation of such a material readily causes excitation to an excited state and partial demotion to the metastable state. But the electrons are then trapped in the metastable states. The faint glow demonstrates that a few electrons demote per second; the glow persists for as long as it takes for all the electrons to reach the ground state.

How do neon lights work?

Gas discharge lamps and photon emission

The French scientist Georges Claude was one of the first to experiment with different types of lamp. In 1910, he first made a glass tube filled with low-pressure neon. Within less than a decade, signs were being fashioned of glass tubes bent to form words and designs that glowed red or green or blue when the gases inside them were subjected to an electric field.

Figure 9.25 shows a typical neon lamp. It consists of a thin glass tube with an electrode at either end. A neon light requires an extreme voltage, which is provided by a so-called *neon-sign transformer*. Neon atoms are neutral, and cannot conduct electricity, but ionization of the gas-phase neon forms Ne^+ ions (Equation (9.24)), so the tube contains a mixture of electrons and ions:

$$Ne^0{}_{(g)} \longrightarrow Ne^+{}_{(g)} + e^- \qquad (9.24)$$

> A plasma is a conducting mixture of ions and electrons.

We say the gas *discharges* under these conditions, meaning that it can conduct, i.e. an electric current can flow through it.

Collisions involving the mobile electrons are generally elastic. They bounce, like a ball off a wall. But a tiny fraction of the electrons undergo an *inelastic* collision with un-ionized neon atoms, causing a fraction of the electron's internal energy to transfer to the neon atom. The electron subsequently moves away after the collision. It has less energy, and so is slower.

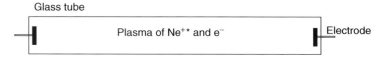

Figure 9.25   A neon lamp comprises two electrodes, one at either end of a thin glass tube. An extreme voltage ionizes the neon to form a plasma of Ne^{+*} and e^-. Inelastic collisions between Ne^* and Ne^+ allow for the release of energy as visible light

The product of the inelastic collisions is a 'hot' atom, Ne*:

$$Ne + energy\ from\ electron \longrightarrow Ne^* \qquad (9.25)$$

The excited-state neon atom is denoted with an asterisk, as Ne*. This hot atom must emit energy to return to the ground state, via the reverse of the reaction in Equation (9.25). The excited-state structure of Ne* has an electron in a orbital of higher energy (which was previously the LUMO). Subsequent relaxation of the electron releases a photon of light. It is this emission that we see. A neon light emits a bright, pink–red glow.

Gas-discharge lamps similarly containing a small amount of the rare gas krypton, emit a green glow; and helium-based lamps emit a pale blue glow. Lamps containing argon emit light in the near infrared, so no colour is visible to the eye.

When we look closely at a neon lamp, we should see that it is the gas itself that emits the light, and not the electrodes.

> The reaction in Equation (9.25) occurs in the gas phase, but the '(g)' subscripts have been removed for clarity.

How does a sodium lamp work?

Initiating a discharge lamp

The first high-intensity sodium lamp was introduced in Europe in 1931. Figure 9.26 shows a schematic view of a sodium lamp; it comprises a glass shell containing sodium vapour at low pressure, metal electrodes to generate a current, and neon gas. The pressure inside the tube is at a relatively low pressure of ~30 Pa, so some of the sodium evaporates to become a vapour. The inner side of the lamp is coated with the remainder of the metallic sodium as a thin film.

An electrode is positioned at either end of the tube, and a large voltage applied. When current first passes between the electrodes, the neon is ionized to form a plasma, and starts to glow (as above), which explains why a sodium lamp first emits a pink shade before it glows with its characteristic orange colour.

> Atmospheric pressure is 1.01×10^5 Pa, so the pressure inside the tube is almost a vacuum.

> Sodium lamps glow pink before orange because of the neon they hold, which 'kick starts' the sodium emission process.

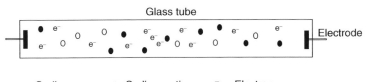

o Sodium vapour ● Sodium cations e⁻ Electron

Figure 9.26 A sodium streetlight. The pressure inside the tube is the relatively low pressure of ~30 Pa, so some of the sodium evaporates to become a vapour; the remainder of the sodium lines the inner side of the lamp as a thin metallic film. Soon after operation commences, 'hot' neon atoms help generate a plasma of sodium

When the lamp gets hotter, energy from the neon is transferred to the remainder of the sodium, which then vaporizes. Atoms of sodium in the gas phase are ionized to form a plasma of Na^+ and e^-. As with the neon lamp (above), some inelastic collisions between electrons and neon atoms generate excited-state Na^*, which must return to the ground state with the emission of a photon (cf. the reverse reaction to Equation (9.25)). The energy of this photon corresponds to light of wavelength 592 nm, which is why the sodium lamp appears orange. The energy release is extremely efficient.

The lamp above is more properly called a *low-pressure sodium lamp*. Such lamps are ideal for street and road illumination, but the monochromatic nature of the emission makes seeing in colour impossible. An adaptation which emits a *range* of colours is the high-pressure sodium-vapour lamp, which is similar to that described above but contains a mixture of mercury and sodium. Such lamps emit a whiter light and are useful for extra-bright lighting in places such as road intersections, car parks and sports stadia.

How do 'fluorescent strip lights' work?

Indirect discharge

Care: the two words 'fluorescence' and 'phosphorescence' are employed interchangeably in everyday life, but in fact relate to different photon pathways.

Strip lighting in a classroom, hospital, business hall or kitchen is often called *fluorescent lighting*, although in fact it is a phosphorescent process, as above. Each bulb consists of a thin, hollow glass tube that is sealed at both ends. It contains gas such as helium, argon or krypton, and a drop of liquid mercury (about 0.5 mg of mercury per kilogram of lamp, or 0.5 parts per million). Like the neon and sodium lamps above, the pressure inside the tube is about 30 Pa, so the mercury evaporates to become a vapour. It is the mercury that yields the light, albeit indirectly.

Again, like the sodium lamp above, application of voltage causes ionization and a plasma of Hg^{2+} ions and electrons soon form. An excited-state ion forms during inelastic collisions between electrons and mercury atoms (cf. Equation (9.25)): this excited-state ion must emit light to return to a lower energy state. A mercury lamp emits this energy in the form of UV light. Because there are only two energy levels that are energetically accessible (i.e. the ground and first excited state), the frequency of the UV emitted is almost monochromatic at 254 nm.

Unfortunately, the emission of UV light is not *useful* for a lamp because our eyes do not respond to UV. Even if they did, the glass of the tube is not transparent to light of wavelength 254 nm. The light must, therefore, be converted to *visible* light. To this end, the inside glass surface of the tube is coated with a thin layer of phosphor which is bombarded with photons of UV light. Each time a photon is absorbed it is re-emitted at a longer wavelength, which we see as visible light.

The reasons why the photon can be re-emitted at a different wavelength were discussed above when we introduced the topic phosphorescence. The exact composition of the phosphor depends on the 'colour type' of the bulb. Some phosphors emit a light that is almost blue, lending a cold, almost clinical atmosphere to a room. More modern bulbs have different phosphors, which emit a more natural light, i.e. with more red and yellow frequencies. The most common colour types are 'cool white', 'warm white', and 'deluxe cool white'. In each case, the phosphors are a mixture comprising varying proportions of calcium halophosphate, calcium silicate, strontium magnesium phosphate, calcium strontium phosphate, and magnesium fluorogermanate. The crystals are doped with impurities such as antimony, manganese, tin, and lead.

> In fact, most 'neon' lamps are mercury lamps in which the inside of the tube is coated with a phosphor. To see if the lamp is truly neon-based, look at the bulb before it glows: a real neon lamp needs no phosphor coating, so the glass is clear and without 'frosting'.

9.5 Other optical effects

Why is the mediterranean sea blue?

Light scattering

The Indian physicist Sir Chandrasekhara Venkata Raman was on a cruise on the Mediterranean Sea in 1921. Some reports suggest it was his honeymoon. Others say the beauty of its deep blue opalescence captivated him. Whatever the reason, he dedicated the rest of his life to understanding its colour and discovered the so-called *Raman effect.*

> *Opalescence* means the ability to show different colours, like the gemstone *opal*. We see different colours when viewing from different angles.

The Raman effect occurs when a beam of light is deflected. When a beam of light impinges on a sample of a chemical compound, a small fraction of the light emerges in directions other than that of the incident beam. The overwhelming majority of this scattered light is of unchanged wavelength, but a small proportion possesses wavelengths different from that of the incident light. This is the *Raman effect*, and earned him the Nobel Prize for Physics in 1930.

> The sky is also blue because of light scattering.

Light impinging on the surface of the Mediterranean Sea is scattered. Of this light, a small proportion is scattered in such a way that the frequency changes, causing it to look more blue than was the incident light. This shift in frequency causes the blue colour of the Mediterranean Sea.

> Raman's studies of the Mediterranean Sea's colour led to the phrase 'blue-sky research', because his work at the time had no obvious contemporary application.

The Raman effect

The Raman effect relates to scattering of light. Raman found that illuminating a transparent substance such as water causes a small proportion of the light to emerge

Raman spectroscopy is an *inelastic* light-scattering technique, in which an analyst directs a monochromated laser beam onto a sample, and determines the frequency and intensity of the scattered light.

Rayleigh scattering is named after its discoverer, John William Strutt (1842–1919), third Baron Rayleigh, an English physicist. He also did important work on acoustics and black-body radiation

The Raman effect is tiny. At most, only one photon per 10^5 collides in an inelastic manner, the exact number depending on the energy (and hence frequency) of the incident light.

at right angles to the direction of illumination. Furthermore, some of this light has a *different* frequency from that of the incident light.

Consider a beam of incident light, each photon of which has the same energy E (so it is monochromatic). These photons strike the sample. Most collisions are elastic, so the energy (and hence frequency) of the scattered photons does not change. Light that has the same frequency as the incident beam is said to have undergone *Rayleigh scattering*; see Figure 9.27.

But the collision is occasionally inelastic, so a molecule takes or gives energy to a photon. These photons are scattered, and lose energy to the sample, so the scattered light has a lower frequency. The shift in frequency Δv exactly matches the vibrational frequencies of molecules in the sample $v_{(vibration)}$; so, by measuring the frequency shifts between the incident and scattered light (v_0 and $v_{(scattered)}$ respectively), we obtain a graph of the proportion of collisions that are inelastic, i.e. the intensity (as 'y') against vibrational energy (as 'x'). We obtain a *Raman spectrum*. Figure 9.28 shows a Raman spectrum of malachite pigment, the sample coming from the famous Lucka Bible, which was made in Paris but now resides in the Czech Republic, at Znojmo.

Figure 9.28 has the appearance of an infrared spectrum, but the y-axis is absorbance (with peaks pointing upwards), whereas infrared spectra are drawn with a y-axis of transmittance, with peaks pointing downwards.

The Raman technique allows us to determine the intensity of Raman bands, and thereby to quantify the concentration of the chemical components in a complicated mixture (a Beer's law calibration graph of intensity against concentration is advisable; see Section 9.1).

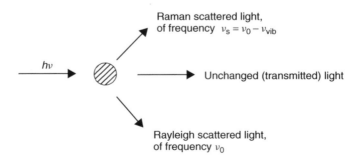

Figure 9.27   In Raman spectroscopy, light from a laser is shone at a sample. It is monochromated at a frequency of v_0. Most of the light is transmitted. Most of the scattered light is scattered elastically, so its frequency remains at v_0; this is Rayleigh scattered light. Raman scattered light has a frequency $v_{(scattered)} = v_0 - v_{(vibration)}$. The sample is generally in solution

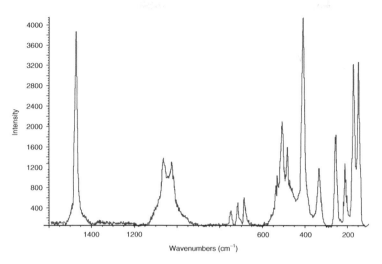

Figure 9.28   Raman spectrum of green malachite pigment (basic copper(II) carbonate $CuCO_3 \cdot Cu$ $(OH)_2$ taken from the Lucka Bible, now in the Czech Republic. The sample came from an initial 'I' from Genesis chapter 1, '*In principio* ...' ('In the beginning ...'). Reproduced with permission by Professor Robin Clarke FRS, University of London

In general, though, Raman spectroscopy is concerned with vibrational transitions (in a manner akin to infrared spectroscopy), since shifts of these Raman bands can be related to molecular structure and geometry. Because the energies of Raman frequency shifts are associated with transitions between different rotational and vibrational quantum states, Raman frequencies are equivalent to infrared frequencies within the molecule causing the scattering.

Pure rotational shifts are small and generally quite difficult to observe, unless the molecules are small and analysed in the gas phase. Rotational motions are hindered for liquid samples, so discrete rotational Raman lines are not observed.

> The intensity of the Raman lines is enhanced significantly if a light beam of high intensity (or better, a laser) is the incident beam.

> Raman spectra may facilitate qualitative analysis, 'asking *what* is it?' and also quantitative analysis – asking 'how much?'

Do old-master paintings have a 'fingerprint'?

Raman spectra

A new, novel application of Raman spectroscopy has emerged recently. In the Middle Ages, most artists prepared their own pigments, e.g. by grinding up coloured earth and binding the powder with linseed oil to make 'burnt umber', or binding soot in sunflower oil to make 'lamp black'. Since the ingredients to hand differed according to geographical region, the chemical composition of the pigments differed from artist to artist; so, analysing a painting's pigments can function almost like a 'fingerprint'.

Until now, many great paintings have been labelled as being 'by an unknown artist', but Raman spectroscopy has been shown to be a powerful technique for identifying the chemical constituents within the pigments of a painting or illuminated manuscript. Raman spectroscopy is not destructive, so none of the Lucka Bible was removed prior to analysing for the malachite shown in the spectrum in Figure 9.28, and none of the page was damaged.

The compositions of pigments is also allowing us to put a precise date to a manuscript, because inks and paints changed with time, becoming more sophisticated as the Middle Ages came to an end.

10

Adsorption and surfaces, colloids and micelles

Introduction

This chapter introduces the topic of adsorption, giving examples of both physical adsorption and chemical adsorption, and discusses the similarities and differences between the two. The standard nomenclature of surface science is given from within this context. The energetics of adsorption are explained in terms of the enthalpies of bond formation $\Delta H_{(ads)}$. Next, isotherms are discussed.

Colloids are introduced in the second half of the chapter. The various classifications of colloid types are discussed, together with ways of forming, sustaining and destroying colloids, i.e. colloid stability. Finally, association colloids ('micelles') are discussed.

10.1 Adsorption and definitions

Why is steam formed when ironing a line-dried shirt?

Adsorption

We often generate steam when ironing clothes made of cotton. The steam forms even if the cloth previously felt dry. But the steam is only formed if the shirt was dried outside, e.g. on a clothes line. We see no steam if the shirt was dried on a radiator before ironing.

Anyone who irons a lot will also notice how the texture of the cotton changes following ironing: beforehand, the cotton felt coarse and somewhat starched, but afterwards it feels more pliable and softer.

Water (in the form of vapour) sticks readily to many surfaces. We say it *adheres* or *adsorbs*. The cause of the *adsorption* is the

> Substances partition preferentially during *adsorption*, leaving the gaseous or liquid phases and forming a new layer on the surface of a substrate.

We call the adsorbing species the *adsorbate*. An adsorbate is atomic, ionic or molecular; solid, liquid or gas.

The adsorbate adheres to the *substrate*, which is the layer beneath. The substrate must be a *condensed* phase, either a solid or a liquid.

The word 'substrate' comes from two roots: the Latin *stratum* means layer, and *sub* is the Latin for 'less' or 'lower than'.

The word 'adsorption' derives from the Latin word *sorbere*, which means 'to suck in'.

formation of a *bond* between molecules of water and a surface, such as the carbohydrate backbone of the cotton fibre. To be a *surface*, it must be a condensed phase, i.e. solid or liquid; clearly, a gas cannot be a condensed phase. This underlying surface to which the *adsorbate* is attached is termed the *substrate* (or, occasionally, *adsorbent*).

The adsorption bond can be quite strong, and is characterized by an energy known as the 'enthalpy of adsorption' $\Delta H_{(ads)}$. Since a *hot* iron is needed to remove the creases in the cotton, we realize that the iron must have supplied sufficient heat to the *adsorbed* water to overcome and hence break the adsorptive bond joining it to the underlying cotton substrate. The steam we experience is adsorbed water released in this way.

A shirt dried previously on a radiator does not form steam while ironing because the heat from the radiator was itself sufficient to break the adsorptive bonds between the water and the cotton. Any adsorbed water was lost *before* the process of ironing commenced.

Steam is again not formed when ironing synthetic fibres, such as nylon or polyester. Several reasons explain the absence: firstly, the temperature of the iron is usually much lower with such synthetic fabrics, to avoid melting them. And we would not necessarily overcome the enthalpy of adsorption $\Delta H_{(ads)}$ with a less energetic iron. But, secondly, the strength of the adsorptive bond formed between water and artificial fabrics is smaller, so the thermal energy of room temperature might be sufficient to dislodge most of the water molecules before ironing commenced.

Also, in being a plastic, the surface of the nylon fibre is smooth, in contrast to most natural fibres such as wool or cotton, which are porous. Cotton, therefore, has a larger surface area. Since adsorption only occurs at the *surface* of a substrate, there is likely to be more water adsorbed per unit mass of cotton than on the same mass of nylon.

Figure 10.1 is a schematic representation of an adsorbate during adsorption onto a substrate.

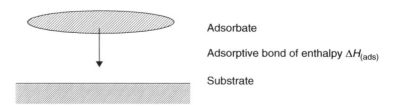

Adsorbate

Adsorptive bond of enthalpy $\Delta H_{(ads)}$

Substrate

Figure 10.1   Adsorption of an adsorbate onto a substrate. The charge necessary to form the adsorptive bond comes from the charge *centroid* of the adsorbate

Why does the intensity of a curry stain vary so much?

Adsorption isotherms

An orange–brown stain forms on the inside of a pan while cooking a curry. The colour represents molecules adsorbed from the constituent spices of the curry.

A strong curry is generally dark red–brown in colour, whereas a milder curry, such as biryani, is paler in hue. And the difference in the intensity of the curry stain arises from the varying concentration of the coloured components in the curry sauce. As an example, we will discuss the red and spicy tasting compound *capsaicin* (**I**), which is the cause of both the hotness of a chilli and contributes to its red colour.

> *Care*: note the spellings here: 'ADsorption' does not relate to the better-known word 'ABsorption'. Absorption is the uptake of something, like a sponge taking water into itself, or a chromophore removing frequencies from a spectrometer beam.

(I)

We must first appreciate that a coloured species such as capsaicin has indeed adsorbed; otherwise, there would be no layer to see. And a higher concentration of capsaicin yields a more intense colour as a straightforward manifestation of the Beer–Lambert law. We discern a relationship between the strength of the curry (by which we mean the concentration of the spices it contains) and the colour of the adsorbate, with a strong curry containing more spice and imparting a more intense colour. Conversely, the amount of dye adsorbed on the pan will be relatively slight after a mild curry (which is more dilute in the amounts of 'hot' compounds it contains).

> *Care*: to a chemist, the words 'strong' and 'weak' do not mean 'concentrated' and 'dilute' respectively, but relate to the *extent of ionic dissociation*; see Chapters 4 and 6.

Adsorption occurs on the microscopic level, with molecules of adsorbate sticking to the atoms on the *surface* of the substrate (in this case, the pan). In practice, it is rare for each and every adsorption site to have a dye molecule adsorbed to it. The coverage is only fractional. We give the name *isotherm* to the fraction of the total adsorption sites occupied by adsorbate. An isotherm is denoted with the Greek symbol θ (theta).

Next, we realize how intensifying the orange stain of capsaicin confirms these variations in the magnitude of θ. A small value of θ tells us that a small proportion of the possible adsorption sites are occupied by adsorbed molecules of capsaicin, and

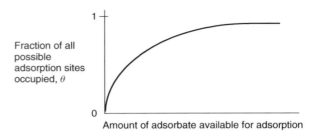

Figure 10.2 Schematic isotherm for the simplest cases of chemical adsorption from solution onto a solid substrate. The amount of adsorbate available to adsorb is best gauged by the concentration c

> A greater number of adsorbed molecules appears as a more intense stain to the eye in consequence of the Beer–Lambert law.

> This argument only holds if each pan-load of curry is allowed to reach its thermodynamic *equilibrium*.

a higher value of θ means adsorbate adheres to a larger proportion of the adsorption sites. And the magnitude of θ changes because the metal of the pan (the *substrate*) is more likely to come into contact with the capsaicin dye (the *adsorbate*) when the curry contains a more concentrated solution of the capsaicin.

Imagine we need to investigate further the intensity of the stain. We would start by making several curries, each made in an identical pan and cooked for an identical length of time, although each containing a different concentration of red capsaicin. We expect the molecules of capsaicin to occupy a larger proportion of possible adsorption sites when there is more of it in solution; and we expect θ to be smaller when there is less capsaicin in solution. This is indeed found to be the case.

A graph of the proportion θ against the (equilibrium) concentration of adsorbate is also called an *isotherm*. (This dual usage of the word can cause some confusion.) A typical adsorption isotherm is shown in the schematic diagram in Figure 10.2, and shows how the proportion θ of occupied sites increases quite fast initially as the concentration [capsaicin] increases. Above a certain concentration of adsorbate, however, the amount of capsaicin adsorbed does not increase but remains constant. In this example, the maximum value of θ is about unity. In other words, *all* the possible adsorption sites are bonded to a molecule of capsaicin – but only if the concentration of the capsaicin is huge.

> Notice how the value of θ is zero at the origin of the isotherm; this should be no surprise, since there is no adsorbate in solution to adsorb!

We might have predicted a value of $\theta = 1$ at high concentrations of capsaicin adsorbate. We would say (simplistically as it turns out), 'if each adsorption site is occupied, then no further adsorption could occur'. We have formed an adsorbed *monolayer*, with the substrate bearing a full complement of adsorbate. Notice also from Figure 10.2 how the value of θ at the origin of the isotherm is zero. This ought not to surprise us, because no adsorbate can adsorb if none of it is in solution.

Similarly, most gases readily adsorb from the gas phase to form a *bound* gas, as a direct analogy to the adsorption of capsaicin

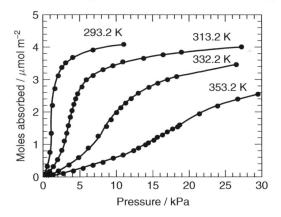

Figure 10.3 An isotherm of amount of gaseous cyclohexane adsorbed (as '*y*') against pressure *p* (as '*x*'), depicted as a function of temperature. The substrate is a catalyst comprising a mixture of metal oxides, called Stirling-FTG. (Figure reproduced by permission of Pergamon from the paper 'Towards a general gas adsorption isotherm', G. M. Martinez and D. Basmadjian, *Chem. Eng. Sci.*, 1996, **51**, 1043)

from solution. (In effect, the molecules of gas changed phase.) Figure 10.3 shows the isotherm constructed with data for gaseous cyclohexane adsorbing onto a solid substrate. The graph in Figure 10.3 shows θ (as '*y*') as a function of gas pressure *p* (as '*x*').

The graph only ever goes through the origin at zero pressure. We have discovered that the only way to have a completely 'clean' substrate (one with no adsorbate on it, with $\theta = 0$) is to subject the surface to an extremely low pressure – in effect, we have subjected the substrate to a strong vacuum. Effectively, the vacuum 'sucks' the adsorbate away from the substrate surface. We give the name *desorption* to the removal of adsorbate.

The extent of adsorption is a function of temperature *T*, as implied by the term 'isotherm', so the construction of an isotherm graph should be performed within a thermostatted system. When adsorbing from solution, the value of θ also depends on the solvent; generally, if the solvent is polar, such as water or DMF, then the extent of adsorption is often seen to decrease because molecules of solvent will occupy sites on the substrate in preference to molecules of solute.

Finally, note how the isotherm can alter dramatically if adsorption occurs in the presence of other adsorbates. Curry is actually a mixture of great complexity, so the example here of capsaicin is somewhat artificial.

> The removal of an adsorbate is termed *desorption*.

> The shape of an isotherm depends on the choice of adsorbate, substrate, temperature *T* and, in a solution-phase system, the solvent.

> The arguments here are simplified because all foods contain a wide assortment of physically disparate molecules. In a more rigorous set of experiments, each solution should contain a *single* solute.

Why is it difficult to remove a curry stain?

The strength of adsorption and the magnitude of $\Delta H_{(ads)}$

Each molecule, ion or atom of solute is bonded to the surface of the adsorbate. If we wish to remove the adsorbate from its substrate then we must overcome the enthalpy of adsorption $\Delta H_{(ads)}$. So, removing adsorbate requires the input of energy. If $\Delta H_{(ads)}$ is large, then more energy is needed to overcome it than if $\Delta H_{(ads)}$ is small. It is difficult, therefore, to remove a layer of adsorbed curry stain without expending energy, which we experience physically as a difficulty in removing the stain; the red colour won't just 'wash off'.

> The two extremes of 'physical adsorption' and 'chemical adsorption' are abbreviated to *physisorption* and *chemisorption* respectively.

There is a wide range of adsorption enthalpies $\Delta H_{(ads)}$, ranging from effectively zero to as much a 600 kJ per mole of adsorbate. The adsorptive interaction cannot truly be said to be a 'bond' if the enthalpy is small; the interaction will probably be more akin to van der Waals forces, or maybe hydrogen bonds if the substrate bears a surface layer of oxide. We call this type of adsorption *physical adsorption*, which is often abbreviated to *physisorption*.

At the other extreme are adsorption processes for which $\Delta H_{(ads)}$ is so large that real chemical bond(s) form between the substrate and adsorbate. We call this type of adsorption *chemical adsorption*, although we might abbreviate this to *chemisorption*.

> A *centroid* is the location of a physicochemical phenomenon or effect or quantity. A *charge centroid*, therefore, represents the part of a molecule or ion having the highest charge density.

Now, to return to the orange stain, formed on the surface of a pan by adsorption of capsaicin from a solution (the curry). Such organic dyes are usually unsaturated (see the structure **I** above), and often comprise an aromatic moiety. The capsaicin, therefore, has a high electron density on its surface. During the formation of the adsorption bond, it is common for this electron cloud to interact with atoms of metal on the surface of the pan. Electron density flows *from* the dye molecule via the surface atoms *to* the conduction band of the bulk metal. The arrows on Figure 10.4 represent the direction of flow as electron density moves from the *charge centroid* of the dye, through the surface atoms on the substrate, and thence into the bulk of the conductive substrate.

As a good generalization, most aromatic and unsaturated species adsorb readily. Their adsorption is facilitated if the substrate is electronically conductive, being either

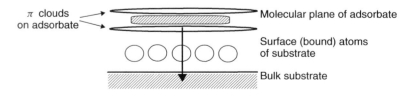

Figure 10.4 Schematic representation of a dye molecule adsorbing on a substrate. The large arrow indicates the movement of charge from the charge centroid to form an adsorptive bond. Notice the way charge delocalizes, once it enters the substrate

metallic or semiconducting, because the charge donated during the formation of an adsorption bond can readily conduct away from the precise location of adsorption. We say the donated charge is *delocalized*. No such delocalization of the charge donated during adsorption occurs if the substrate is non-conductive or poorly conducting. The adsorption of *additional* adsorbate onto a poor conductor is, therefore, rendered less likely because *electronic repulsions* would be induced between the substrate and incoming molecules of adsorbate. Such repulsions make the donation of more charge to a surface already bearing an excess surface charge unfavourable from electrostatic considerations.

> The charge donated during adsorption conducts away from the immediate site of adsorption, and delocalizes, albeit partially. In fact, most of donated charge resides close to the adsorption site.

A further difficulty arises: when additional molecules of adsorbate approach the substrate, causing the proportion of sites filled θ to approach unity, it becomes difficult for additional adsorbate to find a suitable angle of approach that allows for successful adsorption. It is like finding a seat on bus: it is easier to see the seats available in a partially empty bus. It is not only difficult to *see* the empty seats in a full bus: it can also be difficult even to squeeze through the crowd to get to them. The schematic diagram in Figure 10.5 represents such a situation. The molecule of adsorbate must approach the substrate from within the cone-shaped area above the vacant adsorption site. Otherwise, *steric restriction* will prevent successful adsorption. Any attempts to approach the substrate from an angle outside this cone will cause the incoming adsorbate to strike a molecule that is adsorbed already, causing the attempt at adsorption to fail.

The surface of the substrate is not homogeneous, which presents two problems. First, both electronic repulsions and then steric restrictions promote adsorption at lower θ and make it more difficult at higher θ. The influence of these problems

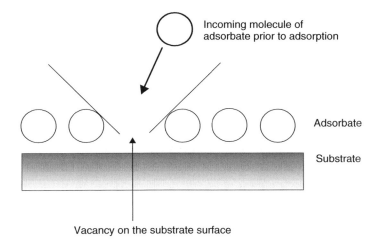

Incoming molecule of adsorbate prior to adsorption

Adsorbate

Substrate

Vacancy on the substrate surface

Figure 10.5   Schematic diagram to show the steric problems encountered when θ tends to unity. The incoming molecule of adsorbate must approach from within the confines of the cone shape if adsorption is to succeed

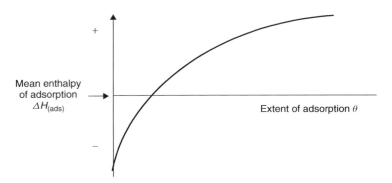

Figure 10.6   The enthalpy of adsorption $\Delta H_{(ads)}$ is a function of θ. The position of the x-axis represents the *mean* enthalpy; the magnitude of the vertical deviation from the x-axis is $\Delta\Delta H_{(ads)}$

Care: note the double 'Δ' symbol in $\Delta\Delta H_{(ads)}$, which represents the *change* in $\Delta H_{(ads)}$ from its mean value rather than the enthalpy of adsorption itself.

is seen in the way the mean enthalpy varies: we tend to talk in terms of $\Delta\Delta H_{(ads)}$, where the double Δ implies we are looking at variations in ΔH. Figure 10.6 shows how the value of $\Delta H_{(ads)}$ varies as a function of θ. The range of $\Delta H_{(ads)}$ values depends markedly on the identities of adsorbate and substrate, but the difference in $\Delta H_{(ads)}$ between $\theta = 0$ and 1 can be as large as 20 kJ mol^{-1}.

Why is iron the catalyst in the Haber process?

Physisorption and chemisorption

The usual choice of catalyst for the Haber process to produce ammonia is metallic iron, often mixed with various other additives to alter subtly the magnitude of the activation energy E_a (see Chapter 8). In the Haber process, a mixture of gaseous nitrogen and hydrogen are heated to 600 °C. The gases are compressed to a pressure of about 6 atm to facilitate efficient pumping through the plant, and forcing them over the iron catalyst. The reaction in Equation (10.1) occurs at the catalyst's surface:

$$N_{2(g)} + 3H_{2(g)} \longrightarrow 2NH_{3(g)} \tag{10.1}$$

The strength of the N≡N bond decreases after charge is donated to the surface of the iron.

The physical conditions to effect a satisfactory extent of reaction are fairly severe, but are needed to overcome the enormous strength of the nitrogen triple bond. The role of the iron is essential: chemisorptive adsorption of nitrogen occurs on the surface of the iron, with charge being donated *from* the N≡N bond *to* the surface of the iron. As a result, less electron density remains between the nitrogen atoms, sufficiently weakening the bond. Furthermore, nitrogen gas is *immobilized* while adsorbed. While adsorbed, the probability of a hydrogen molecule striking the nitrogen molecule is enhanced. In a similar way, it is easier to throw a ball at a stationary athlete than one who runs fast.

> Molecules of nitrogen are *immobilized* by adsorption, thereby facilitating collisions with a molecule of hydrogen.

But why do we *heat* the mixture to effect reaction? We will perform a thought experiment: imagine immersing iron into nitrogen at different temperatures. We will see the effects of the relative differences between physisorption and chemisorption.

(1) *Very cold.* Immersing a clean iron surface into liquid nitrogen at 77 K ($-196\,^\circ$C) yields a weak physisorptive bond. The N≡N molecule is probably aligned parallel to the metal surface, with electron density donating from the centroid of the triple bond directly to iron atoms on the surface of the metal via a van der Waals type of interaction. The experimental value of $\Delta H_{(ads)}$ is small at about 1.5 kJ mol^{-1}.

(2) *Room temperature.* Now imagine the iron is removed from the liquid nitrogen and allowed to warm slowly up to room temperature (at, say, 298 K). The energy from the room is always equal to $\frac{3}{2}R \times T$ (see p. 33), so the energy of the room increases with warming. The value of RT at 298 K is 2.2 kJ mol^{-1}. As RT is greater than $\Delta H_{(ads)}$ at this temperature, the molecules of nitrogen shake themselves energetically until breaking free. In summary, increases in temperature are accompanied by desorption of nitrogen from the iron until the iron is bare.

(3) *Very hot.* We finally consider the case of roasting metallic iron in a nitrogen atmosphere at about 600 $^\circ$C. Such roasting occurs, for example, during the Haber process for making ammonia. When hot, both nitrogen atoms from molecular dinitrogen form a different type of bond (a *chemical* bond) to the surface of the metal, binding to adjacent iron atoms. The extent of adsorption is thereby increased. The enthalpy change $\Delta H_{(ads)}$ is greater than 500 kJ mol^{-1}.

Following adsorption, electron density seeps from the inter-nitrogen bond, decreasing its bond order from three to perhaps one and a half. Figure 10.7 depicts schematically the chemisorption of the nitrogen molecule onto iron. To complete the picture, Figure 10.8 shows how diatomic hydrogen gas readily adsorbs to the surface-bound layer of 'complex' on the iron surface – a structure sometimes called a 'piggy back'.

As a simple proof that a complex-like structure forms on the surface of metals immersed in a mixture of nitrogen and hydrogen gases, try immersing a piece of red-hot bronze in an atmosphere of ammonia. The surface of the metal soon forms a tough, impervious layer of bronze–ammonia complex, which imparts a dark-brown colour to the metal. The brown complex reacts readily with moisture if the metal is iron and is impermanent, but the complex on bronze persists, thereby allowing the colour to remain. These ammonia complexes explain why so many bronze statues and medals have a dark-brown colour.

> Roasting a bronze artifact in dry NH_3 imparts to it a deep-brown appearance, caused by a chemisorbed *adlayer* of ammonia–bronze complex, which persists long after removing the bronze from the furnace.

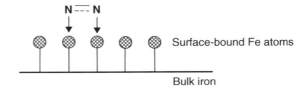

Figure 10.7 During the adsorption of molecular nitrogen onto iron metal, the two nitrogen atoms donate an increment of charge to adjacent atoms of iron on the metal surface, as depicted by the vertical arrows. The N≡N triple bond cleaves partially, with a resultant bond order of about 1.5

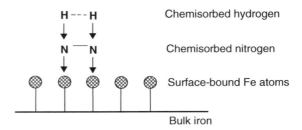

Figure 10.8 Since the nitrogen is immobilized, hydrogen gas can adsorb to the surface of the bound –N–N– structure, thereby effecting a further decrease in the bond order between the two nitrogen atoms. The bond order between the two hydrogen atoms is also less than one

Why is it easier to remove a layer of curry sauce than to remove a curry stain?

Multiply adsorbed layers

> We mean $\theta = 1$ when we say θ is 'unity'.

> There can only be one *chemisorbed* layer on a substrate because, after bonds have formed with the substrate, there is no more substrate with which to form bonds.

The idea of θ having a maximum value of unity is simplistic. Try this quick experiment: take the pan used to prepare a curry and try to wash it clean. There will be a stain adjacent to the metal, above which is a layer of curry sauce. The stain is thin – possibly even transparent – which we call the adsorbed layer. The layer of stain has a minimal thickness and it is difficult to remove. Conversely, we call the overlayers of curry sauce 'bulk material', which can take any thickness and are relatively easy to remove, even with a fingernail.

There are layers on layers – we call them *multiple layers* (or *multilayers*). A chemisorbed layer is formed by the creation of chemical bonds. For this reason, there can only be a single chemisorbed layer on a substrate. Conversely, it is quite likely that a material can adsorb *physically* (or *physisorb*) onto a previously formed *chemisorbed* layer, either on more of the same adsorbate or even on a different adsorbate.

This concept is illustrated by the example of curry on a saucepan. A chemisorbed layer forms on the pan, and physisorbed curry can adhere to the chemisorbed layer. Similarly, the hydrogen gas in the previous example adsorbs on nitrogen gas chemisorbed on iron.

How does water condense onto glass?

Physisorbed layers adsorbing on chemisorbed layers

We wake up to find it is a bitterly cold day outside. We wish to wash, so we fill the bathroom sink with hot water; but, as soon as we look up, we see how water condenses on the cold glass of the mirror.

As soon as steam emanates from the water in the sink, it will rise (owing to eddy currents; see Chapter 1). A tiny fraction of the airborne water (i.e. steam) will condense on the mirror, and soon a strongly bound chemisorbed layer forms to cover its whole surface. The layer is microscopically thin, making it wholly invisible to the eye. We will call this layer 'LAYER 1'. Each molecule of water in LAYER 1 is now physically distinct from normal water, since charge has been donated to the substrate. Each water molecule in LAYER 1 is, therefore, slightly charge deficient, compared with normal water.

But the water in the sink is still hot, causing more steam to rise. The air in the bathroom is cold, causing the airborne water molecules to lose energy as they attain thermal equilibrium. No more water can form a chemical bond with the glass of the mirror because there are no adsorption sites available. So they form a physisorptive bond onto the chemisorbed layer of water, to form 'LAYER 2'. Each molecule as it forms a physisorbed layer donates charge to water in LAYER 1, which we recall is slightly charge deficient. Physisorption causes the newly bound water molecules in LAYER 2 to be slightly charge deficient, though certainly not as deficient as those in LAYER 1.

> The layer immediately adjacent to the substrate need not be physisorbed.

Further layers can physisorb, one above another. The extent of charge deficiency will decrease until, after about five or six layers have physisorbed, they are energetically indistinguishable. By the time we have about a dozen layers adsorbed on the glass, we are no longer able to speak about layers, and start to talk about bulk condensate of water. At this point, more water condenses onto this mass of water, causing its weight to increase, and eventually it runs down the mirror as warm condensation.

Another example: consider the adsorption of bromine on silica. We start by placing elemental bromine and silica at either end of a long sealed vessel. Having a relatively high vapour pressure, bromine volatilizes and molecules of bromine soon chemisorp onto the surface of the silica. When the chemisorbed monolayer is complete, successive layers of bromine form by physisorption. The first physisorbed layer is unique, since its under-layer is chemisorbed, but the second physisorbed layer has physisorbed layers both on its top and bottom. In fact, after several adsorbed layers have been deposited, such layers become virtually indistinguishable, and may be considered to be identical in nature to 'normal', i.e. liquid, elemental bromine.

How does bleach remove a dye stain?

Reactivity and properties of adsorbed layers

Before we start, we assume that the dye stains the surface by forming an adsorptive interaction. The majority of dyes in the kitchen come from vegetables. A particularly intense dye is *β-carotene* (**II**), which colours carrots, the golden leaves of autumn and some flowers. The intensity of the *β*-carotene colour arises from the extended conjugation.

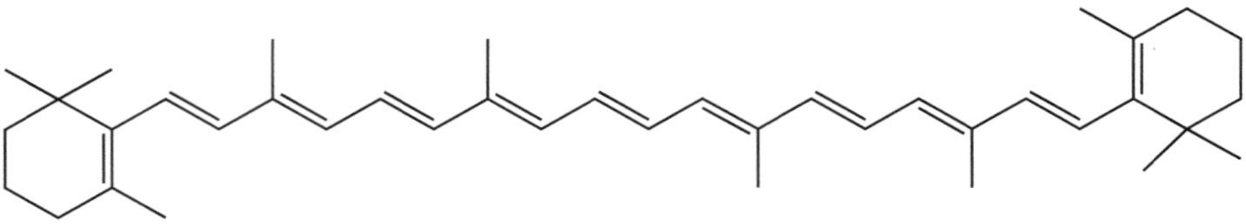

(II)

Wiping the stained portion of work surface with bleach quickly removes the colour of the adsorbed dye as a result of the adsorbed material *reacting* with the bleach. One of the active components within household bleach is hypochlorous acid, HOCl, in equilibrium with molecular chlorine. Cl_2, H^+ and ClO^- add across the double bonds in the dye (see Figure 10.9) – usually near the middle of the conjugated portion, because such bonds are weakened as a result of their conjugation. The colour is seen to vanish because the conjugation length decreases.

> The conjugation strength is also decreased because some of the electron density has been donated to the surface.

Secondly, by decreasing the extent of conjugation, the electron density within the adsorbate decreases. Since a large electron density was possibly a major cause of the adsorptive bond being so

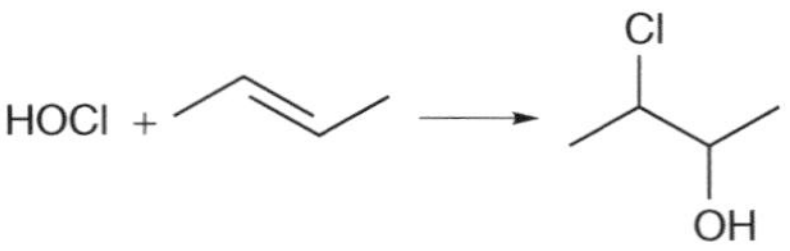

Figure 10.9 Hypohalite addition across one of the weak, central double bonds of *β*-carotene. The colour of an adsorbed material is lost when exposed to bleach because of *hypohalite addition* across a double bond, thereby decreasing the extent of conjugation, if not removing it altogether

strong, and the charge from that double bond is no longer available, the adsorption strength decreases greatly. We should also expect the solubility of previously insoluble products of hypohalite to increase, allowing them to wash away.

How much beetroot juice does the stain on the plate contain?

The Langmuir adsorption isotherm

The dye responsible for the red–mauve colour of beetroot juice is potent, because a small amount imparts an intense stain on most crockery. It has a large extinction coefficient ε (see Chapter 9). The dye is fairly difficult to remove, so we safely assume the adsorption bonds are strong and chemisorptive by nature.

We can determine how much of the dye (the 'adsorbate') adsorbs on a plate (the 'substrate') by devising a series of experiments, measuring how much adsorbate adsorbs as a function of concentration. We then analyse the data with the *Langmuir adsorption isotherm*:

$$\frac{c}{n} = \frac{c}{n_m} + \frac{1}{bn_m} \qquad (10.2)$$

where c is the *equilibrium* concentration of coloured adsorbate in contact with the substrate, n is the amount of adsorbate adsorbed to the plate (the substrate) and n_m is the amount adsorbed in a *complete* monolayer. Finally, b is a constant related to the equilibrium constant of adsorption from solution.

We perform a series of experiments. We purchase a stock of identical plates and prepare several solutions of beetroot juice of varying concentration. One plate is soaked per solution. The chemisorptive bond between the dye and plate is strong, and equilibrium is reached after only a few seconds. The excess juice is decanted off for analysis, e.g. by means of optical spectroscopy and the Beer–Lambert law, provided we know the extinction coefficient ε for the juice.

We will find a decrease in the amount of dye remaining in solution because some of it has adsorbed to the plate surface. The amount adsorbed is readily determined, since the initial concentration was known and the equilibrium concentration is readily measured.

> We obtain the amount of dye n in a monolayer as '$n_{(initially)}$ − $n_{(remaining\ in\ solution)}$'.

From Equation (10.2), we expect a plot of $c \div n$ (as 'y') against c (as 'x') to yield a straight-line graph of intercept $1/(bn_m)$ and with a gradient $1/n_m$.

Worked Example 10.1   The following data refer to the adsorption of the red–mauve dye from beetroot juice on porcelain at 25 °C. (1) Show that the data obey the Langmuir adsorption isotherm. (2) Demonstrate that 1.2×10^{-8} mol of dye adsorb to form a monolayer. (3) Estimate the area of a single dye molecule if the radius of a plate was 17.8 cm (we assume the formation of a complete monolayer).

Equilibrium concentration c/mmol dm^{-3}	0.012	0.026	0.047	0.101	0.126
Amount adsorbed n/nmol	2.94	4.98	7.94	9.00	9.59

Answer strategy

> Notice the tiny numbers here: mmol means milli-moles (m $= 10^{-3}$) and nmol means nano-moles (n $= 10^{-9}$).

(1) If data obey Equation (10.2), then a plot of $c \div n$ (as 'y') against c (as 'x') should yield a straight-line graph of gradient $1/n_m$. Any non-linear portion(s) of the graph represent concentration ranges not following Equation (10.2), and hence not following the Langmuir adsorption isotherm. Figure 10.10 shows the Langmuir plot constructed with the data above. The graph is linear, indicating that the data obey the Langmuir isotherm.

(2) We then determine the gradient of the line. Its reciprocal has a value of n_m, which is the number of moles of beetroot juice in a monolayer. The gradient of the graph is 0.0796×10^6 mol^{-1}. The reciprocal of its gradient equates to n_m, so $1 \div$ (gradient) $= n_m = 1.256 \times 10^{-5}$ mol.

> China or porcelain is made from clay, itself made up of *aluminosilicate*.

This is a tiny value, and reveals how a small amount of beetroot dye will stain a standard aluminosilicate plate; in other words, it demonstrates how large is the extinction coefficient ε of the beetroot juice chromophore.

(3) We calculate the number of molecules by multiplying n_m by the Avogadro number L. We already know the area over which the monolayer forms via πr^2, so we calculate the area per molecule with the following simple calculation:

$$\text{area per molecule} = \frac{\text{area of the substrate}}{\text{number of molecules}} \tag{10.3}$$

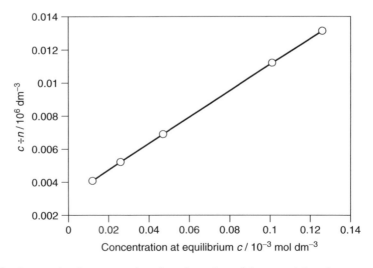

Figure 10.10   Langmuir plot concerning the adsorption of beetroot-juice dye onto a porcelain substrate at 25 °C: graph of $c \div x$ (as 'y') against c (as 'x')

We obtain the number of molecules as $n_m \times L$, where L is the Avogadro number:

$$\text{number of molecules} = 1.256 \times 10^{-5} \text{ mol} \times 6.023 \times 10^{23} \text{ molecules mol}^{-1}$$

so

$$\text{number of molecules} = 7.56 \times 10^{18} \text{ molecules}$$

We calculate the area A of the plate as πr^2 (in terms of SI units) so $A = 3.142 \times (17.8 \times 10^{-2} \text{ m})^2 = 0.093 \text{ m}^2$. Then, inserting numbers into Equation (10.3), we obtain:

$$\text{area per molecule} = \frac{0.093 \text{ m}^2}{7.56 \times 10^{18} \text{ molecules}}$$

$$A = 1.32 \times 10^{-19} \text{ m}^2 \text{ molecule}^{-1}$$

Aside

One of the major components within aluminosilicate china is kaolin clay. Kaolin is used in the treatment of stomach ache, often being found for example in 'kaolin and morphine'. Just like the beetroot juice in this example, biological toxins adsorb strongly to the surface of the kaolin and are immobilized thereafter, and leave the body during defecation. Many animals in the wild, such as elephants in Africa, spider monkeys in South America and brightly coloured macaws, eat naturally occurring kaolin clay at the end of a meal in order to prevent stomach ache or outright poisoning.

Background to the Langmuir model

The Langmuir model first assumes the adsorption sites are energetically identical. Actually, this assumption is not borne out when adsorption occurs predominantly by physisorption. The spread of $\Delta H_{(ads)}$ values between the various sites can be as high as 2 kJ mol^{-1}, which is often a significant fraction of the overall enthalpy of adsorption when physisorption is the sole mode of adsorption. By contrast, energetic discrepancies between sites can be ignored when adsorption occurs by chemisorption.

The second assumption follows from the first: to ensure that the adsorption sites are energetically identical, we say the adsorption occurs wholly by chemisorption.

Third, we assume the adsorption process is dynamic. Adsorption and desorption occur to equal and opposite extents at equilibrium, provided the mass of material adsorbed remains constant.

Justification Box 10.1

For the purposes of the derivation below, we will consider the process of adsorption from the *gas* phase. A simple example of a system involving adsorption of gases is the Haber process, in which $N_{2(g)}$ and $H_{2(g)}$ adsorb to the surface of metallic iron.

There are N possible number of adsorption sites on the substrate, the fraction of which bearing adsorbed material is θ. By corollary, the fraction of empty sites is $(1 - \theta)$; the number of filled sites will thus be $N\theta$ and the number of empty sites will be $N(1 - \theta)$.

The rate constant of adsorption is k_a and the rate constant characterizing the way sites lose adsorbate is k_d. From simple kinetics (see Chapter 8), the rate of adsorption depends on the number of sites available $N(1 - \theta)$, the rate constant of the process k_a, and the amount of adsorbate wanting to adsorb, which is proportional to pressure p. Overall, the rate of adsorption is $N \times (1 - \theta) \times k_a \times p$.

Clearly, desorption can only occur from a site already bearing adsorbed material. The rate of desorption is proportional to the number of such sites, with a proportionality constant of k_d.

The net rate of adsorption is the difference between the rate of adsorption and the rate of desorption:

$$\frac{d\theta}{dt} = \boxed{N(1 - \theta)\,pk_a} - \boxed{N\theta k_d}$$

(10.4)

net rate   rate of   rate of
adsorption   desorption

The fraction θ stays constant at equilibrium, so the *rate of change* of θ will be zero, so

$$N(1 - \theta)pk_a - N\theta k_d = 0$$

implying

$$N(1 - \theta)pk_a = N\theta k_d \qquad (10.5)$$

The N terms cancel, so

$$(1 - \theta)pk_a = \theta k_d \qquad (10.6)$$

Dividing by k_d yields

$$(1 - \theta)p\frac{k_a}{k_d} = \theta \qquad (10.7)$$

> The ratio of rate constants yields the equilibrium constant of adsorption K.

The ratio of rate constants $(k_a \div k_d)$ yields the equilibrium constant of adsorption K; so, multiplying the bracket gives

$$pK - pK\theta = \theta \qquad (10.8)$$

Grouping the θ terms:

$$pK = \theta(pK + 1) \qquad (10.9)$$

So the fractional coverage of the total adsorption sites θ becomes

$$\theta = \frac{pK}{pK + 1} \qquad (10.10)$$

We have not yet expressed Equation (10.10) in its final, useful form. The fraction θ can be replaced if we realize how θ is the ratio of the volume of gas adsorbed $V_{(ads)}$ divided by the volume of gas adsorbed evenly to form a complete monolayer V_m. Substituting for θ yields

> *Care*: the product of K and p could be written as '$K\,p$' (i.e. $K \times p$), which could be mistaken for a gas-phase equilibrium constant K_p.

$$\frac{V_{(ads)}}{V_m} = \frac{pK}{pK + 1} \qquad (10.11)$$

Rearranging Equation (10.11) gives

$$\frac{p}{V_{(ads)}} = \frac{p}{V_m} + \frac{1}{K V_m} \qquad (10.12)$$

which is the *Langmuir adsorption isotherm*.

Accordingly, a plot of $p \div V$ (as 'y') against p (as 'x') should yield a straight-line graph of intercept $1 \div (K \times V_m)$ and gradient $1/V_m$. And knowing V_m from the gradient allows for calculation of K.

We obtain Equation (10.2) when Equation (10.12) is written in terms of concentration c.

Why do we see a 'cloud' of steam when ironing a shirt?

Deviations from the Langmuir model

We saw on p. 487 how ironing a shirt with a hot electric iron was usually accompanied by the release of adsorbed water. But the amount of water adsorbed within a monolayer is tiny – no more than about 10^{-8} mol, as shown by Worked Example 10.1. Even a moment's thought tells us too much steam is released when heating the cloth beneath a hot iron. In fact, the water does not adsorb according to a simple Langmuir model with, for example, multiple layers adsorbing to form a 'sandwich'. We find several layers of water physisorbing, one on top of another, with the ultimate substrate being an underlayer of chemisorbed water. This way, the amount of water can be many times greater than that comprised within a chemisorbed monolayer.

For this reason, we sometimes construct Langmuir plots and find they are not particularly linear, particularly at the right-hand side, which represents higher concentrations. This deviation from

> The gradient of a Langmuir plot is always positive, since it is clearly impossible for a monolayer to contain a *negative* amount of material.

linearity generally implies layers of physisorbed material have formed on top of the chemisorbed monolayer at high c. Less frequently, the deviations suggest the system lies energetically on the borderline between chemi- and physisorption.

SAQ 10.1 The following data refer to the adsorption of butane at $0\,°C$ onto tungsten powder (area $16.7\ m^2\,g^{-1}$). Calculate the number of moles adsorbed in a monolayer, and hence the molecular area for the adsorbed butane (at monolayer coverage); and compare it with the value of $32 \times 10^{-20}\ m^2$ estimated from the density of liquid butane.

Relative pressure $p/p^{\ominus}$	0.06	0.12	0.17	0.23	0.30	0.37
Volume adsorbed (at s.t.p.) $V_{(ads)}/$ cm^3 g^{-1}	1.10	1.34	1.48	1.66	1.85	2.05

SAQ 10.2 When nitrogen was adsorbed onto a 213 mg sample of graphitized carbon black, the pressures necessary to maintain a given degree of adsorption were 15 kPa at 312 K and 17 kPa at 130 K.

(1) Calculate the *mean* enthalpy of adsorption $\Delta H_{(ads)}$.

(2) Why 'mean' in this context?

[Hint: the process of adsorption is a condensation reaction, so we can employ the Clausius–Clapeyron equation (Equation (5.5)) to answer part (1). Be careful, though, because ΔH for a condensation reaction will be negative.]

10.2 Colloids and interfacial science

Why is milk cloudy?

Particle suspensions and phase dispersal

Milk contains both organic and inorganic components. A majority of the milk is water based, and contains water-soluble solutes such as calcium compounds. But milk also contains as much as 15 per cent by mass of water-insoluble, fat-based compounds.

These two extremes of oil and water do not mix, and remain as separate *phases* (see Chapter 5). Mixing these oil- and water-based components causes two distinct layers to form in a process called *phase separation* – much as we see when mixing oil and vinegar when preparing French dressing, or when shaking water and toluene solu-

Two *immiscible* liquids do not mix.

tions within a separating funnel. We say the water- and fat-based phases are *immiscible*. But, unlike the two distinct layers that form when mixing two immiscible liquids in a separating funnel, the fat particles in milk retain their microscopic size, and remain *suspended* or *dispersed*. We say the water-based phase is the *dispersal medium*, and the fat-based phase is the *dispersed medium*.

We see no particles after dissolving an ionic solute in water, so aqueous solutions of potassium chloride or copper nitrate are crystal clear. In fact, a beam of light can pass straight through such a solution and emerge on the other side of a flask or beaker, with no incident light *scattered* by the particles. Scattering would only occur if the particles were similar in size to, or larger than, the wavelength λ of visible light (350–700 nm). The average diameter of a solvated ion is 10^{-9}–10^{-8} m, and so is much smaller than the wavelength λ.

But the microscopic fat particles suspended in milk have an average diameter in the range 10^{-7} to 10^{-5} m, i.e. much larger than λ of visible light. A beam of incident light is *scattered* rather than transmitted by a suspension of particles – a phenomenon known as the *Tyndall effect*.

In summary, milk appears cloudy because the suspended fat particles scatter any incident light.

> The continuous phase is said to be the *dispersAL medium* (or *phase*), and the suspended particles are the *dispersED medium* (or *phase*).

> The study of the scattering of light by colloidal systems has a long history. The *Tyndall effect* describes the scattering of light by suspended particles. In fact, the first rigorous theory was that of Rayleigh in 1871.

Aside

Tyndall scattering also causes the blinding effect of shining a car headlight directly into a thick bank of fog or mist; it also yields the beautiful iridescent colours on the wing of a butterfly or peacock tail, and an opal and mother of pearl. A good potter can reproduce some of these optical effects with so-called iridescent glazes, which comprise colloidal materials.

The light is scattered in a so-called scattering pattern, which is a function of the wavelength of the light λ and the angle θ between the incident and scattered light. Although no precise stipulation relates the size with the light scattered, particles that cause light scattering generally have diameters that are ten times larger than λ, or more.

Incidentally, these different scattering patterns explain why a car headlamp causes more scattering with a quartz–halogen bulb than with a standard tungsten filament, because of their differing wavelengths and power.

What is an 'aerosol' spray?

Colloids and aerosols

Squirting a solution of perfume through a fine tube, near the end of which is a slight constriction, generates a fine spray of perfume comprising tiny particles of liquid. The spray remains airborne for a short time, and then settles under the influence of

> A *colloid* is one phase suspended within another.

gravity. We call this suspension an *aerosol*, which is one form of a *colloid*.

A colloid is a broad category of mixtures, and is defined as one phase suspended in another. A perfume spray is made up of a liquid (the perfume) *dispersed* in a gas (the air). The principle underlying the perfume atomizer is the same as the nozzle on a can of polish, and the jets within the carburettor in the internal combustion engine. In each case, the colloid formed is an aerosol.

Aside

> The word 'colloid' comes from the Greek root *coll*, meaning 'glue'.

In 1860, the Scottish chemist Thomas Graham noticed how substances such as glue, gelatin, or starch could be separated from other substances, like sugar or salt, by dialysis across a semi-permeable membrane, such as cellulose or parchment (made from treated animal skin). He gave the name *colloid* to substances unable to diffuse through the membrane – because the particles were too large. In fact, colloidal particles are larger than molecules, but are too small to observe directly with a standard optical microscope.

What is 'emulsion paint'?

Emulsions and colloid classification

> Emulsion paint comprises pigment bound in a synthetic resin such as urethane, which forms an emulsion with water.

Emulsion paint is easy to apply, aesthetically good looking, and forms a hard wearing and waterproof coating. Like milk in the previous example, the paint comprises one phase dispersed within another. It is a colloid, but this time a liquid finely dispersed in another liquid, which we call an *emulsion*.

In emulsion paint, the dispersion phase is the liquid of the paint and is generally water-based for emulsion paints. The water is buffered (p. 269) with ammonia or simple amines to yield a slightly alkaline solution. The dispersed phase in most emulsion paints comprises small particles of immiscible urethane.

Water evaporates from the paint during drying, with two consequences. Firstly, molecular oxygen *initiates* a polymerization reaction. And, secondly, as water evaporates from the layer of paint, so the volume of paint decreases, causing the urethane particles to come together more closely, thereby encouraging *propagation* of the polymerization reaction, causing juxtaposed urethane particles to join and hence form a continuous layer. The eventual paint product is a solid, polymeric layer of poly(urethane).

Classification of colloids

A colloid is defined as a non-crystalline substance consisting of ultra-microscopic particles, often of large single molecules, such as proteins, usually dispersed through a second substance, as in gels, sols and emulsion.

There are eight different types of colloid, each of which has a different name according to the identity of the *dispersed* phase and the phase acting as the *dispersion* medium.

A gas suspended in a liquid is called a *foam*. Obvious examples include shaving foam (the gas being butane) and the foam layer generated on the surface of a warm bath after adding a surfactant, such as 'bubble bath'. The gas in this last example will be air, i.e. mainly nitrogen and oxygen.

A gas suspended in a solid is also called a *foam*. This form of colloid is relatively rare in nature, unless we stretch our definition of 'solid' to include rock, in which case pumice stone is a colloidal foam. Synthetic foams are essential for making cushions and pillows. There is also presently much research into forming metal foams, which have an amazingly low density.

A liquid dispersed in a different liquid is called an *emulsion*, as above. In addition to emulsion paint, other simple examples include butter, which consists of fat droplets suspended in a water-based dispersion medium, and margarine, in which water particles are dispersed within an oil-based phase.

> The word 'emulsion' comes from *emulsio*, Latin for the infinitive, 'to milk'.

A liquid dispersed in a solid is called a *solid emulsion*. Few examples occur in nature, other than pearl and opal, the solid phase of which is based on chalk.

A solid or liquid suspended in a gas is called an *aerosol*. A good example of a solid-in-gas aerosol is smoke, either from a fire or cigarette. An example of a liquid-in-gas aerosol is the liquid coming from a can of polish or paint, or the perfume emerging from an atomizer.

A solid suspended in a liquid is called a *sol*, although a high concentration of solid is also called a *paste*. Some paints are sols, particularly those containing particles of zinc to yield weatherproof coatings. Toothpaste is also a sol. Many simple precipitation and crystallization processes in the laboratory generate a sol, albeit usually a short-lived one, since the solid settles under the influence of gravity to leave a clear solution above a layer of finely divided solid. Try mixing solutions of silver nitrate and copper chloride; this generates a cloudy sol of white silver chloride, which soon settles to form a white powder and a clear, blue solution. Simple sols are rarely stable unless the liquid is viscous, or if chemical interactions bind the dispersed phase to the dispersion phase.

A solid suspended in a solid is called a *solid suspension*. Pigmented plastic – particles of dye suspended in a solid polymer – is a simple example. Freezing a liquid–liquid emulsion, such as milk, also yields a solid suspension. A layer of partially set paint, in which the urethane monomer has been consumed but the polymer has yet to form long chains, can also be thought of as representing a solid suspension.

> Short, imperfectly formed polymer chains are properly called *oligomers*, from the Greek *oligo*, meaning 'small'.

Table 10.1 Types of colloid dispersion

Dispersed phase	Dispersion medium	Name	Example
Gas	Gas	—[a]	—
Liquid	Gas	Liquid aerosol	Fog, perfume spray
Solid	Gas	Solid aerosol	Smoke, dust
Gas	Liquid	Foam	Fire extinguisher foam, shaving foam
Liquid	Liquid	Emulsion	Milk, mayonnaise
Solid	Liquid	Sol, paste	Toothpaste, crystallization
Gas	Solid	Solid foam	Expanded polystyrene, cushion foam
Liquid	Solid	Solid emulsion	Opal, pearl
Solid	Solid	Solid suspension	Pigmented plastics

[a]A gas-in-gas system is not a colloid, it is a *mixture*.

Finally, there are no gas-in-gas colloids: we cannot *suspend* a gas within another gas, since it is not possible to have gas 'particles' of colloidal dimensions. Introducing one gas to another generates a simple mixture, which follows the thermodynamics of mixtures, e.g. Dalton's law (p. 221).

We summarize these colloid classifications in Table 10.1.

Aside

Ice cream is simultaneously both an emulsion and a partially solidified foam, so it comprises three phases at once. The ice cream would be too solid to eat without the air, and too cold to eat without discomfort. The air helps impart a smooth, creamy consistency. The solid structure is held together with a network of globules of emulsified fat and small ice crystals (where 'small' in this context means about 50 μm diameter).

Why does oil not mix with water?

Interfaces and inter-phases

The plural of 'medium' is *media*, not 'mediums'.

Not all colloid systems are stable. The most stable involve solid dispersion media, since movement through a solid host will be slow. Emulsions also tend to be stable; think, for example, about a glass of milk, which is more likely to decompose than undergo the destructive process of phase separation. Aerosols are not very stable: although a water-based polish generates a liquid-in-air colloid, the particles of liquid soon descend through the air to form a pool of liquid on the table top. Smoke and other solid-in-gas aerosols are never permanent owing to differences in density between air and the dispersed phase.

In fact, we generate the most unstable colloids immediately before we need them. A good example is oil and water, which explains why we shake the bottle before serving French dressing to a salad.

The problems causing the observed instability of colloids stems from the *interface* separating the two phases. In French dressing, for example, each oil particle is surrounded with water. The instability of the water–air interface causes the system to minimize the overall area of contact between the two phases, which is most readily achieved by the colloid particles *aggregating* and thereby forming two distinct phases, i.e. oil floating on water.

Aside

'Milk fat' comprises lipids, which are solid at room temperature. If they were liquid, we could correctly call them 'oils'.

The melting temperature of milk fat is 37 °C, which is significant because 37 °C is the body temperature of a cow, and milk needs to be a liquid at this temperature. As well as the obvious dietary properties imparted by the fat, milk fat also imparts lubrication, which is why cream has a 'creamy' feel.

.3 Colloid stability

How are cream and butter made?

Colloid stability

Ordinary milk displays an amazingly diverse range of colloid chemistry. It owes its white colour to multiple light scattering from globules of fat suspended in an aqueous phase (see p. 504); protein aggregates supplement the scattering efficiency. The colloidal milk coming straight from a cow is stable except from slight gravitational effects, so centrifuging the milk separates the cream from the milk, thereby forming two stable colloidal systems, both of which comprise fat and water phases.

> Such separation occurs naturally, since cream is less dense than milk, but such separation is too slow for commercial cream production.

Mechanical agitation of the cream – a process called *whipping* – creates a metastable foam (i.e. it contains much air). Further whipping causes this foam to collapse: some water separates out, and the major product is yellow butter. Incidentally, butter is a different form of colloid from milk, since its dispersed medium is water droplets and its dispersal phase is oil (milk is an oil-in-water colloid). Forming butter from milk is a simple example of *emulsion inversion*.

Similarly, the protein in milk is very rich in colloidal chemistry. Most of the protein is bound within aggregates called *casein* micelles (see p. 512). The colloids in milk are essentially stable even at elevated temperatures, so a cup of milky tea, for

In the dairy industry, milk is warmed to kill potentially dangerous bacteria such as *E. coli, lysteria*, and *salmonella* – a process known as 'pasteurization' after Louis Pasteur (1822–1895) who discovered the effect in 1871.

example, will not separate, although boiled milk sometimes forms a surface 'skin' of protein. This skin chemisorbs to many metals and ceramics, thereby explaining why it is generally so difficult to clean it off a saucepan used to boil the milk. The formation of such skin is termed *fouling* in the dairy industry, and represents a potential health risk.

But most colloids are intrinsically unstable, and can be 'broken' by applying heat, changing the pH or adding extra chemicals, such as salt, addition of which causes a large change in ionic strength. All colloids are thermodynamically unstable with respect to the bulk material, but the kinetics of destroying the colloid are often quite slow. Colloidal milk will persist almost indefinitely before other biological processes intervene (we say the milk 'goes off').

How is chicken soup 'clarified' by adding eggshells?

The 'breaking' of colloids

An expert cook starts the process of making chicken soup by boiling the carcass of a chicken in water. The resultant *broth* is cloudy. The broth is allowed to cool, and the chicken bones are removed. The chef then adds pieces of pre-washed eggshell to the soup. The shells of a dozen or so eggs will be sufficient to remove the cloudiness of one pint of soup: we say the soup has been *clarified*. We can finish making the soup by adding vegetables and stock, as necessary, after removing the eggshells.

We say the colloid is *broken* when the dispersion medium no longer suspends the dispersed medium.

Chicken broth is cloudy because it is colloidal, containing microscopic particles of chicken fat suspended in the water-based soup. Like milk, cream or emulsion paint, the cloudy aspect of the soup is a manifestation of the Tyndall effect. Adding the eggshells to the colloidal solution removes these particles of fat, thereby removing the dispersed medium. And without the dispersed medium, the colloid is lost, and the soup no longer shows its cloudy appearance. We say we have *broken* the colloid.

In this example, a simple mechanism for breaking a colloid was chosen. The eggshells are made of porous calcium carbonate, their surface covered with innumerable tiny pores. The particles of fat in the broth accumulate in these small pores. Removing the eggshells from the broth (each with oil particles adsorbed on their surfaces) removes the dispersed medium from the broth. One of the two components of the colloid is removed, preventing the colloid from persisting.

How is 'clarified butter' made?

The thermal breaking of colloids

Clarified butter was once a popular delicacy, though it is no longer in vogue to the same extent. It is ideal for making dishes that benefit from a buttery flavour but need to be cooked over a strong flame, such as omelets, since it does not burn as easily as ordinary butter. We clarify normal butter by removing its milk solids, which incidentally increases its *smoking point* temperature.

> The *smoking point* of butter is the temperature at which partial combustion starts, yielding smoke. An oil or fat is only safe to cook with below its smoke point.

To prepare it, small nuggets of unsalted butter are melted slowly in a deep saucepan. The water in the butter evaporates, causing the milk solids to sink to the bottom of the pan, and a foam rises to the surface and is skimmed off. The clear, yellow melted butter is then poured off to leave the milk solids at the bottom of the saucepan, to be discarded.

On cooling, the butter has essentially the same taste as conventional butter, but the appearance has changed to an almost waxy consistency. And the butter is clear, because the colloidal system of water in oil is broken, hence Tyndall light scattering is no longer possible. For this reason, the process is often called *clarification*. Perhaps the weight-conscious need to know its calorific value *per unit mass* is greater than normal butter as a consequence of losing most of its water and milk solids.

We break the colloid by heating. In fact, the only temperature-independent colloidal systems are those involving a solid dispersion medium, such as lava. As with all thermodynamic quantities, the equilibrium constant associated with forming a colloid depends strongly on temperature. In general, warming a colloid results in two possible outcomes: either the dispersed medium dissolves into the dispersion medium (they become miscible), or the dispersion and dispersed media separate to form two separate layers. The crystallization of small particles of solute from solution is a good example of the former situation, when crystals form as a solution cools; the solution is clear and homogeneous at higher temperatures.

Why does hand cream lose its milky appearance during hand rubbing?

The mechanical breaking of colloids

Most hand creams are colloidal, and generally have a thick, creamy consistency. The majority of hand creams are formulated as a liquid-in-liquid colloid (an emulsion), in which the dispersion medium is water based, and the dispersed phase is an oil such as palm oil or 'cocoa butter'. These oils are needed to replenish in the skin those natural oils lost through excessive heat and work.

> Most oil-in-water emulsions (like cream or hand cream) feel creamy to the touch, and most water-in-oil emulsions (like margarine) feel greasy.

The hand cream is opaque as a consequence of the Tyndall effect. The milky aspect is lost soon after rubbing the cream into the skin to yield a transparent layer on the skin. Furthermore, the hands feel quite damp, and cool. We notice with pleasure how our hands feel softer after the water evaporates.

The mechanical work of rubbing and kneading the hand cream breaks the colloid. The oil enters the skin – as desired – while the water remains on the skin surface before evaporating (hence the cooling effect mentioned above).

The mechanical breaking of colloids is also essential when making butter from milk: the solid from soured cream is churned extensively until phase separation occurs. The water-based liquid is drained away to yield a fat-rich solid, the butter.

Why does orange juice cause milk to curdle?

Colloid stabilization and emulsifiers

> The word 'vitamin' is an abbreviation of the two words 'vital mineral'. Vitamins were once considered to be those minerals that were vital for a healthy life. The modern meaning is somewhat more comprehensive.

We have already seen how milk is an emulsion comprising oil as a dispersion medium in a water-based dispersion medium. Milk fats also form colloids. The aqueous component of milk contains many vitamins, especially the salts of calcium, which baby mammals need to produce strong teeth and bones.

Colloidal proteins in milk are chemically unstable, so adding an acidic solution such as orange juice to a glass of milk will 'break' it, causing the milk to *curdle*. The milk rapidly separates into two separate layers, termed *curds* and *whey*. The orange juice supplements the aqueous phase, so the bottom layer in the glass will look orange. The upper layer will remain somewhat cloudy, and comprises the fats and organic components within the milk. Drinking the curdled mixture is likely to taste revolting.

> The word 'curdling' comes from *curd*, the coagulated solid formed by adding acid or rennet to milk. Curds are the precursors of most cheeses.

Milk is an unusual colloid in comprising oil particles suspended in water. Adding, say, olive or sunflower oil to water will not produce a stable colloid. Two layers will re-form rapidly even after vigorous shaking, with the oil floating above the water. Milk is stable because it contains an *emulsifier*, i.e. a compound to promote the formation of a colloidal emulsion.

The difference in stability between milk and other colloidal oil-in-water colloids, such as French dressing, is the naturally occurring protein casein (see p. 509), which milk contains in tiny amounts. Casein is a phospholipid also found in egg yolk (from whence it gets its name), animals and some higher plants. Molecules of casein stabilize the colloid by *adsorbing* to the interface separating the minute particles of oil dispersed in the water-based dispersion

> The emulsifier casein *adsorbs* to the interface between the oil and water.

medium. The usual repulsions experienced between oil and water are overcome by surrounding the oil particle in this way, thereby promoting the persistence of oil particles as a suspension.

We now look more closely at the structure of casein. It is a long molecule with different ends: one end is polar and the other is non-polar. In milk, the polar group (ending with a phosphate group) is positioned to face the polar water, and the non-polar end faces the oil. In effect, each particle of oil has a *double* coating: the inner

> Each colloid particle is surrounded with an *electric double layer*.

layer is the non-polar end of the casein emulsifier, and the outer layer is a sheath of polar phosphate groups.

Table 10.2   Vocabulary concerning the breaking of colloids

Term	Definition	Example
Aggregation	The reversible coming together of small particles to form larger particles	A suspension of micro-crystals forming soon after nucleation
Coagulation	The irreversible formation of the thermodynamically stable phase in bulk quantities	The precipitation of crystals
Flocculation	Adding fibrous or polymeric materials to entrap solid particles of colloid	Adding silicate to turbid drain water causes the solution to clear
Coacervation	Separation of a liquid-in-liquid colloid into its two separate phases	French dressing forms an upper layer of oil and a lower layer of aqueous vinegar

But we must appreciate how the phosphate group acts much like the anion from a weak acid (see Chapter 6), so its exact composition will depend on the pH of solution. The pH of cow's or human milk is about 7 (see Table 6.4). If the pH decreases much below about 6 (e.g. by adding an acid in the form of orange juice), the phosphates become protonated. The emulsifying properties of casein cease as soon as its structure changes, causing the milk to separate.

Table 10.2 contains additional vocabulary relating to the thermodynamics of breaking colloids.

How are colloidal particles removed from waste water?

Aggregation, coagulation and flocculation

Waste water from a drain or rain-overflow usually contains sediment, including sand, dust and solid particles such as grit. But smaller, colloidal particles also pollute the water. Water purification requires the removal of such particulate matter, generally before disinfecting the water and subsequent removal of any water-soluble effluent.

> The adjective 'particulate' means 'in the form of particles'.

Larger particles of grit and dust settle relatively fast, but colloidal solids can require weeks for complete sedimentation (i.e. colloid breaking) to occur completely. Such sedimentation occurs when microscopic colloid particles approach, touch and stay together because of an attractive interaction, and thereby form larger particles, and sink under the influence of gravity. We call this process *aggregation*.

But time is money. The waste industry, therefore, breaks the colloid artificially to remove the particulate solid from the water. They employ one of two methods. Firstly, they add to the water an inorganic polymer such as silicate. The colloid's thermodynamic stability depends on the *surface* of its particles, each of which has a slight excess charge. As like charges repel (in consequence of Coulomb's law;

The word 'flocculate' comes from the Latin *floccus*, meaning an aggregate (originally of sheep, hence their collective noun of 'flock').

see p. 313), the colloid particles rarely come close enough for any attractive interactions to develop. But the colloid particles *are* attracted strongly by ionic charges along the backbone of the silicate (principally, existing as pendent groups of $-O^-$). The Coulombic attraction between the microscopic colloid particles and the silicate polymer increases the weight of the chains, promoting its settling under the influence of gravity. We call this process *flocculation*.

Secondly, the water-treatment companies add aluminium compounds, such as aluminium sulphate, $Al_2(SO_4)_3$, to break the colloid. The aluminium ion $Al^{3+}_{(aq)}$ has a large surface charge, and adsorbs strongly to the surface of any colloidal particle, particularly those possessing a negative surface charge. The net surface charge of such a 'modified' colloid particle is neutralized, thereby obviating the repulsions between colloid particles, and enhancing the rate of sedimentation. This second method of breaking a colloid is termed *coagulation*. Changing the solution pH will also change the surface charge of colloid particles, again breaking the colloid.

Coagulation is thermodynamically irreversible, but flocculation is reversible.

Aside

It is fairly common to employ the words 'flocculation' and 'coagulation' instead of 'aggregation'. Strictly, aggregation means particles of, for example, colloid coming together without external assistance.

Conversely, *flocculation* implies those aggregation processes effected by the intertwining of fibrous particles, for example in the wool trade, or the entrapment of silt particles in foul water, as above.

10.4 Association colloids: micelles

Why does soapy water sometimes look milky?

Association colloids

Soapy water often looks milky. This milky appearance indicates that a colloid has formed, with one phase suspended in another. But soapy water introduces another complexity: whereas water containing a lot of soap does indeed have a cloudy aspect, dilute solutions of soap are not cloudy, but clear. We see this behaviour when washing our face in the sink (yielding a concentrated and, therefore, milky soap solution) or washing in a larger volume of water, such as a bath, when the water can remain clear. Whether or not a soap solution forms colloid depends on its concentration.

We are dealing here with a new type of colloid: the *micelle*. A micelle forms when molecules aggregate to form particles suspended in solution (i.e. a colloid). A micelle is often called an *association colloid*, because it forms colloid by the association of a discrete number of components. Such colloids form by the self-organizing (or 'self-assembly) system. The most common micelles form from detergents and surfactants, but alcoholic drinks containing absinth (*Pernod, Ricard*, Ouzo, etc.) also form micelles, thereby explaining why Ouzo becomes cloudy after adding water.

> The word 'micelle' comes from the diminutive of the Latin word *mica*, meaning 'crumb'.

Although the micelle particle is an aggregate, it behaves like a liquid; indeed, it is often convenient to regard these micelle aggregates as a separate phase. For this reason, we usually class micelles as a liquid-in-liquid colloid.

> The colloid is held together with interactions of various types, as described on p. 517.

The micelle is an aggregate containing several molecules: we will call them 'monomers', M. In this example, each monomer is a molecule of soap. In general, the micelle 'particle' forms via the *step-wise* addition of n monomer molecules:

$$M + M_{(n-1)} \longrightarrow M_n \qquad (10.13)$$

It is impossible to specify, let alone quantify, all the equilibrium steps, so we formulate our ideas in terms of approximate *models*.

One of the models best able to describe the properties of micellar colloid solutions is the *closed-association model*. In it, we start by assuming the colloid comprises n molecules of monomer. We approximate by saying the colloid forms during a single step:

$$nM \longrightarrow M_n \qquad (10.14)$$

> The long-forgotten English chemist Cooper (1759–1839) was the first to give a satisfactory explanation of colloid formation (including the CMC, as below).

The equilibrium constant of colloid formation (Equation (10.14)) is given by

$$K_{(\text{micellation})} = \frac{[M_n]}{[M]^n} \qquad (10.15)$$

As n increases, so the value of $K_{(\text{micellation})}$ changes quite dramatically with concentration. Being an equilibrium constant, the value of $K_{(\text{micellation})}$ is fixed (at constant temperature), so there is effectively no micelle at low concentration but, at a sharply defined concentration limit, just about all of the monomer M becomes bound up within the micelle.

This concept is well illustrated in Figure 10.11, which shows the way n dictates the proportion of monomer units incorporated into the micelle colloid. As the value of n increases, the steeper the graph in Figure 10.11 becomes. For example, a micelle formed from 30 monomer units has a very steep concentration dependence, i.e. almost a 'step', but below a certain concentration – which is characteristic of the temperature and the monomer – about 10 per cent of the colloid is micellar. Above this limit, however, more than 90 per cent is bound up within the micelle.

The concentration range over which the transition 'monomer → micelle' is effected decreases significantly as the value of n increases.

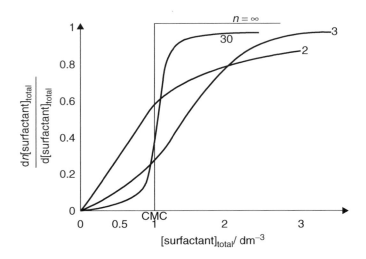

Figure 10.11 As the aggregate number n increases, so the fraction of the added surfactant that goes into the micelle (as 'y') varies more steeply with total concentration of surfactant monomer (as 'x'). The critical micelle concentration (CMC) is the midpoint of the region over which the concentration of the micelle changes (Reproduced by permission of Wiley Interscience, from *The Colloidal Domain* by D. Fennell Evans and Håkan Wennerström)

The graph in Figure 10.11 shows well how a micelle would forms at a single concentration, rather than over a range. Although a single concentration is a theoretical construct, the concentration range can be remarkably narrow.

We call the centre of the concentration range the *critical micelle concentration* (CMC). As an over-simplification, we say the solution has no colloidal micelles below the CMC, but effectively all the monomer exists as micelles above the CMC. As no micelles exist below the CMC, a solution of monomer is clear – like the solution of dilute soap in the bath. But above the CMC, micelles form in solution and impart a turbid aspect owing to Tyndall light scattering. This latter situation corresponds to washing the face in a sink.

The number of monomers in a typical micelle lies in the range 30–100, which is large enough for a reasonably well defined CMC to show. The value of n depends on the choice of monomer, with large monomer molecules generally forming larger micelles, according to Table 10.3.

The properties of the micelle are very well defined; so, for example, the maximum micelle radius is a simple function of the hydrocarbon chain in the monomer.

> The CMC relates to the total concentration of monomer.

> In fact, a truly micellar solution will comprise micelles having a statistical spread of n values and hence a range of molar masses. For example, a sulphonic acid with a chain of 14 CH_2 units produces a micelle of $n = 80$, albeit with a standard deviation $\sigma = 16.5$

Table 10.3 A micelle comprises n monomer units. The value of n depends on the length of the monomer chain

Surfactant	CMC/mol dm^{-3}	Temperature/$^{\circ}$C	n
$C_6H_{13}SO_4Na$	0.42	25	17
$C_7H_{15}SO_4Na$	0.22	25	22
$C_{12}H_{25}SO_4Na$	8.2×10^{-3}	25	64
$C_{14}H_{29}SO_4Na$	2.05×10^{-3}	40	80

What is soap?

The physical nature of colloids

One of the principal causes of the English Civil War was the sudden imposition of steep taxes on soap, in 1637. The people could not afford to clean themselves, and rioted.

But how does soap work; why did the people want soap rather than water alone? To explain the mode by which soap works, we start by describing the structure of soap molecules in water. A good example of such a soap is the dual-nature molecule, sodium dodecyl sulphate (SDS, **III**), with its long, snake-like structure. Its head is an ionized sulphonic acid group $-SO_4^-Na^+$, ion paired with a sodium cation to yield a salt. The remainder of the molecule is a wholly non-ionic straight-chain alkyl group. We give such molecules the general name of *surfactant*.

> The word *surfactant* is an abbreviation of 'surface-active agent'.

(III)

The sulphonic acid 'head' is capable of electrostatic interactions just like any other charged species; the 'tail' can only interact weakly via induced dipoles of the London dispersion force type (see p. 47). For this reason, the head and tail of SDS behave differently, and often in a contradictory manner. In Chapter 5 we saw how polar and non-polar species tend not to associate. As an example, mixing benzene and water forms separate layers, i.e. forming separate phases, with the organic layer floating above the aqueous. In a similar way, the hydrocarbon ends of the SDS molecules seek to avoid contact with water. We say they are *hydrophobic*. The sulphonic acid groups differ and seek solvation by water: they are *hydrophilic*. In common language, we might say they are 'water hating' and 'water loving' respectively.

> The words 'hydrophobic' and 'hydrophilic' derive from Greek roots: 'hydro' comes from *hudoor* meaning water and 'phobic' comes from the Greek *phobos* meaning fear or hate (hence the English word 'phobia'). The Greek *philos* means love or friendship.

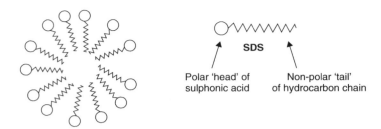

Figure 10.12   Micelles of sodium dodecyl sulphate (SDS) comprise as many as 80 monomer units. The micelle interior comprises the hydrocarbon chains, and is oil like. The periphery presented to the water of solution is made up of hydrated hydrophilic sulphonic acid groups

The interactions between SDS and water represent a compromise between the extremes of complete phase separation (as happens when benzene and water mix) and molecular dispersion (SDS in dilute aqueous solution). A micelle forms. To minimize the energetically unfavourable interactions with water, SDS molecules aggregate to form a variety of microscopic structures, such as the 'dandelion head' in Figure 10.12.

We can summarize the principal properties of these aggregates, saying: they form spontaneously at a well-defined concentration, the CMC (see p. 516); and adding more monomer to the solution yields more micelles, each colloidal particle having the same size, and ensuring the concentration of free monomer does not change.

> The association forces between juxtaposed surfactant monomers is physical, not chemical, so the motion of the hydrocarbon tails within a micelle is similar to the local motion in a sample of pure hydrocarbon.

We now consider the two competing forces affecting the formation of these micelles. On the one hand, the hydrophobic hydrocarbon end of the molecule is taken away from the polar environment of water into the oil-like interior of the micelle – a process which provides the driving energy of the micellation process. On the other hand, repulsions are minimized between the charged heads of each monomer as they come together during micellation, so the hydrophilic sulphonic acid groups are spatially well separated. These two processes compete: the first has the effect of increasing the micelle size, and the second has the effect of decreasing it. The aggregation process stops when the micelle reaches its optimum size (e.g. see value of n in Table 10.3). This competition helps explain why adding more monomer does not augment the existing micelles, but merely generates further micellar colloid in solution.

The properties of SDS in dilute solutions generally remind us of a simple ionic solute much like NaCl or KNO_3. But above a concentration of 8×10^{-3} mol dm^{-3} the tensions inherent in the two distinct natures of SDS become apparent, and the SDS becomes a soap.

Why do soaps dissolve grease?

Detergency

We are now able to describe the way soaps clean the skin. Soaps were first mentioned on p. 239, when we introduced the action of aqueous alkali on the skin, reacting *chemically* with oils, fats and greases in a process we call 'saponification'. But the best soaps are not inorganic alkalis, they are organic salts of long-chain carboxylic acids. They are also milder on the skin.

We also need soaps for cleaning the crockery after a meal of chips, pizza or greasy sausages. Such cleaning can be difficult and time-consuming unless we first add to the water an effective soap or detergent such as 'washing-up liquid'.

Each micelle has a polar periphery and an oil-like core. When molecules of monomer collide with the solid surface of, say, a dirty plate, the non-polar ('hydrophobic') end adsorbs to the non-polar grease. Conversely, the polar ('hydrophilic') end readily solvates with water. Soon, each particle of oil or grease is surrounded with a protective coating of surfactant monomer, according to Figure 10.13.

Having 'disguised' each particle of oil or grease, it can readily enter solution while sheathed in its water-attracting 'overcoat' of surfactant. And if the oil particles enter the solution, then the oil is removed from the plate, and is cleaned.

After leaving the plate, the grease particle remains encapsulated within the micelle, surrounded with the oil-like hydrocarbon chains of the soap monomers. The soap cleans the plate by allowing the grease to enter solution.

> A compound able to 'dissolve' grease by forming micelles is called a *detergent*. Naturally occurring detergents are also called *soaps*.

> We often describe the structure of this coating as a *bi-layer*, with the inner (oil-facing) part made up of water-repelling hydrocarbon chains, and the outer (water-facing) layer comprising the sulphonic acid groups.

> A detergent forms a micelle from an otherwise insoluble phase.

Why is old washing-up water oily when cold but not when hot?

The Krafft point temperature

Washing up the dishes after a meal demonstrates how the temperature affects the colloid's stability. The dirty water is cloudy after washing up because it contains

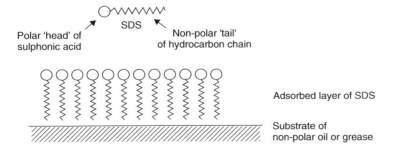

Figure 10.13 The non-polar tail of an SDS molecule readily adsorbs to the surface of non-polar oil or grease on a plate or hand. The polar sulphonic acid heads point toward the solution, and are hydrated

micelles of surfactant; each micelle particle contains a tiny particle of grease. But, after cooling the washing water below about 35 °C, an oily scum forms on its surface and the water beneath becomes clear. Re-warming the water causes the oily compounds that make up the scum to re-enter the water, to re-form the colloid.

Micelles only form above a crucial temperature known as the *Krafft point temperature* (also called the *Krafft boundary* or just *Krafft temperature*). Below the Krafft temperature, the solubility of the surfactant is too low to form micelles. As the temperature rises, the solubility increases slowly until, at the Krafft temperature T_K, the solubility of the surfactant is the same as the CMC. A relatively large amount of surfactant is then dispersed into solution in the form of micelles, causing a large increase in the solubility. For this reason, IUPAC defines the Krafft point as the temperature (or, more accurately, the narrow temperature range) above which the solubility of a surfactant rises sharply.

In reverse, the surfactant precipitates from solution as a hydrated crystal at temperatures below T_K, rather than forming micelles. For this reason, below about 20 °C, the micelles precipitate from solution and (being less dense than water) accumulate on the surface of the washing bowl. We say the water and micelle phases are *immiscible*. The oils re-enter solution when the water is re-heated above the Krafft point, causing the oily scum to *peptize*. The way the micelle's solubility depends on temperature is depicted in Figure 10.14, which shows a graph of [sodium decyl sulphate] in water (as '*y*') against temperature (as '*x*').

> To peptize means a bulk phase enters solution as a colloid (here, as a micelle).

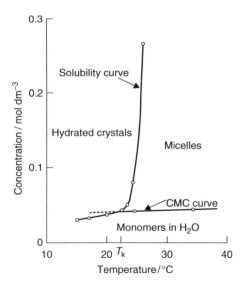

Figure 10.14 Graph of [surfactant] (as '*y*') against T (as '*x*') for sodium decyl sulphate in water. The Krafft temperature is determined as the intersection between the solubility and CMC curves, yielding a T_K of about 22 °C. At lower temperatures, the micelles convert to form hydrated crystals, which we might call 'scum' (Reproduced by permission of Wiley Interscience, from *The Colloidal Domain* by D. Fennell Evans and Håkan Wennerström)

Table 10.4 Krafft point temperatures[a] for sodium alkyl sulphates in water

Number of carbon atoms	10	12	14	16	18
Krafft temperature/°C	8	16	30	45	56

[a]Data reproduced with permission from J. K. Weil, F. S. Smith, A. J. Stirton and R. G. Bristline, *J. Am. Oil. Chem. Soc.*, 1963, **40**, 538.

The value of T_K is best determined by warming a dilute solution of surfactant, and noting the temperature at which it becomes clear. Table 10.4 lists the Krafft points for a series of colloidal systems based on aqueous solutions of sodium alkyl sulphate (cf. structure **III**).

Aside

Tap water in some parts of the country contains large amounts of calcium and magnesium ions. The calcium usually enters the water as it seeps through limestone and chalk, since these rocks are sparingly soluble.

One of the simplest ways to tell at a glance if our tap water contains much Ca^{2+} – we say it is 'hard' – is to look at the bath after letting out the water. Surfactants form micelles with the calcium di-cation at temperatures above T_K, i.e. in a hot bath. But after cooling to a temperature of about 20 °C, the micelles precipitate to yield hydrated crystals – which we observe as a ring of 'scum' along the waterline.

The *brown* colour of this bath ring generally derives from occlusion of dirt and particles of skin within the crystals.

A quick glance at Figure 10.14 suggests a simple way of saving time when cleaning away the bath ring is to scrub with *hot* water since the solid crystals convert back to water-soluble micelles above T_K. The micelles can flow away from the bath after removing the plug.

Why does soap generate bubbles?

Surfactants and surface tension

A surfactant yields bubbles by decreasing the surface tension γ of the solution. Most detergents are surfactants. A detergent such as a long-chain alkyl sulphonic acid dramatically decreases the surface tension γ of water by adsorbing at the air–water interface. Stated another way, the sulphonyl groups attach to the meniscus of the water via strong hydrogen bonds. The polar sulphonyl group points toward the water, while the long alkyl chain points away, into the air above the water. Figure 10.15 shows a schematic representation of the structure, which looks much like a long-pile carpet in miniature.

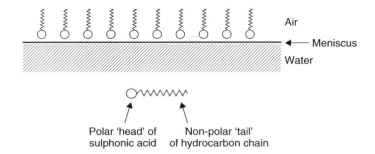

Figure 10.15   Molecules of surfactant adsorb at the air – water interface, thereby decreasing the surface tension γ. The polar sulphonic acid heads point toward the solution, thereby ensuring some extent of hydration

> The water forms *beads*, rather than any other shape, because a sphere has the lowest possible surface area per unit volume, thereby minimizing the surface area presented to the atmosphere.

The first effect of decreasing the surface tension is to make it easier for the water to adhere to a surface, be it a crockery bowl or waste food on a plate; we say the 'wetting properties' of the water are enhanced. Of course, the phrase 'adsorbed water' is an over-simplification. As soon as a dish enters a detergent solution, the molecules of detergent residing on the surface of the water realign to form a layer *between* the water and the dish. This multi-layer structure helps explain detergency: water forms a thin, continuous layer on a surface if detergent is added; otherwise, the water forms small, spherical beads.

Why does detergent form bubbles?

Surfactants and bubbles

> A surfactant is a soap.

The air in water colloid formed during the turbulent flow of water, such as water passing over a dam or running water into a bath, is a particularly unstable colloidal system. The bubbles rise to the surface within a few seconds to leave a clear, homogeneous liquid. But adding a soap or detergent completely changes the properties of the water, allowing the ready formation of an air-in-water colloid. Why is there a difference?

Detergents aid the removal of dirt. Commercial synthetic detergents were first developed in the 1950s. Detergents act mainly on the oil-based films that trap dirt particles. Most detergents have an oil-soluble portion (usually a hydrocarbon chain), and a water-soluble portion, which is generally ionic.

Detergents act as emulsifiers, breaking the oil into tiny droplets, each suspended in water. The disruption of the oil film allows the dirt particles to become solubilized (they *peptize*). Soap, the sodium or potassium salt of long-chain fatty acids, is a good detergent, although it often forms insoluble compounds with certain salts found in hard water, thus diminishing its effectiveness.

Detergents are classified into four groups:

(1) *Anionic*, or negatively charged, e.g. soaps.

(2) *Cationic*, or positively charged, e.g. tetra-alkyl ammonium chloride (used as fabric softeners).

(3) *Non-ionic*, e.g. certain esters made from oil (used as degreasing agents in industry).

(4) *Zwitterionic*, containing both positive and negative ions on the same molecule.

Surface tension is the tendency of liquids to reduce their exposed surface to the smallest possible area. A single drop of water – such as a rain drop – tries to take on the shape of a sphere. We attribute this phenomenon to the attractive forces acting between the molecules of the liquid. The molecules within the liquid bulk are attracted equally from all directions, but those near the outer *surface* of the droplet experience unequal attractions, which cause them to draw in toward the centre of the droplet – a phenomenon experienced as a *tension*.

> The high *surface tension* of a water meniscus explains why various small insects are able to skate across the surface of a pond.

Adding a surfactant such as decadodecylsulphonic acid to the solution changes the magnitude of the surface tension.

Answers to SAQs

1 Introduction to physical chemistry

1.1 If the resistance per degree is $6 \times 10^{-6} \; \Omega \, °C^{-1}$ and $R = 3.0 \times 10^{-4} \; \Omega$ at $0 \, °C$, then $T = 140 \, °C$.

1.2 'length' is the variable, '3.2' is the number, 'k' is the factor and 'm' is the unit.

1.3 $T = (273.16 + 30) \; K = 303.16 \; K$.

1.4 $(287.2 - 273.16) \; K = 14.04 \, °C$.

1.5 Draw a graph of volume V (as 'y') against temperature T (as 'x').

1.6 From Equation (1.8), $V_2 = 1.17 \; dm^3$.

1.7 From Equation (1.11), $V_2 = 0.4 \; dm^3$.

1.8 We rearrange Equation (1.6) to yield $\dfrac{V_1}{V_2} = \dfrac{T_1}{T_2}$, so a tenfold rise in temperature results in a tenfold increase in volume.

1.9 From Equation (1.13) and remembering that $1 \; dm^3 = 10^{-3} \; m^3$, $p_2 = 13.7 \times 10^6 \; Pa$.

1.10 From Equation (1.14), $T_2 = 508 \; K$.

1.11 The area inside the cylinder is $297 \; cm^2$ (i.e. $297 \times 10^{-4} \; m^2$). Then, using Equation (1.15), $F = 1.68 \times 10^7 \; Pa$.

1.12 The 'room temperature energy' is $\frac{3}{2} \times R \times T$, so energy $= 3.3 \; kJ \, mol^{-1}$.

2   Introduction to interactions and bonds

2.1 $V = 1.5 \times 10^{-2}$ m^3 = 15 200 cm^3.

2.2   In each case, we assume the bond will be polar if the separation between the two electronegativities is large, and non-polar if the separation is small. (a) HBr is polar; (b) SiC is non-polar; (c) SO$_2$ is non-polar; (d) NaI is very polar, indeed, fully ionized.

2.3   302 K.

2.4 $p = 2.18 \times 10^6$ Pa; p is about 0.5 per cent lower when calculated with the ideal-gas equation, Equation (1.13).

2.5   The molar mass of water is 18 g mol^{-1}, so 21 g represents 1.167 mol. The energy liberated is then 1.167 mol $\times$ 40.7 kJ mol^{-1} = 47.5 kJ.

3   Energy and the first law of thermodynamics

3.1   The molar mass of water is 18 g mol^{-1}. C_V in J K^{-1} g^{-1} $\times$ 18 g mol^{-1} = 75.24 J K^{-1} mol^{-1}. (Note how the units of g and g^{-1} cancel here.)

3.2   The rise in temperature is 80 °C. 1.35×10^3 g of water represents 75 mol. The energy necessary is, therefore, 80 °C $\times$ 75 mol $\times$ 4.18 J K^{-1} mol^{-1} = 25.08 kJ.

3.3   Inserting values into energy is $V \times I \times t$ (Equation (3.10)) yields 1920 J.

3.4   The energy produced by the electrical heater (Equation (3.10)) is 108 kJ; we call this value C. The molar mass of anthracene is 178 g mol^{-1}, so 0.40 g represents 2.25×10^{-3} mol. The molar enthalpy liberated is, therefore, 4806 kJ mol^{-1}.

3.5   From Equation (1.13), 0.031 m^3 is the same as 31 dm^3.

3.6   The change in volume ΔV is 1.99 dm^3 = 1.99×10^{-3} m^3. Using Equation (3.12), $w = 199$ J.

3.7   The fully balanced equation is $1C_2H_6 + 3.5O_2 \rightarrow 2CO_2 + 3H_2O$. From Equation (3.22), the value of ΔC_p is 144.9 J K^{-1} mol^{-1}. (By 'per mole' here we mean *per mole of reaction*.) Then, from Equation (3.21), the value of $\Delta H_c^{\ominus}$ at 80 °C is -1550.8 kJ mol^{-1}.

3.8 $\Delta H_r^{\ominus} = H_m^{\ominus} (NO_2) - \{2H_m^{\ominus} (NO) + H_m^{\ominus} (O_2)\}$

3.9   The fully balanced equation is $C_6H_{12}O_6 + 9O_2 \rightarrow 6CO_2 + 6H_2O$. Using Equation (3.29), the value of $\Delta H_c^{\ominus}$ is -2804.8 kJ mol^{-1}.

3.10   Having drawn a suitable Hess-law cycle, $\Delta H_r^{\ominus} = -0.21$ kJ mol^{-1} (or -210 J mol^{-1}).

3.11 $\Delta H_{(\text{lattice})} = -1185$ kJ mol^{-1}.

4 Reaction spontaneity and the direction of thermodynamic change

4.1   The liquid water will have a lower entropy than the gaseous water, so the process will *not* be thermodynamically spontaneous.

4.2   The number of moles decreases from 1.5 mol of gas to 1 mol. Assuming equivalent entropies per mole, the entropy decreases. ΔS will be negative, so we do not expect a spontaneous process.

4.3   From Equation (4.10), $\Delta S = 39 \text{ J K}^{-1} \text{ mol}^{-1} \times \ln\left(\dfrac{273 \text{ K}}{258 \text{ K}}\right) = 2.2 \text{ J K}^{-1}$ mol^{-1}.

4.4 $\Delta S = 25.8 \ln\left(\dfrac{350 \text{ K}}{300 \text{ K}}\right) + 1.2 \times 10^{-2} (350 \text{ K} - 300 \text{ K}) = 4.58 \text{ J K}^{-1}$ mol^{-1}.

4.5 $\Delta H_r - 230.6 = \text{kJ mol}^{-1}$, $\Delta S_r = -423.2 \text{ J K}^{-1} \text{ mol}^{-1}$, and hence $\Delta G_r = -104.5 \text{ kJ mol}^{-1}$.

4.6   From Equation (4.21), $\Delta G_r - 4.09 = \text{kJ mol}^{-1}$.

4.7   (1) Taking straight ratios yields $p_{(\text{vacuum})} = 3682 \text{ Pa}$. (2) From Equation (4.39), $\Delta G = -8.2 \text{ kJ mol}^{-1}$.

4.8   The left-hand side is a logarithm and, therefore, has no units. ΔH_r has units of J mol^{-1} and R has units of J K^{-1} mol^{-1}, so $\Delta H_r \div R$ has units of K. The units of the reciprocal temperature in the right-hand bracket is K^{-1}, which cancels with the K from $\Delta H_r \div R$; therefore, the right-hand side is also dimensionless.

4.9   Remembering to convert from kJ to J, $K = 127$.

4.10   (1) $\Delta S = 560 \text{ J K}^{-1} \text{ mol}^{-1}$. Its positive value implies a thermodynamically spontaneous process. (2) From Equation (4.62), ΔG_r at 332 K is -220.8 kJ mol^{-1}.

4.11   Using Equation (4.74), $\Delta G^{\ominus}$ is negative only above 985 K. Maintaining this high temperature explains why synthesis gas is not economically viable.

4.12   Using Equation (4.78), $\Delta H = 14.7 \text{ kJ mol}^{-1}$.

4.13   Draw a van't Hoff graph of $\ln K$ (as 'y') against $1/T$ (as 'x'), then multiply the gradient by $-R$ to obtain $\Delta H = 36.7 \text{ kJ mol}^{-1}$.

5 Phase equilibria

5.1   About 100 mmHg, which equates to $1.31 \times 10^4 \text{ Pa}$ ($0.13 \times p^{\ominus}$).

5.2   A fivefold increase in pressure means $p_2 - p_1 = 5p^{\ominus} - p^{\ominus} = 4p^{\ominus}$. Using $4p^{\ominus}$ as the 'dp' term in Equation (5.1) yields $\Delta V_m = 75.6 \times 10^{-6} \text{ m}^3 \text{ mol}^{-1}$.

5.3 The pressure $p = $ force $\div$ area (Equation (1.15)), so $p = 1000$ N $\div$ 1 m^2 (he has two snow shoes) $= 10^3$ Pa. Inserting values again into Equation (5.1) yields d$T = -7.3 \times 10^{-5}$ K, i.e. unnoticeable.

5.4 From Equation (5.5), and remembering to convert from kJ to J, $T_2 = 352.9$ K ($= 79.8\,^{\circ}$C).

5.5 We start by writing the equilibrium constant K as '[sucrose]$_{(water)}$ $\div$ [sucrose]$_{(CHCl_3)}$', which equals 5.3. So for every 6.3 portions of sucrose, one portion dissolves in the chloroform and 5.3 in the water. One portion out of 6.3 is 15.8 per cent.

5.6 Remember to convert the volume of water into a mass of 900 g. From Equation (5.17), $\Delta T = 0.305$ K and $T_{(freeze)} = 273.45$ K.

5.7 10 g of N$_2$ is 0.357 mol and 15 g of CH$_4$ is 0.938 mol. $p_{(CH_4)} = 0.357 \div (0.357 + 0.938) = 0.275$ or 27.5 per cent.

5.8 From Equation (5.22), $p_{(total)} = 0.286 \times p^{\ominus}$.

5.9 From $p_{(total)} = 0.286 \times p^{\ominus}$, the vapour comprises 69 per cent toluene.

6 Acids and bases

6.1 (a) base = carbonate and acid = bicarbonate; (b) base = H$_2$EDTA^{2-} and acid = H$_3$EDTA$^-$; (c) acid = HNO$_2$ and base = NO$_2^-$.

6.2 pH $= 0.7$.

6.3 [HCl] $= 5 \times 10^{-7}$ mol dm^{-3} when pH $= 6.31$ and 4.79×10^{-7} mol dm^{-3} when pH $= 6.32$.

6.4 [HNO$_3$] $= 6.3 \times 10^{-2}$ mol dm^{-3}.

6.5 The exponent is -5, so the pH is $+5$.

6.6 pH $= 11.8$.

6.7 [H$_3$O$^+$] $= 10^{-9}$ mol dm^{-3}, so [OH$^-$] $= 10^{-5}$ (because $K_w = 10^{-14}$).

6.8 pH $= 12$.

6.9 $K_a = K_w \div K_b = 5.75 \times 10^{-10}$.

6.10 $$K_{a(1)} = \frac{[\text{H}_3\text{EDTA}^-][\text{H}^+]}{[\text{H}_4\text{EDTA}]} \qquad K_{a(2)} = \frac{[\text{H}_2\text{EDTA}^{2-}][\text{H}^+]}{[\text{H}_3\text{EDTA}^-]}$$

$$K_{a(3)} = \frac{[\text{H}_1\text{EDTA}^{3-}][\text{H}^+]}{[\text{H}_2\text{EDTA}^{2-}]} \qquad K_{a(4)} = \frac{[\text{EDTA}^{4-}][\text{H}^+]}{[\text{H}_1\text{EDTA}^{3-}]}$$

6.11 Using Equation (6.43), $V_{(alkali)} = 8.6$ cm^3.

6.12 With $s = \frac{1}{2}$ and using Equation (6.46), $V_{(alkali)} = 12.5$ cm^3.

6.13   Using Equation (6.50), pH $= 9.04$.

6.14   The pH is 9.22 before adding any alkali. 8 cm^3 of this alkali contains 8×10^{-4} mol of NaOH, so 8×10^{-4} mol of NH$_3$ is formed and 8×10^{-4} mol of NH$_4^+$ is consumed. Accordingly, there are 0.0508 mol of NH$_3$ and 0.0492 mol of NH$_4^+$. Then, from Equation (6.50), pH $= 9.23$.

7   Electrochemistry

7.1   1 mol of electrons $= 1\ F$, so $10^{-9}\ F$ are passed, which represent 6.022×10^{14} electrons.

7.2   If $n = 1$, then $1\ F$ generates 1 mol, so 10^{-10} mol are formed.

7.3 $\frac{2}{3}$ mol.

7.4   The iron couple is the more positive, so $\ominus$ Co$_{(s)}$|Co^{2+}$_{(aq)}$||Fe^{3+}$_{(aq)}$|Fe$_{(s)}$ $\oplus$.

7.5   Using Equation (7.15), $\Delta G_{(cell)} = -289$ kJ mol^{-1}.

7.6 $\dfrac{\mathrm{d}(emf)}{\mathrm{d}T} = 2.79 \times 10^{-4}$ V K^{-1} using Equation (7.20), so $\Delta S_{(cell)} = 26.8$ J K^{-1} mol^{-1} using Equation (7.18).

7.7 $emf = 0.0648$ V at 610 K.

7.8 $emf = 0.0271$ V at 660 K, so $\Delta S_{(cell)} = -146$ J K^{-1} mol^{-1} with $n = 2$.

7.9 $\Delta G_{(cell)} = -212$ kJ mol^{-1}, $\Delta S_{(cell)} = 67.5$ J K^{-1} mol^{-1} and $\Delta H_{(cell)} = -192$ kJ mol^{-1}.

7.10 $E_{Co^{2+},Co}$.

7.11 E_{Br_2,Br^-}.

7.12 $\gamma_\pm = \sqrt[5]{\gamma_+ \times \gamma_+ \times \gamma_- \times \gamma_- \times \gamma_-}$

7.13 $I = 6 \times c$.

7.14   At low ionic strengths, the $\sqrt{I}$ term in the denominator becomes negligible, so $(1 + b\sqrt{I})$ tends to unity, yielding the limiting Debye–Hückel law.

7.15 $I = 4 \times c$. $\gamma_\pm = 0.064$, i.e. decreases by 93 per cent.

7.16   Using Equation (7.41), $a(Cu^{2+}) = 4.17 \times 10^{-4}$.

7.17   Using Equation (7.43), $E_{AgCl,Ag} = 0.251$ V.

7.18   Using Equation (7.48), $emf = 0.037$ V.

7.19   Using Equation (7.49), $K = 0.017$ V.

7.20 $emf = 0.153$ V.

7.21   The numerical value of $\dfrac{2.303RT}{F}$ at 298 K is 0.059 V or 59 mV.

7.22   Use Equation (7.53). Subtract E_j from both sides of the equation, then manipulate the value of $E_{(\text{right-hand side})}$, i.e. $E_{Ag^+, Ag}$, to obtain the activity as normal.

8   Chemical kinetics

8.1   rate $= k_2[CH_3COOH][CH_3CH_2OH]$

8.2   rate $= k[Cu^{2+}_{(aq)}][NH_{3(aq)}]^4$

8.3   rate $= k[Ce^{IV}][H_2O_2]$. If each concentration doubles, then $[Ce^{IV}]_{(new)} = 2 \times [Ce^{IV}]_{(old)}$ and $[H_2O_2]_{(new)} = 2 \times [H_2O_2]_{(old)}$. Inserting the new concentrations into the rate expression will increase the new rate fourfold.

8.4   (1)   rate $= k \times p(NO_2) \times p(N_2O_4)$

        (2)   rate $= k \times p(SO_2)^2 \times p(O_2)$

        (3)   rate $= k \times [Ag^+][Cl^-]$

        (4)   rate $= k \times [HCl] + [NaOH]$

8.5   (1) Second order; (2) first order; (3) third order.

8.6   Equation (8.12) is rate limiting because it is the slow step. Its rate is $k \times [\text{ethanal}] \times [O_2]^2$.

8.7   From Equation (5.19), partial pressure $=$ mole fraction, $x \times p^{\ominus}$, so $p(O_2) = 0.21 \times 10^5$ Pa, so $p(O_2) = 2.1 \times 10^4$ Pa.

8.8   rate$_{(\text{forward})} = k \times p(H_2) \times p(Cl_2)$ and rate$_{(\text{backwards})} = k \times p(HCl)^2$.

8.9   rate $= k \times [I^-] \times [S_2O_8^{2-}]$, so rate $= k_2 = 0.03$ dm^3 mol^{-1} s^{-1}.

8.10   Using Equation (8.24), and converting to seconds, $[A]_t = 7.39 \times 10^{-3}$ mol dm^{-3}.

8.11 $k_1 = 1.27 \times 10^{-4}$ s^{-1}.

8.12   Calculate the number of seconds in a year $(= 365.25 \times 24 \times 60 \times 60 = 3.156 \times 10^7$ s year$^{-1})$, then divide 1.244×10^4 year^{-1} by it.

8.14   Using Equation (8.27), $[MV^{+\bullet}]_t = 0.25 \times 10^{-3}$ mol dm^{-3}.

8.15   If 15 per cent is consumed then 85 per cent remains, so $[A]_t = 0.85 \times 10^{-3}$ mol dm^{-3}, yielding $k_2 = 0.238$ dm^3 mol^{-1} s^{-1}.

8.16   Draw a graph of $1/[A]_t$ (as 'y') against time t (as 'x'). The gradient equals k_2, so $k_2 = 6.36 \times 10^{-4}$ dm^3 mol^{-1} s^{-1}.

8.17   Because it is a 1:1 reaction, if $[A]_t = 0.05$ after 0.5 h then
$[B]_t = 0.15$ mol dm^{-3}. $t = 1800$ s. Using Equation (8.30),
$k_2 = 2.25 \times 10^{-3}$ dm^3 mol^{-1} s^{-1}.

8.18   The length of time represents two half-lives. 10 g halves to 5 g after one
half-life; this mass halves again after a second half-life.

8.19   If ^{14}C remaining is 88 per cent, then the age is 1031 years. If ^{14}C
remaining is 92 per cent, then the age is 670 years. Quite a difference!

8.20   From Equation (8.27), $1/[A]_t = 1/0.5[A]_0$. Accordingly, algebraic
addition of the two concentration terms as '$1/[A]_t - 1/0.5[A]_0 = kt_{1/2}$'
yields '$1/[A]_0 = kt_{1/2}$', which, after dividing by 'k' yields $t_{1/2} = \dfrac{1}{[A]_0 k_2}$.

8.21 $t = 600$ s and $k' = 1.33 \times 10^{-3}$ s^{-1}. $k_2 = k' \div [K_3[Fe(CN)_6]]$
$= 2.266$ dm^3 mol^{-1} s^{-1}.

8.22   Equating Equation (8.49) to K, $K = 1200$.

8.23   Equation (8.50) includes $\ln \left[\dfrac{(1 - 0.4) \text{ mol dm}^{-3}}{(0.7 - 0.4) \text{ mol dm}^{-3}} \right]$. The units cancel. If
the mathematics is performed erroneously, the log becomes
$\ln(1 \div 0.7) = 0.357$. If, however, we perform the mathematics in the two
brackets first, we obtain $\ln(0.6 \div 0.3) = 1.20$.

8.24   Using Equation (8.55), and saying the left-hand side is merely $\ln(2)$, and
converting each temperature to kelvin, we calculate $E_a = 41.6$ kJ mol^{-1}.

8.25   Using Equation (8.55), $k_2 = 4.23 \times 10^{10}$ dm^3 mol^{-1} s^{-1} at 330 K, i.e.
about 5 per cent faster.

8.26   (1)   Using Equation (8.55), $E_a = 53.6$ kJ mol^{-1} and $A = 1.98 \times 10^{12}$ dm^3
mol^{-1} s^{-1}.

       (2)   Using Equation (8.56), $\Delta H^{\ddagger} = 50.9$ kJ mol^{-1}; using Equation (8.61),
$\Delta G^{\ddagger} = 56.3$ kJ mol^{-1}, and $\Delta S^{\ddagger} = -18.1$ J K^{-1} mol^{-1}.

       (3) $E_a - \Delta H^{\ddagger} = 2.7$ kJ mol^{-1}, which is very close to the theoretical
value of 2.4 kJ mol^{-1}.

9 Physical chemistry involving light: spectroscopy and photochemistry

9.1   From Equation (9.2), $\lambda = 3 \times 10^{-4}$ m, which is a microwave photon.

9.2   From Equation (9.3), $\nu = 1.51 \times 10^{15}$ Hz.

9.3   From Equation (9.4), 1.0 MJ mol^{-1} equates to an energy per photon of
1.66×10^{-18} J, and describes a frequency $\nu = 2.5 \times 10^{15}$ Hz.

9.4   From Equation (9.6), $T = 0.63$.

9.5   From Equation (9.7), $A = 1.39$.

9.6   From Equation (9.21). (1) $\mu = 1.614 \times 10^{-27}$ kg; (2) 2.947×10^{-27} kg.

9.7   From Equation (9.21), $\mu = 1.577 \times 10^{-27}$ kg. From Equation (9.20), $\omega = 89\,118$ m$^{-1} = 891$ cm^{-1}, in the infrared region of the spectrum.

9.8   From Equation (9.2), $c = \lambda \times \nu$, $\lambda = 0.1224$ m $= 12.2$ cm.

10 Adsorption and surfaces, colloids and micelles

10.1 (1) We obtain the number of moles adsorbed in a monolayer as the reciprocal of the Langmuir plot's gradient: $V_m = 8.49$ cm^3 g^{-1}. From the ideal-gas equation (Equation (1.13)), this volume equates to 3.74×10^{-4} mol g^{-1}.

      (2) This number of moles adsorbs to form a monolayer of area 16.7 m^2, so 1 mol has the almost incredible area of $44\,650$ m^2. 1 mol comprises 6.022×10^{23} molecules, so one molecule has an area of $44\,650$ m$^2 \div 6.022 \times 10^{23} = 7.42 \times 10^{-20}$ m^2.

10.2 (1) $\Delta H_{(ads)} = -232$ J mol^{-1} (or -0.23 kJ mol^{-1}). The mass of sample is irrelevant. (2) This enthalpy is a *mean* because all enthalpies depend on temperature (see Equation (3.19)), and the temperature was varied over 143 K.

Bibliography

The books and comments contained in this bibliography are each cited at the recommendation of the author alone.

Any web address (URL) is prone to alteration without warning. All the URLs here were correct when the bibliography was completed in Easter 2003. Please be careful with upper and lower cases when typing in the URL on your browser.

General bibliography

The best selling textbook of physical chemistry in the world is undoubtedly *Atkins's Physical Chemistry*. The latest edition is the seventh by P. W. Atkins and Julio de Paula, Oxford University Press, Oxford, 2002. Many students will find it rather mathematical, and its treatment is certainly high brow. Its 'little brother' is *Elements of Physical Chemistry* (third edition), P. W. Atkins, Oxford University Press, Oxford, 2001, and is intended to overcome these perceived difficulties by limiting the scope and level of its parent text. Both are thorough and authoritative.

Walter Moore's two books *Physical Chemistry*, Longmans, London, 1962, and its abridgement *Basic Physical Chemistry*, Prentice Hall, London, 1983, both deserve their popularity. Although Moore is rigorous, he never loses sight of his stated aim to *teach* his subject, perhaps explaining why his prose is just as informative as his mathematics.

Physical Chemistry (third edition) by G. W. Castellan, Addison Wesley, Reading, MA, 1983, is not much in vogue these days, in part because many of his symbols do not conform to the IUPAC system. Castellan's strength is his explanations, which are always excellent. His mathematical rigour is also notable.

Worked examples

Several texts approach the topic by means of worked examples. *Physical Chemistry* (2nd edition), C. R. Metz, McGraw Hill, New York, 1989, is a member of the Schaum 'out-line' series of texts, and *Physical Chemistry*, H. E. Avery and D. J. Shaw, Macmillan, Basingstoke, 1989, is part of the 'College Work-out Series'. Both books are crammed with worked examples, self-assessment questions, and hints at how to approach typical questions. Avery and Shaw is one of the few general textbooks on physical chemistry that a non-mathematician can read with ease.

Molecular chemistry texts

Many books approach physical chemistry from the *molecular* level, looking first at atoms, ions and molecules in isolation, and only then analysing the way such species interact one with another. In this approach, macroscopic properties are shown to relate to the underlying microscopic phenomena. Such books are strong on quantum mechanical and statistical mechanical theory, implying a far more conceptual approach. They are always far more mathematical than texts majoring on phenomenological chemistry. Their approach is, therefore, the exact opposite to that taken in this book. Some (possibly most) students will struggle with them, but those whose appetites have been whetted will find ultimately the 'microscopic to macroscopic' route is a far more powerful way to understand the world and its physical chemistry. *Physical Chemistry: A Molecular Approach*, Donald A. McQuarrie and John D. Simon, University Science Books, Sausalito, CA, 1997, is a good example of this kind of approach, and comes highly recommended. *Principles of Chemistry*, Michael Munowitz, W. W. Norton, New York, 2000, comes from the same conceptual stable, and is slightly less mathematical. *Physical Chemistry*, John S. Winn, HarperCollins, New York, 1995, is also quite good.

As an adjunct, *Theoretical and Physical Principles of Organic Reactivity*, Addy Pross, Wiley, New York, 1995, presents a penetrating analysis of the way reaction profiles help the understanding of the physical chemist.

Experimental chemistry

Accurate and reliable determination of physicochemical data lie at the heart of our topic, but good books addressing this aspect are surprisingly difficult to find. *Experimental Physical Chemistry: A Laboratory Textbook* (second edition), Arthur M. Halpern, Prentice Hall, Upper Saddle River, NJ, 1988, is one of the better books. It does presuppose a thorough understanding of fundamental methods of analysis, but it includes all the necessary theory for each of its 38 experiments.

Also worth a glance is *Practical Skills in Chemistry*, by John R. Dean, Alan. M. Jones, David Holmes, Rob Read, Jonathan Weyers and Allan Jones, Prentice Hall, Harlow, 2002. Being a general text, the techniques required by the experimental *physical* chemist occupy a relatively small proportion of the book.

Popular science

Many 'fun' science books are now available. For example, try Len Fisher's *How to Dunk a Doughnut: The Science of Everyday Life*, Weidenfield and Nicholson, London,

2002. Gary Snyder's *The Extraordinary Chemistry of Ordinary Things* (third edition), Wiley, New York, 1998, has the associated Website *http://www.wiley.com/college/ extraordinary*. Louis A. Bloomfield's book *How Things Work* (second edition), Wiley, New York, 2001, aims more at an understanding of physics; it can be investigated electronically at the Website *http://www.wiley.com/college/howthingswork*. This latter site has a message board, so you can ask him your own questions. For a related physics-based Website, try *http://www. howstuffworks.com.*

The history of chemistry

The late J. R. Partington's four-volume work, *A History of Chemistry*, Macmillan, London, completed in 1970, must be counted as *the* work on the subject. Volume IV recounts the history up to about 1960. He is enormously learned and erudite. He is also authoritative; the book is well supported with footnotes.

For another masterly introduction, try *The World of Physical Chemistry* by Keith Laidler, Oxford University Press, Oxford, 1995. Laidler is himself a proficient physical chemist. It represents a complete survey of the scientific development of physical chemistry from the 17th century to the present day. Excellent.

Laidler's next book *To Light Such a Candle: Chapters in the History of Science and Technology*, Oxford University Press, Oxford, 1997, explores the links between science and technology. Laidler follows the Victorian historian Carlyle's advice that 'history is the biography of great men', so we find chapters on Maxwell, Faraday, Planck and Einstein, among others. It is a well-presented book, and highly informative.

Most recently, Laidler published *Energy and the Unexpected*, Oxford University Press, Oxford, 2003. This book falls more readily into the popular science genre, but it does have a historical emphasis.

The Website *http://www-groups.dcs.st-and.ac.uk/~history/Mathematicians* is hosted by St Andrew's University in Scotland, and cites biographies of several hundred scientists, including most of the 'greats' mentioned in this text.

Finally, The Royal Society of Chemistry's magazine *Chemistry in Britain* has an article on the history of chemistry in most issues.

Introduction

Julian of Norwich

One of the best versions of Julian of Norwich is *Revelations of Divine Love*, Penguin, Harmondsworth, 1966, with Clifton Wolters's masterly introduction. *Julian's Way* by Rita Bradley, HarperCollins, London, 1992, brings out the allusions to the interconnectedness of things.

Definitions

Several sources were used regularly as sources of the definitions found in this text. The IUPAC Website *http://www.iupac.org/publications/compendium/index.html*

defines several hundred terms and concepts. Peter Atkins's little book *Concepts in Physical Chemistry*, Oxford University Press, Oxford, 1995, is presented in the form of a pocket dictionary, and is invaluable not just for definitions. *The Oxford English Reference Dictionary*, Oxford University Press, Oxford, 1995, is surprisingly rich in scientific information. Finally, try the *Oxford English Dictionary*, available free at *http://www.oed.com*.

Linguistic introduction

Minerva, publishes Umberto Eco's masterpiece *The Name of the Rose* (1994) in a superb English translation by William Weaver. A 'must read'.

Anyone wanting a gentle, humorous, introduction to the topic should read Bill Bryson's *Mother Tongue: Our Language*, Penguin, Harmondsworth, 1990. More detailed is the gem *Our Language* by Simeon Potter, Pelican, Harmondsworth, 1958, which glistens with detail and is always a delight to read.

For the budding etymologist, all good dictionaries include etymological detail. *English from Latin and Greek Elements* by Donald M. Ayers, University of Arizona Press, Tuscon, AZ, 1985, is a must for any budding etymologist, and *The Oxford English Reference Dictionary* and the *Oxford English Dictionary* (available as above) are both invaluable.

1   Introduction to physical chemistry

General reading about variables and relationships

The conceptual frameworks underpinning physical chemistry are discussed well in *The Foundations of Physical Chemistry*, in the Oxford Primer Series, Oxford University Press, Oxford, 1996. Each of its authors, Charles Lawrence, Alison Rodger and Richard Compton, is an experienced teacher of physical chemistry. Its sister volume, *Foundations of Physical Chemistry: Worked Examples*, by Nathan Lawrence, Jay Wadhawan and Richard Compton, Oxford Primer Series, Oxford University Press, Oxford, 1999, is a treasure trove of worked examples.

The zeroth law of thermodynamics

The zeroth law is described at *http://www.sellipi.com/science/chemistry/physical/thermodynamics/zero.html*, although not in any great depth.

Ink-jet printers

Peter Gregory's short article, 'Colouring the Jet Set' in *Chemistry in Britain*, August 2000, p. 39, concerns the fabrication of inks for PC printers. He also discusses the underlying mechanical features allowing a printer head to function. Alternatively, the 'How Stuff Works' Website at *http://www.howstuffworks.com* features the relevant pages at, ~/question163.htm, ~/inkjet-printer.htm and ~/inkjet-printer2.htm.

King Edgar

Details concerning the reign of this remarkable and wise man are mentioned at the informative site *http://www.chrisbutterworth.com/hist/edgar.htm*, although it does not mention the legend of how the first foot rule was made. For more detailed information about King Edgar, try *The Saxon and Norman Kings*, Christopher Brooke, Fontana, London, 1963, chapter 8, or Sir Frank Stenton's magisterial volume in the Oxford History of England, *Anglo-Saxon England* (third edition), Oxford University Press, Oxford, 1971.

Measuring temperature

For a superior introduction to this difficult topic, try Peter Rock's now classic book, *Chemical Thermodynamics*, Oxford University Press, Oxford, 1983. The treatment in *Temperature Measurement* (second edition), by Ludwik Michalski, Joseph McGhee, Krystyna Eckersdorf and Jacek Kucharski, Wiley, New York, 2001, is aimed at engineers manufacturing temperature-measuring machines, such as electrical and optical sensors, but some of its introductory material might help.

Thunder and lightning

The sites *http://www.usatoday.com/weather/thunder/wlightning.htm* and *http://www. nssl.noaa.gov/edu/ltg* both discuss the fascinating topic of thunder and lightning. Neither extrapolates in order to explore the validity of the gas laws. The site *http://bcn.net/ ~lti/f_sets/links.html* has more scientific content, but is more difficult to navigate.

Système Internationale

One of the best scientific sites on the web is IUPAC's own site, *http://www.iupac.org/ publications/compendium/index.html*, which defines several hundred terms and concepts.

The Maxwell–Boltzmann law

For biographies of these great men, try *http://www-groups.dcs.st-and.ac.uk/~history/ Mathematicians/Boltzmann.html* and *http://www-groups.dcs.st-and.ac.uk/~history/ Mathematicians/Maxwell.html*.

2 Introducing interactions and bonds

General reading

The Chemical Bond, by J. N. Murrell, S. F. A. Kettel and J. M. Tedder, Wiley, Chichester, 1978, is an excellent text, and describes the interactions inherent in all bond formation. It is a very mathematical read, and tackles the topic in terms of valence bond and molecular orbitals: not an easy read, but well worth a try.

The content of *The Forces Between Molecules*, by Maurice Rigby, E. Brian Smith, William A. Wakeham and Geoffrey C. Maitland, Oxford University Press, Oxford, 1986, is more explicitly about interactions than formal bonds. Again, it will be a fairly austere and mathematical read. In the Oxford 'Primer' series, try *Energy Levels in Atoms and Molecules* by W. G. Richards and P. R. Scott, Oxford University Press, Oxford, 1994. It's easier than the two books above, and again helps provide some of the background material to the subject. It is still mathematically based.

One of the better books – precisely because it is less mathematical – is Jack Barrett's *Structure and Bonding*, RSC, Cambridge, 2001, in the new Royal Society of Chemistry 'tutorial chemistry texts' series. Although the layout and style were designed to make the topic accessible, be warned that even this book can be very rigorous and looks a bit daunting.

Finally, for organic chemists wanting to look at bonds and interactions, try *Mechanism and Theory in Organic Chemistry* (third edition) by Thomas H. Lowry and Kathleen Schueller Richardson, Harper & Row, New York, 1987, which looks at the physical chemistry underlying so much of organic chemistry – a topic sometimes known as 'physical–organic chemistry'. A similar approach is apparent in *Orbital Interactions Theory of Organic Chemistry* by Arvi Rauk, Wiley, New York, 1994, although *all* his discussions are cast in terms of orbitals.

Liquid crystals

The literature on liquid crystals and LCDs is simply vast. A good start is *Thermotropic Liquid Crystals*, by George Gray, Wiley, Chichester, 1987. Gray was one of the principal pioneers in the early days of liquid crystals.

Alternatively, those wanting more material could consult Dietrich Demus, who, with his colleagues, has published numerous texts on the subject. The introductory chapters in volume I, *Handbook of Liquid Crystals: Fundamentals*, by Dietrich Demus, John W. Goodby, George W. Gray, Hans W. Spiess, Volkmar Vill, Wiley–VCH, Weinheim, 1998, are all high-brow but encompass the topic with much useful descriptive prose. The site *http://www.eng.ox.ac.uk/lc/research/introf.html* gives a beautifully clear good lay-man's introduction, and *http://www.eio.com/lcdhist.htm* contains dozens of super links.

For those wishing to follow up the topic and are interested in coloured LCDs, try Chapter 5 of Peter Bamfield's *Chromic Phenomena*, Royal Society of Chemistry, Cambridge, 2001.

Nucleation

Nucleation is a difficult topic. The physical chemistry of small particles attracting and aggregating is described in Chapters 9–11 of D. H. Everett's authoritative text, *Basic Principles of Colloid Science*, Royal Society of Chemistry, Cambridge, 1988.

Hydrogen bonds and the structure of water

Martin Chaplin of South Bank University, London, has produced an extensive series of Web pages concerning water, its structure and properties. The home page is located at *http://www.sbu.ac.uk/water*. The structure of water is discussed at *http://www.sbu.ac.uk/water/clusters.html*.

The site *http://www.nyu.edu/pages/mathmol/modules/water/info_water.html* hosts a nice discussion of water, including two short video clips: (1) the quantum-mechanically computed movement of two water molecules united by means of a single hydrogen bond, at *http://www.nyu.edu/pages/mathmol/modules/water/dimer.mpg*; (2) a short film of several hundred water molecules dancing within a cube at *http://www.nyu.edu/pages/mathmol/modules/water/water_dynamics.mpg*.

Cappuccino coffee

The story about the Capuchin monk Marco d'Aviano can be found at *http://www.mirabilis.ca/archives/000710.html* or the sometimes tongue-in-cheek site *http://www.sspx.ca/Angelus/2000_January/A_Politically_Incorrect_Monk.htm*.

DNA and heredity

Organic Chemistry: A Brief Introduction by Robert J. Ouellette, Prentice Hall, New Jersey, 1998, contains a super introduction to the history of DNA and heredity. Stephen Rose's now classic book *The Chemistry of Life*, Penguin, Harmondsworth, 1972, goes into more depth, and includes a good discussion of H-bonds in nature and DNA. The sites *http://www.dna50.org.uk/index.asp* and *http://www.nature.com/nature/dna50/* have good pictures and links.

The site *http://www.netspace.org/MendelWeb/Mendel.html* reproduces the full text of Mendel's original research papers.

London dispersion forces

A good start is the long and well-referenced review article 'Origins and applications of London dispersion forces and Hameker constants in ceramics' by Roger J.

French in *Journal of the American Ceramics Society*, 2000, **83**, 2117. Its introductory sections are excellent; its application sections are not particularly germane to us but are interesting. It's worth downloading an electronic copy of this paper from *http://www.lrsm.upenn.edu/~frenchrh/download/0009jacersdispersionfeature.pdf*.

For a brief biography of London, go to *http://onsager.bd.psu.edu/~jircitano/London.html*, or read *Fritz London: A Scientific Biography* by Kostas Gavroglu, Cambridge University Press, Cambridge, 1995.

Critical and supercritical fluids

Steve Howdle's short article 'Supercritical solutions' in *Chemistry in Britain*, August 2000, p. 23, is a good introduction to the topic.

The Web page *http://www-chem.ucdavis.edu/groups/jessop/links.html* contains many good links to other sites. Anyone wanting a more in-depth look at the topic should consult the bi-monthly *Journal of Supercritical Fluids*.

Etymology of the word 'Molecule'

The etymological detail concerning Faraday and Cannizzaro come from Everett's *Basic Principles of Colloid Science*, Royal Society of Chemistry, Cambridge, 1988. The other citations in the chapter come from the Oxford English Dictionary: G. Adams, *Nat. & Phil.*, 1794, **I**, iii, 79; Kirwen, *Geol. Ess.*, 1799, 478; W. Wilkinson, *Oult. Physiol.*, 1851, 9; and Tyndall, *Longm. Mag.*, 1882, **I**, 30.

Nitrogen fixation

Most biochemistry texts discuss the so-called 'nitrogen cycle' in some depth. Rose (above), for example, touches on the topic a few times. Additional information may be found at the *http://academic.reed.edu/biology/Nitrogen*, *http://www.infoplease.com/ce6/sci/A0860009.html* and *http://helios.bto.ed.ac.uk/bto/microbes/nitrogen.htm* Websites.

The van der Waals equation

The page *http://www.hull.ac.uk/php/chsajb/general/vanderwaals.html*, at Hull University's Website, includes an interactive page – the 'van der Waals calculator' – to determine values for real and ideal gases, with the van der Waals equation. The site *http://antoine.frostburg.edu/chem/senese/javascript/realgas.shtml* includes a different calculator with more variables, but is not quite so easy to use.

The page *http://www.chuckiii.com/Reports/Science/Johannes_van_der_waals.shtml* contains a brief biography of van der Waals. The page *http://www.hesston.edu/*

academic/FACULTY/NELSONK/PhysicsResearch/Waals mentions other detail, and cites a few interesting links.

Linus Pauling

For a brief life of this brilliant man, try the Web page of the Swedish Academy of Science, who awarded him two Nobel prizes: go to *http://www.nobel.se/chemistry/laureates/1954/pauling-bio.html*. Alternatively, his authorized biography, *Linus Pauling*, is by Anthony Serafini, Paragon House, 1991.

Etymology of the word covalent

Langmuir defined 'covalent' in *Proc. Nat. Acad. Sci.*, 1919, **V**, 255. His quote (reproduced on p. 68) relates to sodium chloride.

Electron affinity and ionization energy

It is surprisingly difficult to find reliable values of I and $E_{(ea)}$. Probably the most extensive collection of data is *Bond Energies, Ionization Potentials and Electron Affinities* by V. I. Vedeneyev, V. L. Gurvich, V. N. Kondrat'yev, Y. A. Medvedev and Ye. L. Frankevich, Edward Arnold, London, 1966. The *Chem Guide* Website has several good pages, e.g. look at *http://www.chemguide.co.uk/atoms/properties/eas.html*.

3 Energy and the first law of thermodynamics

General reading

Gareth Price's book *Thermodynamics of Chemical Processes* in the affordable Oxford Primer series, Oxford University Press, Oxford, 1998, describes much of the background material discussed here, and goes into some depth.

Chapter 2 of E. Brian Smith's book *Basic Chemical Thermodynamics*, Clarendon Press, Oxford, 1990, is a superb introduction to the topic. His Chapter 1 discusses concepts such as reversibility and the broader question, 'Why do we need thermodynamics?' His Chapter 5 covers the measurement of thermodynamic parameters.

In the new RSC 'Tutorial Chemistry Texts' series, Jack Barrett's *Structure and Bonding*, RSC, Cambridge, 2001, is another superb resource because it covers many of the thermodynamic aspects featured in this chapter. Be warned, though: the overall treatment is a rigorous and somewhat mathematical tour of the topic.

Sweat and sweating

The Website *http://www.sweating.net* contains generous chunks of relevant information. Alternatively, try the page *http://www.howstuffworks.com/sweat.htm* at the 'How Stuff Works' site. The page *http://www.howstuffworks.com/sweat3.htm* introduces the necessary thermodynamics, and the page *http://www.howstuffworks.com/sweat2.htm* describes how sweat is made.

Lord Kelvin

The site *http://scienceworld.wolfram.com/biography/Kelvin.html* has a relatively short but informative biography of Kelvin, the man. Little science is mentioned. Kelvin also appears often in all the biographies of Joule (e.g. see below).

James Joule

For a brief but lively biography of Joule the man, try 'Heated exchanges' by Leo Lue and Les Woodcock, in *Chemistry in Britain*, August 2001, p. 38. The article well describes the confusion surrounding the formulation of the laws of thermodynamics, explaining why the first law was defined before the zeroth, etc. The site *http://scienceworld.wolfram.com/biography/Joule.html* also has a photo of the great man. D. S. L. Cardwell has written a more detailed biography of this remarkable man: see *James Joule: A Biography*, Manchester University Press, Manchester, 1989. Cardwell's account contains sufficient scientific detail to place Joule's work in its proper context.

Kirchhoff's Law

The Website *http://www-groups.dcs.st-and.ac.uk/~history/Mathematicians/Kirchhoff.html* contains a substantial biography and several photographs of this great but long forgotten scientist. The shorter biography at *http://scienceworld.wolfram.com/biography/Kirchhoff.html* concentrates more on his discoveries in physics than in thermochemistry.

Be warned that a Web search for 'Kirchhoff' will yield dozens of pages on Kirchhoff's rules, which relate to electronic circuits.

Conversion of diamond into graphite

There is a wealth of information on this thermodynamic transformation, e.g. in most textbooks of physical chemistry. The site *http://chemistry.about.com/library/weekly/aa071601a.htm* has copious links, while the short Web page *http://members.tripod.com/graphiteboy/Graphite_Diamond.htm* cites a few nice details

Enthalpies of reaction

The short, fun Website *http://schools.matter.org.uk/Content/Reactions/BE_enthalpy NO.html* has a few examples that might amuse.

Enthalpies of combustion

Any edition of the *Handbook of Chemistry and Physics*, CRC Press, Boca Raton, FL, will contain an authoritative array of data, and all standard texts on physical chemistry publish tables of $\Delta H_c^{\ominus}$ data. The Website *http://www.innovatia.com/Design_Center/rktprop2.htm* is relatively mathematical, but good quality. Alternatively, try the site *http://www.geocities.com/CapeCanaveral/Launchpad/5226/thermo.html*, which has a fairly good treatment.

Bond enthalpies

Any edition of the *Handbook of Chemistry and Physics*, CRC Press, Boca Raton, FL, will cite some data, and all standard texts on physical chemistry will publish tables of $\Delta H_{BE}^{\ominus}$ data. Also, try the Web page *http://www.webchem.net/notes/how_far/enthalpy/bondenthalpy.htm*.

Hess's law

The site *http://www.chemistry.co.nz/hess_law.htm* from New Zealand has a short biography of Hess, including photographs. The page *http://dbhs.wvusd.k12.ca.us/Chem-History/Hess-1840.html* has an .html copy (in English translation) of Hess's original paper, dating from 1840.

Peppermint and cooling agents

It is relatively difficult to find details of cooling agents, but try T. P. Coultate's, *Food: The Chemistry of its Components* (fourth edition), Royal Society of Chemistry, Cambridge, 2002, p. 241.

4 Reaction spontaneity and the direction of thermodynamic change

One of the better books on the topic is Peter Atkins's *The Second Law: Energy, Chaos, and Form*, P. W. Atkins, Scientific American, New York, 1994, which explains the complicated and interpenetrating truths underlying this topic. Its 'popular science book' style may irritate some readers, but will undoubtedly help many, many more.

Clausius

The Website *http://www-groups.dcs.st-and.ac.uk/~history/Mathematicians/Clausius. html* contains a short biography of Clausius.

Entropy as 'the arrow of time'

The idea of entropy being 'the arrow of time' has attracted a huge following from mystical poets through to Marxists (see *http://www.marxist.com/science/arrowoftime.html*). The outline of the science in the Marxist.com site is actually very good.

The site *http://www-groups.dcs.st-and.ac.uk/~history/Mathematicians/Eddington. html* contains a brief biography of Sir Arthur Eddington, who first proposed the phrase' arrow of time'.

Most people seem to think T.S. Eliot's poem *The Wasteland* (1922) contains the lines 'not with a bang ...' but in fact they conclude the poem *The Hollow Men* (1925). Incidentally, the 'whimper' is an obscure reference to Dante's description of a newborn baby's cry upon leaving one world to enter another. The text of *The Hollow Men* is available at *http://www.aduni.org/~heather/occs/honors/ Poem.htm*.

Gibbs function

A short biography of Gibbs may be found in *Physical Chemistry: A Molecular Approach*, p. 924. A longer version is available at *http://www-groups.dcs.st-and.ac. uk/~history/Mathematicians/Gibbs.html*. To read some of his delightful quotations, try *http://www-groups.dcs.st-and.ac.uk/~history/Quotations/Gibbs.html*.

Le Chatelier's principle

Le Chatelier's original work was published in the journal of the French Academy of Sciences, as H. L. Le Chatelier, *Comptes rendus*, 1884, **99**, 786. It's in French, and even good translations are hard to follow.

The best biography is available at the French Website *http://www.annales.com/ archives/x/lc.html*. The short article 'Man of principle' by Michael Sutton in *Chemistry in Britain*, June 2000, 43, also includes a nice introduction to the man and the background to his science. Websites bearing the same title as Sutton's article proliferate: two of the better ones (which include photographs) are *http://www.woodrow.org/ teachers/ci/1992/LeChatelier.html* and *http://www.stormpages.com/aboutchemists/ lechatelier.html*.

The example of chicken breath was inspired by the article, 'From chicken breath to the killer lakes of Cameroon: uniting seven interesting phenomena with a single chemical underpinning' by Ron DeLorenzo, *Journal of Chemical Education*, 2001, **78**(2), 191. The article also discusses boiler scale, the way that carbon dioxide partitions between fizzy drink and the supernatant gases (see p. 165), and stalactites and stalagmites.

5   Phase equilibria

General reading

In the affordable Oxford Primer series, *Thermodynamics of Chemical Processes* by Gareth Price, Oxford University Press, Oxford, 1998, describes some of the background material discussed here, albeit in modest depth.

Tin and Napoleon

Whether true or not, the story about Napoleon is quite well known, and is mentioned on the Websites *http://www.tclayton.demon.co.uk/metal.html#Sn* and *http://www. corrosion-club.com/tinplague.htm*.

Thermodynamic data for tin may be found in *Tin and its Alloys and Compounds*, B. T. K. Barry and C. J. Thwaits, Ellis Horwood, Chichester, 1983. The International Tin Research Institute has an informative Website at *http://www.tintechnology.com*, the 'library' page of which is particularly good.

Supercritical fluids

For the underlying science of supercritical fluids, try Steve Howdle's short article 'Supercritical solutions' in *Chemistry in Britain*, August 2000, p. 23, which represents a useful introduction to the topic. For more applications of such fluids, try the short review article 'Some applications of supercritical fluid extraction', by D. P. Ndiomu and C. F. Simpson in *Analytica Chimica Acta*, 1988, **213**, 237. The article is somewhat dated now but readable. A look at the contents list of *The Journal of Supercritical Fluids* will be more up-to-date: go to *http://www.umecheme. maine.edu/jsf*.

Decaffeinated coffee

The article 'Caffeine in coffee: its removal: why and how?' by K. Ramalakshmi and B. Raghavan in *Critical Reviews in Food Science and Nutrition*, 1999, **39**, 441 provides an in-depth survey of the physicochemical factors underlying decaffeination of coffee with supercritical CO_2.

D. J. Adam, J. Mainwaring and Michael N. Quigley have described a simple experiment to remove caffeine from coffee with a Soxhlet apparatus: see *Journal of Chemical Education*, 1996, **73**, 1171. Their solvent was a chlorinated organic liquid rather than supercritical CO_2. The abstract is available at *http://jchemed.chem.wisc.edu/journal/issues/1996/Dec/abs1171.html*.

Chemical potential

Chemical potential can be a very difficult topic to grasp, but any standard textbook of physical chemistry will supply a more complete treatment than that afforded here. A particularly useful introduction to the thermodynamics of solutions and mixtures is Chapter 6 of *Basic Chemical Thermodynamics*, by E. Brian Smith, Oxford University Press, Oxford, 1990.

Be careful when searching the Web, though, because typing the phrase 'chemical potential' will locate a great many pages that relate to solid-state and atomic physics: both disciplines use the term to mean different concepts.

Cryoscopy and colligative properties

Perhaps the best short description of cryoscopy and the uses of colligative properties is Michael Sutton's delightful article 'One cool chemist', *Chemistry in Britain*, 2001, June, p. 66, concerning François-Marie Raoult. Sutton sketches a brief portrait of Raoult before widening the scope to include the physicochemical factors underlying the depression of freezing point. He gently guides the reader through the early history of how our Victorian forebears made ice cream at a time predating refrigerators, before explaining the physicochemical principles underlying the action of salt on a frozen path. Highly recommended.

A slightly different approach will be found in Mark Kurlansky's fascinating book *Salt: A World History*, Walker Publishing Company, New York, 2002, which is more than a 'popular science' book, and will be enjoyed as well as being informative.

The site *http://www.northland.cc.mn.us/Chemistry/colligative_properties_constants.htm* cites many $K_{(cryoscopic)}$ and $K_{(ebullioscopic)}$ constants.

Anaesthesia

Michael Gross has written a beautifully clear article 'The molecules of pain' describing how the sensation of pain originates. He also discusses a few recently discovered molecules that act as chemical 'messengers' to the brain: see *Chemistry in Britain*, June 2001, p. 27.

The site *http://www.oyston.com/history* has a fascinating history of the topic, mentioning such early anaesthetics as ether, chloroform and nitrous oxide. Local anaesthetics are often injected in the form of liquids or solutions, see the article 'pharmacology of local anaesthetic agents' by a British anaesthetist, Dr J. M. Tuckley, may be found at *http://www.nda.ox.ac.uk/wfsa/html/u04/-u04_014.htm*. General anesthetics are more usually administered as a gas: see the article, 'General anaesthetic agents' at *http://www. mds.qmw.ac.uk/biomed/kb/pharmgloss/genanaesth.htm*, again from a British hospital.

The anaesthetic properties of halothane are described in some depth at the French Website *http://www.biam2.org/www/Spe2391.html*, which includes health and safety data.

Solid-state gas sensors

The complicated topic of solid-state electrical conductivity is well described in *Solid State Chemistry and its Applications*, A. R. West, Wiley, Chichester, 1984, although it does not explicitly discuss sensors. Those wanting more depth should look at *Transition Metal Oxides*, P. A. Cox, Clarendon Press, Oxford, 1992, which provides a readable account of the conduction of ions and electrons through solids.

For a background introduction to the doping of semiconductors, try the fun Website *Britney Spears' guide to semiconductor physics* at *http://britneyspears.ac/lasers.htm*, which is actually quite good in parts.

Petrol

The 'press centre' on the Website of British Petroleum plc (BP) is a good source of information, at *http://www.bpamoco.com/centres/press/index.asp*. For example, the page *http://www.bpevo.com/bpevo_main/asp/evo_glo_0003.asp* lists the terms used by most petrol companies, and *http://www.bp.com/location_rep/uk/bus_operating/ manu_ops.asp* cites the amounts of the known carcinogen, benzene, found naturally in petrol.

The page *http://www.radford.edu/~wkovarik/lead* cites some fascinating sidelights on the history of petrol and its additives, and its links are carefully chosen. For example, the excellent page *http://www.chemcases.com/tel* summarizes the history of tetraethyl lead and the attendant controversies over its addition to petrol. Though quite small, the site *http://www.exxon.com/exxon_productdata/lube_encyclopedia/knock.html* contains useful information on the anti-knock properties of petrol and its additives.

Incidentally, as our discussion initially focused on the way we *smell* petrol, Natalie Dudareva's article 'The joy of scent' is a good introduction. She concentrates on flowers, but her theory and references are germane: see *Chemistry in Britain*, February 2001, p. 28.

Finally, one of the better books to analyse the environmental impact of petrol is *Green Chemistry: An Introductory Text* by M. Lancaster, Royal Society of Chemistry, Cambridge, 2002.

Steam distillation

The best book describing essential oils and their extraction is *The Essential Oils: Individual Essential Oils of the Plant Families* (in six volumes) by Ernest Guenther, 1948 (reprinted 1972–1998).

For discussions concerning the extraction of essential oils, e.g. for aromatherapy, see the sites *http://www.origanumoil.com/steam_distillation.htm*, *http://www.distillation. co.uk* and *http://www.lifeblends.com/howitworks5.html*. For a more in-depth description of oil extraction, see Guenther (above) or *http://www.fatboyfresh.com/essentialoils/ extraction.htm*. The site *http://www.chamomile.co.uk/distframe.htm* has some super graphics of steam distillation.

The Chemistry of Fragrances, D.H. Pybus and C. S. Sell, Royal Society of Chemistry, Cambridge, 1999, gives an extremely balanced assessment of aromatherapy as a scientific discipline. See also 'The joy of scent', above.

6 Acids and bases

General reading

Aqueous Acid–Base Equilibria and Titrations, Robert de Levie, Oxford University Press, Oxford, 1999, is a good resource, whose scope extends far beyond this book. Its particular emphasis is speciation analyses, which are discussed in an overtly mathematical way. The maths should be readily followed by anyone acquainted with elementary algebra. *Water Chemistry*, by Mark M. Benjamin, McGraw Hill International Edition, New York, 2002, is a longer book and covers the same material as de Levie but in substantially greater depth.

The dissociation of water

One of the best resources for looking at the way equilibrium constants K vary with ionic strength is the Web-based resource Joint Expert Speciation System (JESS) available at *http://jess.murdoch.edu.au/jess/jess_home.htm*. Notice the way that values of any equilibrium constant (K_a, K_w, etc.) changes markedly with ionic strength I.

Chlorine gas as a poison gas

The information concerning the Second Battle of Ypres is embedded within a disturbing account of man's scope for evil, *World War One: A Narrative* by Philip Warner, Cassell Military Classics, 1998. The ghastly effects of poisoning with chlorine gas are recounted at *http://www.emedicine.com/EMERG/topic851.htm*.

Sulphuric acid formed in the eyes

The paper 'The design of the tears' by Jerry Bergman in *The Technical Journal* 2002, **16**(1), 86, is one of the best short introductions to the operation of the tear duct, and mentions SO_3 in the vapours emanating from onions; or go to *http://www. answersingenesis.org/home/area/magazines/tj/TJ_issue_index.asp*. The book *Clinical Anatomy of the Eye*, R. Snell and M. Lemp (second edition), Blackwell, Boston, 1998, discusses tears and onions on p. 110.

Acid rain

The literature on acid rain is simply vast. One of the better introductory texts is *Acid Rain: Its Causes and its Effects on Inland Waters*, by B. J. Mason, Clarendon Press, Oxford, 1992. Dozens of Websites supplement and update Mason's book. For a general but more widely ranging survey of pollution and its legacy, try *Pollution: Causes, Effects and Control* (fourth edition), edited by Roy M. Harrison, Royal Society of Chemistry, Cambridge, 2001.

The Websites *http://library.thinkquest.org/CR0215471/acid_rain.htm* and *http:// www.geocities.com/CapeCanaveral/Hall/9111/ACIDRAIN.HTML* simplify the issues (some might add 'over-simplify'). The Website of *Greenpeace* at *http://www. greenpeace.org* can be sensational at times, but its information is usually reliable. For more interesting sidelights on the topic, try the Canadian Government's site at *http://www.atl.ec.gc.ca/msc/as/as_acid.html*.

Bases in nature

The Websites *http://www.consciouschoice.com/herbs/herbs1308.html* and *http://www. homemademedicine.com/insectbiteandbeesting.html* cite the many naturally occurring plants having interesting medicinal properties.

Neither of these sites contains much scientific theory. A more in-depth survey of the phenomenon of pain may be found at *http://www.chic.org.uk/press/releases/pain2.htm*, produced by the Consumer Health Information Centre (CHIC); and Michael Gross' short article 'The molecules of pain' describes the sensation of pain and the mode(s) by which pain killers operate: see *Chemistry in Britain*, June 2001, p. 27.

The Website *http://www.ediblewild.com/nettle.html* lists many more properties of stinging nettles – particularly impressive being the number of ways to consume nettles in the form of edible foodstuffs, drinks and beers.

Carbolic acid and Lord Lister

A good introduction may be found at the Websites *http://web.ukonline.co.uk/b.gardner/ Lister.html* and *http://limiting.tripod.com/list.htm*. The former site also contains some fascinating links.

Indigestion tablets and neutralization reactions

The Internet site *http://www.picotech.com/experiments/antiacid/results.html* investigates the chemistry behind several well-known tablets.

Concerning the effects of aluminium and the onset of Alzheimer's disease, the site *http://www.alzscot.org/info/aluminium.html* gives good information – but it can be a bit scary! The site *http://www.hollandandbarrett.com/healthnotes/Drug/Sodium_ Bicarbonate.htm* describes the action of $NaHCO_3$ (an aluminium-free anti-acid product) in indigestion formulations.

Finally, the site *http://www.naplesnews.com/today/neapolitan/a84826n.htm* gives good advice on diet, aimed at avoiding the problem before we need to ingest an anti-acid tablet.

The colours of flowers and indicators

The fascinating book *Sensational Chemistry*, in the Open University's 'Our Chemical Environment' series, OU Press, Milton Keynes, 1995, is an excellent introduction to the topic. The Web has several hundred relevant sites, most of which are simple to follow: just go to a reputable search engine, and type 'acid–base indicator'.

Finally, *Food Flavours: Biology and Chemistry* (fourth edition) by C. Fisher and T. Scott, Royal Society of Chemistry, Cambridge, 2002, supplements many of the titles above.

7 Electrochemistry

General reading

There are a large number of books on electrochemistry. *The* electrochemistry text is undoubtedly *Electrochemical Methods: Fundamentals and Applications* (second edition) by A. J. Bard and L. R. Faulkner, Wiley, New York, 2001. This book is unfortunately a mathematical read, but it contains absolutely everything we need at our level; and the prose is generally a model of clarity.

In the Oxford Primer series, the book *Electrode Potentials* by Richard G. Compton and Giles H. W. Sanders, Oxford University Press, Oxford, 1996, is an introduction. It is intended for the absolute novice, but develops themes to a satisfactory level. Its treatment of the Nernst equation is both thorough and straightforward. It contains copious examples and self-assessment questions.

Fundamentals of Electroanalytical Chemistry, by Paul Monk, Wiley, Chichester, 2002, is intended to be an easy read. It was written for those learning at a distance. Its style is non-mathematical, and involves a series of 'discussion questions'. It's also packed full of worked examples and self-assessment questions.

Dynamic electrochemistry

One of the better articles is 'Electrochemistry for the non-electrochemist' by Peter T. Kissinger and Adrian W. Bott, *Current Separations*, 2002, **20**(2), 51. This brief but informative article outlines 'nine of the most fundamental concepts of electrochemistry which must be mastered before more advanced topics can be understood'. The article can be downloaded from *http://www.currentseparations.com/issues/20-2/20-2d.pdf*.

Also, try *Electrode Dynamics*, by A. C. Fisher, Oxford University Press, Oxford, 1996, which is another title in the Oxford primer series. Its early chapters discuss the transport of analyte through solution and the various rates inherent in a dynamic electrochemistry measurement. It is a readily affordable and readable introduction and highly recommended.

Electrolysis and pain

Perhaps the best site is that of *Amnesty International* at *http://www.amnesty.org*. Its page *http://web.amnesty.org/library/eng-313/index* cites many dozens of case studies, a significant fraction of which involve electricity. Or type 'electricity' into their search engine (almost hidden at the foot of the page), and be shocked at the number of so-called 'friendly' and 'civilized' countries that employ torture: Israel, Indonesia, the USA, and many of the 'new' countries of Europe. Be warned: some of the details are horrific – electricity is a very efficient way of generating pain.

Hair removal ('electrology')

The commercial sites *http://www.electrology.com/info.html* (The American Electrology Association) and *http://www.betterhairremoval.com/electrolysis.htm* contain copious information about the effect of electrolysis on follicles.

Fuel cells

By far the best book is *Fuel Cell Systems Explained* by James Larmanie and Andrew Dicks, Wiley, Chichester, 2000. It's an expensive gold mine of a book.

A simpler 'first glimpse' at fuel cells comes from the brief layman's article in *The Economist*, 'The fuel cell's bumpy ride', published in the 24 March 2001 edition. Its scientific content is good, although its general approach centres on economics. A similarly easy read, but for chemists, is Rob Kingston's 'Powering ahead', in *Chemistry in Britain*, June 2000, p. 24.

Scientific American published a special issue on fuel cells in July 1999. Alternatively, A. John Appleby's introductory article 'The electrochemical engine for vehicles' is a good read; the Internet sites *http://www.chemcases.com/cells/index.htm*, *http://www.fuelcells.org* and *http://www.howstuffworks.com/fuel-cell.htm* and ~fuel-cell2.htm are also worth a try.

The *Campaign for a Hydrogen Economy* (formerly *The Hydrogen Association of UK and Ireland*, HEAUKI) is a treasure-trove of information and advice and may be contacted at 22a Beswick Street, Manchester M4 7HR, UK. It is a non-profit organization that 'works for the world-wide adoption of renewably generated hydrogen as humankind's universal fuel'. CREC can be contacted at co-ordinator@hydrogen-heauki.org.

Standard cells

A brief discussion of standard cells, together with a short biography of Edward Weston, may be found at *http://www.humboldt.edu/~scimus/HSC.54-70/Descriptions/ WesStdCell.htm*. Alternatively, *http://www.humboldt.edu/~scimus/Instruments/Elec-Duff/StdCellDuff.htm* is also useful, although its symbolism does not always adhere to SI or IUPAC.

Corrosion

The 'Corrosion Doctors' site at *http://www.corrosion-doctors.org* contains lots of interesting case studies.

Electrochromism

The best introductory text on electrochromism is *Electrochromism: Fundamentals and Applications* by P. M. S. Monk, R. J. Mortimer and D. R. Rosseinsky, VCH, Weinheim, 1995, although the introductory sections of *Handbook of Inorganic Electrochromic Oxides*, by C. G. Granqvist, Elsevier, Amsterdam, 1995, are also invaluable. For a shorter but nevertheless thorough introduction, read *Chromic Phenomena* by P. Bamfield, Royal Society of Chemistry, Cambridge, 2001. The popular article 'Through a glass darkly' by Paul Monk, Roger Mortimer and David Rosseinsky in *Chemistry in Britain*, 1995, **31**, 380 is very short, but might represent a less threatening introduction.

The Website *http://jchemed.chem.wisc.edu/Journal/Issues/1997/Aug/abs962.html* explores simple laboratory demonstrations of electrochromism, whereas the commercial sites *http://www.refr-spd.com/hgi-window.html*, *http://www.ntera.com/nano.pdf* and *http://www.saint-gobain-recherche.com/pages/angl/materia/mtintell.htm* are more product oriented. Some contain pictures and video sequences.

Eau de Cologne and perfumes

The Chemistry of Fragrances, by D. H. Pybus and C. S. Sell, Royal Society of Chemistry, Cambridge, 1999, contains all the amateur needs to know on the subject. Additionally, the site *http://www.eau-de-cologne.com* (in German) describes the history of the perfume, and the site *http://www.farina1709.com* contains additional information for the interested novice.

Activity and ion association

For a short biography of G. N. Lewis, the early giant of 20th century American chemistry, visit the site *http://www.woodrow.org/teachers/ci/1992/Lewis.html*.

Ionic Solution Theory, H. L. Friedman, Wiley, New York, 1953, is a standard work in the field, but is a bit mathematical and can be difficult to follow. An easier book to follow is *Ions, Electrodes and Membranes* (second edition) by Jiři Koryta, Wiley, Chichester, 1992, and is an altogether more readable introduction to the topic. It can also be trusted with details of pH electrodes and cells. Its examples are well chosen, many being biological, such as nerves, synapses, and cell membranes. It is probably the only book of its kind to contain cartoons.

Easier still is John Burgess' book, *Ions in Solution*, Ellis Horwood, Chichester, 1999. Though it does not go into great detail about activity coefficients γ, its treatment of ionic interactions and solvation is excellent.

Activity coefficient and solubility products

The example on p. 319 comes from *Equilibrium Electrochemistry*, The Open University, The Open University Press, Milton Keynes, 1985. The book contains many other excellent examples.

Reference electrodes

The classic book on the topic remains *Reference Electrodes* by D. I. G. Janz and G. J. Ives, Academic Press, New York, 1961. This book is still worth consulting despite its age. One of the best articles is 'Reference electrodes for voltammetry' by Adrian W. Bott, *Current Separations*, 1995, **14**(2), 33. The article can be downloaded from *http://www.currentseparations.com/issues/14-2/cs14-2d.pdf*. Also, *Electrochemistry* by Carl H. Hamann, Andrew Hamnett and Wolf Vielstich, Wiley–VCH, Weinheim, 1998, has extensive discussions on reference electrodes.

pH electrodes and concentrations cells

Try looking at *The Elements of Analytical Chemistry*, by Séamus P. J. Higson, Oxford University Press, Oxford, 2003. It is an affordable addition to an analyst's library, since it covers all the basics.

Batteries

Understanding Batteries by R. M. Dell and D. A. Rand, Royal Society of Chemistry, Cambridge, 2001, represents a good introduction to the topic, as does *Electrochemistry* (above). Both books discuss battery design, construction and operation. One of the better books is J. O. Besenhard, *Handbook of Battery Materials*, Wiley, Chichester, 1998.

For an easier read, try the feature article in the occasional 'Chemist's Toolkit' series, entitled 'Batteries today' by Ron Dell, in *Chemistry in Britain*, March 2000, p. 34. It goes into greater depth than here, and is informative and relevant. The recent paper 'The lead acid battery' by A. A. K. Vervaet and D. H. J. Baett, *Electrimica Acta*, 2002, **47**, 3297, discusses the recent developments of this, the world's best-selling battery.

The site *http://antoine.frostburg.edu/chem/senese/101/redox/faq/lemon-battery.shtml* cites details of how to make a battery from a lemon.

Ancient batteries

The 'battery' found near Baghdad is mentioned in a footnote in *Modern Electrochemistry*, *J*. O'M. Bockris and A. K. N. Reddy, Macdonald, London, 1970, p. 1265.

8   Chemical kinetics

General reading

Probably the best book on the topic is *Chemical Kinetics and Mechanism*, edited by Michael Mortimer and Peter Taylor, Royal Society of Chemistry, Cambridge, 2002. This book is really excellent, and comes complete with a superb interactive CD. The other good book on kinetics currently in print is *Fundamentals of Chemical Change*, by S. R. Logan, Longmans, London, 1996. Its mathematical treatment may appear daunting at times, but its prose is good. Its coverage also includes the kinetics of electrode reactions, photochemical reactions, and discusses catalysis to a relatively deep level. *Modern Liquid Phase Kinetics*, by B. G. Cox, Oxford University Press, Oxford, 1994, in the Oxford Primers series, is a relatively cheap text. Its illustrations are not particularly good, but the book as a whole is readable.

Although long out of print, a super book for the complete novice is *Basic Reaction Kinetics and Mechanisms* by H. E. Avery, Macmillan, London, 1974. Its author is a superb teacher, whose insights and style will benefit everyone. Although quite old now, the general approach and the theory sections in *Fast reactions* by J. N. Bradley, Oxford University Press, Oxford, 1975, is still of a high standard.

Ozone depletion

A vast library waits to be read on ozone depletion. The best book by far is 'Topic study 1: the threat to stratospheric ozone' in the *Physical Chemistry: Principles of Chemical Change* series, published in the UK by the Open University, Milton Keynes, 1996. From the UK's Royal Society of Chemistry come *Climate Change*, Royal Society of Chemistry, Cambridge, 2001, *Green Chemistry*, M. Lancaster, Royal Society of Chemistry, Cambridge, 2002, and *Pollution, Causes, Effects and Control* (fourth edition), edited by Roy M. Harrison, Royal Society of Chemistry, Cambridge, 2001. Each gives a clear exposition.

The magazine Website *http://www.antarcticconnection.com* has much up-to-date information, as does *http://www.ciesin.org/TG/OZ/oz-home.html* and its links.

The full text of the 1985 Vienna Convention for the 'Protection of the Ozone Layer' is available at *http://www.unep.org/ozone/vienna_t.shtml*, and contains much of the primary data that alerted the scientific community to the threats of CFCs, etc. It is somewhat dated now, but the Web page of the United Nations 'Ozone Secretariat' is more reliable: *http://www.unep.org/ozone/index-en.shtml*.

The 'popular science' pages of 'Beyond Discovery' are compiled by the American National Academy of Sciences and are generally highly readable, and written for the intelligent novice. They may be accessed at *http://www.beyonddiscovery.com*. Do a search using 'ozone' as a key word. On a lighter note, the fun site at 'Fact Monster' *http://www.factmonster.com/ipka/A0800624.html* displays much 'pre-digested' data.

Viologens

The viologens are mentioned a couple of times in this chapter, and are formally salts of 4,4′-bipyridine. The only book dedicated to them is *The Viologens: Physicochemical Properties, Synthesis and Applications of the Salts of 4, 4′-Bipyridine* by P. M. S. Monk, Wiley, Chichester, 1998, and is up-to-date and comprehensive. Although badly out of date, a shorter work is 'Electrochemistry of the viologens', C. L. Bird and A. T. Kuhn, *Chem. Soc. Rev.*, 1981, **10**, 49.

Radiocarbon dating

A good layman's guide to radioactive dating may be found in *Searching for Real Time*, by Richard Corfield, *Chemistry in Britain*, January 2002, p. 22. It presupposes only slight background knowledge of radiochemistry. For its worked examples, it considers the age of the Turin Shroud and the stone of which sedimentary rocks are made, i.e. dating the movements of tectonic plates. The excellent book *Archeological Chemistry* by A. Mark Pollard and Carl Heron, Royal Society of Chemistry, Cambridge, 1996, will reward a thorough reading.

The Website *http://www.c14dating.com* comprises a good treatment of the topic and hosts a huge amount of information. The academic journal *Radiocarbon* always contains examples of radiocarbon dating, as well as dating using the isotopic abundances of other elements (abstracts of its papers may be accessed online, at *http://www.radiocarbon.org*).

The Turin Shroud

The site *http://www.shroudstory.com/c14.htm* is less informative, but does discuss all the evidence underpinning the claims of a botched radiocarbon dating of the

Turin Shroud, e.g. via contamination with other sources of carbon. Its treatment is informative and generally applicable.

Ötzi the Ice Man

For the non-scientist, the fascinating book *The Man in the Ice*, by K. Spindler, Weidenfeld and Nicholson, London, 1993, is well worth a look.

The BBC Website has a page dedicated to Ötzi the Iceman, with some nice pictures: go to *http://www.bbc.co.uk/science/horizon/2001/iceman.shtml*, which also reproduces the full transcript of an hour-long documentary.

The journal *Radiocarbon* regularly publishes fascinating technical papers and case studies of radiocarbon dating. For example, the original data concerning Ötzi derive from the paper, 'AMS-^{14}C dating of equipment from the Iceman and of spruce logs from the prehistoric salt mines of Hallstatt', W. Rom, R. Golser, W. Kutschera, A. Priller, P. Steier and E. M. Wild, *Radiocarbon*, 1999, **41**(2), 183.

For a briefer scientific analysis of Ötzi, *Archeological Chemistry* (above) devotes a small case study to Ötzi, see its p. 251 ff. And several Websites are devoted to him, including *http://info.uibk.ac.at/c/c5/c552/Forschung/Iceman/iceman-en.html* and the French site *http://www.archeobase.com/v_texte/otzi/corp/cor.htm*.

Walden inversion reactions

Walden first published his observations on inversion in *Berichte* 1893, **26**, 210; 1896, **29**, 133; and 1899, **32**, 1855, long before the inversion mechanism was proposed by Ingold in *J. Chem. Soc.*, 1937, 1252. The idea that the addition of one group could occur simultaneously with the removal of another was first suggested by Lewis in 1923, in *Valence and Structure of Atoms and Molecules*, Chemical Catalog Company, New York, 1923, p. 113. Olsen was the first to propose that a one-step substitution reaction leads to inversion, in *J. Chem. Phys.*, 1933, **1**, 418.

Catalytic converters

The hugely complicated chemistry of the catalytic converter is described in some depth in the Open University's excellent book *Physical Chemistry: Principles of Chemical Change*, 'Topic study 2'. Its part 1 is entitled 'the three-way catalytic converter', Open University, Milton Keynes, 1996. It covers the composition of the catalytic surface and the role of each dopant, the actual chemical reactions occurring, and details of the current legal situation regarding atmospheric pollution.

The 'How Things Work' Website is worth a glance: look at pages *http://www.howstuffworks.com/question66.htm* and *http://www.howstuffworks.com/question482.htm*.

Finally, *Green Chemistry*, by Lancaster (above), discusses these converters to some small extent in Chapter 4, especially pp. 107–109.

Photochromic lenses

A layman's guide to the action of photochromic lenses may be found at the 'How Stuff Works' Website, at *http://www.howstuffworks.com/question412.htm*. The Website of Britglass, *http://www.britglass.co.uk/publications/mglass/making2.html*, also cites a few interesting facts. A more thorough treatment is available in the excellent text *Chromic Phenomena*, Peter Bamfield, Royal Society of Cambridge, 2002, section 1.2 of his Chapter 1.

Hormones

The topic of hormones is horribly complicated, so read widely and with care. Most Websites that advertise information about hormones are too simplistic and they should be avoided – particularly those sites that sell hormones, and start with a paragraph or two of 'pseudo science'. Most textbooks of physiology and medicinal chemistry contain sufficient detail. For example, *Medicinal Chemistry*, F. D. King (ed.), Royal Society of Chemistry, Cambridge, 2002, has a good introduction.

Chocolate

The Science of Chocolate by Stephen Beckett, Royal Society of Chemistry, Cambridge, 2000, is probably the best general introduction to the subject, and contains sufficient scientific content to stimulate, while avoiding being overtly technical. Highly recommended.

Activation energy

Addy Pross's book, *Theoretical and Physical Principles of Organic Reactivity*, Wiley, New York, 1995, is an invaluable tool for understanding the way constructing a reaction profile can help the physical chemist to predict the outcome of a chemical reaction. Lowry and Richardson's *Mechanism and Theory in Organic Chemistry* (above) is also germane.

9   Physical chemistry involving light: spectroscopy and photochemistry

General reading

Several books describe the background to this topic. Perhaps the best general introductions come from the Royal Society of Chemistry: *Colour Chemistry* by R. M. Christie, RSC, Cambridge, 2001, is written for the beginner, but does extend to some depth. It is particularly good at describing chromophores of every describable type. *Chromic Phenomena* by P. Bamfield, RSC, Cambridge, 2001, is written more for the specialist, and concentrates on colour changes for device applications. Nevertheless, its introductory material is superb.

Useful books from the Oxford University Press 'primer' series include: *Foundations of Spectroscopy*, by Simon Duckett and Bruce Gilbert, OUP, Oxford, 2000; *Introduction to Organic Spectroscopy*, by Laurence M. Harwood and Timothy D. W. Claridge, OUP, Oxford, 1997; and *Energy Levels in Atoms and Molecules*, by W. G. Richards OUP, Oxford, 1996. Each, particularly the last, represents a clear and lucid introduction.

More in-depth treatments include *Ultraviolet and Visible Spectroscopy* (second edition) by Michael Thomas, Wiley, Chichester, 1997. This book was written as part of a distance-learning course within the Analytical Chemistry by Open Learning (ACOL) series, so it contains a good number of examples and sample questions. Its typical ACOL format will probably annoy some readers.

The two books *Modern Spectroscopy* (second edition), by J. Michael Hollas, Wiley, Chichester, 1992, and *Fundamentals of Molecular Spectroscopy*, by C. N. Banwell (third edition), McGraw Hill, Maidenhead, 1983, are each highly recommended. Each contains instrumental detail, in addition to the theory and applications of the spectroscopic techniques discussed. Both will look too mathematical to many readers, but they are both authoritative and clear.

A simpler version of Hollas's book is now available in the Royal Society of Chemistry's new Tutorial Chemistry Texts series: *Basic Atomic and Molecular Spectroscopy*, J. Michael Hollas, RSC, Cambridge, 2002. It gives a super introduction, and its academic level is well gauged, although it does require a knowledge of molecular orbitals and maths.

Ink-jets

Peter Gregory's article, 'Colouring the jet set' in *Chemistry in Britain*, August 2000, p. 39, introduces the chemistry of colour, and the colour science of inks for PC printers. He places the subject in context, and also discusses the chemical composition of the many of the most common inks. *Essays in Ink (for Paints and Coatings too)* by J. Kunjappa, Nova Science Publishers, New York, 2002, provides a good introduction for the non-specialist who may be curious about the world of ink.

Pigments, paints and dyes

Philip Bell's text *Bright Earth: the Invention of Colour*, Viking, 2001, is a beginner's guide intended for those with absolutely no background in science. The book commences with a history of the subject, detailing the prehistoric painting found in caves such as Lascoux, before moving on to the Aztecs, Incas and other races peopling the history of colour. The sections on theory should be readily followed by absolutely everyone.

A slightly more in-depth study of colour is afforded by *The Physics and Chemistry of Colour* by Kurt Nassau, Wiley, Chichester, 2001. The author describes many everyday examples of colour, from peacock tails through to the Northern Lights, *Aurora Borealis*. Its Chapter 1 is an overview, and is probably a little highbrow at times, but overall is a fascinating read.

Chromophores in nature

There are many case studies and examples from everyday life in *Light, Chemical Change and Life: A Source Book for Photochemistry*, edited by J. D. Coyle, R. R. Hill and D. R. Roberts, Open University Press, Milton Keynes, 1988, and *Our Chemical Environment, Book 4: Sensational Chemistry*, Open University Press, Milton Keynes, 1995. Both are outstanding and well worth a read.

The sixth chapter of *Food: The Chemistry of its Components* (fourth edition), by T. P. Coultate, Royal Society of Chemistry, Cambridge, 2002, revels in the colours of nature, as well as synthetic chromophores.

Melanin in the skin

There is an embarrassing wealth of material concerning melanin, and the interactions of light on the skin. The Website *http://omlc.ogi.edu/spectra/melanin* by Steven Jacques of the Oregon Laser Centre introduces the topic well, with a readable and highly informative discussion.

Although badly dated now, *The Pigmentary System*, edited by J. J. Norlund, Oxford University Press, Oxford, 1988, is still worth reading, particularly its chapter 'The chemistry of melanins and related metabolites', by G. Prota, M. D'Ischia and A. Napolitano.

Several companies now produce sunglasses with melanin pigment embedded within the lenses; the intention is for the commercial lens to mimic the way the eye blocks some wavelengths, i.e. forming 'natural sunglasses'. The informative Website *http://www.fcgmelanin.com* has both 'history' and 'science' pages.

A more in-depth study of the photochemistry of melanin – natural or synthetic – is available at *http://www.findarticles.com*. Type in 'melanin' and look for the article by Corinna Wu of 18 September 1999.

Finally, as an adjunct, try Peter Bamfield's book *Chromic Phenomena* (above) for a brief discussion of hair dyes, on its pp. 110 ff.

Franck–Condon

The recent book *Conformational Analysis of Molecules in Excited States*, by Jacek Waluk, Wiley, New York, 2000, describes the way we can experimentally determine the shapes of molecules in the ground and excited states. It can be a little high brow at times, and is clearly written for physicists (perhaps explaining its overtly mathematical treatment), but it will be useful for those wanting to study a little further.

Incandescent lights

Louis A. Bloomfield's entertaining book *How Things Work: The Physics of Everyday Life* (second edition), Wiley, New York, 2001, discusses neon bulbs and fluorescent strip lighting, see pp. 395–399. For a more scientific look at fluorescent dyes, try Chapter 3 of Peter Bamfield's *Chromic Phenomena* (above), especially pp. 182–184.

Bioluminescence

This intriguing subject is introduced by Tony Campbell in his article 'Rainbow makers' in *Chemistry in Britain*, June 2003, p. 30.

Raman spectroscopy

The Raman effect' by Neil Everett, Bert King and Ian Clegg in *Chemistry in Britain*, July 2000, p. 40, is a good general introduction, written for scientists with no prior experience of Raman spectroscopy. Each of the books cited above under 'general reading' discuss Raman spectroscopy, but in greater depth.

Raman spectroscopy and pigments

This new topic is well described in the article 'A Bible laid open', by Stephen Best, Robin Clark, Marcus Daniels and Robert Withnall, *Chemistry in Britain*, February 1993, 118. The article shows the Raman spectra of several pigments. A more technical description may be found at S. P. Best, R. J. H. Clark and R. Withnall, *Endeavour*, 1992, **16**, 66.

Robin Clark's group operates from University College in London; their Website *http://www.chem.ucl.ac.uk/resources/raman/speclib.html* is informative, and contains links to many spectra. The Raman spectrum of malachite is interpreted in M. Schmidt and H. D. Lutz, *Physics and Chemistry of Minerals*, 1993, **20**, 27.

For more information about the pigments themselves, try *A History of Lettering*, N. Gray, Phaidon, Oxford, 1986, or *The Icon: Image of the Invisible*, Egon Sendler, Oakwood, 1988. Chapters 12–15 of the latter book are a mine of detail concerning the manufacture of pigments.

The colour of water

A simple yet useful answer to the question 'Why is water blue?' may be found in the article by Charles L. Braun and Sergei N. Smirnov, *J. Chem. Educ.*, 1993, **70**,

612. Alternatively, try 'The colours of water and ice', by Terence Quickenden and Andrew Hanlon, *Chemistry in Britain*, December 2000, p. 37.

10 Adsorption and surfaces, colloids and micelles

General reading

Perhaps the best introduction for the novice is the recent article 'The impact of colloid science' by Mike Garvey in the 'Chemists' Toolkit' series in *Chemistry in Britain*, February 2003, p. 28.

Also for the novice, and also very good, are *Introduction to Colloid Science*, D. J. Shaw, Butterworths, London, 1980, and *Basic Principles of Colloid Science*, by D. H. Everett, Royal Society of Chemistry, Cambridge, 1988. Both are rigorous: the former is more pedagogical in style, and an easier read; the latter is very thorough but somewhat mathematical.

Robert J. Hunter has written two good books on colloid science: the magisterial *Foundations of Colloid Science* (second edition), Oxford University Press, Oxford, 2000, is surely the benchmark text on this topic, but it is not cheap. Its smaller offspring is *Introduction to Modern Colloid Science*, Oxford University Press, Oxford, 1993. This latter text loses none of the rigour but is much shorter and cheaper.

The Colloidal Domain: Where Physics, Chemistry, Biology, and Technology Meet (second edition) by D. Fennell Evans and Håkan Wennerström, Wiley, New York, 1999, is a superb book which satisfactorily demonstrates the interdisciplinary nature of the topic. Its biological examples are particularly good. They also present a nice discussion on pp. 602–603 of how colloid science contributed to the growth of several, disparate strands of science.

In the American Chemical Society 'ACS Professional Reference Book' series comes, *The Language of Colloid and Interface Science – A Dictionary of Terms*, Laurier L. Schramm, American Chemical Society, Washington, 1998; and a related text is *Dictionary of Colloid and Interface Science*, Laurier L. Schramm, Wiley, New York, 2001. Both were compiled to be read as dictionaries, but are thorough and well presented.

The book *Principles of Adsorption and Reaction on Solid Surfaces* by Richard I. Masel, Wiley, New York, 1996, concentrates on mechanistic detail and adsorption at the solid–gas interface, but, within its self-imposed limitations, it is a superb book.

Any in-depth study of adsorption requires us to know something of the surfaces involved. One of the better books describing how we characterize a surface is *Surface Chemistry*, by Elaine M. McCash, Oxford University Press, Oxford, 2001. It is both affordable and well paced, and consistently concentrates on concepts rather than diverting the reader's attention to experimental details. Recommended.

Finally, the journal *Langmuir* usually contains relevant papers. Go to its homepage at *http://pubs.acs.org/journals/langd5*.

Adsorption of biological toxins on kaolin

The nature section of the BBC Website describes several examples of animals eating kaolin clay to immobilize toxins by adsorption on the clay's surface. For example, see the first entry on the page *http://www.bbc.co.uk/nature/weird/az/mo.shtml*. Alternatively, read *The Life of Mammals* by Sir David Attenborough, BBC Books, 2002, p. 170.

Food chemistry

A great many of the examples in this chapter involve food. The best chemistry text on the subject is probably *An Introduction to Food Colloids* by E. Dickinson, Oxford University Press, Oxford, 1992, and is an absolute gold-mine of a book. For further reading, though, try *Food: The Chemistry of its Components* (fourth edition) by T. P. Coultate, Royal Society of Chemistry, Cambridge, 2002, or *Food Microbiology* (second edition) by M. R. Adams and M. O. Moss, Royal Society of Chemistry, Cambridge, 2002, both of which are filled with suitable examples. *Food Flavours: Biology and Chemistry* by C. Fisher and T. Scott, Royal Society of Chemistry, Cambridge, 1997, supplements these titles.

Emulsions

The short Web page at *http://www.surfactants.net/emulsion.htm* has a few interesting links. *Food: The Chemistry of its Components* by Coultate (as above) discusses emulsifiers on p. 114 ff.

Pasteurization

The French site *http://www.calixo.net/braun/conserve/pasteurisation.htm* and the Canadian site *http://www.foodsci.uoguelph.ca/dairyedu/pasteurization.html* both cite pertinent details.

Butter

Many Websites willingly divulge the secrets of butter making, such as *http://www.foodfunandfacts.com/milk.htm*. The site *http://www.culinarycafe.com/Eggs-Dairy/Clarified_Butter.html* describes how (and why) to make clarified butter.

Absinthe

The paper 'Absinthe: enjoying a new popularity among young people?' by C. Gambelunge and P. Melai, *Forensic Sci. Int.*, 2002, **130**, 183 gives a brief history of absinthe-based drinks and their effect on 'the artistic muse'. See also, 'Absinthe: behind the emerald mask', D. D. Vogt and M. Montagne, *Int. J. Addiction*, 1982, **17**, 1015 for a little more social background; its description of the effects of absinthe addiction are salutary.

Milk

There is a colossal amount of information on the Web concerning milk. For example, see the page *http://www.sciencebyjones.com/MILK_NOTES.HTM* or the 'Dairy Chemistry and Physics page' at *http://www.foodsci.uoguelph.ca/dairyedu/chem.html*, hosted by the Canadian University of Guelph.

Ice cream

Most of the information about ice cream comes from Dickinson (above), but *Food: The Chemistry of its Components* by Coultate (above) is also worth a glance.

Soap and Civil War

The soap tax was one of the 'trigger points' leading inexorably to the Great English Civil War. This tax is but one example of the foolishness of the British King Charles I, although it is wise to appreciate how many people aggressively capitalized on his foolishness. The best account of the causes of the Civil War is *The King's Peace 1637–1641*, by C. V. Wedgewood, Penguin, Harmondsworth, 1955. The soap tax is mentioned on p. 160; the introductory background to the King's ill-advised policy commences on p. 157.

The Tyndall effect

'Blue skies and the Tyndall effect' by M. Kerker in *J. Chem. Educ.*, 1971, **48**, 389 is a nice introduction. Alternatively, try Chapter 7 'Some important properties of colloids: II scattering of radiation' in Everett (above), which is extremely thorough.

The history of how the science of light scattering grew is well delineated in the fascinating book, *The Scattering of Light*, M. Kerker, Academic Press, 1967, in which Kerker outlines how the work of various scientists has intertwined.

The Krafft temperature

One of the most cited papers is J. K. Weil, F. S. Smith, A. J. Stirton and R. G. Bristline, *J. Am. Oil Chem. Soc.*, 1963, **40**, 538. One of the better sources of data is Shaw, *Introduction to Colloid Science* (see above) p. 87, and Evans and Wennerström, *The Colloidal*

Domain: Where Physics, Chemistry, Biology, and Technology Meet (see above) p. 11. The paper, 'Use of quantitative structure–property relationships in predicting the Krafft point of anionic surfactants' by M. Jalali-Heravi and E. Konouz, *Internet Electronic Journal of Molecular Design*, 2002, **1**, 410, has a nice introduction and useful references. It can be downloaded at *http://www.biochempress.com/av01_0410.html.*

The journal *Colloid and Polymer Science* occasionally has relevant papers: go to its homepage at *http://link.springer.de/link/service/journals/00396.*

Water treatment

One of the best books on the topic is *Green Chemistry: An Introductory Text* by M. Lancaster, Royal Society of Chemistry, Cambridge, 2002. Its treatment is thorough, but not always in great depth. It is clearly aimed at the complete novice. Chapter 5 of *Pollution, Causes, Effects and Control* (fourth edition), edited by Roy M. Harrison, Royal Society of Chemistry, 2001, describes the material in a little more depth; do not be confused by the careless title of the book *Basic Water Treatment* by C. Binnie, M. Kimber and G. Smethurst, Royal Society of Chemistry, Cambridge, 2002, which is neither an elementary text nor a treatise on treating alkaline water.

Index

Natural materials cited in the text: *flora*, *fauna*, **food, and natural chemicals**

Chemicals cited in the text: elements, compounds, alloys and materials
This list does not include all compounds and materials cited in tables.

Etymologies, definitions and the meanings of special words
The majority of these entries relate to etymologies. The remainder relate to precise scientific usage.

General index
Tables are indicated as (T)

Printed and bound by CPI Group (UK) Ltd, Croydon, CR0 4YY